THE SOLAR OUTPUT AND ITS VARIATION

THE SOLAR OUTPUT AND ITS VARIATION

Edited by

Oran R. White
*High Altitude Observatory of the
National Center for Atmospheric Research*

COLORADO ASSOCIATED UNIVERSITY PRESS
BOULDER

Copyright 1977 Colorado Associated University Press
Boulder, Colorado
International Standard Book Number: 0-87180-07105
Library of Congress Card Catalog Number: 76-15773
Printed in the United States of America
Designed by Dave Comstock

THE SOLAR OUTPUT AND ITS VARIATION

PREFACE   *Oran R. White and Gordon Newkirk, Jr.*   vii

CHAPTER I. Solar Variability: A Summary   1
    Editor's Comments   1

    Section I. Terrestrial Implications of a Changing Solar Input   3
    Introduction, *Stephen Schneider*   3
    The Magnetosphere, *Thomas Holzer*   5
    The Thermosphere-Ionosphere, *Raymond Roble*   8
    The Stratosphere-Mesosphere, *Paul Crutzen*   13
    The Troposphere, *Stephen Schneider*   16

    Section II. The Evidence for a Variable Solar Output   19
    Paleoclimatic Evidence for the Solar Cycle and Its Variation,
        Murray Mitchell   19
    The Integrated Solar Flux, *John Eddy*   22
    The Solar Spectrum, *Richard Donnelly*   25
    Plasma Flow from the Sun, *Arthur Hundhausen*   36
    Energetic Particles from the Sun and Solar Modulation
        of Galactic Cosmic Rays, *Robert Lin*   39
    The Theory of the Solar Interior and Its Predictions
        of Solar Variability, *Roger Ulrich*   41
    Activity and Variability in Other Stars, *Oran White*   44

CHAPTER II. Historical and Paleological Evidence for Solar Variability   49
    Editor's Comments   49
    Historical Evidence for the Existence of the Solar Cycle, *J. A. Eddy*   51
    Climatic Change and the Possible Influence of Variations
        in Solar Input, *J. D. Hays*   73

CHAPTER III. The Integrated Solar Flux   91
    Editor's Comments   91
    Contemporary Measures of the Solar Constant, *C. Fröhlich*   93
    1976 Rocket Measurements of the Solar Constant and Their
        Implications for Variation in the Solar Output in Cycle 20,
        R. C. Willson and J. R. Hickey   111
    Variations in Insolation Caused by Changes in Orbital Elements
        of the Earth, *A. D. Vernekar*   117

CHAPTER IV. The Solar Spectrum   131
    Editor's Comments   131

The Solar Spectrum Above 1 mm, *F. I. Shimabukuro*	133
The Solar Spectrum Between 10 and 1000 μm, *W. G. Mankin*	151
The Solar Spectrum Between .3 and 10 μm, *A. K. Pierce and R. G. Allen*	169
The Solar Spectrum Between 1200 and 3000Å, *D. F. Heath and M. P. Thekaekara*	193
The Solar Spectrum at Lyman-Alpha 1216Å, *A. Vidal-Madjar*	213
The Solar Spectrum Between 300 and 1200Å, *J. G. Timothy*	237
The Solar Spectrum Between 10 and 300Å, *J. E. Manson*	261
Comment on the Solar Spectrum Between 10 and 300Å, *A. B. C. Walker*	279
The Solar Spectrum Below 10Å, *R. W. Kreplin, K. P. Dere, D. M. Horan, J. F. Meekins*	287
The Availability and Development of NBS Radiometric Standards, *R. P. Madden*	313
Problems and Controversies in the Measurement of the Solar Spectral Irradiance, *R. F. Donnelly*	317
Models of the Solar Atmosphere, *E. H. Avrett*	327
CHAPTER V. The Solar Plasma and Energetic Particles	**349**
Editor's Comments	349
Plasma and Magnetic Fields from the Sun, *W. C. Feldman, J. R. Asbridge, S. J. Bame, J. T. Gosling*	351
Measures of Energetic Particles from the Sun, *L. J. Lanzerotti*	383
Particle Output of the Sun in the Past, *D. Heymann*	405
Solar Induced Variations of Energetic Particles at One AU, *P. E. Damon*	429
CHAPTER VI. The Theory of the Solar Variation	**449**
Editor's Comments	449
Theoretical Predictions of Variations in the Solar Output, *D. O. Gough*	451
CHAPTER VII. Stellar Variability	**475**
Editor's Comments	475
The Solar Output and Variability Viewed in the Broader Context of Stellar Activity, *J. L. Linsky*	477
APPENDIX I. List of Participants and Contributors in the Solar Output Workshop, Boulder, April 26-28, 1976.	517
INDEX	**523**

PREFACE

Oran R. White and Gordon Newkirk, Jr.
High Altitude Observatory

The total output of the sun of matter, radiation, and magnetic fields comprises a fundamental body of information from which we have gained much of our present knowledge of the sun as an energy-producing stellar body; therefore, exact knowledge of those outputs is of great importance to aeronomy, meteorology, atmospheric physics, stellar physics, and solar physics. The total radiant flux from the sun is one of the critical empirical boundary conditions that permit construction of models of the solar interior and the specification of nuclear processes as the ultimate source of all of the energy emitted by the sun. Likewise, the global solar output and its relation to the extreme variations observed from point to point on the solar surface provide the essential means for comparison with other stars, whose surface structures are poorly, if ever, resolved. The variation of the solar output with time that occurs as a result of solar activity is another important link to other stars where magnetic activity is difficult to measure directly. Because of the sun's central role in determining the energy budget and physical state of the earth's atmosphere, exact specification of the solar output is essential for progress in our quantitative understanding of the structure of the terrestrial and other planetary atmospheres as well as of the solar-terrestrial interaction. Given this broad need for accurate knowledge of the solar output, the goal of this volume is to set forth the best information available on the total output of the sun as it is measured outside of the earth's atmosphere and to bring attention to the uncertainties that exist in this view.

Several components of the total solar output show large variations with time; consequently their specification at only one time will yield an incomplete picture of the sun. Solar variability over time scales ranging from seconds to decades is documented by modern observations for those outputs associated with solar activity. However, the changes occurring over epochs of more than a few hundred years can only be inferred from early scientific records and other historical data. For the longer spans, over millions or billions of years, our only tools for studying the possibility of solar change are the record of exposure to high energy particle fluxes in extraterrestrial samples and the paleoclimatic record of the earth. The results from these two approaches, however, are sometimes ambiguous and may be influenced by assumptions—about the nature of the earth's atmosphere or the stability of the lunar surface, for example.

The goal of this volume is to provide a broad perspective of our current knowledge of the sun's output and to furnish our colleagues in climatology, aeronomy, stellar physics, planetary physics, and solar physics with a compendium of the quantitative material now available. We hope that this collection will stimulate an interdisciplinary conversation that will lead to answers to such questions as the role of solar variation in the climate of the earth and the other planets; the relationship between solar and laboratory plasmas; the physical processes that produce stellar chromospheres, coronae, and mass loss; and the inference of physical conditions in the solar interior and outer atmosphere from measurements of the solar output.

The idea of producing such a volume began in discussions with Walter Orr Roberts on problems associated with past and future measurements of the variability of the sun. It became clear that existing observing technology presents the stellar astrophysicist and the atmospheric physicist with measurements of many different physical parameters, each of which is only a part of the total solar output. It also became clear to us that the scientific community in general did not have any reference giving either an overall picture of the state of the measurements or a coherent understanding of all of the outputs from the sun. Therefore, in an effort to provide a contemporary view of these outputs and their variation, we undertook the preparation of this book by means of an interdisciplinary workshop attended by colleagues actively engaged in research programs relevant to measurement and interpretation of the solar output.

Our knowledge of solar variability is, of necessity, incomplete at this time; and some topics were judged to lie beyond our purposes of specifying the output of both the active and quiet sun as well as describing the behavior of the solar output over time intervals from seconds to billions of years. For example, solar flare emission is not discussed in great detail, nor have we attempted a review of the theory of the solar cycle and the solar dynamo. Other such areas where knowledge is most lacking are indicated throughout the book. It is our hope that this work will provide a productive guide to those problems that are both important and amenable to solution during the coming years.

Preparations of the contributions to this volume culminated in the Solar Output Workshop held in Boulder, Colorado, from 26 to 28 April 1976. The workshop was attended by most of the authors and organizers as well as by a small number of consultants. The organizing committee for the workshop and the editorial board for this book were Richard Donnelly, John Eddy, Donald Heath, Arthur Hundhausen, Gordon Newkirk Jr., A. Keith Pierce, Stephen Schneider, Adrienne Timothy, Roger Ulrich, William Wagner, Harold Zirin, Julius London, and Oran White. The purposes of the workshop were to examine each contribution critically and to develop a summary of our knowledge of the solar output through discussions among scientists with a broad range of interests covering solar physics, atmospheric physics, stellar astronomy, and paleoclimatic research. Our workshop benefitted from the previous meeting held at Big Bear Solar Observatory in June 1975 to discuss the "solar constant" problem.* Members of the scientific community who attended the Solar Output Workshop and/or participated in the preparation of this volume are listed in Appendix I.

Although this project began within the High Altitude Observatory of the National Center for Atmospheric Research,** several institutions joined in supporting the workshop itself and the publication of the book. Those organizations that made direct contributions to the project were: the High Altitude Observatory, the Advanced Study Program, and the Atmospheric Analysis and Prediction divisions of the National Center for Atmospheric Research; the Environmental Research Laboratories of the National Oceanic and Atmospheric Administration; the Sacramento Peak Observatory of the United States Air Force; and the National Aeronautics and Space Administration. Support of the workshop or publication of this volume by these agencies is not to be construed as endorsement of the volume's contents as the official policy of any of these agencies.

We are deeply indebted to the thoughtful people in these organizations and to the National Science Foundation for their enthusiastic encouragement of this project as well as for their direct assistance in making this volume a reality, accessible to practicing scientists and students requiring quantitiative data on the output of the sun.

Special Acknowledgments by the Editor

The completion of a volume such as this one draws on many skills ranging from the professional judgment of one's scientific associates to the equally professional skills in manuscript preparation and publication. I am personally indebted to Dick Donnelly, Jack Eddy, Julius London, Gordon Newkirk, Steve Schneider, and Bill Wagner for their patience in helping me prepare the summary Chapter I. The workshop and collection of the material would not have been possible without the careful and professional assistance of Jane Aschenbrenner of HAO. Likewise, the technical editing of the final manuscripts was expertly done by Merry Maisel and her associates in the NCAR Publications Office. The final production was a collaborative effort among all of us at NCAR. Even though it is small compensation

---

* Zirin, H., Moore, R.L., and Walters, J. (eds.), 1976, Proceedings of the workshop: The solar constant and the earth's atmosphere, *Solar Phys. 46*, 347.

** The National Center for Atmospheric Research is operated by the nonprofit University Corporation for Atmospheric Research and sponsored by the National Science Foundation.

for much thoughtful, hard work, I express my sincere personal thanks to each person who helped complete this project.

Oran R. White

Addendum to the Second Printing

Because of the interest in the solar output and its role in the solar-terrestrial interaction by the research community in the Soviet Union, this volume is being translated into Russian by MIR Publishers, 2 Pervy Rizhsky pereulok, Moscow, I-110, GSP, USSR. The Russian language edition should be available during 1980 as a printing of 5000 copies. It is, indeed, rewarding to the contributing authors and me to see such genuine scientific interest in our collection of quantitative data on the total solar output.

I. SOLAR VARIABILITY: A SUMMARY

A brief glance through this volume immediately shows the large variety of subjects covered by the authors and the level of detail used to discuss the solar output quantitatively. In view of the breadth and complexity, a summary chapter touching briefly on each general area is included to emphasize the state of our understanding of the solar output and its variation and to raise some provocative questions about this variation and its influence at the earth.

This summary chapter is the combined effort of several colleagues present at the Solar Output Workshop. What we seek in this chapter is not consensus on how well we can specify the solar output and its variation, but rather a definition of the empirical uncertainties and how they affect disciplines in which quantitative knowledge of the sun is necessary for progress on current problems. Beginning in Chapter II, the reader will find the detailed discussions of the solar output problem presented by the individual reviewers.

Oran R. White

I. TERRESTRIAL IMPLICATIONS OF A CHANGING SOLAR INPUT

Introduction
Stephen Schneider

The earth's climate is fundamentally controlled by the output of the sun. Therefore, any fluctuations in the sun's output of radiation, fields, or particles should be accompanied by a response in the terrestrial climate system, which is itself comprised of the atmosphere, the oceans, the land, the cryosphere (snow and ice component) and, to some extent, the biota (see references below for GARP and NAS publications giving summaries of climate theory in 1975). The atmospheric component is usually taken to include only the first 30 km or so above the earth's surface, that is, the troposphere and lower stratosphere. Most of the atmosphere's energy is contained in this region, and influences from the higher atmospheric layers and the sun itself are regarded as imposed external forcing conditions on the climatic system. A change in the radiant energy output of the sun (the solar "constant" or total solar irradiance) could provide a direct energy input to the climatic system in an amount sufficient to cause some of the observed climatic changes. However, the less energetic (but better established) variations in the solar output of particles or fields—those that are energetically too small to cause detectable climatic changes—could indirectly influence the upper atmosphere enough to create a response that cascades downward and causes a detectable response in the lower atmosphere. Much of the statistical evidence connecting solar variability with a climatic response offers correlations that can be explained only by such a cascade. Some amplification mechanism is required to explain how the small energy perturbations associated with solar variability might be responsible for a terrestrial response containing much more energy. The search for such physical amplification mechanisms is an important component of solar-terrestrial research, since statistical correlations in the absence of causal mechanisms give little physical insight or understanding of the solar-terrestrial system.

The interacting components of the global climate system respond to changes in external forcing functions with differing time scales and magnitudes. Therefore, a chief problem of terrestrial climate theory is to determine whether past fluctuations in the states of the components of the climatic system were caused by fluctuations in external forcing functions, such as the solar output or volcanic dust veils, or were due to "internal oscillations" within the climatic system itself. Since the climate near the earth's surface has a response time dominated by the relatively long thermal inertia of the upper layers of the oceans, any significant variability in the climate at the earth's surface forced by variations in solar emission requires that solar variability occur over times generally much greater than a month. Since solar variability over scales of years appears to be small relative to the total energy input to the troposphere, connections between solar variability and surface climate are still hotly debated issues.

On the other hand, the neutral and ionized upper atmosphere, consisting of the stratosphere-mesosphere (about 15-80 km), the thermosphere-ionosphere (80-500 km), and the magnetosphere ($>$ 500 km), has a relatively rapid response time. There is proof that variability in solar emissions of radiation, particles, and fields can be related to measurable and predictable responses in the upper atmosphere. However, the extent to which upper atmospheric variability can be correlated with surface phenomena is an issue of debate, as mentioned earlier.

The different atmospheric regions discussed here have significant similarities and marked contrasts. From the troposphere to the thermosphere, the atmosphere is primarily controlled by absorption of electromagnetic radiation. However, the absorption influences very different physical processes in the different atmospheric regions. These differences occur because the absorption of solar photons takes place in significantly different energy ranges in the spectrum.

In the lowest region, the troposphere, the gases are relatively transparent to solar radiation at wavelengths longer than 3200 Å, where we find the bulk of the solar radiative energy, i.e., the near-UV, visible, and near-IR radiation. The most notable absorption is due to the CO_2 and H_2O bands in the near-infrared. Because most of the visible incident solar radiation reaches the earth's surface, a large fraction of its energy is inserted into the atmosphere by the evaporation-precipitation cycle of water vapor. In general, the hydrological cycle processes are the most significant solar-driven physical processes in the troposphere.

The thermosphere absorbs almost all EUV flux at wavelengths shorter than 1000 Å, i.e., the photons capable of dissociating the primary atmospheric molecules (O_2 and N_2) and the 1000-1900 Å radiation that dissociates O_2. Consequently, thermospheric processes involve the more abundant molecules, atoms, and ions; the sources and sinks of these species; and their transport through the atmosphere.

The stratosphere-mesosphere absorbs photons mainly in the 1900-3200 Å range. This absorption is determined by small cross sections of O_2 over the lower half of the range and by large cross sections of several important trace gases, especially ozone, at various wavelengths within the range. In contrast to thermospheric absorption, the major gaseous species no longer dominate. Through the very slow dissociation of O_2 and the rapid dissociation of trace gases, the solar flux in the 1900-3200 Å band drives many important gas phase chemical reactions. Therefore, the stratosphere-mesosphere system is often referred to as the chemosphere.

The very outer part of the earth's atmosphere, the magnetosphere, interacts with the sun in a completely different physical regime than exists in the lower layers, since photon absorption processes are not directly important. Instead, the ionization state, magnetic field structure, and dynamics of the magnetosphere are the result of interaction with the particle and magnetic field flux known as the solar wind. The state of the magnetosphere is only indirectly coupled to the solar photon flux through the dynamical processes that connect it to the lower atmosphere where that radiation is so important. However, the transfer of energy downward from the magnetosphere by precipitating particles and electric fields is a source of the general excitation of the ionosphere-thermosphere system, while the photon contribution is significant but secondary in comparison.

These general background statements give only a glimpse of the current scientific interest in understanding the physics of the solar-terrestrial interaction and the terrestrial response to any change in the solar output. The remainder of this introductory section presents more detailed descriptions of the terrestrial response. We begin with the outermost part of the earth's atmosphere, the magnetosphere, where large responses to energetically small solar fluctuations are observed; we then progress downward to the troposphere, where the physical coupling between the solar output and the earth's atmosphere is most difficult to establish. It is in this lowest part of the earth's atmosphere that the social and economic consequences of any solar-induced variability in the earth's climate are of interest.

This introduction to the solar-terrestrial interaction problem is written as a guide and motivation to those colleagues interested in pursuing the detailed properties of the solar output as described in this volume. However, our knowledge of the solar output is of equal interest for understanding the atmospheres of the other planets in the solar system, the properties of the interplanetary medium, and the relationship between the sun and other stars.

The Magnetosphere
Thomas Holzer

When the solar wind plasma impinges upon the terrestrial magnetic field, the earth is enclosed in a cavity from which the solar wind is largely excluded and within which the terrestrial magnetic field is largely contained. This entire cavity is called the magnetosphere, and its outer boundary layer, between the earth's magnetic field and the solar wind plasma, is called the magnetopause. The magnetospheric cavity has a shape reminiscent of a comet in that it extends a relatively short distance (about ten earth radii) from the earth toward the sun but is drawn out into a long tail away from the sun. In the lower magnetosphere near the earth, the magnetic field differs only slightly from an undistorted terrestrial dipole field; but in the outer magnetosphere, where solar wind effects are important, the field configuration is quite deformed, being strongly compressed on the sunward side and becoming greatly extended in the tail.

The magnetosphere can be looked upon as a dynamic system. It transfers from interplanetary space to the terrestrial atmosphere part of the energy, momentum, and mass that is carried outward from the sun by the solar wind and solar energetic particles. The transfer efficiency for this energy, momentum, and mass is quite variable and is determined almost entirely by the stress exerted by the variable solar wind on the magnetosphere at the magnetopause. It is convenient to think of this stress as being composed of two components, one normal to and one tangential to the magnetopause. The normal stress is controlled predominantly by the ram pressure (proportional to the bulk flow energy density) of the solar wind, whereas the tangential stress is apparently determined by magnetic field line reconnection occurring on the sunward face of the magnetopause.

Together, these two components of the stress determine the number of geomagnetic field lines that are connected directly to the interplanetary magnetic field. These connecting field lines intersect the terrestrial atmosphere at high geomagnetic latitudes and provide paths of direct access to the atmosphere by energetic particles from interplanetary space. The area of the atmosphere into which energetic particles can enter is somewhat larger than the area intersected by interconnected magnetic field lines, but both of these areas vary in size with changes in the magnitude of the stress exerted by the solar wind at the magnetopause. Solar energetic particles alter the ionization and chemistry of the mesosphere and stratosphere at high geomagnetic latitudes, and because of these changes in the auroral zones there can be, at all latitudes, a modification of the radiation-to-heat conversion in the mesosphere, stratosphere, and upper troposphere, as well as a modification of the absorbing properties of the atmosphere in the wavelength range of ultraviolet radiation that can be harmful to living organisms (e.g., Reid et al., 1976). Another consequence of energetic particle ionization in the lower ionosphere is the production of conductivity enhancements that play a role in the electrical coupling between the ionosphere and the lower atmosphere. The magnitude of the effects of energetic particles on the atmosphere is sensitive to the energy flux density and the energy spectrum of the particles and to the area of the atmosphere to which the particles have access.

Lower energy (solar wind) particles also have access to the magnetosphere along interconnected magnetic field lines, but most of the energizing and deposition of these particles occurs during the catastrophic magnetospheric relaxation process known as the magnetospheric substorm. The magnetospheric substorm acceleration process may be their most important means of entry into the lower atmosphere. The energy released in this process has been transferred from the solar wind to the magnetosphere through the tangential stress at the magnetopause and has been stored in the tail of the magnetosphere. On the average, substorms occur every four to five hours, and they endure for one or two hours. Most of the particles energized in a substorm have relatively low energies (about 10 keV), but there is a significant fraction produced with much higher energies (characteristic of energetic solar particles). Some of the higher energy particles are immediately deposited in the atmosphere, while others populate the radiation belts and are deposited in the atmosphere at subauroral latitudes at some later time. The atmospheric effects of these high-energy particles accelerated in the magnetosphere are similar to those of the energetic particles from both solar and galactic sources.

The lower energy particles deposited in the atmosphere by substorms produce ionization and chemical changes, mainly in the thermosphere, and their primary effect on the atmosphere is the high-latitude heating of the thermosphere in the auroral zones. Thermospheric heating can also be brought about as a direct result of tangential stress at the magnetopause. The tangential stress induces Lorentz (J × B) forces throughout the magnetosphere and ionosphere, and these forces drive dissipative currents (primarily in the ionosphere at high latitudes), which result in the so-called Joule heating of the thermosphere. In addition, the Lorentz forces drive plasma flows in the magnetosphere and the high-latitude ionosphere, and momentum from these ionospheric flows is transferred directly to the neutral atmosphere through frictional drag. The heat and momentum added to the high-latitude thermosphere through particle precipitation and through the action of Lorentz forces can produce significant effects at lower latitudes and perhaps at altitudes lower than 80 km.

Of course, there are certain well-known practical consequences of particle precipitation and of enhanced ionospheric currents associated with magnetospheric substorms. The modification of the electrical conductivity of the ionosphere by particle preicpitation can significantly affect radio transmission, and precipitation events can thus seriously disrupt communication systems at high latitudes. Currents flowing in the ionosphere also induce currents in the earth, in transmission lines, and in any other good conductor relatively near the seat of the ionospheric currents; thus such currents associated with a substorm can do serious damage to transmission lines and can increase the eddy-current deterioration of structures built at high latitudes, such as the Alaska pipeline.

There are several solar particle and field parameters of major importance in determining the transfer of energy, momentum, and mass from interplanetary space to the terrestrial atmosphere through the magnetosphere. The most important of these are the solar wind mass density, bulk flow speed, and magnetic field (magnitude and direction), and the energy flux density and spectrum of solar energetic particles. It should be emphasized, however, that the important physical processes in the magnetosphere are not so well understood that the accurate specification of these parameters enables an accurate estimate of the energy, momentum, and mass inputs into the atmosphere from the solar wind. Of course, it is clear that the energy, momentum, and mass inputs through the magnetosphere are quite small in comparison with the energy, momentum, and mass content of the lower atmosphere; therefore, any significant effects of magnetospheric processes on the lower atmosphere must be indirect, as in the case mentioned above in which energetic particle precipitation modulates EUV absorption through a chemical modification of the stratosphere.

A good guide to recent scientific literature in the field of magnetospheric physics can be found in the U.S. National Report to the IUGG (1975, pp. 872-1048).

References

Reid, G.C.; Isaksen, I.S.A.; Holzer, T.E.; and Crutzen, P.J., 1976, *Nature 259*, 177.
U.S. National Report to the International Union of Geodesy and Geophysics, 1975, *Rev. Geophys. Space Phys. 13*, 1-1106.

The Thermosphere-Ionosphere
Raymond Roble

Solar electromagnetic energy at wavelengths shorter than 2000 Å interacts strongly with the major neutral gases within the earth's thermosphere, the atmosphere above 80 km. For wavelengths shorter than 1025 Å, virtually all solar photons are absorbed by the major neutral constituents in photoionization processes to produce the ionosphere, a weakly ionized plasma that is embedded within the thermosphere (see, for example, the review by Evans, 1975). Both the magnitude of the ionization and thermal structure of the daytime ionosphere are maintained by the solar EUV radiation ($\lambda < 1025$ Å).

The solar EUV radiation, besides creating the ionosphere, is the most important energy source for heating the neutral gas within the earth's thermosphere above 150 km. Of the solar EUV energy absorbed in a typical ionization event, approximately 50% appears as chemical energy of the ion production and 50% appears as kinetic energy of the ejected photoelectron. The energy in these thermal and chemical channels cascades through a chain of chemical and physical processes involving both atoms and molecules, described by Stolarski *et al.* (1975). They showed that of the solar EUV absorbed in the ionizing event, 40 to 50% of the energy is reradiated from the thermosphere as UV airglow; 20% is transported as chemical energy by atomic oxygen to the lower thermosphere below 90 km, where it is regained through recombination; and 30 to 40% appears as direct kinetic heating of the neutral gas. Below 150 km, solar energy absorption in the Schumann-Runge continuum (1300-1750 Å) and Schumann-Runge bands (1750-2100 Å) becomes the dominant heat source for the neutral atmosphere, again with a heating efficiency of 30 to 40% relative to the solar energy absorbed. Thus, over a wide range of altitude within the thermosphere, the neutral gas heating can be determined by calculating the amount of solar energy that is absorbed and multiplying it by a heating efficiency of about 33%.

The interaction between physicists modeling the ionosphere and those measuring the solar EUV flux points to the need for accurate solar flux values. In the recent literature there has been considerable discussion of whether or not the magnitude of the solar EUV flux as measured by Hinteregger (1970) is sufficient for modeling ionospheric processes (e.g., Swartz and Nisbet, 1973; Roble and Dickinson, 1973; Prasad and Furman, 1974; Heroux et al., 1974; Nisbet, 1975; Heroux et al., 1975). Some authors argue that a doubling of Hinteregger's flux measurements is necessary to obtain agreement between ionospheric theoretical calculations and observations; others disagree. There have been other measurements of the solar EUV flux by Heroux et al. (1974), Timothy et al. (1974), and Schmidtke (1975); however, they were made at different times and levels of solar activity as measured by the solar radio emission at 10.7 cm. The Hinteregger flux values were obtained on 11 March 1967, when the solar F10.7 value was 144×10^{-22} Wm^{-2} Hz^{-1} and the integrated energy below 1025Å was 2.46 erg cm^{-2} s^{-1}. Heroux et al. (1974) measured an integrated energy of 3.27 erg cm^{-2} s^{-1} on 23 August 1972, when the F10.7 value was 120×10^{-22} Wm^{-2} Hz. Timothy et al. (1972) obtained about 6.10 erg cm^{-2} s^{-1} on 3 April 1969, when the F10.7 value was 189×10^{-22} Wm^{-2} Hz^{-1}, and Schmidtke (1974) obtained 3.8 erg cm^{-2} s^{-1} on 2 March 1973, when the F10.7 value was 101.6×10^{-22} Wm^{-2} Hz^{-1}. Recently, Hinteregger (1975) has discussed the various solar EUV flux measures and also has published solar EUV flux values obtained from the *Atmosphere Explorer* satellite during 1974-1975. His measurements give solar flux values essentially similar to those of Heroux et al. (1974); however, they were obtained during solar minimum, when the F10.7 values were typically 80×10^{-22} Wm^{-2} Hz^{-1}.

Roble (1976) has examined the ionospheric properties determined over the incoherent scatter radar station at Millstone Hill, Massachusetts (42.6°N, 71.5°W). According to these calculations, the Hinteregger (1970) values are sufficient for modeling the ionospheric properties during solar minimum, when the F10.7 values are typically 80×10^{-22} Wm^{-2} Hz^{-1}. However, modeling the ionospheric properties for solar maximum (F10.7 $\sim 160 \times 10^{-22}$ Wm^{-2} Hz^{-1}) requires twice these values. It appears, from ionospheric modeling considerations, that the solar EUV output is linearly related to the solar F10.7 cm output, with the Hinteregger (1970) values adequate for modeling solar minimum while twice that is required for solar maximum, in solar cycle 20. More direct measurements, however, of the solar EUV output over a solar cycle and also during various levels of solar activity are required to study the earth's ionospheric properties. It would be desirable to have long-term solar EUV flux measurements accurate to ±10% for ionospheric modeling purposes.

The thermosphere is also the region where the bulk of the solar wind energy is deposited within our atmosphere through auroral processes at high latitudes.

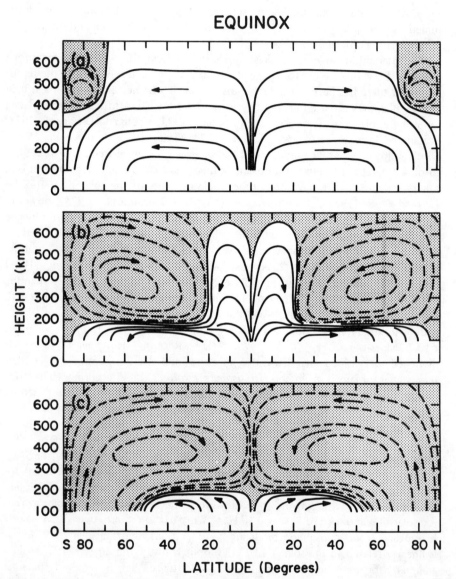

Fig. 1 Schematic diagram of the mean thermospheric circulation during equinox. The upper panel shows the circulation during quiet geomagnetic conditions, the middle panel is for average conditions, and the bottom panel is for geomagnetic storms.

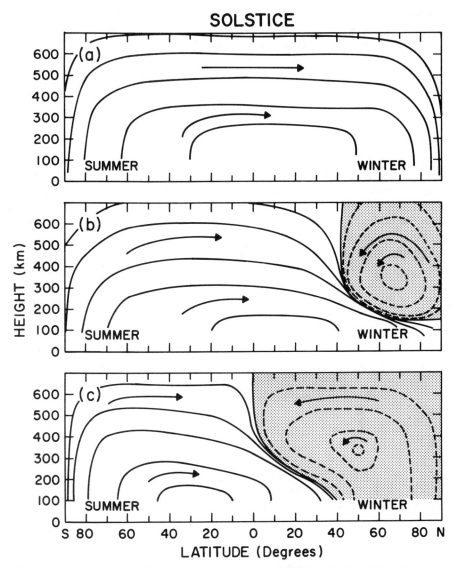

Fig. 2 Schematic diagram of the mean thermospheric circulation during solstice. The upper panel shows the circulation during quiet geomagnetic conditions, the middle panel is for average conditions, and the bottom panel is for geomagnetic storms.

Locally, this energy can be greater than the absorbed solar electromagnetic radiation, thereby making the thermosphere the part of our atmosphere that is competitively forced by the two forms of solar energy interacting with the earth's outer atmosphere. These interactions make the thermosphere a dynamically active region with large variations about a basic state that are alternately being forced and relaxed due to the changing auroral conditions. Nowhere within our neutral atmosphere is this interaction of the two forms of solar forcing more apparent than it is in the thermosphere.

Dickinson *et al.* (1975) have recently examined the mean meridional circulation and latitudinal structure of the thermosphere produced by the interaction of the two forms of solar energy forcing, and their results are summarized in the accompanying figures, which represent the longitudinally averaged mass flow within the thermosphere for two seasons. The upper panel of Figure 1 shows the mean meridional circulation during equinox for very quiet geomagnetic activity. A Hadley circulation exists, with rising motion over the subsolar point, poleward transport, and sinking motion over the poles. The middle panel shows the mean circulation during average geomagnetic activity, where the high-latitude heat source from solar wind energy drives a reverse circulation to midlatitudes, a flow which opposes the radiatively driven circulation. During geomagnetic storms (bottom panel), the high-latitude heat source reverses the circulation throughout most of the thermosphere.

Figure 2 shows the circulation for the same geomagnetic activity sequence, but in solstice conditions. The high-latitude heat source is again present in both hemispheres, but in the summer hemisphere it reinforces the radiatively driven summer-to-winter pole circulation, whereas in the winter hemisphere it opposes the circulation. The position and strength of the reverse circulation depends upon the level of geomagnetic activity, and thus the mean circulation can undergo considerable variation with solar change. Recent studies suggest that for average geomagnetic conditions it is necessary to have about 2.5 times more solar wind energy entering the high-latitude summer hemisphere than the winter hemisphere during solstice. It also appears that 5 times more solar wind energy is being deposited in the thermosphere by auroral processes during solar maximum than during solar minimum. To understand the thermospheric structure and circulation, it is important to have not only an accurate prescription of the solar electromagnetic flux below 1900 Å, but also a prescription of the high-latitude heat and momentum sources produced by the interaction of the solar wind with the earth's magnetic field.

References

Dickinson, R.E., Ridley, E.C., and Roble, R.G., 1975, *J. Atmos. Sci. 32*, 1737.
Evans, J.V., 1975, *Rev. Geophys. and Space Phys. 13*, 887.
Heroux, L., Cohen, M., and Higgins, J.E., 1975, *J. Geophys. Res. 80*, 4732.
_____, _____, and _____, 1974, *ibid. 79*, 5237.
Hinteregger, H.E., 1970, *Ann. Geophys. 26*, 547.
_____, 1975, paper presented at the XVI IUGG General Assembly, Grenoble, France.
Nisbet, J.S., 1975, *J. Geophys. Res. 80*, 4770.
Prasad, S.S., and Furma, D.R., 1974, *ibid. 79*, 2463.
Roble, R.G., 1976, *ibid. 81*, 26.
_____, and Dickinson, R.E., 1973, *ibid 78*, 249.
Schmidtke, G., 1976, *Geophys. Res. Lett.*, in press.
Stolarski, R.S., Hays, P.B., and Roble, R.G., 1975, *J. Geophys. Res. 80*, 2266.
Swartz, W.E. and Nisbet, J.S., 1973, *ibid. 78*, 5640.
Timothy, A.F., Timothy, J.G., Willmore, A.P., and Wager, J.H., 1972, *J. Atmos. Terr. Phys. 34*, 969.

The Stratosphere-Mesosphere
Paul Crutzen

As mentioned earlier, the most significant solar-driven processes in the stratosphere and mesosphere involve the chemistry of neutral trace species. In particular, the ozone layer is maintained by the balance between odd oxygen production by photodissociation of molecular oxygen (O_2) and an array of chemical loss processes. Ozone (O_3) is generally accompanied by atomic oxygen (O) in a ratio that is determined by the balance between the three-body reaction of O with O_2 to form O_3 and the photodissociation of O_3 back to its constituents. This absorption of solar photons is also the dominant means by which the sun heats the stratosphere and mesosphere, and it is in balance with CO_2 infrared cooling. These chemical exchanges, together with dynamic processes, determine the temperature and wind structure in the stratosphere-mesosphere system.

At wavelengths less than 3100 Å the atomic oxygen (O) produced by ozone photodissociation is in the excited electronic state $O(^1D)$. The chemical energy available in the excitation of $O(^1D)$ drives many important chemical transformations. In particular, it converts the nitrous oxide (N_2O) transported up from the lower atmosphere into nitric oxide (NO). This is the major source of the NO and NO_2 that provides the dominant loss process for odd oxygen. Also of major importance is the oxidation of water (H_2O) and methane (CH_4) by $O(^1D)$ into hydroxyl radicals (OH). The OH, in turn, is a major component of several important reactions.

Two additional nitrogen cycle photodissociative processes are notable. First, most of the N_2O is dissociated to N_2 and O, thus decreasing the amount of NO and NO_2 available to reduce the ozone concentration. Second, above 50 km, NO is photodissociated, allowing the free nitrogen atom to react with another NO to convert two odd nitrogens to N_2. The rates of these two processes are determined mainly by solar irradiance in the 1800-2400 Å region. Since the NO and NO_2 concentration controls the ozone abundance, one may speculate that an increase in the solar output in this wavelength band will lead to increased concentrations of ozone in the stratosphere by a combination of the following effects (Crutzen, 1973):

(a). More primary production of ozone.

(b). More destruction of NO. Although this process takes place above about 40 km, it may have a significant effect on NO in the ozone layer because nitric oxide is a long-lived species in the stratosphere.

(c). Less production of NO, leaving less N_2O in the stratosphere for oxidation to NO.

It is estimated by means of model calculations that a 20% increase in the solar irradiance between 1800 and 2400 Å would yield an increase of about 3% in the total ozone column. Because of the role of ozone as a shield against UV radiation at ground level and the possibility of man's activities affecting the ozone layer, it is important to know the empirical boundary condition set by the variability of solar UV radiation near 2000 Å.

On the other hand, an increase in the solar output in the 2400-3100 Å region will have the principal effect of increasing the production of $O(^1D)$ atoms, which will lead to enhanced production of NO and, consequently, a lower ozone concentration. This effect may be increased by greater solar heating of the stratosphere, which would tend to lower the concentration of ozone in the upper stratosphere because of the temperature dependence of the ozone-destroying reaction

$$O + O_3 \rightarrow 2 O_2 \quad .$$

In addition, above 25 km and below 40 km, larger concentrations of OH, resulting from more intense production of $O(^1D)$ atoms, will tend to lower the concentration of atomic oxygen and ozone through catalytic reactions involving hydrogen and other species (i.e., H and HO_2). All these effects, leading to reduction in the odd oxygen concentration, are slightly counteracted by the fact that larger production of OH would also favor the conversion of catalytic NO_x (NO + NO_2) to noncatalytic HNO_3.

From this brief description of the chemistry of the stratosphere-mesosphere system, the complexity of the simplest steady-state chemical model should be very evident; but the chemical specification of this part of the earth's atmosphere is even more difficult in the real case, where the vertical and horizontal flow of atomic species and a variable solar input are present. One obvious step toward modeling the chemical structure and energy balance of the stratosphere successfully would be the empirical specification of the time variation in the solar spectral irradiance at wavelengths important in the oxygen-ozone chemistry.

The chemical composition of the stratosphere and mesosphere is also influenced by the flux of solar and galactic cosmic rays, mainly protons. Although their average long-term energy input into the stratosphere and mesosphere is very small compared to the energy supplied by absorption of solar UV radiation, the chemical effects resulting from the interaction of the solar and galactic proton fluxes with the air in the stratosphere and mesosphere may be very significant. One important effect of this interaction is the production of nitric oxide. On occasion, the quantity of nitric oxide in the stratosphere and mesosphere produced by solar proton events may be so great as to more than double the content of nitrogen oxides at high geomagnetic latitudes (Crutzen, Isaksen, and Reid, 1975). For example, the additional input of NO_x in the stratosphere resulting from the solar proton event of August 1972 caused depletion in stratospheric ozone concentration of about 20% at altitudes of 35-40 km (Heath, Krueger, and Crutzen, 1976). A sudden change in ozone concentration may result in large perturbations of stratospheric and mesospheric dynamics. It is important to note that the photochemical lifetime of NO in the stratosphere is very long and, therefore, a single energetic solar proton event may cause perturbations in the stratospheric composition and energy balance that last for several years.

Solar proton events cause an immediate but short-term effect on the mesospheric radiative heating rates; a sharp decrease in ozone content occurs due to enhanced production of OH and larger catalytic destruction of ozone. The reason for only a short-term effect in the mesosphere is the short photochemical removal time of H, OH, and HO_2. It seems unlikely, therefore, that the global effects of a solar proton event can be as important in the mesosphere as in the stratosphere.

In the stratosphere and most of the troposphere, the steady ionization of the air is caused primarily by galactic cosmic ray protons. There is a well-known variability in ionization rates that follows the 11-y sunspot cycle, with low cosmic ray flux (i.e., low ionization rates) occurring during the maximum phase of the solar cycle. It has been postulated that the variable input of NO resulting from these varying ionization rates should be the cause of the cyclic change in the total concentration of ozone detected at the high latitude station at Tromso, Norway (±5% with maximum values occurring about 36 months after the maximum of the solar cycle, according to Angell and Korshover, 1973). Although such fluctuations in ozone content could be explained qualitatively by the variable input of NO in the lower stratosphere (Ruderman and Chamberlain, 1975), detailed two-dimensional calculations by the author show that this can only account for no more than 10% of the observed total ozone variability. There must, therefore, be another reason for the observed periodicity in the total ozone concentration— possibly variability of the output of solar ultraviolet radiation in the 1800-2400 Å band. The cause for these variations in total ozone is not necessarily to be found in stratospheric processes. For instance, more intense poleward transfer of ozone from the equatorial regions and/or a lower average location of the tropopause during the maximum phase of the solar cycle could cause more ozone to be collected in the lower stratosphere in polar latitudes. This implies the existence of sun-weather relationships in the troposphere, causal relationships which are still highly speculative.

Statistical evidence for such relationships has been claimed by King (1975). Unfortunately, physical mechanisms that would explain such sun-weather relationships are not well established. A new approach may be to explore the possible influence of the variable ionization on the earth's electric field properties, which may effect thunderstorm frequency, the hydrological properties of the atmosphere and, thus, the meridional transfer of latent heat versus sensible heat. Other interesting speculations on solar activity and weather involve cirrus cloud formation and aerosol formation. For more detailed treatment of some of these possible mechanisms see discussions by Markson (1973), Roberts and Olson (1973), and Dickinson (1975). Reck (1976) discusses some of the possible climatic effects that may be caused by changes in atmospheric composition and, in particular, by changes in the ozone distribution.

References

Angel, J.K., and Korshover, J., 1973, *Mon. Wea. Rev. 101*, 426.
Crutzen, P.J., 1973, *Pure Appl. Geophys. 106-108*, 1385.
─────────, Isaksen, I.A.A., and Reid, G.C., 1975, *Science 189*, 457.
Dickinson, R.E., 1975, *Bull. Am. Meteorol. Soc. 56*, 1240.
Heath, D.F., Kreuger, A.J., and Crutzen, P.J., 1976, submitted to *Science*.
King, J.W., 1975, *Astronaut. Aeronaut. 13*, 10.
Markson, R., 1973, *NASA SP-366*, 171-178.
Reck, R.A., 1976, *Science 192*, 557.
Roberts, W.O., and Olson, R.H., 1973, *Rev. Geophys. Space Phys. 11*, 731.
Ruderman, M.A., and Chamberlain, J.W., 1975, *Planet. Space Sci. 23*, 247.

The Troposphere
Stephen Schneider

Solar variability is a controversial, yet often invoked hypothesis to explain tropospheric changes. These can be classified into two categories: short-term or weather effects (lasting a few days to a few weeks) and longer term climatic effects (occurring on time scales greater than a few weeks). In addition, both categories of terrestrial responses have been associated with solar variability whose energy is much less than the energy of the terrestrial response, thereby requiring some amplification mechanism to explain an observed correlation (e.g., see Olson,

Roberts, and Zerefos, 1975, for a postulated statistical solar-weather connection and see Dickinson, 1976, for an example of a postulated physical amplification mechanism).

The most obvious physical mechanism that might explain tropospheric variability in terms of solar forcing is the possibility that such changes are imposed by variations in the total solar irradiance, a physical process that need not necessarily invoke an amplification mechanism to make the terrestrial response energetically consistent with the postulated variation in solar output.

To date, no *long-term* record of the total flux of solar radiation reaching the earth's orbit has been compiled from an extraterrestrial platform. Many attempts have been made to measure the solar "constant" (which we will henceforth call the total solar irradiance, S), and the difference between individual measurements often exceeds the stated accuracy of each measurement (see Frohlich's article in Chapter III). Furthermore, many of these data were taken within the earth's atmosphere, which increases the chance that the record was contaminated by fluctuations in atmospheric turbidity. However, two well-known studies, by a group at the Smithsonian Astrophysical Observatory and a Soviet group, have related S to the mean number of sunspots, R. The findings of both groups suggest that the value of S increases with sunspot number but eventually reaches a maximum at R = 100 and subsequently decreases. The Kondraytev and Nikolsky (1971) result, for example, suggests that S increases more than 2% from no sunspot activity to moderate activity where the sunspot number is about 80 and then decreases back to its lowest values for higher activity (R $\sim$ 200).* Recent examination of the Smithsonian data by Eddy (1976) suggests that S may vary with the envelope of the solar cycle, not the average value of the sunspot number.

Although the validity of these relationships between R and S is not yet established, the potential climatic consequences of such a large variation in the solar irradiance, if it occurs, are sufficiently compelling to warrant testing what such a variation in solar output might mean. In order to provide a physically tangible example of the potential climatic response to solar constant variation, Schneider and Mass (1975) used a simple climatic model of the earth's heat balance.

This model leads to the following analysis. The sensitivity β_S of the *global* surface temperature T_S to changes in total solar irradiance is defined as

$$\beta_S = S \frac{\partial T_S}{\partial S} .$$

Various values for β_S can be obtained by using different physical and mathematical models to compute the relationship between T_S and S. The simplest approach,

---

*Editor's Note: Recent results by Willson and Hickey, in this volume, show the solar cycle variation of S to be much smaller, $\Delta S < 0.75\%$, from 1969 to 1976.

however, is to use not T_s but the planetary radiative equilibrium temperature T_p. Then the planetary radiation balance relation,

$$S(1-\alpha) = \sigma T_p^4$$

where α is the earth's albedo and σ is the Stefan-Boltzmann constant,* can be used to estimate the sensitivity

$$\beta_p = S \frac{\partial T_p}{\partial S}.$$

For mean earth conditions of $\alpha = 0.3$ and $T_p \simeq 255K$, the sensitivity, β_p, is $\sim 65K$. That is, a 1% decrease in S would lower T_p by 0.65K. But the global surface temperature T_s is about 287K, and its sensitivity to changes in S depends on the changes in absorbing terrestrial gases that might occur simultaneously with changes in T_s. For example, in the one-dimensional radiative-convective model by Manabe and Wetherald (1967), it is assumed that the relative humidity of the earth's atmosphere is nearly constant, and this assumption leads to an estimate of $\beta_s \simeq 120K$, nearly double the estimate for β_p.

If the positive feedback effect of ice, temperature, and albedo were included, β_s could be increased by as much as a factor of four; and if negative climatic feedbacks were included, it might be reduced severalfold (e.g., see the survey by Schneider and Dickinson, 1974, or NAS, 1975). However, since the uncertainties in the present state of the art cannot resolve even the algebraic sign of all climatic feedback mechanisms, it is sufficient for our purposes to point out that a state-of-the-art, order-of-magnitude estimate of β_s is

$$50 \lesssim \beta_s \lesssim 500°K$$

This estimate is derived from the range of values encountered in the literature of climatic modeling.

Should the lower limit, i.e., 50K, for β_s prove to be closer to reality, then a variation in S of some 2% ($\simeq 27$ Wm^{-2}) would be needed to explain the $\sim 1K$ *global* surface temperature variations that have occurred over the past few thousand years. However, if the higher limit were closer to reality, then a change of only 0.2% in S ($\simeq 2.7$ Wm^{-2}) sustained over several years or longer could explain such observed climatic fluctuations.

---

*We are, for simplicity, considering values of S and α integrated over wavelength in this case; whereas in reality the integrated product

$$S\alpha = \int_0^\infty S_\lambda(\lambda) \, \alpha_\lambda(\lambda) \, d\lambda$$

is what describes the amount of solar energy reflected to space. Thus, it is not only the variability of the total solar irradiance that is of interest to climatic theory, but also the corresponding change in the shape of the solar spectrum, since the earth's albedo is not uniform with λ.

Unfortunately, climatic theory is not yet able to provide much more than this rough, factor-of-ten estimate of the sensitivity of the global surface temperature to long-term variations in solar energy input, and it is imperative for terrestrial climate theory that both observational and theoretical investigations of the possible variability of S be undertaken. This need to establish the climatic sensitivity, β_s, from a known change in total solar irradiance is of more than academic interest. A knowledge of the terrestrial climatic response to a known solar variation would provide a gauge against which climatic models could be calibrated. These models, in turn, are the only tools available for society to estimate the seriousness of the growing influences human activities may have on climate, and some present climate models suggest that the climatic consequences of some energy-related human activities will soon produce detectable and possibly irreversible climatic changes (e.g., see the discussion in Chapters 5 and 6 of Schneider and Mesirow, 1976).

REFERENCES

Dickinson, R.E., 1975, *Bull. Am. Meteorol. Soc. 56*, 1240.
Eddy, J.A., 1976, *Science 192*, 1189.
GARP, 1975, *Report of the Global Atmospheric Research Program Study Conference*, GARP Publication Series No. 16, WMO, Geneva, Switzerland.
Kondratyev, K. Ya., and Nikolsky, G.A., 1970, *Q.J.R.M.S. 96*, 509.
Manabe, S., and Wetherald, R.T., 1967, *J. Atmos. Sci. 24*, 241.
NAS, 1975, *Understanding Climatic Change: A Program for Action*, National Academy of Sciences, Washington, D.C.
Olson, R.H., Roberts, W.O., and Zerefos, C.S., 1975, *Nature 257*, 113.
Schneider, S.H., and Dickinson, R.E., 1974, *Rev. Geophys. Space Phys. 12*, 447.
─────────, and Mass, C., 1975, *Science 190*, 741.
─────────, with Mesirow, L.E., 1976, *The Genesis Strategy: Climate and Global Survival*, Plenum, New York.

II. THE EVIDENCE FOR A VARIABLE SOLAR OUTPUT

Paleoclimatic Evidence for the Solar Cycle and its Variation
Murray Mitchell

The paleoclimatic record on Earth deserves close inspection for possible clues to the long-term behavior of the sun. A number of points can be made.

First, there is no evidence that, during at least the last 3 billion years of the earth's history, the climate of the earth as a whole ever ranged below the ice point, or that it ever approached the boiling point of water. This conclusion follows from evidence of life during all of that period and from theoretical considerations that indicate that an ice-covered earth would be unlikely ever to recover from such a condition, even in the face of a solar constant several percent higher than its present-day value. All this suggests that the solar constant is unlikely to have varied during the past 3 billion years by more than 10 or 15%, at least for periods of time longer than a few months or years. A caveat is needed here, however, and that concerns the possibility that the earth's biota is capable of modulating the planetary temperature over geologically long periods, through biogenic feedback mechanisms such as oceanic plankton blooms (which affect ocean albedo) or changes of atmospheric CO_2 levels (which alters the "greenhouse" warming effect on the atmosphere). In such a case, past long-term variations of the solar constant over a range in excess of 10 or 15% might be possible without resulting in inadmissibly large variations of terrestrial climate.

Secondly, during the past 600 million years of the earth's history there seem to have been a rather large number and variety of climatic changes of a periodic or quasiperiodic character. On the whole, a great many causal mechanisms can be visualized to account for such climatic changes, and an explanation for most of them in terms of solar variability is unnecessary. The principal quasiperiodic variations of climate are identified below in terms of the characteristic wavelength and possible origins of each.

Nature of Variation	Characteristic Wavelength	Most Plausible Origin(s)
Interval between major ice ages	$\sim 3 \times 10^8$ y	Tectonic plate motions (thermal isolation of polar continents)
Lesser long-term climatic changes	~ 3 to 6×10^7 y	Mountain building episodes; marine transgression epochs; changing vulcanism (?)
Changes of ice volume and general temperature levels during present ice age	$\sim 100,000$ y $\sim 40,000$ y $\sim 20,000$ y	Earth orbital variations (Milankovitch parameters)
Little Ice Age cycle during past 10,000-y postglacial period	$\sim 2,500$ y	Solar variability (?)
Lesser general climatic changes (?)	$\sim 200, 400, 800$ y	Solar variability (?)

| Relatively minor climatic changes on time scales of years and decades | ~ 10, 20, 80 y | Internal variability of climatic system; volcanic activity; solar variability (?) |

It may be noted that the greatest reliance on solar variability as the plausible origin of climatic changes involves changes of the order of centuries and millennia. These are the same ranges of time scale for which ^{14}C variations and other evidence of solar variability are clearly suggested by nonmeteorological data, implying real solar/climate relationships on such time scales.

Many of the above-mentioned points are further explored in the paper by Hays in this volume. He concludes that whether or not actual climatic changes are controlled to a large extent by solar activity, a better understanding of climatic changes of the past can be expected to clarify for us the sensitivity of the climatic system to *postulated* changes of solar output with a specified magnitude and on a specified time scale of solar change. Advances in the numerical modeling of climate, now progressing at a healthy pace, will ultimately provide more quantitative insights into solar variability effects on climate—or so we have every reason to believe.

As evident in the contribution by Vernekar, the history of the earth's orbital variations (orbital eccentricity, obliquity of the ecliptic, and precession of the earth's axis) is now quite precisely known for at least the past million years. These variations do not greatly affect the solar radiation received by the earth overall during the year, but lead to alterations in its latitudinal and seasonal distribution on time scales of 10^4 to 10^5 y, locally reaching magnitudes on the order of 10%.

Data on historical variations of climate, ^{14}C, and other indicators of possible solar variability on time scales of decades, centuries, and millennia are presented in the papers by Eddy and Damon. From these data it is evident that the rhythmical character of solar variability seen in the modern record since 1700 AD, including the 11-y and longer cycles, is not necessarily characteristic of the sun in the last 7,000. At least temporarily, the 11-y solar cycle may not have been operating prior to 1700 AD (specifically, in the Maunder Minimum period, from 1640 to 1715 AD), and the possibility must be recognized that the sun could again lapse into such an "anomalous" state of behavior (perhaps one in which the 11-y cycle is suppressed) at any time in the future. Variations of solar constant—to whatever extent such variations are real—may conceivably follow not the 11-y sunspot cycle but some other pattern, possibly the envelope of the 11-y sunspot cycle. These matters cry out for better observational documentation and theoretical understanding.

The Integrated Solar Flux
John Eddy

The "solar constant," S, is the amount of solar radiative energy of all wavelengths received per unit time and unit area at the top of the earth's atmosphere, corrected to mean sun-earth distance. The most probable modern value, from measurements made by modern techniques in the last 10-y, is

$$S = 1373 \pm 20 \text{ W m}^{-2}.$$

This result is a "most probable" value derived from a number of different and disparate experiments made above the surface of the earth: six from aircraft, three from balloons, and three from spacecraft. It was derived for this compendium by Claus Frohlich, who critically evaluated each experiment, allowed for likely systematic differences, and reduced each measurement to the same radiometric scale. A more thorough discussion of the error assessment and of evidence for secular change during the last decade is contained in Frohlich's review in Chapter III.

Most modern efforts at determining the solar constant have been directed at determining its average or instantaneous value, as above, rather than possible short- or long-term variability. This is more than anything a reflection of the recent growth of aircraft, rocket, and space technology: the opportunity to measure S from above the atmosphere has come only recently, and the measurements of the past decade or so have been largely exploratory attempts using new vehicles. Only very recently have we had the opportunity to make continuous, calibrated measurements of S from orbiting platforms.

In any case, it is sadly true that at present the limits on temporal change in S are poorly known. Ground-based measurements have proved of only marginal value in this regard. Early in this century, the Smithsonian Astrophysical Observatory (SAO) began a series of mountaintop measurements of S that continued for nearly 50 years. This program, under the direction of Charles G. Abbot, and before him, Samuel P. Langley, reported changes in S at the 0.1 to 1% level that were deemed to be of possible significance. There was also an apparent long-term increase in the average value of S of about 0.25% in 50 years. Current assessment of the SAO data suggests that most or all of the changes reported are probably indistinguishable from uncertainties in atmospheric transmission, for which corrections were necessarily large, and from changes in instruments, calibration, and technique. As Frohlich points out in his review, the atmospheric transmission at noon on Abbot's Table Mountain (California) site was only about 76%, requiring a correction of 24% in the deduced solar constant. Even on the highest modern astronomical stations, such as Mauna Kea or Mt. Evans, the midday transmission is only 82%. Broad band measurements of the solar constant from mountaintops are thus highly limited, if useful at all. Calibrated measurements of specific spectral regions from the ground are a different matter and will continue to be important in the evaluation of S and its possible variation.

A modern reevaluation of the SAO historic data is now under way to establish the limits of reliability of the data and the reality of any long-term effects. Until this is done the SAO data should probably be interpreted as indicating that changes in S over the first half of this century were within the limits of about ±1%.

More promising measurements of possible variability of S come from spacecraft data. The first continuous measurement of S from above the terrestrial atmosphere was made by the Jet Propulsion Laboratory experiment on Mariners VI and VII in 1969. The instrument was an absolute pyroheliometer, with measurement accuracy not less than ±1%. Data were obtained for a period of about five months, although vitiated somewhat by problems of instrument drift. Even though the Mariner solar constant measurement was made with a device designed for another purpose, the result is of crucial importance in solar-terrestrial studies, for it indicated that S did not vary by more than the limits of measurement precision, about ±0.25%, near the maximum of the solar cycle, during which time the daily sunspot number went through extremely high and low values. There was some indication of an interesting and as yet unresolved variability of a few tenths of 1% with a period of about 14 d.

A different solar monitor is now in operation on the Nimbus 6 spacecraft, launched in midsummer 1975. The instrument, a wire-wound thermopile, measures total radiation and spectral components in nine wavelength channels. The situation is not ideal, for the Nimbus instrument views the sun only about 5 min per orbit in a wide (18°) field, without solar pointing. Throughout the first 18 months of operation, during which time solar activity was very low, Nimbus 6 found S constant to within its measurement precision of about ±0.2%.

Two early results that linked variation of S with sunspot number R are frequently cited and deserve modern clarification. These are the balloon measurements by Kondratyev and Nikolsky, made in a series of more than 20 flight attempts between 1961 and 1967, and an analysis of some of the earliest SAO data by Anders Ångstrom in 1922. Both found evidence that S varied with R, first rising to some maximum value and then falling at very large sunspot numbers. Kondratyev and Nikolsky determined that S increased by more than 2.5% between low values of R up to R = 100, then fell by a similar amount with further increase in R. Ångstrom had found a similar relation, though reduced in amplitude and shifted in time from the early mountaintop data. He later retracted it as of likely atmospheric origin. The Kondratyev and Nikolsky result seems uncertain, since the variability which they found was close to their limit of measurement and smaller than their flight-to-flight repeatability. Moreover, as noted in Frohlich's review in Chapter III, the Mariner data, which were taken at about the same time and under a similar daily range of sunspot number, did not show the variability found in the balloon data. Modern measurements of S from space should clear up the uncertainty surrounding this point. We may hope that the Nimbus 6 experiment will answer the question as solar activity increases in the new solar cycle, 21.*

Photometric observations of Uranus and Neptune, made in the blue spectral region (4400 Å) at the Lowell Observatory from 1955 to date, when corrected for the effects of changing distance and solar phase angle, reveal a highly correlated, smooth, slow variation with a peak-to-peak amplitude of 2%. The variations appear to repeat after about 11-y, but it would be premature and highly speculative to conclude that a periodicity exists in the variations on the basis of only 1.5 cycles of observation.

A second series of Lowell Observatory observations, made from 1971 to 1976 at 4700 Å and 5500 Å, reveals monotonic linear increases in the brightness of Titan (a satellite of Saturn), Uranus, and Neptune. Each of these bodies has an atmosphere. The total brightening amounts of 6%, 3%, and 2%, respectively, averaged over the two colors. The accuracy of the annual mean magnitudes is about ±0.3% for each object and color, so that the reality of the brightening phenomena seems well established.

While the first set of observations does not exclude the possibility of a 2% variation of the sun's visible output, such a variation has never been detected in other solar measurements. Because of the significantly different *rates* of brightening of the three objects, a mechanism is implied by which some variable component of the solar radiation or particle flux influences planetary albedos. If this turns out to be the case, then planetary brightness changes reveal solar variability in the nonvisible spectrum. This seems to be the only acceptable explanation for the highly-correlated variations, which have been observed now for nearly 20 y.

With Nimbus 6 now operational and a similar follow-up experiment planned for Nimbus 7, we can have some confidence that the possible variability in S and its spectral components will at last be more definitely known. It is encouraging that solar constant experiments are planned for the NASA Solar Maximum Mission and for the Space Shuttle. The Space Shuttle seems an especially promising method for the task since many missions are planned and since the instruments can be calibrated shortly before and after each flight. Balloon-borne and rocket measurements will also be valuable in the decade or so before the start of the Space Shuttle series. Measurements from high altitude aircraft, such as were carried out in the 1960s, can probably no longer compete. Nor do ground-based measurements of the total integrated flux seem of value, because of problems of variable atmospheric absorption at even the best sites. What seems important and probably crucial from the ground is a long-term program of monitoring solar flux in specific wavelength regions of interest, where atmospheric correction can more reasonably be made and monitored.

---

*Editor's Note: New results by Willson and Hickey in Chapter III show the change in the solar constant from 1969 to 1976 is less than .75%; thus, the Kondratyev and Nikolsky result is not confirmed.

Whether these new programs will be continued for a sufficiently long time is another matter. It is unfortunate that the continuous measurement of the solar constant has never been given priority in science, either by funding agencies or by atmospheric or solar physicists. A program to measure S and its possible variation with solar activity was begun in England by Balfour Stewart in the days of Queen Victoria, but was soon dropped—for reasons that have ever since damped interest in the task: it is a difficult measurement; it is an unexciting measurement; it requires dedication and funding over periods measured not in years but in decades. Modern science does not seem set up to tackle problems of this time scale.

The Solar Spectrum
Introduction
Richard Donnelly and Oran White

When we look at the solar spectral irradiance measurements with the goal of stating the absolute fluxes and their degree of variability, we find the time record to be seriously incomplete, even for one solar cycle. Furthermore, this experimental field is marked by active controversy and quantitative disagreement about the absolute flux levels at all wavelengths of interest in the solar-terrestrial interaction. This situation arises primarily because of the experimental difficulties in absolute radiometry over the enormous wavelength range that extends from the radio region to X-rays. Over the entire solar spectrum at least five different experimental techniques are required for imagery and dispersion of the solar spectrum, not to mention some rather major difficulties in the absolute radiometry in particular spectral regions. The space technology to observe the solar spectrum outside of the earth's atmosphere has been largely developed over the last fifteen years, and we are just now at the stage where high-resolution measurements of the spectral irradiance are feasible on a routine basis from space. The experimenters in this field readily admit their dissatisfaction with the state of the science as far as our view of the spectrum is concerned, and they continue to press the problems in a competitive fashion that should be fruitful over the next decade.

The various spectral regions examined in this study are as follows:

Radio	$\lambda \geq 1$ mm
Far infrared	$1 \text{ mm} > \lambda \geq 10 \text{ }\mu\text{m}$,
Infrared	$10 \text{ }\mu\text{m} > \lambda \geq 0.75 \text{ }\mu\text{m}$,
Visible, or optical	$0.75 \text{ }\mu\text{m} > \lambda \geq 0.3 \text{ }\mu\text{m}$,
Ultraviolet (UV)	$3000 \text{ Å} > \lambda \geq 1200 \text{ Å}$,
Extreme ultraviolet (EUV)	$1200 \text{ Å} > \lambda \geq 100 \text{ Å}$,
Soft X-rays	$100 \text{ Å} > \lambda \geq 1 \text{ Å}$,
Hard X-rays	$1 \text{ Å} > \lambda$

where, of course, 1 m = 10^3 mm = 10^6 μm = 10^9 nm = 10^{10} Å. The absolute flux values characteristic of the various wavelength bands are shown schematically in Figure 1, prepared by Harriet Malitson from data in Chapter IV. The figure clearly shows the shape of the absolute solar spectrum with the dominance, energetically, of the visible spectrum between 3000 and 10,000 Å. The intrinsic variability of the spectrum is difficult to depict in such a figure, but it is clear from Malitson's presentation that the visible spectrum is relatively stable, while the radio, UV, and X-ray regions display large fluctuations associated with solar activity.

Figure 1

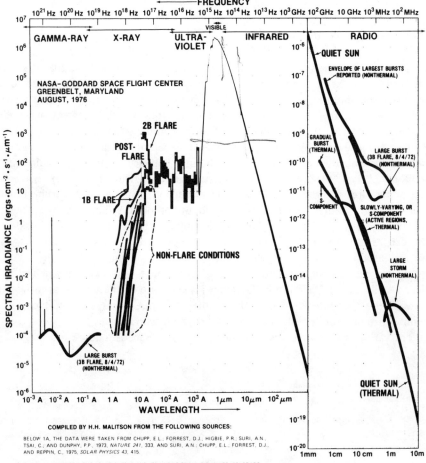

In addition to the specific reviews of each wavelength range in Chapter IV, Robert Madden and Richard Donnelly give their views on the problems of absolute radiometric standards and the comparison of different measurements. Madden describes both the source and detector standards currently available from the U.S. National Bureau of Standards and the research program designed to improve existing standards and develop new ones. As a user of solar spectrum measurements in ionospheric problems, Donnelly speaks directly to the problems of comparing measurements made at different times and with different techniques. As future programs make use of spectral irradiance measurements both from the ground and from space, internal consistency over the entire solar spectrum can only be obtained through use of good absolute calibration procedures that account for atmospheric effects above the instrument and for the sensitivity variation of instruments in space. As Donnelly emphasizes, the combination of different measurements into a single quantitative picture of the solar spectrum can be successful only when the measurements were made at (or can be corrected to) comparable spectral resolutions, similar levels of solar activity, and negligible stray light levels.

The problems of physical units in the discussions in Chapter IV continue to plague us. Although uniform adherence to one system such as the International System of Units (SI) is preferred, individual preferences for cgs units show up in the review papers. In most cases, the reviews had progressed to the point where a change in units was impractical, but the conversion factors are stated in the tables of flux values.

A similar problem exists in comparing percentage errors stated for various parts of the solar spectrum. Absolute measurements have been made at optical wavelengths for a long enough time that experimenters are usually careful in their use of words and error estimates. They speak of precision (repeatability of measurements), relative temporal and spectral accuracy, and standard deviations, which are all well-defined quantities. However, in many measurements from satellites, the term "best estimates of error" is given as the experimenter's estimate of the *total* uncertainty in his results. The lack of use of a uniform, statistically defined set of error estimates often prohibits meaningful comparison of discrepant measurements. In such cases, the ultimate source of the experimental difficulty is concealed from us because the systematic and random errors are not clearly separated.

In addition to summarizing the accuracy and resolution of existing measurements, we also consider the accuracy requirements for new experiments. In many cases the requirements are set by the needs of ionospheric and atmospheric physics rather than by solar physics alone. We hope that colleagues in these disciplines will examine and verify the experimental parameters (absolute radiometric accuracy, relative accuracy, and spectroscopic resolution) indicated for future work and, thereby, supplement the information we gathered from sections 4.4

and 6.2 of *The Physical Basis of Climate and Climate Modeling*, no. 16, GARP publications Series, World Meteorological Organization, 1975.

Radio Wavelengths
Fred Shimabukuro

The solar radio spectrum ($\lambda > 1$ mm) contains an almost negligible fraction of the total solar irradiance, but it does provide valuable information about the solar atmosphere and its variation. The radio flux is usually discussed in terms of the quiet-sun emission, the slowly varying component (or S-component), and radio bursts. The quiet radio flux is the solar output without the contribution of discrete sources, and the quiet radio sun can be represented as a blackbody with an apparent brightness temperature ranging from about 6000K at 1 mm to over a million degrees at the meter wavelengths. The S-component has its origin in localized areas of high electron densities and magnetic fields that rotate and evolve with the associated active regions. The highly transient solar burst emission runs through the entire radio spectrum, having intensities ranging up to values orders of magnitude greater than those of the quiet sun on a time scale from less than a second to days.

Absolute measurements of the solar radio flux are typically made by comparing the intensity received from the sun with a blackbody calibration source. The principal sources of error are the uncertainty of the antenna characteristics, the fact that the sun and the calibration signal are not coupled in exactly the same way into the antenna terminals, and the effect of atmospheric attenuation. An absolute flux determination to 5% can be achieved with specially calibrated antennas and is most easily done at the cm wavelengths. Relative measurements are more precise, and changes of less than 1% in one second can be detected at cm wavelengths. Measurements at meter wavelengths have a basic signal-to-noise problem since the flux levels are low, the sky background brightness temperature is at a relatively high level, and antennas in this wavelength range are generally low gain. The frequency resolution varies from $\nu / \Delta \nu > 10^5$ in spectral line work to $\gtrsim 10^2$ in typical flux measurements.

Solar radio astronomers usually do not measure the solar flux with high absolute accuracy because errors of 10% can easily be tolerated in the models of the solar atmosphere. Accurate relative values are more important in measurements of bursts, limb brightening, and disk structure, where emphasis is on spatial and temporal resolution. At the present time the 10.7 cm flux is monitored with good absolute and relative accuracy as an activity index.

Although the radio flux has little direct impact in the solar-terrestrial interaction, its study is of interest in radio communications (where solar bursts can be a disruptive source of noise), as a diagnostic tool in short-term predictions of solar events emitting high-energy particles, and as an indicator for other solar fluxes, such as the UV and X-ray emissions. It should be pointed out that the 10.7 cm flux is determined to a greater accuracy than measurements of the UV flux; and if there is a failing in correlations between the radio and short wavelength emissions, it is in the application and not the accuracy of the 10.7 cm flux measurements.

The Far Infrared Region
William Mankin and Heinz Neckel

The solar disk spectrum from 10 μm to 1000 μm may be represented by blackbody intensities corresponding to temperatures near 4950K at 10 μm, declining to a broad minimum about 4400K near 150 μm, and rising to near 6000K at 1000 μm. The integrated irradiance in this wavelength region at mean earth-sun distance is 0.802 ± 0.026 Wm^{-2}, which amounts to only .057% of the solar constant. The spectrum is continuous, so the spectral resolution of observations is comparatively unimportant. The long wavelength solar emission from active regions can increase the irradiance at 500-1000 μm by amounts on the order of 1%, but this has no known important consequences for the earth.

There is agreement within observational errors among different observations (with one exception) as well as among observations and model predictions. The typical error given by the experimenters is about ±7%, and the mean level appears to be established with an accuracy ranging from ±3% at 15 μm to ±5-10% between 100 μm and 1000 μm. An increase in absolute accuracy by a factor of two to four seems to be possible with recent techniques. Such improvement would be important in a better determination of the temperature minimum between the solar chromosphere and corona.

The Optical and Infrared Region
Heinz Neckel

The 0.3 to 10 μm region contains 99% of the total solar radiation and thereby strongly weights the solar constant. It is one of the most important spectral regions with respect to both the thermal balance of the earth's atmosphere and our understanding of the solar photosphere and lower chromosphere. It includes the maximum of brightness temperature near 1.6 μm, which is the emission from the deepest layer observable on the sun. Optical and infrared radiation are usually considered to be "quiet sun" emissions. To date there are very few independent measurements of the spectral irradiance in this band, and none that reliably indicate variation in these solar fluxes except for the known changes in certain

Fraunhofer lines affected by the presence of action regions on the solar disk.

The accuracy of observations available today remains a point of controversy because of systematic differences between the different flux determinations. The limit set by the absolute accuracy of radiation standards is about 1-2%; however, occasional calibration errors up to 8% have been reported that obviously affected some recent observations of the solar irradiance. The precision of the comparison between the sun and the radiation standard certainly is not the only factor in determining the overall accuracy. However, high precision is one condition for high absolute accuracy, and the precision obtained should be used as one of the criteria to judge the reliability of the different observations.

Concerning the region above 5000 Å up to 12 μm, a mean error of ±1.5% or even better is representative for the most precise observations compiled by Labs and Neckel. Actually, the continuum intensities derived from these observations scatter around any reliable model prediction with a standard deviation of ±1.3% (see the paper by Avrett in Chapter IV). The total flux agrees with the value of the solar constant given by Frohlich (see Chapter III) within 1.4%. Below 5000 Å, the accuracy becomes less: it is estimated to be ±4-5% around 4000 Å and may rise to about ±10% near 3000 Å.

In the visible and near infrared, an increase of absolute accuracy by a factor of two may be obtained with considerable effort. Such improvement is of academic interest in solar modeling because the theory cannot take such accuracy into account. From the viewpoint of measuring variations of solar irradiance over a solar cycle to determine its effect on the earth's atmosphere, a long-term precision of ±0.5% is needed. Although, under good atmospheric conditions, the error introduced by extinction can be kept around ±0.5%, there is hardly a ground-based site with a sufficiently large number of such clear days to determine variations with periods on the order of days or weeks. For this purpose measurements from outside the atmosphere are required, with a precision on the order of 0.5% maintained over a period of years.

Ground-based measurements of certain important minor constituents in the earth's atmosphere, e.g., ozone, make use of pairs of narrow wavelength bands in the solar spectrum. These measurements are interpreted by assuming that solar flux at these wavelengths is constant. Small variations in the solar flux at these wavelengths could cause errors in the interpretation of ozone temporal variations. It is therefore important that extraterrestrial measurements of temporal variations in the solar spectral irradiance include high-wavelength-resolution measurements at the particular wavelengths used by atmospheric monitors. Wavelength pairs commonly used for monitoring the columnar ozone content are as follows: pair A, 3055 Å and 3254 Å; pair B, 3088 Å and 3291 Å; pair C, 3114.5 Å and 3324 Å; pair D, 3176 Å and 3398 Å; and pair C', 3324 Å and 4536 Å. The bandwidths at these wavelengths range from about 18 to 40 Å (p. 47, G.M.B. Dodson, Part III, Observer's Handbook for the Ozone Spectrophotometer, *Annals of the*

International Geophysical Year V, Pergamon Press, London, 1957; K. Hanson, NOAA).

The Ultraviolet Region
Donald Heath

The solar flux in the 1200 to 3000 Å region is important because of its effects in the upper atmosphere and because of its value in empirically modeling the solar chromosphere. Although this band contains only about 1% of the total solar irradiance, this energy is significant because it is completely absorbed by ozone and diatomic oxygen molecules in the upper atmosphere of the earth. As a result of these selective absorption processes, the values of the solar flux in this band are crucial to the understanding of the oxygen chemistry in the stratosphere (See Fig. 1 and Chapter IV for current flux estimates).

The spectral make-up of this region is complex because at about 1400 Å the solar spectrum makes a transition from a strong continuum broken by Fraunhofer absorption lines at longer wavelengths to an emission-line spectrum superimposed on a weak continuum background at shorter wavelengths. This change occurs because of the decrease in the continuum opacity in the solar atmosphere towards the UV. Below 1400 Å the solar chromosphere is sufficiently opaque for us to see its emission-line spectrum; at longer wavelengths, the continuum emission from the lower photospheric layers dominates the spectrum. Because of this transition property, the UV spectrum is a good spectral diagnostic for inferring physical conditions in the solar chromosphere and low corona.

Absolute accuracies are currently ±10 to 20%, and some experimenters conclude that ±4 to 8% should be achievable. The relative accuracy as a function of either wavelength or time is about ±2% at best. The relative flux accuracy required in studies of the time variation of ozone is ±1%. (For accuracy requirements see the *Physical Basis of Climate and Climate Modeling*, GARP Publ. Ser. No. 16, 1975, International Council of Scientific Unions, World Meteorological Organization; and the chapter on radiation by John Gille in *Planning Document for the Middle Atmosphere Program*, ed. S.A. Bowhill, Atmosphere Physics Programs Committee, Special Committee on Solar Terrestrial Physics, National Academy of Sciences, Washington, D.C.) Absolute flux accuracies of 10% to 20% have been requested for modeling the upper atmosphere above 70 km.

At the wavelengths of the absorption bands of O_2, i.e., 1750-2040 Å, the spectral resolution in total irradiance measurements needed for atmospheric research is about 0.020 Å, a resolution which is attained in current space experiments. For long-term monitoring in this region, a 1 Å resolution appears to fit most atmospheric requirements, but at least one good high-resolution irradiance spectrum is still needed as a reference.

The largest percentage increase observed during a flare in the 1200 to 3000 Å range is in the Lyman-alpha line of neutral hydrogen, e.g., about 16% for a class 3b flare. Similarly, the variations associated with the 27-d rotational period decline exponentially with increasing wavelength, being about 25% at 1200 Å and about 1% at 3000 Å. At 1750 Å, this variability is comparable to that due to the 6% annual variation associated with the changing sun-earth distance. One of the least understood and one of the most important questions is whether or not the solar UV flux changes over the 11-y solar cycle. Some measurements indicate a factor-of-two variation near 1700 Å, but colleagues dispute this (see Chapter IV). Note that it is well established that the solar flux below 1200 Å does vary over a solar cycle, presumably because of the strong effect of solar activity on the emission from the chromospheric-coronal interface.

The need now exists for a concentrated campaign to measure the UV solar spectral irradiance from solar minimum to solar maximum during the next solar cycle using the most advanced absolute spectroradiometric techniques. Such measurements are extremely important for understanding observed variations in the structure of the stratosphere and mesosphere with regard to whether they are produced either by changes in the solar spectral irradiance or by the introduction of anthropogenic constituents. The improved measurement and absolute calibration techniques now permit a program to resolve the question of the large irradiance variation at 1700 Å empirically over the next solar cycle, but the achievable accuracy will probably be less than the 1% level required in the ozone studies.

The Lyman-alpha line at 1216 Å
Elmo Bruner and Alfred Vidal-Madjar

The solar emission in the hydrogen Lyman-alpha line, which contains more radiative energy than at all shorter wavelengths in the solar spectrum, is important as the source of ionization in the upper D-region of the earth's ionosphere at times of low solar activity. As the resonance line of the most abundant chemical element in the solar system, this particular spectral line is of prime importance in understanding the physics of the upper solar chromosphere as well as the atmospheres of the other planets and comets. Despite these important consequences of the selective absorption by neutral hydrogen, the line contributes negligibly to the total radiative energy output of the sun.

The integrated flux in the Lyman-alpha line depends upon the level of solar activity, varying within limits of about 3.3 to 6.5 erg cm^{-2} s^{-1}. This energy is distributed across a complex profile with self reversals of both solar and terrestrial origin. If measurements could be made free of the earth's hydrogen envelope, the specific intensity at the bottom of the solar reversal should lie within the range 2.5 to 8.1 erg cm^{-2} s^{-1} Å^{-1}, again depending upon the level of solar activity.

These profiles are of the order of 1 Å wide and have been observed with a spectral resolution approaching 0.01 Å. Most observers quote accuracies on the order of ±30% in reporting the absolute flux. Experimenters indicated the current state of the art is ±10% in absolute accuracy and ±5 to 10% relative temporal accuracy.

Variations due to solar rotation can be ±15% while the flux can vary by a factor of about two from minimum to maximum of the 11-y solar cycle. Empirical relations have been found between the solar Lyman-alpha flux and the 10.7 cm radio flux or the Zurich sunspot number. These give predictions in agreement with most measurements to within ±10%. The relations were deduced from high or medium solar activity conditions and are probably less accurate during conditions of very low solar activity.

After reviewing the different needs for Lyman-alpha measurements, it appears that absolute flux values accurate to ±10 to 20% are required whereas a ±1% relative accuracy both across the profile and in time will be useful in both solar and terrestrial atmospheric programs. An attempt to eliminate the systematic absorption of the line core by the geocorona should be made from a high-altitude spacecraft operating at about 10,000 km. Long term monitoring of the Lyman-alpha flux and profile from spacecraft will only be successful when we are able to make in-flight, periodic cross-calibrations with rocket or space shuttle payloads.

Extreme Ultraviolet and X-Rays (> 10 Å)
Richard Donnelly

The solar flux in the 10-1200 Å range is important because it is the main ionizing source for the ionosphere (E and F regions). It also includes the resonance lines and continua for recombination to the ground state for all the major solar constituents; consequently, the solar spectrum in this range is very important for studying the upper chromosphere, transition region, and corona. Even though the energy flux below 1200 Å is a negligible portion of the total solar irradiance, this band is of interest in solar-terrestrial physics because its flux level is so critical in determining the state of the ionosphere.

The accuracy of the absolute solar irradiance measurements between 31 and 1200 Å appears to be ±30%. In the 10-31 Å range, the accuracy is at best ±20% and typically about ±50%. For modeling the ionosphere, it would be desirable in the near future to improve the accuracy of the absolute measurements to about ±10% in the 90-911 Å range, and ±20% in the 10-90 Å and 911-1200 Å ranges. Although different experimental groups assign errors of about ±30% to their measurements between 300 and 1200 Å, there appear to be discrepancies in the fluxes of at least a factor of two in particular wavelength intervals. These discrepancies may be associated with different levels of solar activity. However, it is also possible that the discrepancies arise from different treatment of the necessary corrections applied to the data to account for scattered light and higher order interference (see Chapter IV).

At wavelengths greater than 170 Å, the variations of the flux during single solar rotations, excluding flares, are typically ±15% for radiation emitted by the chromospheric and transition-region ions at temperatures less than 10^6 K, and can be as large as a factor of five for coronal lines emitted by ions requiring temperatures above 10^6 K,. SOLRAD observations in the 8-20 Å range exhibit roughly a factor-of-five variation with rotation and two orders of magnitude over a solar cycle.

At wavelengths longer than 300 Å, near-normal-incidence and grazing-incidence grating spectrometers are the main tools, while only grazing-incidence grating spectrometers are practical at wavelengths between 300 and 31 Å. Below 31 Å, Bragg crystal spectrometers are necessary. Broad-band detectors can be used below 60 Å and at wavelengths longer than 170 Å. Very few absolute-flux data are available in the 60-170 Å wavelength range, i.e., there is a "hole" between the X-ray and EUV measurements, except for several rocket flights. There is evidence for long-term variability in the irradiance between 30 and 1200 Å over the 11-y solar cycle, but the magnitude of the variability is disputed. It cannot be determined accurately from the limited data available from the many different types of experiments.

The wavelength resolution achieved for Bragg spectrometer X-ray measurements has been about 0.1 Å for full-disk measurements and about 0.02 Å for observations with a 6 × 6 arcmin field. Grazing-incidence EUV measurements have achieved, at best, about 0.1 Å wavelength resolution for calibrated irradiance measurements. Some photographic data give about 0.03 Å resolution for relative spectral measurements and show that solar line widths are roughly 0.1 Å in the 30-300 Å range.

X-Rays (<10 Å)
Robert Kreplin and Arthur Walker

From a terrestrial viewpoint, X-rays below 10 Å have a different significance than those above 10 Å because the shorter X-rays are an important source of ionization in a different region of the ionosphere, the D region. The 0.5-10 Å X-rays with a flux below about 10^{-6} Wm^{-2} are not terrestrially important because during these quiet solar periods D-region ionization is produced predominantly by Lyman-alpha in the upper D region and by cosmic rays in the lower D region. From a solar physics viewpoint, we should distinguish between soft X-rays (>1 Å), where the spectrum is dominated by emission lines, thermal free-free continua, and recombination continua; and hard X-rays observed during solar flares (<1 Å), where the spectra appear to have a power-law continuum. Measurements of hard X-ray bursts are important primarily for analyzing the energetic electrons in the solar flare plasma. Although the photoionization from intense hard X-ray bursts can exceed that due to cosmic rays for several minutes at altitudes below the bottom of the normal D region, the consequent effects from X-rays below 0.5 Å are currently not known to be important.

The 1-10 Å solar flux varies greatly, so much that the trend has been to measure it all the time and publish it for general use in *Solar Geophysical Data*. The 1-8 Å flux varies by two to three orders of magnitude over a solar cycle. Large variations occur over the solar rotation period, with marked differences from one rotation to another according to the number, size, age, rate of growth, and spatial distribution of active regions. Solar flares produce increases of up to three orders of magnitude in 1-8 Å X-rays, and four orders in 0.5-3 Å. Because the correlation of X-ray burst intensities with optical flare classifications or radio burst intensities is low, X-ray burst measurements are the only way to evaluate the "Sudden Ionospheric Disturbances" (SID) produced by solar X-ray emission.

The monitoring measurements discussed above have been made with broad-band detectors, where our lack of knowledge of the spectral shape at a given instant is probably the main source of error in interpreting the data. The spectral shape assumed in interpreting the broad-band measurements has been improved through the use of a set of adjacent bands to adapt the spectral assumption to the set of measurements. The broad-band photometer current can be measured to within ±3%. The absolute calibration of the photometer can be known to ±10%. Bob Kreplin estimates the error from assuming a spectrum like that of a single-temperature emitting region is probably about ±30% and estimates the overall absolute accuracy in flux below 10 Å to be ±50% at best. From the viewpoint of ionospheric physics, an absolute flux accuracy of at least ±20% in the 1-10 Å range is desired.

Bragg crystal spectrometers can provide wavelength resolution down to 0.001 Å in this range. The observed wavelength distribution depends on the wavelength dependence of the crystal reflectivity and detector efficiency. At present, these can be measured with a relative flux accuracy of ±10%. The absolute accuracy of crystal reflectivity is difficult to measure; consequently, most absolute fluxes from published crystal spectrometer measurements should be used with caution. Art Walker has achieved the best accuracy in rocket measurements with high spectral resolution, and he estimates it is possible to measure the flux in the range 1-10 Å with about ±25% accuracy. Improved accuracy in this range must make use of high-resolution crystal spectrometer techniques. A realistic goal in this wavelength range is ±3-5% in accuracy and extremely good precision (i.e., repeatability). He suggests this is possible using sealed proportional counters and scintillators that are absolute detectors, with in-flight calibration using radio active sources. We suggest that a monitoring system using a combination of broad-band detectors and measurements with high spectral resolution could provide photometric measurements and the spectral distribution. In the range of energies above

10 keV ($\lambda \lesssim 1$ Å), proportional counter and scintillation measurements allow the irradiance to be determined to ±20% accuracy.

Plasma Flow from the Sun
Arthur Hundhausen

Two different experimental methods for obtaining information on the interplanetary plasma and magnetic field are drawn upon in the reviews by Heymann and Feldman in this volume. The first of these examines evidence for the past existence of the solar wind and its variations given by studies of the surfaces of solid objects that have been exposed to the solar wind on a geological time scale. The second describes the detailed characteristics of the interplanetary plasma and magnetic field observed *in situ* during the past 14 y of spacecraft exploration. I will attempt to summarize these two reviews here and to describe additional evidence for solar wind variability given by geomagnetic activity observations over more than a century and cosmic ray data available for several decades.

Ions with the keV and higher kinetic energies typical of the present solar wind are trapped upon impact on a solid surface. Thus the anomalous presence of noble gases in meteoritic and lunar surface material is attributed to exposure to the solar wind. Quantitative conclusions regarding solar wind properties can be deduced from study of such material only with great difficulty because of possible alteration of the thin surface layer where the trapping occurs and uncertainties in dating the time and duration of exposure. Nonetheless, the presence of layers rich in noble gases on a wide variety of lunar and meteoritic material indicates that "a solar-wind-like entity" has been present more or less continuously for about the past 4×10^9 years. These records also suggest (although they do not prove) secular changes in the isotopic composition of some trapped elements, e.g., a 40% increase in the ratio of ^{3}He to ^{4}He in the solar wind.

A body of historical information regarding the solar wind exists in records of terrestrial phenomena related to geomagnetic activity (see, for example, Chapman and Bartels, 1940). These records were collected quantitatively in the early nineteenth century, and three specific associations of solar and geomagnetic activity were recognized and studied at that time:

1. Geomagnetic activity varied with a period of about 11 y, a period readily associated with that of the sunspot cycle. A general tendency existed for the epoch of maximum geomagnetic activity to fall near the maximum in sunspot number.

2. The large, transient brightenings of small portions of the sun known as solar flares were often followed by intense geomagnetic disturbances or geomagnetic storms.

3. Geomagnetic activity tended to recur with a period of about 27 days, the

rotation period of the sun as viewed from the earth.

These associations came to be regarded as evidence for particle emission from the sun as a consequence of solar activity. The transient, flare-associated phenomena were attributed to clouds of material explosively ejected in the flare process, while the 27-d recurrence was attributed to long-lived streams of particles emitted from some localized source and swept past the earth by solar rotation. Both the tendency for recurrence and the intensity of this latter type of geomagnetic activity were found to be strongest on the declining portions of the 11-y sunspot cycles.

A second body of historical information regarding the solar wind is cosmic ray observations made over the past few decades. These observations are reviewed and summarized by Lanzerotti and Damon in this volume. For our immediate purposes, it will suffice to mention the recognition of two characteristic variations in cosmic ray intensity associated with solar activity:

1. A variation with the 11-y period of the sunspot cycle. The cosmic ray intensity is lower near sunspot maximum, higher near sunspot minimum.

2. A decrease in cosmic ray intensity after some large solar flares.

These two effects were again widely interpreted in terms of ejection of plasma and magnetic fields from the sun at times of high solar activity.

In situ interplanetary plasma and magnetic field observations that began in the late 1950s were first made for long periods of time well outside the perturbing influence of the geomagnetic field in 1962 and have been performed almost continuously since 1964. This body of observations gives an extensive, detailed, and generally accurate description of numerous characteristics of the solar wind. Average values and the degree of fluctuation about these averages are well-established. The fluctuations occur on time scales ranging from days to seconds and are in many cases readily interpreted in terms of waves and large-scale structures propagating outward from the sun. In fact, the existence of the two specific phenomena inferred from geomagnetic studies—transient clouds emitted after some solar flares and long-lived "streams" of plasma—is borne out by the in situ observations. However, the association of these solar-wind structures with "solar activity" is not as simple as might have been expected from these historical studies. For example, the detailed association of the transient shock waves, which are observed to produce most geomagnetic storms, with solar flares is far from one to one; and the interplanetary effects of the more common, small coronal transients observed by the Skylab coronagraph remain obscure. Further, long-lived solar wind streams and the related magnetic polarity structures (sectors) appear to have been the dominant form of solar wind variability throughout the epoch of in situ observations, and these streams seem to originate in coronal holes or the polar caps of the sun, regions which are largely devoid of the common forms of solar activity.

This absence of a "simple" association of observed solar wind characteristics with solar activity is exemplified by the situation with regard to the 11-y sunspot cycle. The past 14 years of in situ observation have revealed no large variations of the parameters characterizing the flow of solar wind that are in phase with the solar cycle; in fact, the major changes in solar wind speed, density (as well as it can be determined over this long time interval), mass, momentum, or energy flux seem to be out of phase with sunspot number, in rough accord with the expectations for long-lived streams (producing recurrent geomagnetic activity). While some solar wind properties—helium abundance and the amplitude of transverse magnetic fluctuations—do vary in phase with the sunspot number, it remains to be demonstrated that these are the physical causes of the classical variations in geomagnetic activity or cosmic ray intensity at the earth.

Thus, straightforward attempts to use modern understanding of the solar wind, magnetosphere, or cosmic rays to infer quantitative solar wind properties from the accumulated bodies of geomagnetic and cosmic ray observations remain largely unsuccessful. However, it has been demonstrated that the variations in the ground-level geomagnetic field at polar cap stations is well correlated with the polarity of the interplanetary magnetic field (sector structure) during the epoch for which in situ observations of the latter are available. This correlation has been used by Svalgaard (1972) and Svalgaard and Wilcox (1975) to infer the sector pattern back through several solar cycles and deduce possible solar cycle variations of this pattern; the reader is referred to the original publications for a detailed description of these variations.

The observed variations in the interplanetary stream and sector structure during the declining portion of solar cycle 20 have been extensively studied in the recent Skylab Workshop on Coronal Holes and are found to fit nicely into a simple interpretive framework. The growth of very large, long-lived, recurrent streams of high-speed wind and the simplification of the sectors into a recurrent dipole-like pattern in early 1974 led to a sequence of recurring geomagnetic disturbances. These interplanetary changes, consistent with those inferred from geomagnetic studies at similar phases of earlier solar cycles, apparently reflect an increasing influence of the polar caps of the sun on the solar wind in the ecliptic plane. In fact, the structure of the corona in 1974 is much like that expected if the solar magnetic field were a dipole tilted about 30° from the rotation axis; flow from the tilted polar caps then gives the observed stream-sector pattern. If this phenomenon occurs in every solar cycle, it provides a mechanism for carrying the cyclic 22-y variation of the sun's polar fields into the ecliptic plane and producing a 22-y cycle at the earth (due to the asymmetric response of the terrestrial magnetosphere to northward- or southward-pointing interplanetary magnetic fields). I caution, however, that the amplitude of the recurrent geomagnetic disturbances of 1974 is unprecedented; geomagnetic activity was higher than at the last maximum in solar activity and as high as ever before recorded. Thus, this epoch may not be entirely typical of past or future solar cycles.

REFERENCES

Chapman, S. and Bartels, J., 1940, *Geomagnetism*, Oxford at the Clarendon Press.
Svalgaard, L., 1972, *J. Geophys. Res.* 77, 4027.
−−−−− and Wilcox, J. M., 1975, *Solar Phys.* 41, 461.

*Energetic Particles from the Sun
and Solar Modulation of Galactic Cosmic Rays*
Robert Lin

Large solar proton flares are one of the most energetic manifestations of the active sun. In these flares a total energy of 10^{32} to 10^{33} ergs can be released in a few minutes, and a substantial fraction of this energy appears in the form of atomic particles accelerated to energies up to the GeV range. Spacecraft and ground-based observations over the last three solar cycles show that roughly 40 flares per solar cycle emit ions with energies greater than 10 MeV and that the output of these ions varies from flare to flare by over five orders of magnitude. Electrons of ~ 10-10^2 keV energies are accelerated and emitted much more frequently and by smaller flares. At energies below about 10 MeV, there is substantial evidence for interplanetary acceleration of the ions by shock waves. Furthermore, at energies near about 1 MeV, enhanced proton fluxes are observed nearly continuously during times of solar activity, indicating that continuous acceleration of these particles may be occurring at the sun. Recent measurements of the elemental and isotopic abundances of solar flare ions show that large variations are present in their composition from flare to flare; in particular, large enhancements of heavy ions at low energies, $\gtrsim 10$ meV/nucleon, and anomalously high He^3/He^4 ratios are observed for some flares. The flare particle acceleration and emission processes responsible for these phenomena are not well understood at the present time.

We caution that the particle fluxes observed at 1 AU are not simply related to the output of the sun as are the electromagnetic emissions. The particles are scattered by magnetic irregularities in the solar wind as they propagate in the interplanetary medium. The simplest flux-time profiles observed at 1 AU following a flare show diffusion effects. The total time-integrated flux is dependent on the amount of scattering. In addition, the interpretation of low energy (< 10 MeV/nucleon) ion fluxes may be complicated by energy changes in their propagation through the interplanetary medium. Furthermore, there are many flux variations observed, particularly at low energies, that appear to be due to the organization of the particle emission by coronal and solar wind structures. Thus, the particle output at the sun may be related to the fluxes observed at 1 AU in a very complex way.

Information on solar energetic particles in the ancient past has become available from studies of meteorites and lunar samples. Together, these contain records in the form of radioactive and stable isotope products and tracks that date back to at least 4×10^9 years, and most likely 4.6×10^9 years. The information is quite precise for periods close to the present and becomes increasingly uncertain at the earlier times. The main conclusions of these studies are that the energy spectrum and average flux level of solar flare ions has not changed by more than a factor of two over the past 5×10^6 years and that energetic solar flare ions with similar energy spectra have been present as far back as 4 to 4.6×10^9 years. Over most of the period covered by the available samples, no unambiguous evidence for periods of greatly enhanced flare activity is available, although the limited data do not exclude such activity. However, recently discovered isotope anomalies in certain meteorites can be interpreted as evidence for an extremely active sun at the beginning of the solar system.

The sun also regulates the entry of galactic cosmic rays from beyond the heliosphere. Presumably, magnetic irregularities carried outward by the solar wind scatter the cosmic rays and sweep them out of the heliosphere. The observation of very small gradients in the cosmic ray density by the Pioneer spacecraft to Jupiter indicates that these scattering features must be located predominantly outside the orbit of Jupiter. The mechanism by which the galactic cosmic ray level is modulated at the earth by the solar wind is not well understood at the present time. However, the empirical relationship between the amplitude of the modulation and solar activity (as represented by sunspot number) is well documented over the last three solar cycles, i.e., the galactic cosmic ray flux is smallest at times of highest solar activity.

The collisional interaction of galactic cosmic rays with the earth's atmosphere produces neutrons that in turn produce ^{14}C. The fluctuations in the atmospheric ^{14}C with time can be determined by analysis of the radiocarbon concentration in tree rings, as discussed by Damon in this volume. These analyses show an excess of ^{14}C during the Maunder and Sporer minima in solar activity (see Eddy, Chapter II of this volume), indicating that the solar modulation of the galactic cosmic rays was, as expected, low. But in the preceding 7.5 millennia there are many similar fluctuations in ^{14}C concentration, which implies that variations in solar activity similar to the Maunder Minimum are commonplace over scales of 100 to 1000 years. Meteorite and lunar sample studies of galactic cosmic rays in contrast show that, in general, solar modulation of galactic cosmic rays has not drastically changed over a period of 50×10^6 years; but the time resolution of this technique is less than for the ^{14}C dendrochronological analyses.

The terrestrial effects of the energetic solar particles emitted by a large flare include magnetospheric effects and radio blackouts due to ionization of the earth's atmosphere over the polar cap. Possibly the most important terrestrial effect of the energetic particles is their production of nitric oxides by the dissociation of N_2 in the earth's atmosphere. The increased nitric oxide concentration leads to depletion of ozone. After the large 4 August 1972 solar flare event a 16% decrease in ozone over high-latitude (> 60°) regions was reported. Large flare events may thus have a strong influence on the absorption of solar UV by the ozone in the earth's atmosphere.

Future research should resolve the questions of the nature of the acceleration mechanisms in flares and shocks and the cosmic ray propagation and modulation processes in the interplanetary medium. The intriguing puzzle of the anomalous elemental and isotopic composition of solar particle fluxes must be solved. The low-energy component of solar cosmic rays, from solar wind energies up to about 10 MeV, represents a major new area of research. Rather complex and unique behavior is observed for these low-energy particles. Particle measurements with good elemental, isotopic, and change-of-state resolution extending to low energies are needed. Indirect measurements of solar particles at the sun such as provided by gamma-ray, hard X-ray, and radio observations are also necessary. More information is also needed about the solar and interplanetary environment of these low energy particles.

For studies of solar energetic particle emission in the distant past, more accurate determinations of exposure times for extraterrestrial samples are needed in order to obtain absolute particle fluxes and their variations. In particular, the specification of exposure ages for breccias is required. It is important to determine whether the early sun was extremely active and responsible for producing the isotopic anomalies of certain primitive meteorites. As a basis for the interpretation of primitive records, measurements of present-day ion composition in flare emission, more nuclear cross-section data, and a better understanding of the effects of intense irradiation of matter would be useful. Finally, it would be desirable to apply the new techniques being developed for the investigation of single particles and selected grain surfaces.

It is very important to extend the coverage and range of the radiocarbon analyses, in order to determine whether the types of fluctuations in solar activity observed during the Maunder Minimum are predictable. Increasing the temporal resolution of the ^{14}C analysis would make it possible to identify very large flares, such as that of 4 August 1972, which have strong effects on atmospheric ozone.

The Theory of the Solar Interior
and Its Predictions of Solar Variability
Roger Ulrich

Our knowledge of solar structure and evolution is based on a limited set of observational data as an empirical boundary condition. Some of these data, such as the mass, radius, luminosity, oblateness, and neutrino emission rate, relate directly to integrated properties of the sun. Other data, such as those obtained from solar activity and solar velocity field measurements, relate to dynamical processes occurring on the sun's visible surface. To the extent that we understand these dynamical processes, the data can be used as a probe of solar structure. Our knowledge of the elemental solar composition comes from the analysis of the solar, planetary, and cometary spectra; from the analysis of the solar wind; and from the analysis of meteorites and lunar soil samples. We have no data that tell us the composition of the solar interior directly. There is some hope that the study of stars in binary systems, where mass has been transferred from one star to its companion, may give an indication of the composition in the interior of other stars. Finally, terrestrial paleoclimate records may set some limits on the evolutionary behavior of the sun.

Most of our understanding of stellar structure comes from an application of the basic conservation laws of physics to the calculation of stellar models. These calculations also rely on measurements of the cross-sections of nuclear reactions and on more complicated calculations of photon opacities. The possible presence of large magnetic fields and radial gradients in the rotation rate are usually ignored in stellar model calculations, but such physical effects are clearly present on the sun. The fundamental constants of physics are usually assumed to be independent of time and place.

Because most of our understanding of solar interior structure and evolution is based on deductions and calculations rather than direct observational evidence, the theoretical questions about variations in the total solar output remain open. Calculations, except for those involving a variable gravitational constant, make a clear prediction that the solar luminosity was only 0.7 to 0.8 times its present value shortly after the sun began to burn hydrogen on the main sequence. A small primordial solar luminosity has certain drastic implications for the earth's climate, but as Hays and Mitchell point out in this volume, the paleoclimate record does not show the luminosity increase.

Solar activity and the solar wind are related to the strength of the sun's rate of rotation and its magnetic field. Observations of the strength of chromospheric activity in star clusters of various ages show that young stars rotate more rapidly and are more active than old stars. Young stars may also have stronger stellar winds. These observations are consistent with the idea that the solar magnetic field and rotation rate decay with age. A similar decrease in the strength of the sunspot cycle is probably associated with the decay of chromospheric activity, but no dynamo theory calculations have yet been done with parameters that characterize a younger sun.

On a shorter time scale, some forms of solar output are associated with the solar cycle itself, and variations may occur over an 11-y period. During a single rotation period various irregularities in the solar atmosphere, such as active regions and giant convective cells (the existence of the latter is not completely established), are carried across the visible disk of the sun. Thus, variations on a time scale of the solar rotation period are to be expected in the radiation falling on the earth. Although the dynamo theory of the solar magnetic cycle has reached a moderately sophisticated state, it does not yet include most nonlinear feedback effects nor account for long-term trends in the strength of the solar cycle.

Although we do have the Babcock-Leighton picture for the magnetic cycle of the sun, it only describes the cyclic generation and dispersion of magnetic fields in the solar photosphere. This theory is not general enough to predict the global properties of energy distribution throughout the visible solar atmosphere; consequently, it cannot give us insight into the possibility of variations in the physical output of the sun. As Gough's study in Chapter VI indicates, examination of the common time scales used in the existing hydrodynamic theory does not point to 11 or 22 y as a characteristic cycle time for large-scale solar processes that are thought to be capable of influencing the solar output. It is fair to say that our current theoretical understanding is not sufficient to quantitatively predict the occurrence of the solar cycle from first principles; therefore, we are also unable, a priori, to predict variability in output associated with the solar cycle. Given a model of the solar atmosphere at different times in a solar cycle, however, we should be able to compute the emitted solar spectrum by using the methods described by Avrett in Chapter IV. Given an equivalent empirical boundary condition for the hydro-magnetic structure in the solar atmosphere, a similar modeling of the solar wind out to 1 AU can presumably be attempted with success.

Recently, processes that could lead to major variations in the solar constant on a time scale of 10^6 to 10^7 have been suggested. The theoretical and observational basis for these suggestions is not as firm as for the processes discussed above. However, these suggestions deserve consideration because of the potentially major impact on the earth's environment that could result if the proposed processes occur. First, a stability analysis by Dilke and Gough indicates the presence of a slowly growing mode of oscillation. They further propose that mixing is induced in the solar core after the oscillation reaches a finite amplitude. If such mixing were to occur, the luminosity of the sun would drop temporarily. Detailed analysis of solar models suggests that the instability exists in some solar models although the effect of convection in the outer solar layers could easily alter this result. Also, the mechanism by which mixing is induced by the oscillation has not been determined.

Another theoretical model for solar luminosity variations involves the passage of the sun through a dense interstellar cloud. If the dynamical pressure of matter within this cloud is able to overcome the solar wind, then cool, relatively dense matter could fall onto the sun and earth. Interstellar dust would be introduced into the earth's atmosphere at a rate comparable to or greater than the rate associated with volcanic eruptions, but whether the cosmic dust would significantly affect the earth's climate depends on the particle size distribution; large particles would rapidly precipitate out. An accretion shock on the solar surface would generate soft X-rays and EUV line radiation. If the interplanetary matter during such an accretion event is opaque in the Lyman continuum, most radiation from the sun shortward of 912 Å will be converted to Balmer continuum radiation. The net effect of all these solar processes is unclear. The duration of solar passage through such a cloud would be 10^6 to 10^7 y.

Much interest in solar structure has been generated in recent years by the study of new measurements. First is the effort by Davis and co-workers to detect neutrinos from the sun. To date, the 1 σ upper limit for the neutrino flux is about a factor of four lower than expected on the basis of solar model calculations. No proposed explanation of the discrepancy has yet proved satisfactory. A second new measurement under study is that of large-scale solar oscillations. The correct identification of the correspondence between the observed frequencies and the frequencies of normal modes of oscillation in solar models could play a role as important in probing the internal structure of the sun as the study of seismic waves is in probing the structure of the earth. The amplitude of the solar oscillations is too low for any significant energy transport to be associated with these motions.

Although the promise held out by both these new types of data for increasing our knowledge of solar structure is very great, no definite conclusions can yet be drawn as to the possible variation of the ultimate solar energy source, the hydrogen-burning nuclear reaction in the sun's core.

Activity and Variability in Other Stars
Oran White

The output of radiation from the sun is the direct empirical link between stellar astrophysics and solar physics since the total solar luminosity and the properties of the solar spectrum define the sun in relation to other stars. Using the spectral classification for stars, the sun is identified as a G-type dwarf star lying on the stellar main sequence. Stellar evolution theory describes how stars of solar mass and initial solar abundance change their luminosity and effective temperature (and thus their spectrum) with time. In addition, velocity fields, magnetic fields, and related phenomena probably change with time; consequently, the study of solar-type stars permits us to look both forward and back along the evolutionary sequence.

As one example of this approach, consider the evolutionary decay of stellar rotation and magnetic fields. Skumanich (1972) analyzed the strongly correlated stellar parameters of the Ca II emission index (a gross measure of chromospheric conditions) and stellar rotation rate as a function of stellar age, and he determined that these quantities have a characteristic decay time of 1.1×10^9 years. Using the established correlation between the Ca II emission and solar magnetic field, Skumanich's result suggests that both solar rotation and solar activity may decrease at a similar rate.

Our view of the sun in the context of other stars is being improved through detailed comparisons of solar and stellar spectra, as described in Chapter VII. The technique of this work is the identification and analysis of the evidence for solar-type structures such as chromospheres, coronae, and winds in the output of other stars. The emphasis, however, is on the understanding of the physical processes operating in stellar atmospheres. Questions familiar to solar physicists, such as the role of nonequilibrium excitation and ionization, mechanical energy transport, magnetic fields, and coronal-type winds, also arise in the stellar data, but under different physical conditions appropriate to each stellar type. The search for evidence of stellar variability and stellar activity continues naturally in this work, and there is evidence for stellar activity cycles even though the time record of the Ca II index is relatively short.

Detailed spectroscopic analysis of stellar line profiles is as yet possible only for nearby, bright stars where the lines can be measured with high spectral resolution. The spectra of two such stars, Procyon and α CEN A, show evidence for nonradiative heating in their upper photospheres. Given this evidence for departures from radiative equilibrium in stars other than the sun, we naturally ask about the existence of stellar chromospheres, where the requirement for nonradiative energy and transport is even more severe. Using various spectral diagnostics developed in the study of the solar chromosphere, Linsky summarizes the evidence for stellar chromospheres in Table 4 of his review. Although the reader should refer to this table to see the evidence in detail, it is apparent that dwarf stars of types F to M, type G to M giants and supergiants, and long-period variable stars show emission lines characteristic of a chromosphere. Strong emission lines characteristic of solar plages and flares are seen in active chromosphere stars such as close spectroscopic binaries, T Tauri stars, novae, and flare stars. Thus, the energy balance problem that continues to be a focus of work on the solar chromosphere arises in a wide variety of stellar atmospheres. It is important to establish the extent to which different types of solar chromospheres, i.e., quiet sun, plages, chromospheric network, and flares, may be reliable prototypes for chromospheres of other stars.

Continuing an upward progression, what is the evidence for the existence of stellar coronae and chromospheric-coronal transition regions? To date, the UV spectra of only two cool stars, α CMI and α AUR, are well observed. These two stars appear to present clear evidence for transition regions, but the temperatures at the bases of the layers and the thickness of the layers differ from the solar case. Linsky argues that the existence of a transition region implies the existence of a stellar corona, and the stellar X-ray spectra from α AUR show high stages of ionization that require the high temperatures and low densities characteristic of the solar corona. Although the number of stars displaying clearly identified coronal spectra is as yet relatively small, the coming improvements in the sensitivity of X-ray experiments should lead to clearer specification of coronae for the different spectra types. Stellar radio observations are not yet sufficiently sensitive to show emission typical of the quiet solar corona, but highly variable, nonthermal radio emission has been detected from α ORI, α SCO, and the UV Ceti-type flare stars. This constitutes evidence for active stellar coronae.

Given the existence of a high-temperature corona, the outward streaming of matter as a stellar wind is inevitable. To date, there is almost no spectroscopic evidence for such a flow similar to the solar wind in solar-type stars. Since the question of mass flow from stars is important, the properties of stellar winds are inferred theoretically from empirical boundary conditions in the small number of stellar coronae observed to date. The predicted stellar winds from giant and dwarf stars have not yet been observed, even though the flow should be spectroscopically detectable for α AUR and α BOO. Stellar winds have been extensively studied, however, in hot stars and in cool supergiants.

From the above description of solar-type phenomena in the atmospheres of other stars, we have hints of spectroscopic variability similar to that associated with solar activity. Unfortunately, it is extremely difficult to measure magnetic fields in solar-type stars, and existing data do not yet establish the existence of stellar magnetic cycles. The principal evidence for such activity lies in O.C. Wilson's measurements of the Ca II emission index, which in the solar case is strongly correlated with local magnetic fields. Although this work is not yet finished, Wilson described his preliminary results for us. His measurements since 1968 show the existence of stellar cycles in several stars of spectral type later than G0 V. No periodic variation is seen in stars hotter than G0 V. Publication of this definitive study is expected in 1978.

One major component in solar activity, the flare, is unambiguously identified in the variation of UV Ceti-type stars. Although there can be large differences in the flare energy relative to the solar case, stellar flares show many of the same properties of solar flares, i.e., time-varying emission-line spectra, large mass motions, and high temperatures and densities. Since some flare stars show slow periodic brightness changes, it has been suggested that this variation is caused by the rotation of starspots on and off of the visible stellar disk. The flare star data show variations that suggest the existence of cycles with decade lengths and the presence of active regions with lifetimes of months. The polarization and strength of radio emission from such objects requires nonthermal emission from the flare source, and the lag of such emission behind the optical flare can be taken as evidence for a disturbance propagating through a stellar corona.

Based on the above broad view of the stellar variability problem, a look at solar variation in the context of stellar properties allows us to identify solar-like phenomena occurring under physical conditions substantially different from those in the sun's atmosphere. The lack of a long time record of intrinsic stellar variability may be partly compensated for by studying stars similar to the sun and differing only in age. Linsky's work to classify different solar regions as distinctly different chromospheres, coronae, etc., can be a very productive step in the quantitative understanding of the spectra from different types of stars. The yield from this approach is critically dependent on the availability of high-resolution stellar spectra and the application of a physically realistic theory of radiative transfer and stellar atmospheres. As the solar work has pointed the way in this approach to stellar spectroscopy, our knowledge of solar activity can serve as guide in both observation and interpretation of stellar variability. In return, the broader picture of stellar variability may give clues to the slow evolutionary changes in solar activity and, in turn, these clues can help us understand solar variations (or the lack thereof) discussed in other papers in this volume.

REFERENCE

Skumanich, A., 1972, *Ap. J. 171*, 565.

II. HISTORICAL AND PALEOLOGICAL EVIDENCE FOR SOLAR VARIABILITY

These and other recent studies of both historical data and the paleoclimatic record are of great scientific value because they raise questions about the regularity of the solar cycle as well as about long-term variations in solar luminosity predicted by theories of the solar interior. The papers by Eddy and Hays that follow are important in solar physics because they add a new dimension to the empirical characterization of the sun that someday must be described by suitable physical models. Eddy's examination of historical data on solar-terrestrial interaction over the last several hundred years suggests that the degree of solar activity may become quite small for periods as long as one generation. These times of depressed solar activity may occur infrequently, at intervals of 500 to 1,000 years. Eddy's studies have now been extended back in time some 7,000 years through the use of the dendrochronological signature of the ^{14}C production rate in the earth's atmosphere, and it is from this record that he suggests the occurrence of other periods of depressed activity like the "Maunder" and "Sporer" minima in the solar past.

Paleoclimatic research like that described by Hays gives us a record, completely independent of known solar history, that extends back in time at least 3 billion years, the time scale over which living organisms have existed on the earth. The conclusion that there is no paleoclimatic evidence for a change in the solar luminosity of more than 10-20% over the last 3 billion years runs directly counter to the hypothesis of a 30% increase in solar luminosity over the same time period due to a change in the mean molecular weight due to nuclear reactions in the solar core. However, as Mitchell pointed out in Chapter I, the biosphere may have a natural mechanism by which its temperature tends to remain in the range necessary for the support of life despite a systematic change in the insolation. Hays postulates that other factors, such as continental drift and vulcanism, are as important in determining the earth's climate as intrinsic changes in the solar output. A recent article by Hays, Imbrie, and Shackleton (Hays, J.D., Imbrie, J., and Shackleton, M.J., 1976, *Science 194*, 1122) makes a strong argument for associating the periodic glaciations of the earth with variations in the earth's orbit around the sun at 23,000, 42,000, and 100,000 y intervals.

This current work, then, suggests that we may now possess some understanding of the long-term variations in the earth's climate that result from natural changes in the insolation received at the earth. And at the other end of the time scale, in blocks of hundreds of years or less, we now have a working hypothesis that demands attention from both solar physicists and their associates who carefully examine the paleoclimatic record. The value of these works lies not only in their conclusions, but in the questions and challenges they raise for the traditional picture of a slowly varying sun possessing a regular solar cycle.

Oran R. White

HISTORICAL EVIDENCE FOR THE EXISTENCE OF THE SOLAR CYCLE

John A. Eddy
High Altitude Observatory
National Center for Atmospheric Research

The regular 11-y cycle of solar activity, so evident in the record of sunspot numbers, is the basis of much of our understanding of the sun and the foundation for much of solar-terrestrial physics. It is probably true that the sunspot cycle is the sun's best known feature; it is certainly the best documented and quite possibly the most extensively observed phenomenon in all of astronomy. We should recognize, however, that continuous and controlled records of sunspots cover only a little more than one century, that the extensions of the record before about 1850 become increasingly uncertain, and that in the longer perspective we have little if any evidence that today's 11-y cycle is an enduring feature of the sun. Indeed, the available indirect evidence suggests that in the longer term, change and not regularity may rule solar behavior.

In this summary we will review what is known about the solar cycle and solar behavior from historical evidence—that is, from man-made records of roughly the last 1,000 years. We will also make use of the ^{14}C record (discussed in Chapter V in this volume by Paul Damon) for the period of overlap with historical data. A graphical summary of the extent of existing data on solar history is shown in Figure 1.

I. THE SUNSPOT CYCLE

The cycle of solar activity is most evident in a plot of the annual means of the numbers of observed sunspots, as in Figure 2 (Waldmeier, 1961). Sunspots themselves are not particularly fundamental in solar activity or necessarily related to the solar output, nor are they a specific cause of any known terrestrial effect. They are convenient as an index of almost all other activity on the sun, including active regions, plages, flares, intense magnetic fields, prominences, and to some extent, changes in the coronal form and in the solar wind. Sunspots are by far the most easily observed indicator of solar activity and the source of the longest direct record of solar history.

II HISTORICAL AND PALEOLOGICAL EVIDENCE

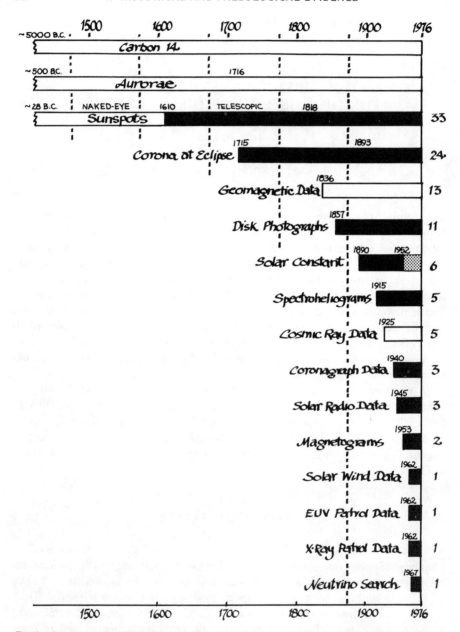

Fig. 1 Summary of available data on past solar behavior. Solid bars, direct data; open bars, indirect data. Horizontal scale is years A.D. Vertical lines are 100-y periods from present. Numbers on right margin are numbers of 11-y cycles for which data exists.

ANNUAL MEAN SUNSPOT NUMBERS FROM 1610 TO THE PRESENT*

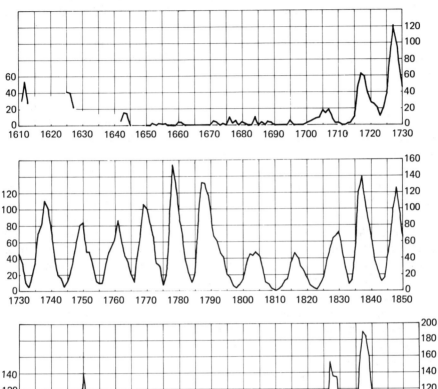

*From Waldmeier (1961) and Eddy (1976a)

Fig. 2 Annual Mean Sunspot Number, A.D. 1700-1975, from Waldmeier (1961) and Eddy (1976).

II HISTORICAL AND PALEOLOGICAL EVIDENCE

The number of spots visible on the sun varies considerably from day to day as a result of solar rotation and the growth and decay of individual spots and groups. These effects are smoothed out in annual averages, where a cycle of about 11-y is clearly evident. Amplitudes of the cycle are far less regular: maxima range from well-observed lows of about 60 to an all-time high of 190 in 1957. At minima, annual average sunspot numbers approach, but seldom reach, zero. The average value at minima of the cycle is about six.

A number of subsidiary features of the sunspot cycle have long been noted: times of rise to maxima are generally shorter than times of fall; heights of maxima sometimes alternate in strength from cycle to cycle; amplitudes of maxima, taken in the long term, seem to follow a longer period of seven or eight 11-y cycles, as first noted by Wolf (1862) and later elaborated by Gleissberg (1944).

All is not regular and predictable in the modern sunspot cycle and it is fair to say that none of the methods of predicting its future behavior have met with unqualified success. A close and critical look at sunspot numbers reveals that each of the cycles is unique and that the record as a whole is not the mark of perfect clockwork. The distribution of observed periods which make up the 11-y average is far from Gaussian (Figure 3) and includes extreme values of 8 and 15-y, nearly a factor of two apart. In examining the curve of annual sunspot numbers we must be warned to avoid our common habit of mentally substituting regularity in an observed, noisy signal; sunspot numbers since about 1850 are probably *perfectly* observed; there is little if any "noise" in the record.

Although sunspots were "discovered" with the telescope in about 1611 and had been observed before that for at least 2,000 years, their cyclic behavior was not known until 1843, when a German amateur, Heinrich Schwabe, pointed out an apparent 10-y periodicity in 17-y of his own observations (Schwabe, 1843). The long delay between the introduction of the telescope and the realization of this now obvious feature of the most easily observed object in the sky has been

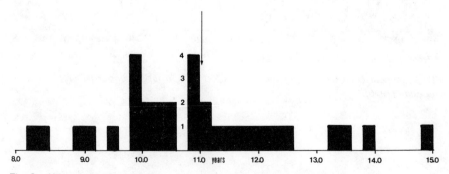

Fig. 3 Histogram of periods between maxima in the sunspot cycle, A.D. 1700 to present. Arrow indicates the mean value, about 11.1 y.

something of an embarrassment in astronomy, although it now seems that the oversight may be partly explained by the suppression or cessation of the cycle during part of the interim.

Shortly after Schwabe's announcement professional astronomers set out to establish whether or not the cycle was a real and continuing effect, through a program of controlled, international observations of the sun that continues today. The leader and organizer of this effort was Rudolf Wolf of the Zürich Observatory, whose definition of "sunspot number" is yet in use. Wolf also carried out extensive searches of past data on sunspots to determine whether the cycle had existed in prior time. After a long and dedicated effort he assembled historical evidence that filled the gap between Schwabe's work and the telescopic discovery of sunspots in the early seventeenth century. Wolf concluded that the 11-y cycle had indeed been present since 1700 and probably earlier, and his reconstructions of earlier sunspot numbers (Wolf, 1856) became truth in all subsequent descriptions of solar history. More than half of the record shown in Figure 2 is the result of Wolf's historical reconstructions.

II. LIMITATIONS OF THE SUNSPOT RECORD

Modern determinations of the daily sunspot number are made on a worldwide basis and compiled with considerable care, including a correction for differences in telescopes, sites, and individual observers. The annual numbers are the mean of 365 of these daily determinations. The historical part of the curve—all numbers before about 1850—is, by contrast, of a distinctly inferior quality, and moreover it degrades considerably with past time. This is due in part to the uncertainty of appropriate correction factors, which should serve warning to take the absolute values of the early record with caution. More serious are the gaps in the daily observing records which made up the historical annual estimates. Wolf was able to find adequate historical observations to reconstruct daily sunspot numbers some 34 years into the past—to 1818—although these of course lacked observing corrections. Beyond that the record worsened. He was able to locate sufficient data to establish *monthly* estimates to 1750, and only enough information to reconstruct *annual* estimates to 1700, at which date his useful numbers stopped. It hardly seems necessary to point out that the "annual means" resulting from these less-than-complete samples of daily numbers are suspect, particularly since observed daily sunspot numbers vary considerably at any time of the sunspot cycle. Also we can guess that sporadic, historical data more likely sample times when the sun was more active and that annual averages deduced from them would be systematically high. In any case we can grade the historical data into three categories of reliability: good from 1852 to 1818, fair from 1817 to 1750, and poor from 1749 to 1700. It is probably wise to question whether Wolf's annual sunspot numbers for the earliest period, 1700 to 1749, represent anything more than a wishful extrapolation to a "normal" cycle. Certainly the amplitudes of this portion of the curve are little more than estimates. Another consideration is whether Wolf's inadvertent bias entered the data, for it seems likely that once he had established the existence of the cycle in the later historical data, he was probably inclined toward finding evidence of it in earlier, less distinct, and less complete records.

Schwabe's announcement and Wolf's extensive study convinced solar astronomers not only of the reality of the cycle, but also of its supposed universality in time. Added support, it is true, has come from the continued operation of the cycle since Schwabe's day. Still, it is an interesting commentary on the ways of science that the wholehearted acceptance of cyclic behavior as a fundamental feature of the sun came as an abrupt switch from equally devoted denial of it in the years before Schwabe (Johnson, 1858), and that after Wolf's work few seemed willing to doubt that the 11-y cycle had always existed on the sun.

III. THE PERIOD 1645-1715 (THE MAUNDER MINIMUM)

Late in the last century Spörer (1887, 1889) and Maunder (1890) called attention to a 70-y period just before 1700 when practically no sunspots were reported. In later papers Maunder (1894, 1922) made the case that during this time, from about 1645 to 1715, the normal sunspot cycle seemed to have either vanished or nearly vanished. He pointed out that the incident, if true, was cause to doubt whether the sunspot cycle of modern time was really a universal solar feature.

A recent and more thorough investigation of the case (Eddy, 1976) has strengthened the reality of Maunder's claim and added further evidence, some of which was unavailable in Maunder's time. In support of the anomaly are original journals and books of the day, auroral records, naked-eye sunspot counts, observations of the sun at eclipse, and the indirect evidence from atmospheric isotopes. A reconstruction of sunspot numbers during the period (Figure 4) seems to confirm Maunder's statement that there were fewer sunspots during the 70-y period than in a normal year of modern solar activity. Throughout the time solar activity seems to have hovered at a level generally lower than at minima of the modern cycle. It seems impossible, however, to distinguish between sunspot numbers of 0 and about 5, which leaves open the important question of whether or not the 11-y cycle was operating at all.

There also seems to be insufficient evidence to establish whether the 11-y cycle existed *before* the onset of what I have called the Maunder Minimum and after the introduction of the telescope, although it does seem true that solar activity was dropping during this time (1611-1645). Thus in the direct, telescopic record of sunspot observation the earliest evidence we have for a well-established

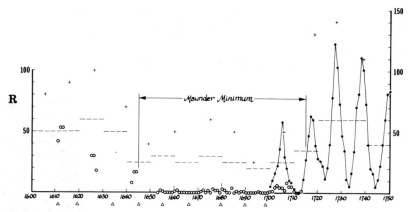

Fig. 4   Estimated annual mean sunspot numbers, A.D. 1610-1750. Open circles from Eddy (1976); connected, solid circles from Waldmeier (1961); dashed lines (decade estimates) and crosses (peak estimates) from Schove (1955). Triangles are Wolf's estimated dates of maxima for an assumed 11.1-y solar cycle.

11-y cycle is about 1700.

In a related study, Eddy, Gilman, and Trotter (1976) have shown that solar rotation was significantly different at the start of the Maunder Minimum from solar rotation in the modern era, in that the surface of the sun was rotating faster at the equator by about 3% and differential rotation was enhanced by about a factor of 3. Until contradictory evidence is shown, we must accept the Maunder Minimum as evidence that the nature of solar activity has changed in modern time and question the long-held tenet of what might be called "solar uniformitarianism."

IV. HISTORY OF THE SUN BEFORE THE TELESCOPE

The existence of the Maunder Minimum suggests a search for evidence of similar periods of solar anomaly in earlier time. Except for the period from 1611 to 1645, during which time the sun was under telescopic study, we are limited to indirect or proxy data (Figure 1). These lines of evidence on historical solar behavior are summarized below.

A. THE AURORAL RECORD

The occurrence of terrestrial aurorae is well correlated with solar activity and sunspot number, and since records of the aurora extend far back in time they provide a long and valuable extension of solar history. Probably the most complete catalog is that compiled over a century ago by Fritz (1873), which covers the period from the start of the Christian era through 1870.

A plot of auroral reports from Fritz's catalog is shown in Figure 5, limited to reports from latitudes south of the polar circle to exclude the regions near the magnetic poles where solar-cycle dependence is less direct. In this presentation we show reports per decade to illustrate long-term effects.

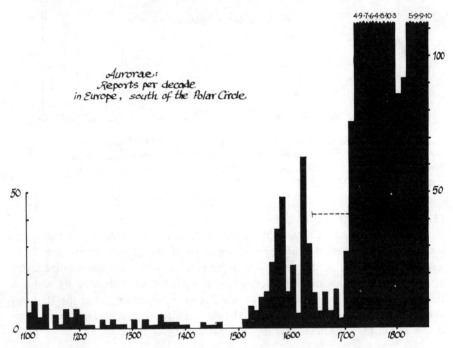

Fig. 5 Reports of the aurorae from the catalog of Fritz (1873), in counts per decade for latitudes 0-66°N. Counts after 1715 must be multiplied by numbers shown at top right of plot. Period of Maunder Minimum in shown as a horizontal line.

The most striking feature in the decade averages is a pronounced rise in the number of auroral reports commencing in the early sixteenth century. This rise is interrupted by a sharp dropout during the Maunder Minimum, after which there is an abrupt rise. If this apparent "auroral turn-on" in the sixteenth through eighteenth centuries were a real solar effect, it would be evidence for a remarkable change in solar behavior. In fact it must be due in part or perhaps wholly to sociological causes and the limits of historical records. It comes at roughly the onset of the age of enlightenment in science. The really steep jump at about 1720 follows a classic paper on the aurora by Edmund Halley (1716), which may have influenced the number of aurorae that were subsequently reported.

Some of the general rise could, of course, be genuine, particularly since the overall trend of auroral reports does not increase monotonically with time. There are more aurorae listed in Fritz's catalog for the twelfth century than for the subsequent three, during which the number fell before the steady increase after the sixteenth century. Other evidence indicates that the period between about 1450 and 1540 was a time of apparent low solar activity, equivalent in character to the Maunder Minimum, and that the twelfth century was a period of apparent high solar activity (Eddy, 1976). These coincidences suggest that there is more than the Renaissance in the auroral curve, although these distinctions remain a subject for more thorough study.

To examine the auroral record for evidence of the 11-y cycle, we must look at individual counts. In Figure 6 we plot annual sums of auroral reports from

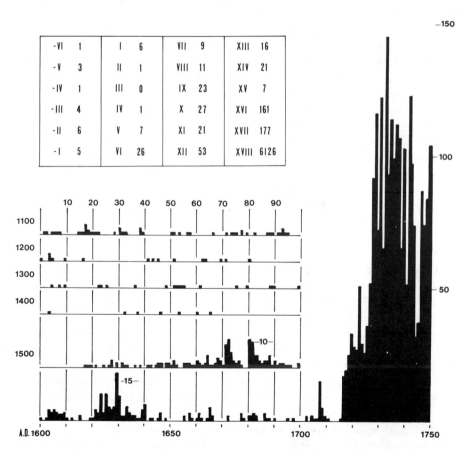

Fig. 6 Reports of aurorae from the catalog of Fritz (1873), in counts per year (first line, twelfth century, etc.) Total reports per century in the catalog are shown at the top, with Roman numeral century numbers.

Fritz's catalog from the twelfth century through the first half of the eighteenth; above it we list reports per century for the entire catalog. We are struck again by the change in their number—the rise, the fall during the Maunder Minimum, and the climb to high, modern values in the early eighteenth century.

More depressing is the paucity of reports from any era before modern time, which speaks for the difficulty of compiling such a catalog and the severe limitations we face in using it. There are only seven aurorae reported for the whole world in the fifteenth century, 19 in the fourteenth century, and 16 in the thirteenth. With a record so thin, it is dangerous to deduce anything about the existence or nonexistence of a solar cycle in early time. Even in the sixteenth and seventeenth centuries, when auroral reports begin to rise, we see no compelling evidence for an 11-y cycle. The two peaks around 1572 and 1580 (11 aurorae per year with a skewed dip between) constitute the only obvious feature that might lead us to think that historical aurora reports show short-term periodicity, and they are an isolated case of a single cycle. Three small peaks in the early twelfth century (three, two, and two aurorae per year separated by 13 and 8 y) might also encourage a cycle-hunter, but the statistics there are too low to stand as significant evidence for regular behavior.

We can hope that better auroral catalogs will be assembled that will provide more illumination on the question of past cyclic behavior, although one suspects that we will always be frustrated in trying to squeeze hard numbers from dim and diffuse records. On the basis of Fritz's catalog we must conclude that the historical aurora record is unable to provide evidence either for or against the existence of an 11-y solar cycle in the past. This is an important point, for we shall find that the auroral data are probably our best hope of resolving the question.

B. NAKED-EYE SUNSPOT REPORTS

Dark spots were seen on the face of the sun with naked eye at least as early as the fourth century B.C. (Bray and Loughhead, 1965), and a long series of reports of them exists in historical records, particularly in the Orient. There is no reason to think that a regular watch was kept for them anywhere at any time, and we must consider all naked-eye sunspot reports as accidental discoveries. Moreover, since only the largest spots or spot groups can be seen, we can assume that these sightings tell only times of relatively high activity. Kanda (1933) has presumed that they therefore identify maxima in a solar cycle. This may be unwarranted since it is common to find short periods of high activity with large spots or groups throughout the modern 11-y cycle, including at times of sunspot minima.

Historical searches have yielded only three or four sightings per century, on the average, and hence their usefulness in establishing patterns or cycles is far more limited than aurorae. There is also inescapable subjectivity in any catalog of them, for there is always doubt whether the feature reported was really associated with the sun. Finally, the frequency of naked-eye sunspot reports may well have been influenced by social pressures; reports may have been suppressed or particularly encouraged in one era or another.

Still, naked-eye sunspot reports can serve several useful functions in establishing solar history: they stand as proof that sunspots existed long before the telescope, and they may help in confirming gross periods of unusually high or low solar activity. Were the sun uncommonly active during some extended period, we might expect to find a greater-than-average number of naked-eye sightings; were it presumed to be inactive for a long time, as during the Maunder Minimum, we should expect a coincident absence of naked-eye reports.

We show in Figure 7 the 53 naked-eye sunspot reports collected by Kanda (1933) from historical records of China, Japan, and Korea for the period between 28 B.C. and the early eighteenth century. In the lower part of the figure the same data are plotted on a compressed scale to emphasize long-term trends. Below that, on the same scale, are plotted reports of Oriental aurorae from Kanda (1933) and Matsushita (1956), which may be considered an appropriate, independent

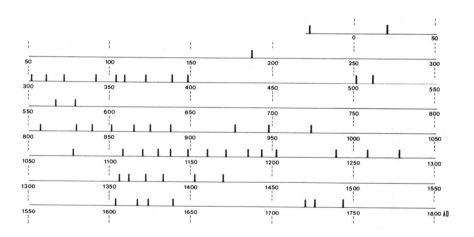

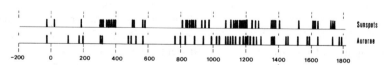

Fig. 7 Reports of naked-eye sightings of sunspots in China, Japan, and Korea, from the catalog of Kanda (1933). Horizontal (time) scale is years A.D. Each sighting is shown as a single, vertical line. Compressed time scale at bottom: top line, same data as above; bottom line, aurorae seen in the Orient as reported by Kanda (1933) and Matsushita (1956).

check. Since China, Japan, and Korea are all at low auroral latitudes, we expect aurorae there only at times of high activity.

We may draw several conclusions. In the long-term (compressed) figures we see a tendency for reports of both sunspots and aurorae to cluster, as in the twelfth century, when 9 were seen, and to coincide in absence for other long periods, such as the seventh and eighth centuries, when for 200 years there is not a single sunspot sighting. The Maunder Minimum is confirmed as a time of absence of both naked-eye sunspots and oriental aurorae.

In the expanded (upper) record in Figure 7 we note that at times of high incidence, naked-eye reports seem to fall as single sightings at intervals very close to the present 11-y sunspot cycle. In the apparently intensely active twelfth century, for example, the spacing between sunspot reports is 10.45 years with standard deviation 1.9 years. We should be cautious in taking this example as evidence for the existence of an 11-y sunspot cycle in the twelfth century, for the picture is too pretty to seem real. It seems highly unlikely that the sun would produce a single burst of activity, like the eruption of Old Faithful, at precisely the same phase of each cycle, and remain relatively dormant for the 11-y between. What seems more likely is that Kanda looked harder at these times for incidents which could be interpreted as sunspot sightings, as proof of his point. But in either case we are in the realm of conjecture.

A more conservative approach is to interpret the naked-eye sunspot record only for its long-term trends, particularly as these are coincident with other, independent evidence of past solar behavior, such as the auroral reports from the West cited earlier or the more objective and continuous evidence from ^{14}C. Here the naked-eye sunspot record comes through rather clearly in support of a very low level of solar activity in the seventh and possibly eighth centuries, a pronounced maximum in the twelfth and early thirteenth, and the Maunder Minimum and another like it (what I have called the Spörer Minimum) in the late fifteenth and early sixteenth, all of which find confirmation in other data.

In Figure 8 we combine oriental reports of sunspots and aurorae (from Kanda and from Matsushita) and show them in 70-y sums to illustrate the apparent long-term trends.

J. Roger Bray of New Zealand has long championed the use of historical records to deduce solar and geophysical history, particularly as those records are coupled with modern evidence from glaciers, forest growth, and atmospheric isotopes (Bray 1965, 1967, 1968, 1971). He concludes a recent essay (1974) on the use of historical Chinese solar observations with this eloquent and provoking summary:

> The recording of sunspots and aurorae is one of the longest series of direct physical observations that the world possesses. Given the present social and technological chaos and the possibility of human extinction, it is conceivable that these data will remain the longest series. Even

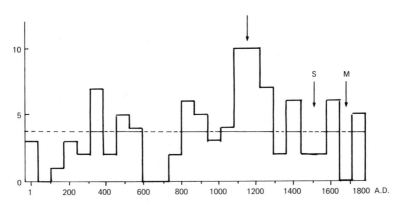

Fig. 8 Number of oriental reports of naked-eye sunspots and aurorae from data of Kanda (1933) and Matsushita (1956), summed over 70-y periods. Horizontal (time) scale is years A.D. Horizontal dashed line shows mean value, about 4. Arrows marked M, S, and GM identify Maunder Minimum, Spörer Minimum, and twelfth century solar maximum found in other data.

short of extinction, the present structure of science, with its formalism and short-term goals, has pretty well stifled all but the professional scientist. The rest of humanity see less and less of the natural world as they fear to examine what their supposed superiors study with a more expensive and refined technology. Perhaps never again will a phenomenon like sunspots be observed and recorded by so many over such a long historical period.

C. APPEARANCE OF THE CORONA AT ECLIPSE

The form and extent of the white-light corona evolves in a well known way with changing solar activity: when sunspot number and solar activity are low the corona is less developed, and confined chiefly to equatorial rays; at maxima of the sunspot cycle (or at times of persistent high activity near the maximum), it is brighter and filled with streamers and rays around the limb of the sun. Since total solar eclipses have long been a source of fascination to man and are fairly well documented in historical records it would seem that we could use early descriptions of the corona to deduce the general state of solar activity at each occurrence. This could be a slightly more reliable signal of maxima or minima in the 11-y cycle than the naked-eye sighting of a single large sunspot, since the coronal form is a more general indicator of solar activity on the whole sun.

Clerke (1894) first pointed out that descriptions of the corona at total eclipses during the 1645-1715 period (Maunder Minimum) suggested that the corona may have been "in abeyance." I have shown (Eddy, 1976) that this seems indeed to have been the case and that first-hand accounts of the eclipsed sun during that time are all consistent with an electron corona much reduced in brightness or absent altogether; there are no descriptions of coronal streamers or any structured form of the corona during the time, even though the eclipsed sun was regularly observed. Where the corona is described it is as a dull, narrow ring of reddish light around the moon, which fits the description of the F-corona or zodiacal light, which would be noted even if the true corona were severely dimmed or absent altogether.

More significant, however, is the fact that we can find no descriptions of a structured corona in any historical record *before* the Maunder Minimum (Eddy, 1976). The first unambiguous account of a structured corona is for the (London) eclipse of May 2, 1715, which is remarkably late in the long history of interest in solar eclipses. It can be hoped that continued work on this intriguing absence will locate some early and clear descriptions, and that historians will take it as a challenge to look. R.R. Newton (1970, 1972) has pursued extensive searches and critical analyses of related ancient and medieval literature and has been unable to turn up a single early account.

The Maunder Minimum seems well enough established through other evidence; thus the absence of the structured corona during the years 1645-1715 may be real. The apparent lack of coronal descriptions *before* 1645 may be explained in several ways. It is possible that no one noticed the corona before the scientific enlightenment of the eighteenth century, that we generally see only what we are trained to look for, or that the structured corona was seen but deemed to be so much like the ordinary aureole around the uneclipsed sun that it did not deserve special description. Those who have seen the spectacle of the modern corona with the naked eye will find this explanation hard to accept. Alternatively, the lack of descriptions could represent a genuine absence of the corona for at least several centuries before the eighteenth—perhaps from the fourteenth or fifteenth centuries through the seventeenth, a period which by other evidence was a time of unusually low activity on the sun, embracing the Spörer and Maunder minima. By this explanation we might excuse the absence of the corona in records before the fifteenth century on the basis of paucity and poor quality of accounts from earlier and earlier time. Finally, the apparent absence could mean that the bright and structured corona is indeed a modern feature of the sun which first appeared at the time of the sun's recovery from the Maunder Minimum, in coincidence with the "auroral turn-on" of the same time. Eddy (1976) has pointed out the possibly significant coincidence of the first description of the chromosphere (1706), the first description of the structured corona (1715), and the dramatic increase in the number of reports of the aurora (about 1720).

D. TREE RINGS

The early hope of A.E. Douglass (1919, 1928) that evidence of past solar cycles could be found in patterns of the rings has been generally disproven in subsequent and more thorough study (LaMarche and Fritts, 1972). It is interesting, however, that Douglass's work, begun for the wrong reason, has proved of such monumental value in the fields of archaeology and climate research. Douglass was able to find some trees where the pattern of ring widths showed 11-y recurrence, but these were isolated examples from restricted regions.

Annual tree-ring widths are primarily determined by local moisture during the growing season. Tree age, type, location, temperature, prevailing sunshine, and the crowding of other trees also figure. An ideally located tree, and particularly one in a high and arid place, will produce annual growth which is proportional to local precipitation. In any case, tree-ring widths are primarily indicators of local climatic conditions, and since local climate has not been shown to be directly related to solar activity, we must discount tree rings as a possible, indirect source of solar history.

It should perhaps be mentioned that long-term histories from tree rings identify clear periods of local and regional drought. Douglass found that trees of the American Southwest gave evidence of a marked drought in the middle seventeenth to early eighteenth century; and before he knew of Maunder's work, Douglass (1919) labeled the period as a time when tree rings of Arizona and California showed singular anomaly. The same period is now known to have been a time of unseasonal cold in Europe and the time of the most severe winters of the so-called Little Ice Age (Gates and Mintz, 1975). Whether and how this marked climatic feature is related to the coincident solar anomaly is at this point conjectural.

E. CARBON 14

The power of terrestrial ^{14}C history as an index of past solar behavior has been pointed out by many authors (see for example Bray, 1967, 1968, 1971; Damon, 1968, 1970; Stuiver, 1961, 1965; and Suess, 1968). In this volume Damon discusses the issue in detail. In review, ^{14}C is formed by an n, p reaction on ^{14}N in the upper atmosphere by the neutrons which result from spallation caused by the impact of galactic cosmic rays. Cosmic rays, in turn, are modulated by changes in overall solar activity. The relationship with sunspot number is inverse: at times of low solar activity the shielding of the earth by the sun's extended field is reduced, we receive more cosmic rays, and ^{14}C production increases. At times of high solar activity the cosmic ray flux and ^{14}C production are suppressed. The strong relationship between cosmic-ray flux and the state of the sun is well illustrated in modern records of the cosmic-ray nucleonic flux, which are surely one of the better, though unappreciated, real-time indices of solar activity. (See, for example, Lanzerotti, Figure 11, in this volume.)

We may presume that changes in solar activity in the past, including the possible 11-y cycle, were also felt on earth as changes in cosmic ray fluxes, and indirectly registered in the upper atmosphere as changes in ^{14}C. CO_2 eventually finds its way to plants where it is assimilated and preserved. Thus by analyzing the $^{14}C/^{12}C$ ratio in tree rings of known age we can determine the relative amount of the isotope in the atmosphere at the time the ring was formed. Unfortunately, as Damon points out (1976; and Damon, Long, and Wallick, 1973), it is difficult to detect changes as short as 11-y, because the radiocarbon reservoirs act as a low-pass filter that very effectively attenuates effects shorter than about 20-y. It has, however, given a fairly clear picture of major, long-term changes in solar behavior. In the ^{14}C record, for example, is the clear and unmistakable signature of the Maunder Minimum, the Spörer Minimum, and the twelfth-century maximum (Eddy, 1976; Damon, 1976). There is also present, as Damon points out, evidence of a more or less cyclic change during the Christian era with a period of about 180-y or about twice the period of the so-called Gleissberg cycle.

Many other mechanisms affect ^{14}C production (Grey and Damon, 1970), and we must be cautious in ascribing any feature in the fossil ^{14}C record to solar cause. The most dominant long-term effect is not solar but terrestrial in origin: a slow and cyclic modulation with a period of about 10,000 years because of the changing magnetic moment of the earth's field (Bucha 1969, 1970). This, too, acts as a variable shield against the incidence of cosmic rays: the stronger the field, the greater the shield. When this is taken out of the ^{14}C record, the excursions which remain are of likely solar origin (Damon, 1976). This seems to be the case, at least at the most recent and best established end of the ^{14}C history, where features match up well with other evidences of solar change from aurorae, naked-eye sunspot observations, and, in the case of the Maunder Minimum, with historical telescopic observations of the state of the sun. Since it can be verified by direct, historical data, the Maunder Minimum has given us the key, in a sense, to the interpretation of solar effects in the ^{14}C history, and provides confidence in ascribing other, similar ^{14}C excursions to probable solar cause. I have shown (Eddy, 1976) that the Maunder Minimum corresponds to a ^{14}C deviation of about 10 parts per mil (see also Damon, 1976, Table 4) and have suggested that this can be used as a yardstick in interpreting earlier fluctuations in the record.

In Figure 9 we show the long-term record of ^{14}C deviation, from a compendium of data kindly provided by Paul Damon (see Damon, 1976). We have here plotted it with increasing ^{14}C downward for agreement in sense with changes in solar activity (higher activity upward). Thus the rapid decrease in ^{14}C abundance in modern times due to fossil fuel combustion (the Suess effect) appears as an upward spike at the right end of the curve. The overall envelope of ^{14}C deviation has been fitted by a best-fit sinusoidal curve which was derived by Lin, Fan, Damon, and Wallick (1975). This curve has about the same period as the best sinusoidal fit to the magnetic field data (Bucha, 1970), although 180° out of phase, to match the sense of the effect on ^{14}C production. There is some uncertainty in the smoothed magnetic curve and therefore in the contrast and relative prominence of residual features. But there can be little doubt as to the presence of the major features of

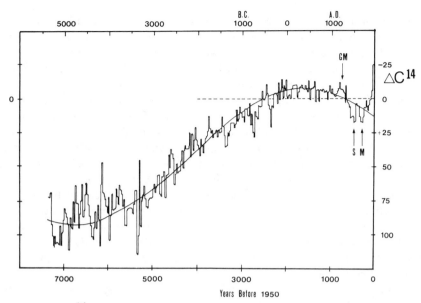

Fig. 9 Measured ^{14}C deviation (relative atmospheric concentration from tree ring analyses) in parts per mil since about 5000 B.C., with smoothed curve of sinusoidal variation in Earth's magnetic moment from Lin, et al., (1975). ^{14}C curve is a compilation from P.E. Damon (this volume). ^{14}C excursions attributed to solar cause are marked with arrows: M, Maunder Minimum; S, Sporer Minimum; GM, twelfth century maximum. The sharp negative (upward) excursion at the modern end of the curve is the Suess effect, due to fossil fuel combustion.

solar history which were described earlier, and of a host of possible earlier changes of equal magnitude.

V. THE SPECTRUM OF TIME PROJECT

In a series of papers over the last 20-y, D.J. Schove has endeavored to bring together evidences of past cyclic behavior of the sun and terrestrial effects in an effort to reconstruct the 11-y solar cycle backward in time. The work, which he has called the Spectrum of Time Project (STP) has produced estimates of peak sunspot numbers (at presumed maxima of the cycle) and decade averaged sunspot numbers to 650 B.C. (Schove, 1947; 1955; 1961a, b; 1962). These have been extensively used in subsequent work which required an estimate of past solar behavior. The STP sunspot numbers were derived chiefly from auroral reports and some of the other evidences cited here.

As pointed out by Eddy (1976) the STP reconstruction may be misleading and must be considered in the light of the *a priori* assumptions under which it was made, and with an appreciation for the necessarily incomplete historical data which were and are available. Since the work was done before our modern understanding of the ^{14}C record, the STP could not have allowed for the major changes in solar behavior which the isotope history now reveals.

The STP reconstruction, like Wolf's early reconstructions of the solar cycle, requires that there be nine solar cycles per century (Schove, 1955) whether or not there is evidence for them. And an implicit assumption of ever-normal solar behavior in the STP causes the peak sunspot estimates and decade average estimates to be more regular and bland than may now seem justified. Since historical data on solar behavior are sketchy and incomplete, Schove endeavored to apply correction factors for spots or aurorae which might have occurred but were not seen (Schove, 1962) and in this process adjusted the raw levels to values typical of the known, modern behavior of the sun. Moreover, implicit in the STP reconstruction are lines of evidence which more recent and critical study would exclude, such as fluctuations in meteorological parameters which are not known to be related to the sun (Schove, 1961*a*).

In the one period of comparison with more recent and detailed reconstruction—the Maunder Minimum—the STP washes out the severity of the effect and supplies regular maxima and minima through the time when contemporary first-hand accounts indicated that no sunspots were seen (Eddy, 1976). The comparison was shown in Figure 4. The differences cause us to question the absolute values of sunspot numbers in other epochs of the STP, and the validity of the 11-y cycle assumption in its earlier periods.

VI. SUMMARY AND CONCLUSIONS

Direct and indirect data on the solar behavior of the last 1,000 y are summarized in Figure 10.

We have long been accustomed to thinking of solar behavior as cyclic and regular and, through training and assumption, to treating the modern course of the solar

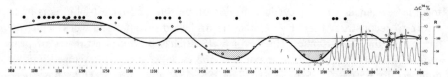

Fig. 10 ^{14}*C variation since 1050 A.D (open circles and heavy curve) compared with other indirect and direct indices of solar activity. Curve at lower right is telescopically-observed sunspot number. Dark circles are naked-eye sunspot observations, from Kanda (1933). Times when* ^{14}*C deviation exceeds* ±*10 parts per mil are shaded.*

cycle as representative of past behavior. Recent and more critical analyses of historical and modern evidence of solar history give cause to doubt these basic beliefs. There is no unambiguous evidence that the 11-y sunspot cycle was operative before about 1700. There is no unambiguous evidence that the sun had a corona, or even a chromosphere, before about the same date. There is cause to think that the incidence of aurorae increased abruptly at about the same time. And there is agreement in all the historical data that in the last 1,000 or 2,000-y the level of activity on the sun passed through three or four severe excursions, each nearly a century long; a probable deep minimum in the seventh century, an almost certain extensive maximum in the twelfth, and two protracted minima, one after the other, in the late fifteenth to early eighteenth centuries. We have no way as yet of knowing whether during these major excursions the 11-y cycle was operating or not.

This suggests that during the last 100 or 200-y, when we have observed the sun most intensively, its behavior may have been unusually regular and benign.

The problems of reconstructing the past behavior of the sun are of more than academic interest and are relevant beyond the field of history. The world's climate in the last 1,000-y, as well as it is known (Gates and Mintz, 1975), has undergone changes which correspond very closely in date and phase and amplitude with the excursions of solar behavior noted above: a warm period in the eleventh to thirteenth centuries (called the "Medieval Climatic Optimum") when solar activity was high, and two severe dips in the Little Ice Age which match the solar Spörer and Maunder minima. Since the end of the Maunder Minimum, the overall envelope of solar activity has been climbing (Figure 11), and during most of this time the world climate has been warming.

It seems possible that long-term changes in solar activity are accompanied by equally ponderous changes in the radiative output of the sun, or its particulate output, either or both of which would exert terrestrial effects. If the changes were in radiation and in the solar constant, the known climatic fluctuations could be explained by no more than a 0.5% per century change in the solar constant—a drift so slow that it lies outside the capability of detection in our present measurements of the solar output and outside our present ability to monitor other similar G stars. Curiously, the only extant long-term data on the solar constant, those of the Smithsonian Astrophysical Observatory, indicate a probable increase of about 0.5% per century between about 1908 and 1955 (Eddy, 1975; Öpik, 1968).

We can answer some of the questions of major solar change by more thorough and critical looks at the historical records of past solar behavior. Or, we can undertake extremely precise measurements of the present variability in all the outputs of the sun—a task for which the papers of this volume can establish a baseline. Probably the best course is both.

II HISTORICAL AND PALEOLOGICAL EVIDENCE

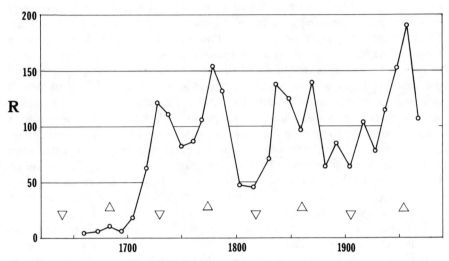

Fig. 11   Annual mean sunspot number R at maxima of the 11-y cycle, A.D. 1645 to present, to demonstrate long-term trends in solar activity. Evident is the 80-year "Gleissberg cycle" (extrema shown as triangles) imposed on a persistent rise since the Maunder Minimum.

REFERENCES

Bray, J.R., 1965, *Nature 205*, 440.

_____ , 1967, *Science 156*, 640.

_____ , 1968, *Nature 220*, 672.

_____ , 1971, *Science 171*, 672.

_____ , 1974, *Scientific, Historical, and Political Essays in Honor of Dirk J. Struik*, ed. R.S. Cohen et al., Boston Studies in the Philosophy of Science, vol. 15., D. Reidel, Dordrecht, 142.

_____ and R.E. Loughhead, 1965, *Sunspots*, Wiley, New York, 1.

Bucha, V., 1969, *Nature 224*, 681.

_____ , 1970, *Radiocarbon Variations and Absolute Chronology*, ed. I.V. Olsson, Almquist & Wiksell, Stockholm, 513.

Clerke, A.M., 1894, *Knowledge 17*, 206.

Damon, P.E., 1968, *Meteorol. Mon. 8*, 151.

_____ , 1970, *Radiocarbon Variations and Absolute Chronology*, ed. I.V. Olsson, Almquist & Wiksell, Stockholm, 571.

_____ , 1976, this volume.

_____ , Long, A., and Wallick, E.I., 1973, *Earth Planet. Sci. Letters 20*, 300.

Douglass, A.E., 1919, *Climatic Cycles and Tree Growth*, Carnegie Institute of Washington Publication 289, vol. 1, Washington, D.C., 102.

——————, 1928, *ibid.*, vol. 2, 125.

Eddy, J.A., 1975, *Bull. Am. Astr. Soc. 7*, 410.

——————, 1976, *Science 192*, 1189.

——————, Gilman, P.A., and Trotter, D.E., 1976, *Solar Phys.*, in press.

Fritz, H., 1873, *Verzeichniss Beobachter Polarlichter*, C. Gerold's Sohn, Vienna.

Gates, W.L., and Mintz, Y., 1975, *Understanding Climatic Change*, Appendix A, National Academy of Sciences, Washington, D.C.

Gleissberg, W., 1944, *Pub. Istanbul Univ. Obs. No. 27*.

Grey, D.C., and Damon, P.E., 1970, *Scientific Methods in Medieval Archaeology*, ed. R. Berger, University of California Press, Berkeley, 167.

Halley, E., 1716, *Phil. Trans. Roy. Soc. London 29*, 406.

Johnson, M.J., 1858, *Mem. Roy. Astr. Soc. London 29*, 196.

Kanda, S., 1933, *Proc. Imp. Acad. Japan 9*, 293.

LaMarche, V.C., and Fritts, H.C., 1972, *Tree-Ring Bull. 32*, 21.

Lanzerotti, L.J., 1976, this volume.

Lin, Y.C.; Fan, C.Y.; Damon, P.E.; and Wallick, E.J., 1975, *14th Int. Cosmic Ray Conf. Munich 3*, 995.

Matsushita, S., 1956, *J. Geophys. Res. 61*, 297.

Maunder, E.W., 1890, *M.N.R.A.S. 50*, 251.

——————, 1894, *Knowledge 17*, 173.

——————, 1922, *J. Br. Astr. Assoc. 32*, 140.

Newton, R.R., 1970, *Ancient Astronomical Observations and the Acceleration of the Earth and Moon*, Johns Hopkins University Press, Baltimore, 39.

——————, 1972, *Medieval Chronicles and the Rotation of the Earth*, John Hopkins University Press, Baltimore, 99, 600.

Opik, E.J., 1968, *Irish Astr. J. 8*, 153.

Schove, D.J., 1947, *Terr. Magn. Atmos. Elect. 52*, 233.

——————, 1955, *J. Geophys. Res. 60*, 127.

——————, 1961a, *Ann. N.Y. Acad. Sci. 95*, 107.

——————, 1961b, *J. Br. Astr. Assoc. 71*, 320.

——————, 1962, *ibid. 72*, 30.

Schwabe, H., 1843, *Astr. Nachr. 20*, No. 295.

Sporer, F.W.G., 1887, *Astr. Ges. Vjschr. Leipzig 22*, 323.

——————, 1889, *Bull. Astr. 6*, 60.

Stuiver, M., 1961, *J. Geophys. Res. 66*, 273.

——————, 1965, *Science 149*, 533.

Suess, H.E., 1968, *Meteorol. Mon. 8*, 146.

Waldmeier, M., 1961, *The Sunspot-Activity in the Years 1610-1960*, Schulthess and Co., Zurich.

Wolf, R., 1862, *Astr. Mitt. Zurich No. 14*.

——————, 1856, *ibid. No. 1*, 8.

CLIMATIC CHANGE AND THE POSSIBLE INFLUENCE OF VARIATIONS OF SOLAR INPUT

James D. Hays

*Lamont-Doherty Geological Observatory of
Columbia University*

The history of earth's climate over the past billion years can provide constraints on the amount of variation of solar flux that may have occurred. It can also provide some information on the sensitivity of the earth's climate to variations of solar input. In such an analysis, it must be realized that a number of factors other than solar variability may cause climatic variations. Among them are changes in the composition of the atmosphere, particularly concentrations of CO_2, and changes in the distribution of continents and oceans. In this respect the flooding of continental surfaces by marine water may also be an important factor. Changes in the seasonal and geographic distribution of insolation as a result of changes in the shape of the earth's orbit are certainly important. Changes in the amount of insolation reaching the earth's surface may also contribute to climatic variability whether they result from screening by interstellar dust or from atmospheric dust injected into the stratosphere by explosive volcanic eruptions.

The purpose of this review is to examine the record of earth's climate over the period of time in which it can be read and to assess which, if any, of the climatic variations the earth has experienced are likely to be a result of variations in solar output. First, we examine the kinds of evidence that give us clues to the past climate of the earth. The data that are available and their reliability are very much a function of age. The most accurate records are for the last million years, for which many climatic indicators can be used. Fossils of extant species can be related to temperature tolerances of living forms and fairly precise reconstructions can be made of our globe's climate for perhaps the last million years (CLIMAP, 1976). However, as we probe deeper into the past, higher and higher percentages of the fossil assemblages are no longer found in living form. Therefore, paleoclimatic interpretations must necessarily be more gross. For rocks of great antiquity (more than 50 million years old), groups of plants and animals are used to indicate environmental conditions. For example, reef-building corals are at present confined to the tropics. We assume that this group required similarly warm conditions to flourish in the past even though none of the species that lived 100 million years ago is currently living. Fossil palm trees in what are now high latitudes are used as evidence for past reductions in the thermal gradient between the equator and the pole.

Care must be taken, however, to consider the paleolatitudes of the site being studied. Fossil palm leaves in Greenland and coal measures in Spitzbergen mean one thing if these areas had their present latitudinal position when the plants were growing but something quite different if they were in a more equatorial position. Great strides have been made over the past two decades in understanding the paleogeography of our planet. Paleomagnetic and geomagnetic data can be combined to provide accurate paleolatitudes and relative paleolongitudes for large continental blocks during the last 200 million years. For earlier times only paleolatitudes can be derived from the paleomagnetic data. The interpretation of these data depends on the assumption that the earth's magnetic poles are always closely associated with the geographic poles.

Other types of evidence for past climatic conditions are sediment types that are peculiar to certain climatic regimes. These include tillites (unsorted consolidated rock debris transported and deposited by glaciers), evaporites and windblown sand deposits that are generally confined to arid regions, and coal, which is usually considered indicative of temperate latitudes. Of these, the tillites are probably the most reliable as they clearly represent glacial conditions, usually of broad regional extent. In this review, I consider the paleoclimatic data in the context of the paleogeography of the planet. The time resolution of the record is also a function of age. Dating errors that are inherent in radiometric techniques in combination with the limited exposures and structural complexity of very ancient rocks severely limit the temporal resolution of climatic trends of the distant past. Thus, the older rocks of the earth contain clues to broad, long-term climatic change measured in tens and hundreds of millions of years, while the more recent deposits (last million years) can be used to measure climatic variations of tens to hundreds of thousands of years.

We are far too ignorant at present to be able to fully explain the causes of all the climatic variance that can be read from the geological record. However, certain possible causes outlined above operate on different time scales. For example, significant changes in continental positions require tens of millions of years, while known variations of the earth's orbit occur in tens of thousands of years. The great unknown is solar variability; however, it may be useful to approach the problem by asking the question: "Does the record require solar variability and, if so, on what time scale and by how much?"

I. EARLY CLIMATES OF THE EARTH: THE FIRST 3.5 BILLION YEARS

The earth is probably about 4.5 billion years old, according to dates obtained on meteorites and lunar rocks. The earliest evidence of life on our planet is about

3.5 billion years old. There is fair evidence that since 2.5 to three billion years ago life has been uninterrupted. This in itself puts important constraints on earth's climate by requiring at least some areas of the planet (probably most) to have temperatures between 0 and 100 °C.

By roughly 600 million years ago, life was highly evolved; the fossil record shows many advanced forms of marine animal life. There is no evidence that these creatures evolved in a world that had a climate significantly different from the climates that have prevailed since. Although minor compositional changes of the atmosphere may have occurred since 600 million years ago, these were small compared with changes that probably occurred before.

Although our knowledge of our planet's environment for the first four billion years of its history is spotty, the little evidence we do have points to relative stability since three billion years ago, with surface temperatures over most of the planet above the freezing point of water and probably well below the boiling point over all of the planet. The record of climate improves greatly after about a billion years ago. During this interval, we have evidence of climatic change. For the earlier part of this record, perhaps the most significant evidence is that of glaciations, which indicate a striking departure from the norm.

II. THE RECORD OF CLIMATIC CHANGE OVER A BILLION YEARS

There have been three well-established glacial intervals in the last billion years. Glaciations prior to one billion years ago are problematical. Precise dating of the late Precambrian glaciation or glaciations is imprecise; however, they seem to fall mostly between 900 and 600 million years ago. Evidence for these glaciations is found in both hemispheres, with tillites reported from northeast Greenland, Scandinavia, Spitzbergen, China, North Korea, Katanga, Zaire, and South Australia (William, 1975). The distribution of glacial evidence in the late Precambrian is so widespread as to suggest to some a very extensive distribution—far greater than any glaciation since. However, caution must be exercised in interpreting these ancient glaciations. First, the criteria for recognizing tillites of this age are not always strictly applied, and second, the vast amounts of time involved, possibly several hundred million years, make it questionable that all these areas were simultaneously glaciated.

Figure 1 shows the South Pole roughly 500 million years ago, over what is now northwest Africa, while the North Pole is located in open ocean. Tillites from about this time have been found in the Sahara. Evidence of rather extensive North African glaciation at this time has been reviewed by Fairbridge (1971).

By Lower Carboniferous time the southern continent had drifted substantially north (Fig. 2) and the only evidence of glaciation at that time is some tentatively identified tillites in southern Argentina. More important is the fact that between

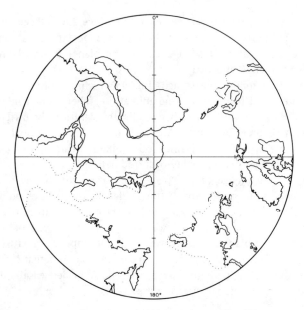

Fig. 1 Global reconstruction of continental positions about 500 million years ago. Locations of tillites are marked with x. (After Briden, Drewry, and Smith, 1974.)

Fig. 2 Global reconstruction of continental positions about 350 million years ago. Tillites are marked with x. (After Briden, Drewry, and Smith, 1974.)

400 million years ago, and about 300 million years ago the pole traced a path across Africa from Morocco to South Africa. However, there is no evidence of glaciation between the Ordovician glaciation of North Africa and the Lower Carboniferous tentative glaciation in Argentina. Apparently the location of a continent over the pole is not the only criterion for glaciation. The distribution of faunas suggests that between these two times the earth had a relatively equitable climate. There is ample evidence of widespread glaciation on the southern continent of Gondwanaland by Late Carboniferous-Lower Permian time 320 million years ago.

This evidence for extensive glaciations in the Permo-Carboniferous has been known for more than a hundred years. Glacial deposits, often resting on scratched rock surfaces similar to striated parts of today's Arctic Canada, are now known from South America, Africa, India, Australia, and Antarctica. Figure 3 shows this distribution together with the directions of motion of the ice. This glacial evidence provided one of the earliest and most compelling arguments for continental drift (note the direction of glacial movement in Argentina).

There is no way that glaciation could have this pattern with the present distribution of continents. However, when the continents are reassembled as they were in the Permo-Carboniferous, the pattern is easily explainable as a large ice cap centered on the pole (Fig. 4).

There is evidence of multiple glaciations (advances and retreats, similar to those of the present interval) of this southern hemisphere ice cap. The earliest glacial advance is placed at about 320 million years ago, with subsequent advances about 300 million and 260 million years ago.

After the breakup of Gondwanaland that occurred in the Mesozoic, the world entered another long warm interval. The distribution of plants and animals during

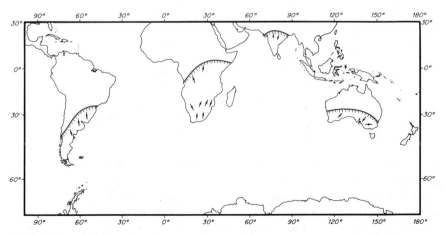

Fig. 3 Present continental configuration showing the distribution of Late Carboniferous (250 million years ago) glaciation. Arrows indicate direction of ice flow. (After Holmes, 1964.)

II HISTORICAL AND PALEOLOGICAL EVIDENCE

Fig. 4 Global reconstruction of continental positions about 250 million years ago (Late Carboniferous). Distribution of tillites of that age is indicated by x. (After Briden, Drewry, and Smith, 1974.)

the Mesozoic and early Cenozoic (250-50 million years ago) strongly suggests that there was probably no permanent ice on the planet with the possible exception of high mountain peaks. Because of erosion, the glacial record of these sectors is usually poor. This 200-million-year interval was characterized by relatively low equator-to-pole thermal gradients. Equatorial temperatures were similar to those at present (25°C), while polar temperatures may have been about 10-15°C and possibly somewhat warmer at times. Evidence from the reconstruction of continental positions during this period is fairly accurate; during much of this time both poles were centered in large oceanic regions (Figs. 5, 6, and 7).

III. CLIMATE OF THE LAST 100 MILLION YEARS

The evolution of climate over the past 100 million years is much better understood than earlier climatic records. This is, in part, due to a more continuous succession of sediments on land, but even more important is the fact that for this period the record of the deep ocean can be studied. In general, this record has the advantage of being far more continuous and less disturbed than the continental record, and the oceanic record more closely approximates global changes in climate than does the record from bits of continents.

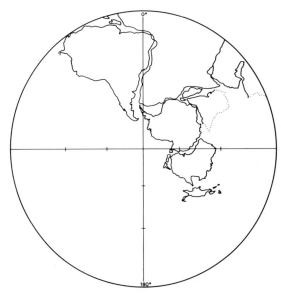

Fig. 5 Global reconstruction of continental positions about 200 million years ago. There is no evidence of permanent ice on the planet at that time. (After Briden, Drewry, and Smith, 1974.)

Fig. 6 Global reconstruction of continental positions about 170 million years ago. There is no evidence of permanent ice on the planet at that time. (After Briden, Drewry, and Smith, 1974.)

Fig. 7 Global reconstruction of continental positions about 100 million years ago. There is no evidence of ice on the planet at that time and high-latitude ocean temperatures were probably 15°C or higher. (After Briden, Drewry, and Smith, 1974.)

Since the earth is currently in a glacial interval and it was not 100 million years ago, the record of the last 100 million years provides an opportunity to examine the evolution of climate from conditions that may have been typical of most of the earth's history during the last billion years to the more unusual conditions that characterize glacial intervals. The interpretation of this record is greatly aided by measurements of O^{18}/O^{16} in the calcareous shells of planktonic (surface living) and benthic (bottom living) marine organisms. Although a number of factors influence this ratio, we can be fairly confident that prior to the development of large polar ice caps these ratios were strongly influenced by the temperature of the water in which the organisms grew. Since the bottom waters of the ocean probably had their origins in high latitudes, the O^{18}/O^{16} record of benthic organisms provides a record of high-latitude cooling.

Figure 8 contains O^{18}/O^{16} data on benthic foraminifera from Douglas and Savin (1975) covering the interval 110-65 million years ago. These measurements were made on a number of North Pacific drill cores. Benthic foraminifera O^{18}/O^{16} measurements covering the interval 55-10 million years ago are from Shackleton and Kennett (1975). These measurements were made on a drill core from south of New Zealand (50°S). The general shape of this curve indicates warming in the Late Cretaceous, peaking somewhere between 90 and 75 million years ago, and a gradual cooling since.

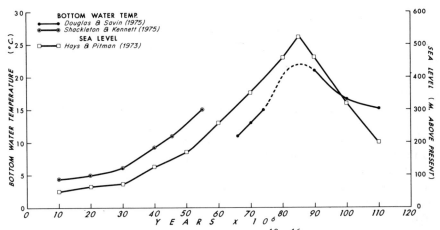

Fig. 8 Bottom water temperatures estimated from the O^{18}/O^{16} ratio of the shells of benthic foraminifera and changes in global sea level calculated from changes in sea floor spreading during the last 110 million years.

Bottom water temperatures were extremely warm in the Late Cretaceous and the possibility that the deep circulation of the ocean at this time was driven primarily by salinity contrasts rather than thermal contrasts must be considered. If this was so, the source of the bottom water was probably in middle latitudes rather than high latitudes.

The general trend of temperature change shown by the oxygen isotopic ratios may be accounted for in part by the movement of continents. However, it is more likely that a second factor is as important or more important. This factor is global sea level change. Much evidence now supports the idea that sea level was significantly higher in the distant past than it is now. This is particularly true of the interval between 110 and 85 million years ago (Hays and Pitman, 1973). We now know that the rate of continental drift or sea floor spreading is not constant. At times in the past, rates were significantly higher than they are now. This is true of the interval between 110 and 85 million years ago. One of the consequences of these higher rates was a general expansion of the volume of the world's oceanic ridge system. This expansion led to a reduction in the volume of ocean basins and the consequent flooding of the continents by shallow seas (Hays and Pitman, 1973). It is likely that this increase in area of shallow seas further contributed to the mildness of global climate and to a reduction of seasonal contrast. I have added to Figure 8 a generalized sea level curve (Hays and Pitman, 1973) computed from changes in spreading rates of ocean ridges. This computed curve agrees well in shape and magnitude with estimates of global sea level change derived from decades of study of marine rocks on the continents and also bears a close resemblance to the paleo-temperature curve.

Figure 9 shows that by Eocene time (~50 million years ago) the North and South Poles were close to their present positions. However, the detailed ocean bottom temperature curve by Shackleton and Kennett (1975) (Fig. 10) shows a gradual cooling. They suggest that the glaciation of Antarctica is marked by the abrupt drop in isotopic values in the mid-Miocene about 10 million years ago. Glaciation of Antarctica by this time or possibly somewhat earlier is supported by much geological data (Craddock, Bastien, and Rutford, 1964). The glaciation of Antarctica marked the beginning of another glacial interval.

Cooling continued with the glaciation of Greenland about 3 million years ago and the formation of high-altitude glaciers in the Sierra Nevada at about the same time. This cooling eventually resulted in the expansion of polar ice into temperate latitudes about one million years ago. Since this time the temperate latitude ice sheets have formed and expanded periodically with gross periods of about 100,000 years.

Fig. 9 Global reconstruction of continental positions about 50 million years ago. There is suggestive evidence that Antarctica may have been glaciated at this time but sea floor temperatures suggest that Antarctic glaciation did not occur until ≃35 million years later, approximately middle to late Miocene.

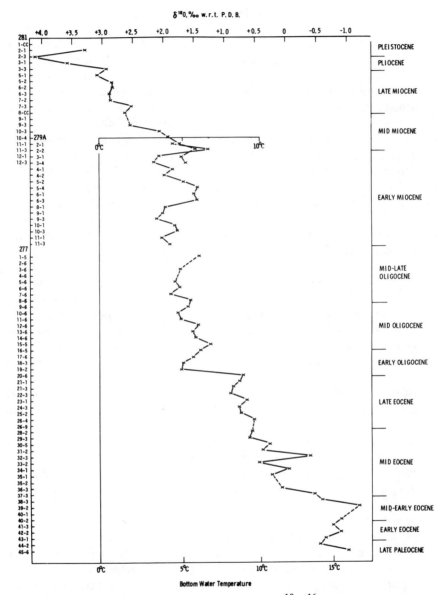

Fig. 10 Estimated bottom water temperatures from the O^{18}/O^{16} ratio of benthic foraminifera recovered from a core drilled at 50° S, south of New Zealand (from Shackleton and Kennett, 1975). Record covers the last 60 million years. Sharp rise in δO^{18} in the mid-Miocene (≃10-15 million years ago) has been interpreted by Shackleton and Kennett (1975) to mark the glaciation of Antarctica. Numbers on the left side of diagram indicate the drill sites and cores sampled.

IV. DISCUSSION

The causes of glacial intervals have been a matter of much speculation. Some have linked them to the rotation period of the galaxy, which is about a quarter of a billion years, because three intervals—the late Precambrian at $\sim$700 million years ago; the Permo-Carboniferous at 300 million years ago, and the most recent at 10 million years ago—have roughly this kind of spacing. However, this spacing is by no means exactly the same as the galactic rotation and other causes seem far more probable. A more likely cause of these long-term changes is the changing distribution of continents and oceans. Reassembly of the continents on the basis of paleomagnetic data indicates that for both the glacial interval 300 million years ago and the present one there was a continent located over the South Pole. The earlier glacial interval shows no evidence of glaciation in the northern hemisphere. In the present glacial interval, the North Pole is over an ocean but this ocean is nearly surrounded by continents and is effectively isolated from thermal advection by ocean currents.

Global atmospheric models are now reaching sufficient levels of sophistication to test the climatic importance of changing continental and ocean positions. One such effort is that of Donn and Shaw (in press), who used the thermodynamic meteorologic model of Adem (1962, 1963, 1964a, b, 1965). Donn and Shaw did not allow for variations in the distribution of shallow seas over the continents. However, they did vary continental positions from 200 million years ago to today in five steps—200 million, 150 million, 65 million, and 35 million years ago and the present.

Just by varying the geometry of continental configurations, the model estimated a gradual decline in northern hemisphere high-latitude temperatures between 200 million years ago and the present. Conversely, equatorial temperatures remained relatively constant. Additional experiments with models that take into account the distribution of epeiric seas and thermal transport by ocean currents will clarify the climatic importance of continental and ocean distributions on global climate. However, it appears likely that continental drift and global sea level changes are important in setting the stage for glacial intervals.

The warmth of interglacial intervals may be enhanced by loss of CO_2 from the oceans. Warming the ocean to as high temperatures as are indicated for the Late Cretaceous would force CO_2 out of solution, thereby enhancing the greenhouse effect. This, of course, would only work to amplify temperature changes caused by other factors. However, it should be emphasized that all evidence suggests that equatorial temperatures have remained fairly constant. The major temperature changes occur in high latitudes.

V. GLACIAL-INTERGLACIAL STAGES— THE RECORD OF THE LAST MILLION YEARS

The causes of climatic variability on the time scale of 10^4-10^5 years has been a subject of great controversy and interest for more than a hundred years, since it was first discovered that extensive middle-latitude ice sheets formed on northern hemisphere continents repeatedly during the past million years. It is beyond the scope of this review to summarize the long history of studies of glacial-interglacial climatic fluctuations. Suffice it to say that our knowledge of the occurrence of glaciations during the last million years has been greatly expanded through the study of deep sea sediments during the past 15 years (Fig. 11). The record recovered with piston cores from the ocean floor is often continuous or nearly continuous for a million years or longer. It has been shown by Shackleton and Opdyke (1973) that, during the last million years, fluctuations of O^{18}/O^{16} in the shells of planktonic organisms preserved in deep sea sediments are primarily a result of changes in the isotopic composition of seawater caused by changing volumes of glacial ice on the continents. As a consequence, the O^{18}/O^{16} record in deep sea cores is a good measure of the waxing and waning of continental glaciers.

Work by Hays, Imbrie, and Shackleton (1976) on two subantarctic cores with very constant accumulation rates for more than 400,000 years demonstrates that the main periods in the climatic record of both the northern and southern hemispheres during this time are ∼100,000, 41,000, and 23,000 years. These periods are nearly precisely the same as the periods of the earth's orbital parameters (eccentricity, 100,000 years; obliquity, 41,000 years; and precession, 23,000 years). There is little doubt that the timing of the glacial-interglacial stages is controlled by seasonal variations of solar insolation, which are related to the changing geometry of the earth's orbit.

Although these climatic variations do not need to be related to variations in solar output, they are caused by seasonal and latitudinal changes in solar input. Therefore, the connection between glacial-interglacial changes and changes in the geometry of the earth's orbit underlines the sensitivity of earth's climate to possible variations of the solar constant.

VI. CLIMATIC OSCILLATIONS LESS THAN 10,000 YEARS AGO

Although demonstrable climatic changes with periods between 1,000 and 10,000 years have not yet been proved statistically, there is suggestive evidence that some may exist. There is good evidence for climatic variability, and climatic variability on these time scales has the best possibility of eventually being linked to solar variability.

The evidence for climatic variability during this interval comes from a variety

86   II HISTORICAL AND PALEOLOGICAL EVIDENCE

MAIN TRENDS IN GLOBAL CLIMATE: THE PAST MILLION YRS.

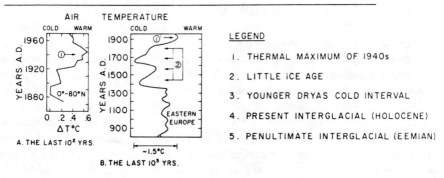

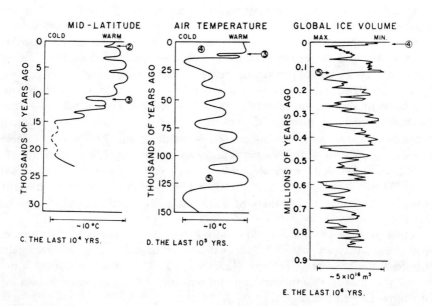

Fig. 11 Main trends in global climate from J. Imbrie, W. Broecker et al., 1975, "A survey of past climates," Appendix A, Understanding Climatic Change: A Program for Action, Panel on Climatic Variation, U.S. Committee for GARP, National Academy of Sciences, Washington, D.C.

of sources. These include the advances and retreats of mountain glaciers (Denton and Karlen, 1973) and changes in isotopic composition of ice from Greenland ice cores. Denton and Karlen made an extensive study of mountain glacier advances in both Alaska and Lapland. These advances were carefully dated with the use of ^{14}C and the investigators concluded that periods of glacier advances in these widely separated areas are highly correlated. Advances occurred between 5800 and 4900

B.P., again between 3300 and 2400 B.P., and finally in what is referred to as the Little Ice Age: 1300-1900 A.D. These times of mountain glacier advance are spaced at about 2500 year intervals.

A review by Denton and Karlen (1973) of ^{14}C dated advances and retreats of the waning phase of the last ice age (from 14,000-10,000 B.P.) suggests a similar temporal spacing of relative advances and retreats of the large European and American ice sheets. These authors also made the connection between past variations in ^{14}C and their own record of climatic change. The following paragraphs draw heavily on the Denton and Karlen paper.

Radiocarbon measurements on tree rings show two types of variations during the last 7,000 years (see, for example, Suess, 1970a; de Vries, 1958). The first is a gradual decrease in ^{14}C activity of about $100°/_{oo}$ between 7000 and 2200 B.P. This change has been attributed to geomagnetic field variation (Grey, 1969; Lingenfelter and Ramaty, 1969; Bucha, 1970; Suess, 1970a). In addition, short-term fluctuations of up to $30°/_{oo}$ occur. These shorter term fluctuations are best known for the past 1,000 years; however, Ferguson (1970) extended the chronology to cover the past 7,000 years. Figure 12 shows these short-term changes plotted as deviations in ^{14}C from a best-fit sine wave curve drawn through the scatter of points representing the long-term change (Suess, 1970a, figs. 1 and 2). Suess (1968) pointed out the close correlation between these variations and climatic change during the past few millennia.

A variety of historical data, including agricultural records of harvest dates and records of lake, river, and North Atlantic sea ice, can be used to reconstruct the trend of climate during this time. These data are summarized by Lamb (1966) and Ladurie (1971). Both authors agree that the northern European climate was warmer than it is now between 1000 and 1200 A.D. Climate then became cooler between 1200 and 1400, with a possible small recovery between 1400 and 1500. This possible recovery was succeeded by a prolonged cold interval known as the Little Ice Age. Mitchell (1961) documented a general warming since the late 1800s that reversed about 1940.

Variations of measured ^{14}C activity closely parallel these historical climatic records (Suess, 1968; Damon, 1968), with high ^{14}C activity corresponding to colder climate. Denton and Karlen (1973) extended this correlation by using Suess' (1970a) ^{14}C measurements and their own record of mountain glacier advances and retreats (Fig. 12). Figure 12 shows that periods of glacier expansion, as far as they are known, correspond with intervals of high ^{14}C activity.

There are now three explanations for these short-term variations of ^{14}C. First, it has been suggested that, like the long-term trend, the short-term fluctuations of ^{14}C activity are a result of variations of the earth's magnetic field. This suggestion does not yet have supporting evidence (Damon, this volume). Second, the ^{14}C variation may be a consequence of climatic change itself. Changes in the exchange rate between ocean and atmosphere caused by changing rates of ocean mixing

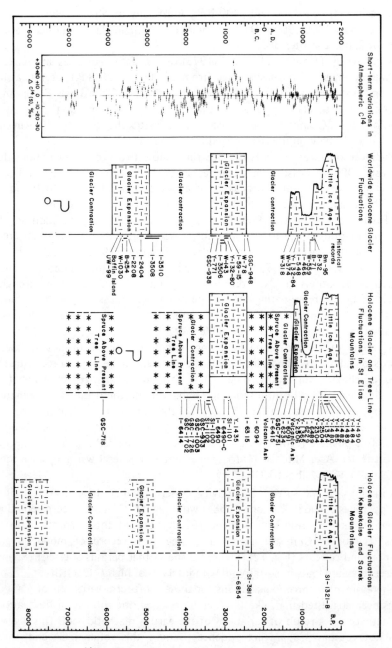

Fig. 12 Variations in ^{14}C activity and mountain glacier fluctuations during last 7,000 years (Denton and Karlin, 1973).

could cause variation in ^{14}C activity. However, the direction and magnitude of the change in response to global cooling is not known.

The third hypothesis relates change in ^{14}C activity to changes in solar activity. Stuiver (1961) suggests that solar activity modulates atmospheric ^{14}C content by influencing the galactic cosmic ray flux in the vicinity of the earth. At times of active sun, ^{14}C production is diminished because the weak magnetic field carried through the solar system by intensified solar wind deflects an increased number of incoming cosmic rays. Conversely, fewer cosmic rays are deflected during times of quiet sun, thereby leading to increased ^{14}C production.

The above hypothesis is strengthened by direct measurements that show an approximate inverse relationship between sunspot (solar flare) activity and cosmic ray flux into the upper atmosphere (Stuiver, 1961; 1965; Lingenfelter, 1963). Stuiver (1961; 1965) showed that this was true back to the 19th century, and de Vries (1958) and later Suess (1968) demonstrated this relationship back into the Middle Ages. In spite of this suggestive evidence that some climatic change during the last 7,000 years may be related to changes in solar activity, it still has not been demonstrated that changes in solar activity are accompanied by changes in solar output, which seems the most obvious way to influence Earth's climate.

VII. CONCLUSIONS

The paleoclimatic record is rich with information that may bear on the constancy of the solar flux with time. The record clearly demonstrates the sensitivity of the present climatic regime to seasonal and latitudinal variations of solar input. However, to unravel the complex influences of changes in other climatic boundary conditions requires much additional effort. With rapidly increasing understanding of the changing relationship of continents and oceans, it should be possible with ocean/atmosphere models to determine whether solar variability is a necessary part of the forcing functions that drive climatic change on time scales of tens of millions of years.

I believe that long-term climatic changes such as glacial intervals, separated by hundreds of millions of years, will eventually be explained by changing ocean and continent geometries. Middle-latitude ice sheet variations on time scales of tens of thousands to hundreds of thousands of years are best explained by seasonal and latitudinal variations of solar input caused by changes in the geometry of the earth's orbit. However, climatic changes on shorter time scales (hundreds to thousands of years) may well be a result, in part, of changes in solar output.

REFERENCES

Adem, J., 1962, *Tellus 14*, 102.

───────── , 1963, *Monthly Weather Rev. 91*, 375.

───────── , 1964a, *ibid. 92*, 91.

───────── , 1964b, *Geofis. Internat. 4*, 3.

───────── , 1965, *Monthly Weather Rev. 93*, 495.

Baxter, M.S., and Walton, A., 1971, *Proc. Roy. Soc. London A 321*, 105.

CLIMAP, 1976, *Science 191*, 1131.

Craddock, C., Bastien, T.W., and Rutford, R.H., 1964, *Antarctic Geology*, Wiley Interscience, New York, 171-187.

Denton, G., and Karlen, W., 1973, *Quaternary Res. 3*, No. 2, 155.

de Vries, H., 1958, *Koninkl. Nederlandse Akad. van Wetenschappen, Amsterdam, Proc. Ser. B. 61*, 94.

Donn, W., and Shaw, D., 1976, *WMO Pub. 421*, 53.

Douglas, R., and Savin, S., 1975, *Init. Rept. Deep Sea Drilling Project 32*, 509.

Fairbridge, R.W., 1971, *Geol. Soc. Am. Bull. 82*, 267.

Grey, D.C., 1969, *J. Geophys. Res. 74*, 6333.

Hays, J.D., and Pitman, W.C., III., 1973, *Nature 246*, 18.

───────── , Imbrie, J., and Shackleton, N.J., 1976, *Science 194*, 1121.

Ladurie, E.L., 1971, *Times of Feast, Times of Famine*, Doubleday, New York.

Lamb, H.H., 1966, *The Changing Climate*, Methuen and Co., London.

Mitchell, J.M., Jr., 1961, *Ann. N.Y. Acad. Sci. 95*, 235.

Shackleton, N.J., and Kennett, J.P., 1975, *Init. Rept. Deep Sea Drilling Project 29*, 743.

───────── and Opdyke, N.D., 1973, *Quaternary Res. 3*, 39.

───────── and ───────── , 1976, *Geol. Soc. Am. Mem. No. 145*, 449-.

Suess, H.E., 1968, *Meteorol. Monog. 8*, 146.

───────── , 1970a, *Radiocarbon Variations and Absolute Chronology*, ed. I.V. Olsson, Wiley Interscience, New York, 303.

Stuiver, M., 1961, *J. Geophys. Res. 66*, 273.

───────── , 1965, *Science 149*, 533.

William, G.E., 1975, *Geol. Mag. 112*, 441.

III. THE INTEGRATED SOLAR FLUX

Of all the components of the solar output, the total solar irradiance or solar constant is by far the largest contributor to the earth's energy budget. This single global measure is the result of the nuclear processes responsible for the sun's existence as a star capable of supporting life on the earth. Since we do not yet have a firm, quantitative measure of the sensitivity of the earth's climate and weather to changes in the solar constant, however, one very important problem in solar physics continues to be the absolute measurement of the solar constant and its variation with time. There is growing evidence that the solar constant has not changed by more than .75% from 1969 to 1976 in the recent solar cycle; but the record is far from continuous, and measurements have not uniformly taken advantage of new techniques introduced over the last 20 years. Frohlich's article makes the state of the science at the present time very clear and shows that we now have the opportunity to make our records of this fundamental solar parameter continuous into the future if we will only undertake a program to determine both the absolute mean value of the solar constant and its relative variation over, say, the next two solar cycles. Such a solar constant measurement program should be combined with a spectral irradiance measurement effort, particularly if the solar constant program is to be conducted from ground-based sites. Without the spectral irradiance data, it is not possible to correct solar constant data for the effects of aerosols, ozone, and water vapor absorption to better than 1%.

Although this book concerns itself principally with intrinsic variations in the output of the sun, the variation of the insolation received at the earth owing to changes in the earth's orbit around the sun appears to be a very significant factor in the past climate of the earth. As Hays, Imbrie, and Shackleton (Hays, J.D., Imbrie, J., and Shackleton, M.J., 1976, *Science 194*, 1122) argue so persuasively, the recurrent periods of glaciation on the earth may be directly correlated with variations in the eccentricity, obliquity, and precession of the earth's track in the solar system. Vernekar's contribution to this volume deals with the methods necessary to compute this variation in insolation from the mechanics of the solar system itself. Thus, given a truly constant sun and sufficiently accurate procedures for handling the celestial mechanics problem, the insolation record at any terrestrial latitude can be computed both forward and backward in time for use both in comparisons with paleoclimatic studies and in predicting climate changes resulting from the very slow variation of the earth's orbit.

III INTEGRATED SOLAR FLUX

Following the summaries of solar constant measurements given by Dr. Frohlich, Hickey, and Willson in the first edition of this volume, experimental programs to measure changes in the solar constant have continued with interesting preliminary results. Three experiments—a balloon-borne thermopile flown by Murcray and Kosters in 1968 and 1978, Hickey's thermopile experiments on both Nimbus 6 and Nimbus 7 satellites operating continuously from 1975 to 1979, and the cavity radiometer package flown on rockets in 1976 and 1978 by Willson, Kendall, and Hickey—all show systematic increases in the solar constant ranging from about .15% for the Nimbus satellite experiments to .4% for the balloon and rocket experiments. These exciting results are just now being published, but the experimenters have generously communicated them to me prior to formal publication.

Of the three experiments, the rocket flights using cavity radiometers appear to give the most precise measures of the solar constant. The first flight in June 1976 carried three radiometers built independently by R.C. Willson (JPL), J.R. Hickey (Eppley), and J.M. Kendall (JPL); and these measurements establish the baseline value of the solar constant at solar minimum between solar cycles 20 and 21. The second flight in November 1978 carried radiometers built by the same experimenters, but only results from the Willson and Kendall instruments are available now. Both experimenters report increases of .4% with an estimated precision of .1%. This then suggests a systematic increase of the solar constant from the time of solar minimum (sunspot number ~ 10) to periods of much greater solar activity (sunspot number ~ 110) just prior to the peak of activity in solar cycle 21.

The Kosters and Murcray (1979)* result of a .4% increase over the longer time period from the maximum of cycle 20 (1968) to the beginning of cycle 21 agrees in the sign of the change of the solar constant, but the longer time interval of 10 years between experiments does not help us resolve the question of the solar cycle variation of the radiative output of the sun. Nevertheless, a precision of .1% is claimed for the balloon experiments; hence, the change appears to be significant.

Hickey's preliminary results from the comparison of thermopile measurements on both Nimbus 6 and Nimbus 7 apply only to UV bands from 275 to 400 nm. A systematic change of 4% is observed in the 275 nm to 360 nm channel, and this translates into a change of .17% in the total solar irradiance. The change was smaller in the longer wavelength channel from 300 nm to 400 nm; thus, Hickey has suggested that the solar UV spectrum below about 300 nm has increased more than at longer wavelengths over the period from 1975 to 1978.

Changes in the solar constant of these magnitudes are sufficiently large to demand attention from both the solar physicist and the climatologist, and in that regard, solar constant variations exceeding .1% over a two year period are quite unexpected. These new results obviously demand confirmation through future experiments such as those planned for the Solar Maximum Mission to begin in November 1979 together with additional rocket and balloon flights conducted throughout Solar Cycle 21.

* Koster, J.J. and D.G. Murcray, 1979, Geophysical Research Letters, 6, No. 4, page 382-383.

CONTEMPORARY MEASURES OF THE SOLAR CONSTANT

C. Fröhlich
Physico-Meteorological Observatory
World Radiation Center
Davos, Switzerland

I. INTRODUCTION

The solar constant S is the amount of total solar energy of all wavelengths received per unit time and unit area at the mean sun-earth distance in the absence of the earth's atmosphere. It is customarily expressed in units of watts per square meter or, before the general introduction of the SI units, in calories per square centimeter per minute. The term "solar constant" is somewhat misleading, because it is not a true constant but seems to fluctuate slightly, going up or down a few tenths of one percent over periods of years. Any direct evidence for intrinsic variability of S with time is at best uncertain, however; until quite recently there have been no attempts to measure S continuously with sufficient accuracy to distinguish a true variation of less than $\sim \pm 1\%$. The instantaneous value of S can be determined either by performing spectral measurements and integrating them over all wavelengths or by measuring the total irradiance directly with a spectrally nonselective instrument. From the ground, only the first method is feasible, since correction must be made for the highly varying atmospheric transmission at each wavelength by extrapolating the spectral intensity to airmass zero. Any ground measurement is limited in accuracy by the strong absorption of atmospheric ozone, water vapor, and carbon dioxide, which block most of the solar radiation below 2950Å and above 2.5 µm. Measurements by the total irradiance method can be made from above the tropopause (at a height of approximately 10 km) and are potentially much more accurate, since the corrections for atmospheric extinction (absorption and scattering) are small or negligible. In the past decade, a number of measurements of S have been made by this method. A critical review of twelve such determinations—six from high-altitude aircraft, three from balloons, and three from satellite and rocket platforms outside the atmosphere—will be given, and the results will be reduced to the same radiometric reference to enable a direct comparison of the measurements.

II. RADIOMETRIC SCALES AND CALIBRATION PROCEDURES

For the atmospheric sciences, the solar constant is the most important input to the earth's energy budget, and the meteorologist is chiefly interested in measurements of total irradiance. The accuracy of the realization of a total irradiance scale is mainly limited by the necessity of changing the radiation power into a measurable quantity. In the modern radiometers this is accomplished by the absorption of the radiation in a blackened cavity and its change into a heat flux, which is measured. The heat flux measurement is then calibrated by comparing the solar flux with that produced by electrical heating of the cavity. The absolute accuracy (maximum possible deviation from a true SI watt) which can be achieved by these radiometers is approximately ±0.3%.

In meteorology, radiometer measurements began at the end of the last century. To overcome the accuracy problems mentioned above and to permit comparison between different instruments and different stations, special radiometric scales were defined. Through the years, these standards or scales have undergone a number of revisions, reflecting improvements in radiometry. Until the middle of this century two such scales were commonly used: the Ångstrom scale of 1905 and the Smithsonian scale of 1913. Before the beginning of the International Geophysical Year a new scale, the International Pyrheliometer Scale (IPS 1956), was defined as a compromise between these two. The details of the specification and use of these radiometric scales are described by Fröhlich (1973a).

The relationship between the IPS 1956 and the absolute scale, represented by different absolute radiometers, is shown in Figure 1. The relationship is used in the following to reduce all solar constant determinations to the same reference. Such a reference can be established from a synthesis of many different measurements. The Solar Constant Reference Scale (SCRS) indicated in Figure 1 is the result of such a synthesis from measurements during the International Pyrheliometer Comparisons (IPC III and IV) at Davos in 1970 and 1975 (WMO, 1976). The maximum possible deviation of a watt, related to the SCRS, from the true SI watt is estimated to be less than ± 0.3%.

Only a few of the determinations of S have been made with electrically calibrated absolute radiometers. Most have been made with radiometrically calibrated instruments, such as Ångstrom pyrheliometers, Normal Incidence Pyrheliometers (NIPs), or actinometers. The calibration is normally performed by comparison with a reference instrument that is based on a commonly used scale, such as IPS 1956. To simulate as accurately as possible the situation during the determination of S, the calibration is performed with the sun as the source at a high-altitude mountain station. For a spectrally nonselective detector, this procedure is straightforward. For instruments with optical windows, however, the interpretation of the results is more difficult because of the nonuniformity of the spectral transmission of the window and the change of spectral distribution of the solar radiation with altitude. Figure 2 shows typical curves of the transmission of optical windows and of the extinction within the atmosphere at different altitudes, together with a solar spectrum. The atmospheric transmissions are calculated by the method described by Selby and McClatchey (1975), with the U.S. standard atmosphere of 1962 as a model atmosphere. The solar spectrum is from Smith and Gottlieb (1974). Table 1 summarizes the results in terms of transmittance integrated over the solar spectrum. Further, the calculated ratios of the readings

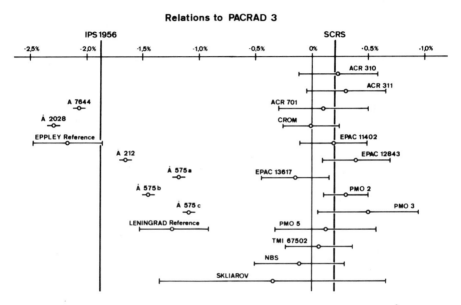

Fig. 1 Relation of different realizations of radiometric scales to PACRAD 3: The Ångström pyrheliometers are referenced by their serial numbers. Å 7644 and Å 2028 are the base for the Eppley Laboratory reference (according to results of IPC III 1970). Å 212, the USSR standard and Å 575, the Leningrad University instrument, are the base for the Leningrad reference (according to results of comparisons at Davos (Å 212, Å 575c), at Byurakan (Å 575a), and Terskol (Å 575b).

The absolute radiometers are referenced by appropriate abbreviations and serial numbers. The results are from IPC IV (1975) except for the NBS instrument, which participated together with PACRAD 3 in IPC III (1970).

The International Pyrheliometer Scale (IPS 1956) is shown as revised after IPC IV (1975). The Solar Constant Reference Scale (SCRS) is the best available representation of the SI scale of total irradiance; its absolute accuracy is estimated to be ± 0.3%.

Details on the instruments can be found as follows: ACR: Willson (1973b), CROM: Crommelynck (1973), NBS: Geist (1972), PACRAD: Kendall and Berdahl (1970), PMO: Brusa and Fröhlich (1975), SKLIAROV: Skliarov (1964). EPAC and TMI are commercial versions of PACRAD, made by Eppley Laboratory, Inc., and Technical Measurement, Inc., respectively.

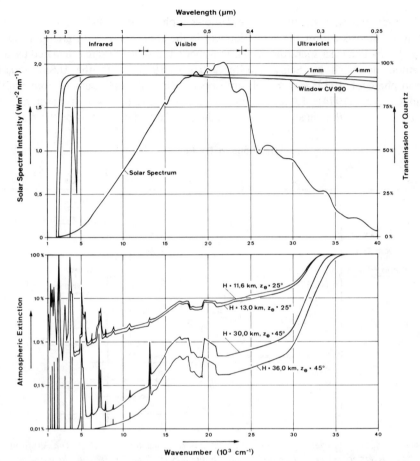

Fig. 2 Solar spectrum after Smith and Gottlieb (1974), atmospheric extinction from model calculations, and transmission of quartz windows as a function of wavenumber. The model calculations use the U.S. standard atmosphere of 1962 (Selby and McClatchey, 1975).

from a nonselective instrument (I_t) to one with a quartz window (I_q) are listed. This ratio is proportional to the calibration factor determined at a given altitude and is used in the following discussion in calculating the corrections for the non-measured spectral part of S blocked by the window. The correction factor is equal to the ratio of I_t/I_q at the calibration site to the ratio calculated for outer space.

For reduction of the individual values to the SCRS, the results of different intercomparisons of reference instruments made during the last years have been used, mainly the ones of the IPC organized by the World Meteorological Organization (WMO) at Davos in 1959, 1964, 1970, and 1975 (WMO, 1973 and 1976), and the Jet Propulsion Laboratory (JPL) comparisons of 1970 and 1972 at Table Mountain organized by Willson and Stallkamp (1971) and Willson (1972).

III. ANALYSIS OF THE SOLAR CONSTANT DETERMINATIONS

The main objective of the following analysis is to reduce recent measurements of S to the same radiometric scale to permit intercomparison. In general, corrections for atmospheric extinction, window transmission, etc., are left as they were made by the individual investigator. Only in cases where the calibration procedure required specific correction of these factors for conversion to the SCRS were they calculated.

A. MEASUREMENTS FROM AIRCRAFT

In 1966 a joint solar constant determination project was initiated by the Eppley Laboratory and the Jet Propulsion Laboratory (JPL). For this purpose, a multi-channel radiometer was developed by Drummond et al. (1967a, 1967b) and calibrated against the Eppley standard instrument at Table Mountain, California (Drummond et al., 1968a). In the following, only the results of the total irradiance channel will be discussed. This channel is equipped with a quartz window (Suprasil W) and has a viewing angle of 15°. As this viewing angle is approximately 10° wider than that of the standard instrument, Drummond et al. applied a correction to the determined calibration constant in order to account for the different amount of circumsolar radiation falling on both detectors. Model calculations (Fröhlich, 1973b) suggest that the correction used was overestimated by about 0.7%. No correction was made for the quartz window of the instrument. From Table 1, the factor is 1.0039. Further, a factor of 1.0236 follows from Figure 1, for the reduction of the Eppley reference to SCRS.

The following table summarizes the factors to be used to correct and reduce the Eppley/JPL solar constant values:

Reduction of Eppley reference to SCRS	1.0236 ± .003
Reduction of the assumed 5-15° correction	0.993 ± .003
Correction for the quartz window	1.0039 ± .002
Total correction factor and rms error	1.0204 ± .003

The mean of the results of six flights with the NASA B-57B jet research aircraft during 1966-67 is S = 1359 ± 13 W m^{-2}, after correcting for atmospheric extinction (Drummond et al., 1968b). These were made at an altitude of about 13 km, where residual atmospheric absorption varied from 5.4 to 6.2%. During these flights the multichannel radiometer was exposed to the sun without any intervening aircraft window. Applying the above correction factor we obtain S = 1387 ± 18 W m^{-2}.

In 1967-68 the Drummond measurements were supplemented by measurements made on seven flights with the NASA CV-990 jet research aircraft. These measurements were performed from the pressurized cabin through an optical window with the Ångström pyrheliometer (Å 9000) (Hickey, *private communication*).* This instrument's response was related to the SCRS by two comparisons. During the flights it was from time to time directly compared with the PACRAD II of Kendall (1973). A correction factor of 1.0187 to SCRS was determined from the comparisons; later tests on Table Mountain (Willson and Stallkamp, 1971) yielded a correction factor of 1.0232. For the present discussion, a mean value of 1.021 ± .003 will be taken. The value S = 1359 ± 15 W m^{-2} given by Drummond (1973) may therefore be corrected to S = 1387 ± 19 W m^{-2}. According to Drummond (1973) corrections were applied for the atmosphere of 5.5 - 5.9% and for the aircraft window of 11.8%.

During the same period a group of scientists from NASA's Goddard Space Flight Center (GSFC) performed measurements of S from a NASA CV-990 jet aircraft at an altitude of about 11.6 km, where residual atmospheric absorption is about 6.4% (Thekaekara, Kruger, and Duncan, 1969). Results of total irradiance measurements were reported for four instruments: a cone radiometer, a Hy-Cal pyrheliometer, and two Ångström compensation pyrheliometers (Å 6618 and Å 7635). Because we were unable to calibrate the Hy-Cal instrument to the SCRS and because some of the Hy-Cal instruments have shown differences of up to 20% (Willson and Stallkamp, 1971, and Willson, 1972), the results of the GSFC Hy-Cal measurements are not included in this summary. The GSFC cone radiometer designed by Kruger has an electrically calibrated cavity receiver. Original reduction of the cone radiometer data, assigning different weighting factors to different sets of the raw results, gave S = 1358 W m^{-2}. As every weighting of results is highly subjective, we prefer using a least squares fit to all data points, which gives S = 1364 ± 24 W m^{-2}. Unfortunately, no result of comparison with other instruments exists, so no direct translation to the SCRS is possible. A critical error analysis of this instrument indicates that the cavity absorptance is overestimated by approximately 0.6% (Fröhlich and Brusa, 1975); and therefore, the GSFC cone radiometer result should be raised to S = 1372 ± 24 W m^{-2}.

The GSFC measurements with the two Ångström compensation pyrheliometers were performed by McNutt and Riley (Å 6618) and by Duncan and Webb (Å 7635). Both are described by Thekaekara, Kruger, and Duncan (1969). The measurements with the Å 6618 instrument give S = 1343 ±30 W m^{-2}. This instrument was calibrated

---

* The preliminary results provided by Hickey for inclusion in this review have now been presented in a more complete form by Hickey and colleagues (Hickey, J.R.; Griffin, F.J.; Helleary, D.T.; and Howell, H.B., 1976, Extraterrestrial solar irradiance measurements from the Nimbus 6 satellite, *Proceedings of the Joint Conference on Sharing the Sun*, Winnipeg, Manitoba, Canada.)

against the Eppley standard after the flights. Therefore, the factor of 1.0236 from Figure 1 can be applied to translate the value to the SCRS, giving S = 1375 $\pm$ 30 W m^{-2}. Duncan and Webb found S = 1349 $\pm$ 40 W m^{-2} from their measurements with Å 7635. They believe that the value is not very accurate because of the unstable performance of the instrument during the calibration against the Eppley standard. The reduction procedure for Eppley calibrations yields a corrected value of 1381 $\pm$ 40 W m^{-2}. The reported instability of Å 7635 is also reflected in comparisons made at Table Mountain with a cavity radiometer of Willson (Willson and Stallkamp, 1971), which show a high standard deviation. The instrumental results can be translated through this comparison to the SCRS; with a correction factor of 1.017, S = 1372 $\pm$ 40 W m^{-2}. For what follows we will take a mean value of S = 1377 $\pm$ 40 W m^{-2} for Å 7635 since it cannot be determined which of the two corrections is the more reliable.

During July and August 1968, Kendall (1973) made some measurements with a PACRAD radiometer from the NASA CV-990, in conjunction with measurements made by the Eppley/JPL group. The instrument PACRAD II is an absolute radiometer with an electrically calibrated cavity receiver (Kendall and Berdahl, 1970) and can easily be related to the SCRS through comparisons made during the IPC III and IV intercalibrations at Davos. The value given by Kendall (S = 1370 $\pm$ 14 W m^{-2}) must be raised by 0.2% in order to be referenced to the SCRS. The final value is 1373 $\pm$ 14 W m^{-2}. Kendall used a window transmission of 0.8907 and assumed an atmospheric extinction of 5.6%. This assumed window transmission is about 1% higher than the one used by Drummond for the same window. Both transmittances were determined independently, and this difference demonstrates the typical variation in accuracy of such determinations.

The mean value of all aircraft measurements amounts to 1378 $\pm$26 W m^{-2} with a single measurement standard deviation of 7 W m^{-2} from the mean. The indicated uncertainty of 26 W m^{-2} is the rms error calculated from the individual errors.

B. HIGH-ALTITUDE BALLOON MEASUREMENTS

At typical balloon altitudes of 30 to 36 km, residual atmospheric absorption is about 1.8 to 1.2%. Early solar constant determinations from high-altitude balloons were performed by Kondratyev and Nikolsky (1970a) over a 7-y period beginning in 1962. In 12 successful ascents, measurements were made by a thermopile actinometer with a quartz window. The corrections for the quartz window were found by model calculations. To calibrate the flight instrument, a thermopile actinometer at Leningrad University was used as a reference. This instrument reads 1.4% lower than the Å 575 actinometer, as was shown later (Nikolsky, 1976). On the other hand, Å 575 was easily related to the SCRS through three different comparisons, as shown in Figure 1. There and in the following table the values of Å 575 have been reduced by 1.4%. The reduction factors to SCRS of Å 575 -

1.4% follow:

1970:	Comparison with Å 7644 at Byurakan (Kondratyev and Nikolsky, 1973)	1.0141
1971:	Comparison with Å 212, the USSR standard, which was translated to the SCRS through IPC III and IV, during comparisons at Terskol (WMO RA VI, 1971)	1.017
1975:	Comparison with PACRAD during IPC IV (WMO, 1976)	1.0131

The results demonstrate the good stability of the Å 575 instrument and yield a mean value of $1.015 \pm .003$, which can be used to reduce S to the SCRS.

Kondratyev and Nikolsky found an apparent variation of S with solar activity, which is discussed below. Their overall value was $S = 1356 \pm 14$ W m^{-2}, which corresponds to the highest value they found. This choice was based on the assumption that the corrections for atmospheric extinction were consistently and systematically underestimated. Moreover, they assumed that most of the apparent variation of S with solar activity was a consequence of variable atmospheric trans-

TABLE 1

Summary of results of model calculations for the integrated transmittance of the U.S. standard atmosphere of 1962 and window materials. The correction factor is used to reduce the calibration factor determined at a mountain station to a corresponding one for outer space and is equal to the ratio of I_t/I_q at the station to I_t/I_q in space.

Place and Altitude	Integrated Atmospheric Transmittance %	Intensity Ratio I_t/I_q	Correction Factor for S
Davos, 1.68 km	65.0	1.0737	1.0049
Table Mountain, 2.25 km	76.1	1.0748	1.0039
Mt. Evans, 4.35 km	81.9	1.0755	1.0033
Aircraft, 11.6 km (behind window)	92.5 83.5		
Aircraft, 13 km	93.5		
Balloon, 30 km	98.2		
Balloon, 36 km	98.8		
Space	100.0	1.0791	

parency rather than a true variation of the solar constant itself (Nikolsky, 1976). A similar conclusion was stated by Ångström (1970) when he reexamined a long series of the solar constant measurements of the Smithsonian Institution. Because of these arguments, we adopt the Kondratyev and Nikolsky highest value, which gives $S = 1376 \pm 18$ W m^{-2} when reduced to the SCRS.

During the end of the period of the USSR measurements, Murcray (1969) also made solar constant determinations from high-altitude balloons with modified NIPs from Eppley. The measurements were performed from altitudes of 31.0 to 35.4 km, where atmospheric extinction is about 1.7%. From four flights he found $S = 1339 \pm 7$ W m^{-2}. Murcray's instrument was originally calibrated against unspecified Eppley Ångström pyrheliometers, and unfortunately no reduction to a standard scale is now available. In 1970, Willson compared Murcray's flight instrument on Mt. Evans, Colorado, with his ACR II, No. 4. The ACR reading was 2.9% higher (Willson, 1973a). As Murcray's pyrheliometers had quartz windows, the correction factor indicated for Mt. Evans in Table 1 of 1.0033 is needed for a calibration. However, if this procedure is applied, the correction for the blocked IR radiation of 11.9 W m^{-2}, which Murcray applied, is no longer needed. The reduction to SCRS is summarized below:

Murcray's value	1339 ± 7	W m^{-2}
Murcray's IR correction	-12	W m^{-2}
	1327 ± 7	W m^{-2}
Reduction to SCRS (1.031)	$+ 41 \pm 4$	W m^{-2}
Quartz correction (1.0033)	$+ 5 \pm 1$	W m^{-2}
Final value	1373 ± 12	W m^{-2}

In 1968-69 Willson made balloon measurements with different Active Cavity Radiometers (ACRs), electrically calibrated cavity receivers (Willson, 1973b). Altitudes were 25 to 36 km, with 2.4 to 1.0% residual atmospheric extinction. The first two flights, in 1968, were made with quartz windows, whose transparency apparently degraded slightly during the flights. Therefore, we will consider only the 1969 ascent, which used a windowless ACR III No. 3. The ACR III No. 3 was compared to the PACRAD III and the ratio was determined to be 1.0001, with the ACR reading higher (Willson, *private communication*). Correction to the SCRS requires a factor of 1.002, giving $S = 1369 \pm 11$ W m^{-2}.

The mean value of the three balloon measurements amounts to $S = 1373 \pm 14$ W m^{-2} with a standard deviation of 3.5 W m^{-2} from the mean.

C. MEASUREMENTS FROM SATELLITES AND ROCKETS

The multichannel radiometer designed by Drummond et al. and described above was primarily developed for the Air Force/NASA X-15 rocket aircraft (Drummond et al., 1967a). Only one successful flight was accomplished, to an altitude of 78-83 km, where residual atmospheric absorption is negligible. The solar constant was measured for 81 seconds. The same corrections as for the B-57B flights has to be applied. Recalculation of the original data yields a value S = 1357 W m^{-2} (Hickey, *private communication*), which, when reduced to SCRS, is S = 1385 $\pm$ 14 W m^{-2}.

As of this writing three satellite measurements are also available: two from Mariners VI and VII (launched in 1969) and one from Nimbus 6 (launched in early summer 1975). The Mariner measurements were intended primarily to provide engineering data on a comparison of the thermal performance of a spacecraft in a simulated test environment against its performance in space, using the incident solar flux as the reference for comparison. The measurement of the absolute value of the solar constant was a secondary objective. The instrument, an electrically calibrated cavity radiometer (known as a Temperature Controlled Flux Monitor, TCFM), was designed by Plamondon and Kendall (1965). The absorptance of the conical cavity was assumed to be 0.997. Measurements on similar cavities show that this value is overestimated, and a value of 0.990 is more realistic (Fröhlich et al., 1975). A description of the TCFM and the results of 150 days of continuous observation have been given by Plamondon (1969). A summary is shown in Figure 3, which includes an appreciable correction applied by Plamondon to compensate for instrumental drift encountered in flight. This is the first continuous record of the solar constant and shows only small fluctuations, which seem to be random. The corrected value is S = 1362 $\pm$ 18 W m^{-2}.

The solar constant Earth Radiation Budget (ERB) experiment on Nimbus 6 is described by Hickey and Karoli (1974). It was launched in July 1975 and at this time is still operating. The instrument contains a total of 10 channels for both total and spectral measurements of the sun's radiation. Table 2 summarizes the spectral bands for all channels. For the solar constant determination, only the channels with no window and with a quartz window will be considered.

The detectors are Eppley thermopiles, calibrated in vacuum against a NIP with simulated solar radiation as source. The window of the vacuum chamber consisted of 25 mm thick quartz glass. The NIP, which is also provided with a quartz window, was calibrated against an Eppley-PACRAD absolute radiometer (EPAC 11402) in Albuquerque, New Mexico. Both the NIP and the absolute radiometer were compared with the PACRAD during IPC IV and may therefore be reduced to the SCRS. As shown in Figure 1, the absolute radiometer agrees with the SCRS, so that no correction is needed for the reference. However, some corrections are needed for the translation of the final thermopile calibration to the SCRS, because of the various spectral distributions used during different steps of the calibration and

TABLE 2

Spectral bands of the ERB solar channels on Nimbus 6 Satellite.

Channel No.	Spectral band (μm)	Filter type
1s	< 0.2 to 5	Sup. W
2s	< 0.2 to 5	Sup. W
3s	< 0.2 to > 40	None
4s	0.530 to 3	OG 530
5s	0.695 to 3	RG 695
6s	0.40 to 0.50	Interference
7s	0.35 to 0.45	Interference
8s	0.30 to 0.40	Interference
9s	0.28 to 0.35	Interference
10s	0.25 to 0.30	Interference

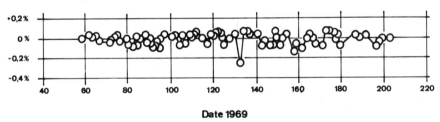

Fig. 3   Temporal variation of solar constant from Mariner VI measurements during 1969, after Plamondon (1969).

afterwards during the solar measurements in space. For both the total and the quartz channel a correction of -0.25% is needed to transfer from the outside calibration of the NIP to the measurements behind the vacuum chamber window. For the quartz channel, which has two windows 2 mm thick, a correction for the unmeasured radiation of 1.0% is needed. Both corrections are evaluated from model calculations similar to the ones described in Section II. The values originally reported by Hickey (1976), of 1392 W m^{-2} for the total channel and 1371 W m^{-2} for the quartz channel, yield S = 1388 $\pm$ 14 W m^{-2} (total channel) and S = 1382 $\pm$ 14 W m^{-2} (quartz channel) after the correction. The determined error is an estimate of the absolute accuracy influenced mainly by the transfer of calibration. Measured temporal deviations of the mean of S for the first five months of Nimbus 6 operation are shown in Figure 4. They amount to less than 0.2%.

The mean of all space determinations amounts to $S = 1379 \pm 15$ W m^{-2} with a rather high standard deviation of 12 W m^{-2}.

D. SUMMARY

Table 3 and Figure 5 summarize all measures of S. The unweighted mean of all these determinations amounts to 1377 W m^{-2} with a standard deviation of 8 W m^{-2}. The originally published values give a mean of 1360 W m^{-2} with a standard deviation of 14 W m^{-2}. The improvement by the reduction to SCRS is evident. However, 7 of the 13 results lie in a very narrow range from 1369 to 1377 W m^{-2}; therefore, the most probable value for the solar constant is proposed to be

$$S = 1373 \text{ W m}^{-2}$$

The calculated rms error from all individual errors associated with each determination amounts to 20 W m^{-2}. The real error of S, i.e., the possible deviation from the true solar constant expressed in true W m^{-2}, is probably smaller. However, the variation from author to author of the applied corrections for the atmospheric extinction and the window transmission as shown in Table 3 indicates that the achieved accuracy of these corrections is on the order of 1-2% for the aircraft measurements. Therefore, most of the indicated 20 W m^{-2} error is probably due to errors in the accounting for atmospheric extinction. In order to improve the accuracy of S to the 0.25% level or better, measurements from spacecrafts, rockets, and high-altitude balloons with accurate in situ atmospheric extinction determinations will be needed. Another accuracy problem is posed by the fact that the proposed value of 1373 W m^{-2} is 1.1% higher than the best available determination of S by integrating spectral measurements, such as the value of 1358 W m^{-2} given by Smith and Gottlieb (1974), as is evident from a critical compilation of measurements made by many different investigators.

IV. TEMPORAL VARIATION

The averaged value of the solar constant is probably better known than its possible variability. Early measurements of S, begun nearly a century ago, established that temporal changes were relatively small—probably less than a few percent. A long series of mountaintop measurements by Charles Abbot of the Smithsonian Institution further defined the range of possible variation: during four solar cycles, from about 1908 until 1952, excursions were within about $\pm$ 1% (Aldrich and Hoover, 1954).

These early measurements—made of necessity from the surface of the earth—were severely limited by uncertainties about the absorption characteristics of the

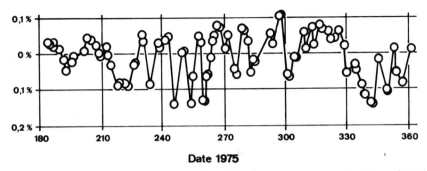

Fig. 4 *Temporal variation of solar constant from Nimbus 6 measurements, after Hickey (1976). Horizontal scale is serial calendar date (January 1 = 1). Some of the slow variations, such as the gradual rise and fall between days 300 and 330, is due to the slowly changing attitude of the spacecraft, for which final corrections are not yet available.*

TABLE 3

Summary of recent solar constant determinations reduced to SCRS. The unweighed mean amounts to 1377 ± 20 W m^{-2} with a standard deviation of 8 W m^{-2}. The mean of the originally published values is 1360 W m^{-2} with a standard deviation of 14 W m^{-2}. The mean corrections indicated are as given in the original papers, except for the quartz window corrections of Drummond and Murcray (see text). Number in col. 1 is for identification in Figure 5.

No.	Author	Platform	Date of Measurements	No. of Measurements	Altitude (km)	Mean Corrections (W m^{-2}) Atmosphere	Mean Corrections (W m^{-2}) Window	SCRS Value (W m^{-2})	Original Value (W m^{-2})
1	Drummond et al.	Aircraft B-57B	1966-67	6	13.0	80	5	1387 ± 17	1359
2	Drummond et al.	Aircraft CV 990	1967-68	7	12.0	79	164	1387 ± 19	1359
3	Kruger	Aircraft CV 990	1967	6	11.6	85	?	1372 ± 24	1364
4	Duncan et al.	Aircraft CV 990	1967	4	11.6	133	?	1377 ± 40	1349
5	McNutt et al.	Aircraft CV 990	1967	5	11.6	95	?	1375 ± 30	1343
6	Kendall	Aircraft CV 990	1968	6	12.2	77	150	1373 ± 14	1370
7	Kondratyev et al.	Balloon	1962-67	12	28 - 34	35	10	1376 ± 18	1356
8	Murcray	Balloon	1967-68	4	31 - 35	24	4	1373 ± 12	1339
9	Willson	Balloon	1969	1	36.0	14	0	1369 ± 11	1366
10	Drummond	Aircraft X-15	1967	1	78 - 81	0	5	1385 ± 14	1361
11	Plamondon	Mariner VI + VII	1969	145 d	1 - 1.4 AU	0	0	1362 ± 18	1352
12	Hickey	Nimbus 6, Ch. 3s	1975	>180 d	1100	0	0	1388 ± 14	1392
13	Hickey	Nimbus 6, Ch. 1s	1975	5	1100	0	14	1382 ± 14	1371

terrestrial atmosphere. This required major correction for wavelengths that were partially absorbed and compensation for parts of the ultraviolet and infrared that were not seen at all. The contribution of these corrections was more than an order of magnitude greater than the temporal variations noted in S. Thus, although variations of a few tenths of a percent were reported as day-to-day, month-to-month, or year-to-year changes, it is quite possible that at least part of the apparent changes was the result of atmospheric uncertainties and instrumental effects. A conservative summary of the Smithsonian measurements is that changes of more than 1% in S were not seen and changes of less than 1% were not capable of being seen.

As described in Section III, Kondratyev and Nikolsky (1970*a*) carried out an extended program of measurements of S from balloons and reported changes in S that they felt were associated with changes in overall solar activity. Their conclusion, often cited, was that S increased and decreased with sunspot number

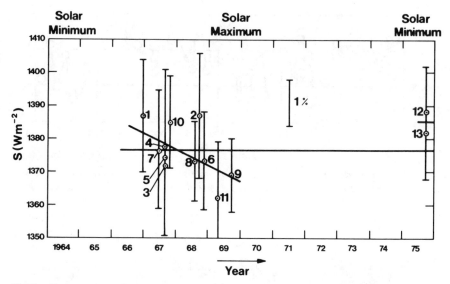

Fig. 5 Measurements of the solar constant (S) described in Section III, as function of date of measurement. Each value has been reduced to the SCRS by author's determination. Horizontal line is the mean value. Least-squares linear regression fit to 1966-1969 data is also shown. See Table 3 for key to numbers identifying individual measurements.

R, rising by more than 2.5% from R = 0 to R = 100, then falling by a similar amount with further increase in R. This seemed to confirm a similar result found by Ångstrom (1922) in an early analysis of some of the Smithsonian data—although the latter was subsequently retracted by Ångström (1970) as more likely an effect of atmospheric change. Moreover, the variability in S found by Kondratyev and Nikolsky was close to their limit of measurement, so that it is difficult to draw definite conclusions from their results.

Spacecraft measurements offer a better hope of resolving the question. Only in the last few years have we had the capability to continuously monitor S from above the atmosphere, where all problems of atmospheric transparency are eliminated. The Mariner spacecraft measurements reported by Plamondon (1969) were historic in continuously sampling the total solar radiation above the atmosphere for 150 days. Analysis, cited earlier, showed that S changed only irregularly during this span—i.e., not in step with the daily variation of sunspot number—and that the total variation was less than about 0.3%. This would seem consistent with the limits of variation reported by Abbot and much smaller than the changes indicated by Kondratyev and Nikolsky, who took their measurements of S at about the same phase of the solar cycle and under very similar conditions of solar activity. Most of the data used by Kondratyev and Nikolsky to establish their S(R) relationship were from five months in 1967; during their balloon flights, the daily sunspot number

R varied from about 20 to 170. During a similar time span for the Mariner observations, in 1969, the daily sunspot number went through about the same range, from 26 to 207. The Mariner experiment required significant correction for apparent instrumental drift, which obviously prevents us from deducing anything about longer-term changes in the solar constant.

More promising is the 10-channel solar constant experiment now on the Nimbus 6 spacecraft, which is still operational at this writing. It is probably too early to draw conclusions from the early results of this important experiment, although, as is evident in Figure 3, variations in S were within limits of about $\pm$ 0.2% in the first six months of operation. During this time—at the minimum of the 11-year cycle—solar activity was very low. Moreover at the time of an isolated peak in sunspot number (R $\approx$ 100 in early August 1975 on day 216) the solar constant measured by Nimbus 6 showed no change greater than that found at times of low sunspot activity (R < 30).

Solar constant measurements compiled in Figure 5 for the 1966-1975 period are of interest in confirming the presence or absence of a longer-term S(R) relationship, for during this time solar activity went through a complete cycle (minimum to minimum). The least-squares linear regression line for the 1966-1970 data and the apparently higher values for 1975 could be associated with the 11-year cycle, with a minimum in S at sunspot maximum (about 1969) and a subsequent increase as the solar cycle progressed toward sunspot minimum. If this were a real change, it would indicate a possible cyclic variation of about 1.5% peak-to-peak amplitude.

The relationship, however, is highly suspect and could as well result from systematic errors either in the original data (which came from varied sources) or in the reduction to the SCRS. We note that it is possible to fit all the data in Figure 5 with a constant value of S when error bars are considered. Resolution of this important point will come with more reliable and continuous measurements through at least one solar cycle.

Finally, we should point out that none of the measurements of S yet attempted has given us the ability to detect the secular variations in S that have been proposed to explain long-term climate changes on the earth (Eddy, 1976)—the ponderous, monotonic drifts that could account for a major glaciation, or the Little Ice Age of the sixteenth and seventeenth centuries, or the global warming which took place between about 1850 and 1950. In order to sense possible changes this subtle, we need instrumentation accurate at the 0.1% level, operational for tens of years, and capable of maintaining calibration over these long periods. That challenge is surely one of the more important and most difficult problems facing experimental science.

ACKNOWLEDGMENTS

The author is indebted to J.A. Eddy, High Altitude Observatory, NCAR, Boulder, for his editorial work and for suggesting Section IV. Appreciation is extended to J.R. Hickey, Eppley Laboratory, Inc., Newport, Rhode Island, for supplying the Nimbus 6 data and for many helpful discussions concerning the Eppley solar constant experiments, to R.C. Willson, Jet Propulsion Laboratory, Pasadena, California, for the results of the many internal JPL comparisons, to G.A. Nikolsky, University of Leningrad, Soviet Union, for clarifying discussions about the traceability of the Leningrad reference scale and to R.W. Brusa of our institute for his contribution to Section II.

REFERENCES

Aldrich, L.B., and Hoover, W.H., 1954, *Ann. Ap. Obs. Smithsonian Inst. 7*, 165.

Angstrom, A., 1922, *Ap. J. 55*, 24.

_____, 1970, *Tellus 22, 2*, 205.

Brusa, R.W., and Frohlich, C., 1975, *Scientific Discussions International Pyrheliometer Comparisons IV*, WRC Davos.

Crommelynck, D., 1973, *I.R.M. Belgique, Pub. Ser. A*, No. 81.

Drummond, A.J., 1973, *The Extraterrestrial Solar Spectrum*, ed. A.J. Drummond and M.P. Thekaekara, Inst. of Environmental Sciences, Mt. Prospect, Ill., 1.

_____ and Hickey, J.R., 1968, *Solar Energy 12*, 217.

_____; Hickey, J.R.; Scholes, W.J.; and Laue, E.G., 1967a, AIAA Fifth Aerospace Sciences Meeting, New York, 23-26 January.

_____; _____; _____; and _____, 1967b, *J. Spacecraft and Rockets 4*, 1200.

_____; _____; _____; and _____, 1968a, *Proceedings 17th International Astronautical Federation Congress*, Vol. 2, Pergamon Press, Elmsford, N.Y., 407.

_____; _____; _____; and _____, 1968b, *Nature 218*, 259.

Eddy, J.A., 1976, *Science 192*, 1189.

Frohlich, C., 1973a, *Proceedings Symposium on Solar Radiation*, Smithsonian Institution, Washington, D.C., 61.

_____, 1973b, *World Meteorological Organization Commission for Instruments and Methods of Observation VI Scientific Discussions*, Helsinki 1973, 28.

_____ and Brusa, R.W., 1975, *Proceedings Workshop on the Solar Constant and the Earth's Atmosphere*, Big Bear Solar Observatory, California, 111.

Geist, J., 1972, *NBS Technical Note 594-1*.

Hickey, J.R., and Karoli, A.R., 1974, *Appl. Optics 13*, 523.

Kendall, J.M., 1973, *Proceedings Symposium on Solar Radiation*, Smithsonian Institution, Washington, D.C., 190.

———— and Berdahl, C.M., 1970, *Appl. Optics 9*, 1082.

Kondratyev, K. Ya., and Nikolsky,G.A., 1970a, *Quart. J. Roy. Meteorol. Soc. 96*, 509.

———— and ————, 1970b, *Atmos.and Ocean. Phys. 6*, 227.

Murcray, D.G., 1969, *Balloon Borne Measurement of the Solar Constant*, Rep. No. AFCRL-69-0070, University of Denver, Denver.

Plamondon, J.A., 1969, *JPL Space Program Summary 3*, 162.

———— and Kendall, J.M., 1965, *ibid. 4*, 66.

Selby, J.A., and McClatchey, R.A., 1975, *AFCRL-TR-75-0255, Environ. Res. Pap. No. 513*, Air Force Cambridge Research Laboratories, Hanscom AFB, Mass.

Skliarov, Y.A., 1964, NASA TT F-321 translation from *Raschet nekotorykh ohshibok bolometricheskogo pirgeliometra 152*, 81-89.

Smith, E.V.P., and Gottlieb, D.M., 1974, *Space Sci. Rev. 16*, 771.

Thekaekara, M.P., Kruger, R., and Duncan, C.H., 1969, *Appl. Optics 8*, 1713.

Willson, R.C., 1972, *Results of the 1972 Table Mountain Radiometer and Radiation Scale Comparisons*, Jet Propulsion Laboratory, Pasadena, Calif.

————, 1973a, *Solar Energy 14*, 203.

————, 1973b, *Appl. Optics 12*, 810.

———— and Stallkamp, J.A., 1971, *Radiometer Comparison Tests*, Rep. No. 900-446, Jet Propulsion Laboratory, Pasadena, Calif.

WMO, RA VI, 1971, *Comparisons Pyrheliometriques de Terskol (URSS)*, WMO internal report.

WMO, 1973, *Final Report of the III. International Pyrheliometer Comparisons at Davos*, Wmo Report No. 362, Geneva.

WMO, 1976, *Report on the International Pyrheliometer Comparisons 1975 at Davos*, in press.

Comment

For the future, it is important to monitor the solar luminosity from outside the earth's atmosphere for a long period in order to see whether there is any systematic variation that relates to the solar rotation period (indicating persistent longitudinal anomalies in the solar flux) or to the longer solar magnetic cycle. The former might be produced by global-scale solar convection, active regions, or both. In addition to its importance for understanding variations in the earth's climate, evidence for variation in solar output over time periods from a few days to a few years (or decades) would be extremely valuable for understanding the structure and dynamics of the sun on a global scale.

Peter A. Gilman
High Altitude Observatory

1976 ROCKET MEASUREMENTS OF THE SOLAR CONSTANT AND THEIR IMPLICATIONS FOR VARIATION IN THE SOLAR OUTPUT IN CYCLE 20*

R.C. Willson, *Jet Propulsion Laboratory*
and
J.R. Hickey, *The Eppley Laboratories*

Since the solar constant is a basic empirical boundary condition in both solar physics and terrestrial climatology, we have pursued a continuing measurement program for this solar output measure through a series of balloon, rocket, and satellite experiments. Using new data from the Earth Radiation Budget (ERB) experiment on the Nimbus 6 spacecraft launched in 1975, one of us (JRH) reported solar constant values ranging from 1388 to 1392 Wm^{-2} in 1975-76 at the minimum of solar activity (see Hickey *et al.*, 1976). An earlier balloon measurement in 1969 near the maximum of solar cycle 20 (see Willson, 1973) gives a solar constant value of 1369 Wm^{-2}, based on the intercomparison by Frohlich in this volume. The apparent increase of the solar constant by 1.5% during the period from 1969 to 1976 was significant in terms of the 1% uncertainty expected for the ERB thermopile data; thus, the possibility of an increase of 1.5% in the solar constant from solar maximum to solar minimum stirred considerable interest in the solar and climate research community.

I. THE 1976 ROCKET EXPERIMENT

Although the ERB thermopile was carefully calibrated prior to launch of Nimbus 6, the instrument is not an absolute detector; and we planned a rocket experiment in

---

*Editor's Note: The data reported by Willson and Hickey were obtained after the Solar Output Workshop, but because of the significance of these preliminary results in setting a 0.75% upper limit on the solar constant variation in the last half of solar cycle 20, I requested that they prepare this note as a valuable extension of the workshop discussions. *ORW*

1976 to make four independent absolute measures of the solar constant as well as to fly a duplicate of the ERB detectors. The four absolute cavity pyroheliometers were a Primary Absolute Cavity Radiometer (PACRAD) described by Kendall and Berdahl (1970), an Eclectic Satellite Pyroheliometer (ESP) described by Hickey (1976), and two Active Cavity Radiometers Type IV (ACR IV) described by Willson (1975).

The rocket instrument package was tested with full sunlight on South Mt. Baldy, N.M. (elevation 3.2 km), late in May prior to launch on 29 June 1976. We included two additional active cavity radiometers, PMO 2 and PMO 5, described by Brusa and Frohlich (1975) as absolute instruments that would remain on the ground for later reference purposes. This preflight measurement and comparison experiment demonstrated that we would obtain agreement between the absolute instruments to within 0.5%.

The instrument package was successfully flown on an Aerobee 170 sounding rocket on 29 June 1976. Solar constant measurements were made by the five instruments over a period of 300 seconds, all at altitudes above 100 km. Although the instruments were recovered intact on 29 June without mechanical damage, the cavities of the PACRAD and ESP radiometers were contaminated by dust during the recovery. This contamination prevents postflight performance comparisons between these two instruments and our other references.

The three types of absolute radiometers were operated in different modes throughout the flight. The PACRAD instrument was operated in a passive mode in which the solar irradiance is determined from the temperature rise in the cavity during the flight. Similarly, the ESP radiometer functioned in a passive mode in which the reference cavity remained in a constant thermal equilibrium state, and the small temperature difference between the solar cavity and the reference cavity yielded the solar constant value. The other two instruments, ACR IV, were operated in the active mode; the solar cavity and heat sink were continuously maintained in thermal equilibrium by a servo system during a shutter "open" and "closed" cycle. This type of operation yields a differential measurement of the cavity heating power between "open" and "closed" conditions. Such fast in situ calibration gave a continuous estimate of the solar irradiance throughout the flight. Because of the different operational modes for the four cavity radiometers, the error analysis is quite different for each type, but we estimate the uncertainty in the solar constant obtained from a combination of results from all four instruments to be less than 0.5%. Our presentation here gives a preliminary result, and the definitive analysis and result for each flight detector will be stated independently by each experimenter in later publications.

II. RESULTS FROM THE 1976 ROCKET EXPERIMENT

Table I gives the measured solar constant values from each of the detectors flown in our June 1976 rocket experiment. The Nimbus 6 ERB output (channel 3), measured simultaneously with the rocket flight, is also given, for comparison with the duplicate ERB detector value. The internal agreement to within better than ±2.5 Wm^{-2} between the four absolute radiometer measurements is evident from the table, and the two ERB detector values also agree with each other. However, the systematic difference between the absolute measurements and the ERB detector (channel 3) is 22 Wm^{-2} or 1.6%. Given the strength of the absolute measurements, the ERB detector (channel 3) measurements should be corrected downward by 1.6% to the absolute value of 1367 Wm^{-2}, which should be compared with Willson's value of 1369 Wm^{-2} measured in 1969. Since these two measurements, one at solar maximum and the other at minimum, agree within the experimental errors for the absolute radiometer data, we conclude that the solar "constant" has remained unchanged to within 0.75% over the last half of solar cycle 20.

TABLE 1

Preliminary Results of the June 1976 Rocket Experiment to Determine the Solar Constant and to Calibrate the Total Irradiance Channel 3 of the Earth Radiation Budget Experiment on the Nimbus 6 Spacecraft.

Experimenter	Platform/Detector	Date	Altitude (km)	Solar Constant (W/m^2)	Uncertainty (W/m^2)
Hickey	Nimbus 6 ERB Ch. 3	1975	1100	1389	-
Hickey	Aerobee ERB Duplicate	1976	100-250	1389	-
Hickey	Aerobee ESP	1976	100-250	1369	± < 7
Kendall	Aerobee PACRAD	1976	100-250	1364	± < 7
Willson	Aerobee ACR 402A	1976	100-250	1368	± < 7
Willson	Aerobee ACR 402B	1976	100-250	1368	± < 7
Average Solar Constant from 1976 Rocket Data				1367	± < 7
Corrected Value from Nimbus 6 (ERB Ch. 3)				1367	± 14

III. IMPLICATIONS FOR FROHLICH'S CONCLUSIONS

Claus Frohlich's review was completed before these new results were known, hence this report is intended to complement his careful study. After a survey of all measurements from aircraft, balloons, and spacecraft, he estimated the most probable value of the solar constant to be 1373 ±20 Wm^{-2}, which is to be compared to a weighted-mean value of 1370 ±1 Wm^{-2} that we obtain from all measurements, including the June 1976 rocket and corrected Nimbus ERB data. Frohlich's 20 Wm^{-2} uncertainty refers to the variance in single measurements of all types, whereas our error estimate of 1 Wm^{-2} is the standard deviation of the weighted mean. When the new 1976 values are included, the range of solar constant estimates is from 1368 to 1379 Wm^{-2}, with the aircraft data showing a large deviation from balloon and rocket experiments.

Aside from the obvious extension of Frohlich's Table 3 by our tabulation of the 1976 results, an important change must be made in his Figure 5, in which the solar constant measurements are plotted as a function of time. Because of the downward correction of the Nimbus 6 ERB (channel 3) values in 1975 and 1976, evidence for a significant systematic variation during solar cycle 20 decreases.

IV. COMMENTS ON THE DIFFERENT TYPES OF SOLAR CONSTANT MEASUREMENTS

As we look to future solar constant experiments, a survey of measurements from aircraft, balloons, and spacecraft gives useful insight into the accuracy limitations inherent in each type. In Table 2, we summarize the average solar constant estimates obtained from the different types of measurement.

The principal difficulty in the aircraft experiments comes from the attenuation due to atmospheric ozone and the aircraft window, and typical transmittance values are 0.89 to 0.94 at 12 km altitude. We estimate that 90% of the attenuation is due to ozone, but this upper atmospheric constituent shows abundance variation in both time and position of ±50% around the mean values used to derive these transmittances (Kaplan, 1975). This observed variation alone would introduce a ±34 Wm^{-2} uncertainty or ±2.5% unless in situ measurements of the ozone attenuation are available for data correction. When all sources of error in aircraft measurements are evaluated in the presence of the new ozone data, we estimate the uncertainties to be in the 3 to 5% range, which is substantially larger than the values usually quoted.

Table 2. Comparison of Solar Constant Results from Different Experiment Types

Observation Type	Solar Constant (Wm^{-2})
Aircraft	1379 (±3) *
Balloon	1371 (±1)
Space	1368 (±2)
Weighted mean result	1370 (±1)

* The figures in parentheses are the standard deviations from the weighted means (not absolute uncertainties in single measurements).

Balloon experiments, however, reach higher altitudes, 20 to 36 km, and penetrate the level of highest ozone concentration at midlatitudes in the northern hemisphere. At such heights, approximately half of the ozone lies below the balloon, thereby decreasing the uncertainty in the attenuation of the solar radiation. At 36 km in altitude, the total atmospheric attenuation is 1% (see Willson, 1973), and the uncertainty in the results from this 1973 balloon measurement is 0.5% without any attempt to account for ozone variations during flight. Even at balloon altitudes, it is necessary to measure the attenuation due to ozone as a function of time and from the same platform if an aggregate uncertainty of ±0.2% in solar constant measurements is to be reached.

The only space flights of absolute radiometers are the two Mariner flights and our 1976 rocket experiment. This latest experiment clearly demonstrates the achievement of the 0.5% level of accuracy without the need for any correction for atmospheric effects. A look at the uncertainties for the rocket and balloon experiments indicates that they are comparable today (both $\sim$ .5%), but our estimates for the rocket experiment are upper limits. Since any correction for even trace atmospheric attenuation requires a good estimate of the solar spectral irradiance as well as specification of absorption coefficients that are not fully known, experiments below 50 km may suffer systematic errors in their results. Our experience suggests that both balloon and rocket experiments can yield solar constant estimates with an accuracy in the 0.2 to 0.5% range today, but rocket and orbiting platforms are probably required for absolute radiometers that will reach the 0.1% accuracy required to determine the climate sensitivity to variation of the solar energy input to the troposphere. Even though solar experiments like those on Nimbus 6 do not use absolute detectors, this class of measurement is very useful for the solar constant problem because it gives an almost continuous record of the relative solar variation. The combination of such relative measurements with *at least one* absolute radiometry experiment above 50 km each year can give a definitive solar constant record in solar cycle 21 until the Solar Maximum Mission flies in 1979, and the Space Shuttle program can provide frequent orbital experiments with recovery of the radiometers in the 1980s.

Acknowledgments

We express our thanks to J. Kendall, Sr., for communicating his latest solar constant result.

References

Brusa, R.W., and Frohlich, C., 1975, *Scientific Discussions IPC IV*, WRC, Davos.

Hickey, J.R., 1976, *Preliminary Design Review of the Eclectic Satellite Pyrheliometer*, AFFE Report, 2 March 1976.

_____ ; Griffin, F.J.; Helleary, D.T.; and Howell, H.B., 1976, Extra-terrestrial solar irradiance measurements from the Nimbus 6 satellite, *Proceedings of Joint Conference on Sharing the Sun*, Winnipeg, Manitoba, Canada.

Kaplan, L.D., 1975, *Solar Constant Workshop*, Big Bear Solar Observatory, California Institute of Technology.

Kendall, J.M., Sr., and Berdahl, C.M., 1970, *J. Appl. Optics 9*, 1082.

Willson, R.C., 1973, *J. Solar Energy 14*, 203.

_____ ; 1975, *Proceedings of the Society of Photo-Optical Instrumentation Engineers Annual Meeting 68*, 31.

VARIATIONS IN THE INSOLATION CAUSED BY CHANGES IN ORBITAL ELEMENTS OF THE EARTH

Anandu D. Vernekar
University of Maryland
College Park, Maryland

We shall assume that any changes in the intensity of solar radiation due to cosmic and astrophysical causes are negligibly small. We shall only consider the variations in the global distribution of solar radiation because of the changes in the earth's orbital elements (eccentricity, obliquity, and longitude of perihelion). Milankovitch (1930) has shown that the solar radiation received at the upper limit of the atmosphere can be represented as a mathematical function of latitude, time, the earth's orbital elements, and the solar constant. Hence the insolation has an annual periodic variation because of the earth's elliptic translational motion around the sun and a long-period variation due to secular variations of the orbital elements. As determined from celestial mechanics, the secular variations result from the perturbations that the eight other principal planets (Mercury, Venus, Mars, Jupiter, Saturn, Uranus, Neptune, and Pluto) exert on the earth's orbit and from the gravitational attraction of the sun and moon on the oblate shape of the earth.

Several investigators (Stockwell, 1875; Harzer, 1895; Michkovitch, 1931) have worked on the solution of secular variations of the orbital elements of the planets in the solar system (excluding the effect of Pluto). A recent solution for the entire solar system (still excluding the effect of Pluto) is given by Brouwer and van Woerkom (1950), which we hereafter refer to as B-W. They have used the most recent information about masses and motions of the planets and also included the second-order effect caused by the long-term inequality in the movement of Jupiter and Saturn. Sharaf and Budnikova (1967) found errors in the initial values of longitude of nodes of Venus and the earth used in the B-W study. They recomputed the integration constants in B-W's solution using the correct initial values. From these results they derived trigonometric formulas for the precession and the obliquity including terms of second degree of eccentricity and inclination. Using these results, Vernekar (1972) calculated the variations of the earth's orbital elements and the insolation for the past two million years. The accuracy of these calculations is limited by the approximations made in solving the differential equations governing the secular variations of the earth's orbital elements. B-W's solution neglects terms of third and higher order in the disturbing masses and terms of third and higher degree of eccentricity and inclination. Recently, Berger (1976a, b) made improvements in the solutions by including terms of third degree of eccentricity and inclination. He also calculated the variations of the orbital elements and the insolation (Berger, 1976c). The difference between the insolation values calculated by Vernekar and by Berger is negligible for the past 115,000 years, but it increases before that time and becomes significant around 300,000 years before the present era.

The following sections discuss the calculations of insolation variations based on B-W's solution for the period over which these calculations are in agreement with Berger's calculations.

I. SECULAR VARIATIONS IN THE EARTH'S ECCENTRICITY, OBLIQUITY, AND PRECESSION

The solutions of the differential equations governing the secular variations of the eccentricity e, the longitude of perihelion from a fixed equinox Π', the inclination of the ecliptic to the invariable plane γ, and the longitude of the ascending node on the invariable plane θ, in terms of trignometric series as obtained by B-W, are

$$\left. \begin{array}{l} e \sin \Pi' = \Sigma\, M_i \sin(-s_i t + \delta_i) \\ e \cos \Pi' = \Sigma\, M_i \cos(-s_i t + \delta_i) \end{array} \right\} \cdots \quad (1)$$

$$\left. \begin{array}{l} \sin \gamma \sin \theta = \Sigma\, N_i \sin(-s_i' t + \beta_i) \\ \sin \gamma \cos \theta = \Sigma\, N_i \cos(-s_i' t + \beta_i) \end{array} \right\} \cdots \quad (2)$$

Here M, N, s, s', δ, and β are numerical constants in the solution of the differential equations obtained when the initial values of the orbital elements of 1900 A.D. and the ecliptic and equinox of 1950 A.D. are used; t is the time in tropical years. The numerical values of the constants for this solution, with correction by Sharaf and Budnikova, are given in Table 1.

The trigonometric formulas for obliquity ε and total precession ψ' derived by Sharaf and Budnikova are

$$\varepsilon = h^* - \Sigma\, c_i N_i \cos\left[(-s_i' + k)t + \beta_i + \alpha\right]$$

$$- \Sigma\, c_{ii} N_i^2 \cos 2\left[(-s_i' + k)t + \beta_i + \alpha\right]$$

$$- \underset{i<j}{\Sigma}\, c_{ij} N_i N_j \cos\left[(-s_i' - s_j' + 2k)t + \beta_i + \beta_j + 2\alpha\right]$$

$$-\sum_{i<j} c'_{ij} N_i N_j \cos\left[(-s'_i + s'_j)t + \beta_i - \beta_j\right] \ldots \quad (3)$$

and

$$\psi' = kt + \alpha + \sum g_i N_i \sin\left[(-s'_i + k)t + \beta_i + \alpha\right]$$

$$+ \sum g_{ii} N_i^2 \sin 2\left[(-s'_i + k)t + \beta_i + \alpha\right]$$

$$+ \sum_{i<j} g_{ij} N_i N_j \sin\left[(-s'_i - s'_j + 2k)t + \beta_i + \beta_j + 2\alpha\right]$$

$$+ \sum_{i<j} g'_{ij} N_i N_j \sin\left[(-s'_i + s'_j)t + \beta_i - \beta_j\right] ,$$

$$+ \sum_{i<j} g''_{ij} M_i M_j \sin\left[(-s_i + s_j)t + \delta_i - \delta_j\right] \ldots \quad (4)$$

TABLE 1

Numerical values of constants in the expressions for eccentricity-longitude of perihelion and inclination-longitude of ascending node of the earth's orbit, referred to the ecliptic of 1950 A.D.

i	M_i	$-s_i$	δ_i	N_i	$-s'_i$	β_i
1	0.0039162	5.463255"	92.25752°	0.0084961	-5.201537"	19.35621°
2	0.0163395	7.344744	196.98320	0.0081159	-6.570802	318.13542
3	0.0104345	17.328323	335.46433	0.0244642	-18.743586	255.16478
4	-0.0148347	18.002327	318.19816	0.0045381	-17.633305	296.51743
5	0.0183400	4.295908	29.60949	0.0275702	0.000004	107.10181
6	0.0028264	27.774064	125.50600	0.0028113	-25.733549	127.00902
7	0.0004427	2.719308	131.98140	-0.0017308	-2.902663	315.02276
8	0.0000115	0.633315	69.02941	-0.0012969	-0.677522	202.28307
9	-0.0001218	-19.182248	-66.28703	–	–	–
10	0.0001471	51.252220	221.40252	–	–	–

The coefficients in (3) and (4) are related to the numerical constants in Table 1 as follows:

$$a_i = \frac{k}{(-s_i' + k)}$$

$$a_{ii} = \frac{1}{4} a_i (a_i^2 - 1)\tanh + \frac{1}{4} a_i^2 \coth$$

$$a_{ij} = \frac{1}{2} \frac{k}{(-s_i' - s_j' + 2k)} [(a_i^2 + a_j^2 - 2)\tanh + (a_i + a_j)\coth]$$

$$a'_{ij} = \frac{1}{2} \frac{k}{(-s_i' + s_j')} [(a_i^2 - a_j^2 - 2a_j - 2a_i)\tanh + (a_i - a_j)\coth]$$

$$b_i = (a_i - 1)a_i \tanh + a_i \coth$$

$$b_{ii} = \frac{1}{4} a_i (a_i^2 + a_i - 1) + \frac{1}{8} a_i^2 (a_i - 1)^2 \tan^2 h + \frac{1}{2} a_i^2 \cot^2 h$$

$$b_{ij} = \frac{k}{(-s_i' - s_j' + 2k)} \{ \frac{1}{2} (a_i^2 + a_j^2 + a_i + a_j - a_i a_j - 2) + a_{ij} \tanh$$

$$- \frac{1}{2} [a_i(a_i - 1) + a_j(a_j - 1)] \tan^2 h + (a_i + a_j) \cot^2 h\}$$

$$b'_{ij} = \frac{k}{(-s_i' + s_j')} \{\frac{1}{2} [5(a_i + a_j) - (a_i^2 + a_j^2) - a_i a_j - 6]$$

$$+ a'_{ij} \tanh + \frac{1}{2} [a_i(a_i - 1) + a_j(a_j - 1)] \tan^2 h\}$$

$$b''_{ij} = 3 \frac{k}{(-s_i + s_j)} \frac{P_o}{\ell}$$

$$g_i = b_i - \coth$$

$$g_{ii} = b_{ii} - (a_i - \frac{1}{2}) \cot^2 h - \frac{1}{2} (a_i^2 - \frac{1}{2})$$

$$g_{ij} = b_{ij} - \frac{1}{2} (a_i^2 + a_j^2 - 1) - (a_i + a_j - 1) \cot^2 h$$

$$g'_{ij} = b'_{ij} - \frac{1}{2}(a_i^2 - a_j^2 - 2a_i + 2a_j)$$

$g''_{ij} = b''_{ij}$

$c_i = a_i - 1$

$c_{ii} = a_{ii} - \frac{1}{2} b_i + \frac{1}{4} \coth$

$c_{ij} = a_{ij} - \frac{1}{2}(b_i + b_j) + \frac{1}{2} \coth$

$c'_{ij} = a_{ij} + \frac{1}{2}(b_i + b_j) - \frac{1}{2} \coth$

$h^* = h - \Sigma N_i^2 \{\frac{1}{2} a_i(a_i - 1) \tanh + \frac{1}{4}(2a_i - 1) \coth\}$.

In the above expressions, $\ell = 54.9066''$ (Newcomb, 1905) and $P_o = 17.3919''$ (Newcomb, 1895) are constants. The numerical values for the constants α, h and k, are

$\alpha = 1.96459°$

$h = 23.40111°$

$k = 50.440174''$.

These numerical values are determined by the use of Newcomb's values for lunar-solar precession $\Psi, \frac{d\Psi}{dt}, \theta$, and ε for 1950 A.D., that is,

$\psi_o = 0$

$(\frac{d\psi}{dt})_o = 50.3733''$

$\theta_o = \varepsilon_o = 23.44579°$.

By use of the numerical values in Table 1, the secular variations of the eccentricity e, obliquity ε, and the longitude of perihelion (based on the moving equinox $\Pi = \Pi' + \psi'$) are calculated from Equations (1), (3), and (4). Figure 1 shows the variations of eccentricity for the past one million years and for one hundred thousand years in the future. Over this time, eccentricity varied from a minimum value very close to zero, which was reached 534,000 y before the present era, to a maximum value of 0.053, which was reached 689,000 y ago. The average period of the variations in eccentricity is about 90,000 y. Figure 2 shows the

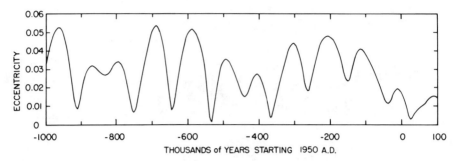

Fig. 1   The variation of eccentricity of the earth's orbit as a function of time for the period from 1,000,000 y before 1950 A.D. to 100,000 y in the future.

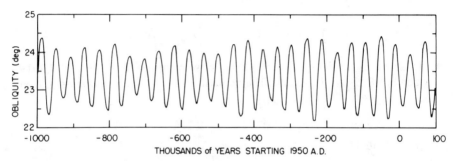

Fig. 2   The variation of earth's obliquity as a function of time for the period from 1,000,000 years before 1950 A.D. to 100,000 y in the future.

variations of the obliquity for the past million years and one hundred thousand years in the future. Over this period obliquity varied between 22.08° and 24.43°, with an average period of about 41,000 y. Figure 3 shows the variations of the longitude of perihelion (based on the moving equinox) for the past half million years and 25,000 y in the future, with an average period of about 21,000 y.

These results are used to calculate the variations in insolation from Milankovitch's formulation.

II. SECULAR VARIATIONS IN GLOBAL DISTRIBUTION OF THE INSOLATION

The intensity of solar radiation received daily by a horizontal surface at the top of the atmosphere is given by

$$W = \frac{\tau S}{\pi \rho^2} (\omega \sin \phi \sin \delta + \sin \omega \cos \phi \cos \delta) \ldots \quad (5)$$

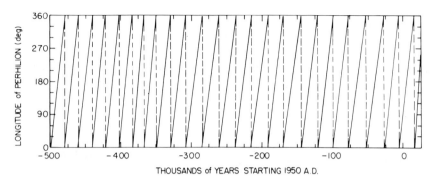

Fig. 3 *The variation of the longitude of perihelion from moving equinox as a function of time for the period from 500,000 y before 1950 A.D. to 25,000 y in the future.*

for geographical latitudes of daily sunrise and sunset,
For latitudes where the sun does not set during the day, the insolation is,

$$w = \frac{\tau S}{\rho^2} \sin \phi \sin \delta \qquad (6)$$

For latitudes where the sun does not rise during the day, the insolation is,

$$w = 0 \quad . \qquad (7)$$

Here S is the solar constant, δ the declination of the sun, ω the hour angle of the sun, ϕ the geographical latitude, τ the duration of the day and ρ the ratio of the magnitude of radius vector of the earth's orbit to the mean distance between the earth and the sun, which can be expressed mathematically as

$$\rho = \frac{1 - e^2}{1 + e \cos \nu}, \qquad (8)$$

where $\nu = 180° - \Pi + \lambda$ and λ is the longitude of the sun in its apparent annual movement around the earth.

From the geometrical considerations, the declination of the sun δ can be expressed in terms of the obliquity ϵ and the longitude of the sun λ

$$\sin \delta = \sin \epsilon \sin \lambda \quad . \qquad (9)$$

Also, the hour angle ω can be expressed in terms of the geographical latitude and the declination of the sun δ

$$\cos \omega = -\tan \phi \tan \delta \quad . \qquad (10)$$

For the present orbital parameters, i.e., $e_0 = 0.0167$, $\epsilon_0 = 23.4458°$, $\Pi_0 = 102°$, or $\nu_0 = 180° - \Pi_0 = 78°$, and the solar constant, $S = 1359\ Wm^{-2}$, the daily insolation calculated from (5), (6), and (7) is shown in Table 2 as a function of geographical latitude and solar longitude. The insolation is given in units of Wm^{-2}. The daily insolation is equal in the two hemispheres at vernal equinox ($\lambda = 0°$) and at the autumnal equinox ($\lambda = 180°$). The insolation values for $\lambda = 90°$ and $270°$ refer to the summer and winter solstices in the northern hemisphere, respectively.

TABLE 2

The intensity of solar radiation (Wm^{-2}) received at the top of the atmosphere during a day of 1950 A.D.

Latitude* (deg.)	Longitude of the sun (deg.)			
	0	90	180	270
90	0	523	0	0
80	76	516	75	0
70	149	492	147	0
60	218	476	215	24
50	280	481	276	86
40	334	482	329	156
30	377	474	372	227
20	410	455	404	295
10	429	425	423	357
0	436	384	430	410
-10	429	334	423	454
-20	410	276	404	486
-30	377	212	372	506
-40	334	146	329	515
-50	280	80	276	514
-60	218	23	215	508
-70	149	0	147	525
-80	76	0	75	551
-90	0	0	0	559

*Positive values are northern latitudes, negative values southern.

The insolation as a function of geographical latitude ϕ and the solar longitude λ is calculated from (5), (6), and (7) by use of the earth's orbital parameters from 35,000 years ago to 5,000 years in the future at 5,000-year intervals. The deviations of the insolation from present values are shown in Table 3 in units of $10^{-1} Wm^{-2}$. Comparing these deviations with the present total daily insolation

TABLE 3

Deviations of daily solar radiation ($10^{-1} Wm^{-2}$) from their 1950 A.D. values

Time in 1000 years	-35				-30				-25				-20			
Long. of sun Lat.	0	90	180	270	0	90	180	270	0	90	180	270	0	90	180	270
90	0	132	0	0	0	-10	0	0	0	-159	0	0	0	-49	0	0
80	1	130	-1	0	-21	-10	21	0	-23	-157	23	0	10	-48	-9	0
70	1	124	-2	0	-42	-10	41	0	-45	-150	45	0	19	-46	-18	0
60	2	167	-3	26	-61	72	61	56	-65	-79	65	56	27	-24	-27	17
50	2	199	-4	5	-79	123	78	46	-84	-38	84	67	35	-12	-34	21
40	2	222	-4	-30	-94	161	93	16	-100	-7	100	61	42	-2	-41	19
30	3	237	-5	-71	-106	190	105	-23	-113	19	113	48	47	6	-46	16
20	3	245	-5	-114	-115	212	114	-67	-123	42	123	30	51	13	-50	10
10	3	245	-6	-157	-120	225	119	-114	-129	62	129	7	54	19	-53	3
0	3	237	-6	-198	-122	230	121	-161	-131	78	131	-18	55	24	-54	-5
-10	3	222	-6	-236	-120	227	119	-207	-129	90	129	-45	54	27	-53	-13
-20	3	200	-5	-269	-115	215	114	-250	-123	97	123	-74	51	30	-50	-22
-30	3	171	-5	-297	-106	195	105	-289	-113	99	113	-103	47	30	-46	-31
-40	2	137	-4	-321	-94	167	93	-326	-100	95	100	-132	42	29	-41	-40
-50	2	98	-4	-342	-79	131	78	-362	-84	84	84	-164	35	25	-34	-50
-60	2	53	-3	-366	-61	81	61	-406	-65	59	65	-205	27	18	-27	-63
-70	1	0	-2	-425	-42	0	41	-501	-45	0	45	-283	19	0	-18	-87
-80	1	0	-1	-446	-21	0	21	-525	-23	0	23	-297	10	0	-9	-91
-90	0	0	0	-453	0	0	0	-533	0	0	0	-302	0	0	0	-92

Time in 1000 years	-15				-10				-5				5			
Long. of sun Lat.	0	90	180	270	0	90	180	270	0	90	180	270	0	90	180	270
90	0	368	0	0	0	552	0	0	0	249	0	0	0	71	0	0
80	20	362	-20	0	-15	544	16	0	-32	246	32	0	16	70	-16	0
70	40	345	-39	0	-30	519	31	0	-62	234	64	0	31	67	-31	0
60	59	307	-57	-34	-44	452	45	-56	-91	191	93	-36	46	103	-46	23
50	76	291	-73	-72	-57	422	58	-111	-117	169	120	-61	59	128	-59	11
40	90	278	-87	-108	-67	398	69	-160	-140	151	143	-79	70	147	-70	-13
30	102	261	-99	-140	-76	370	78	-205	-158	133	161	-92	79	160	-79	-42
20	111	240	-107	-169	-83	335	85	-243	-172	114	175	-102	86	168	-86	-73
10	116	215	-112	-193	-87	295	89	-274	-180	94	184	-108	90	170	-90	-104
0	118	185	-114	-212	-88	249	90	-296	-183	73	186	-109	91	166	-91	-134
-10	116	151	-112	-224	-87	199	89	-309	-180	51	184	-107	90	157	-90	-162
-20	111	115	-107	-230	-83	147	85	-313	-172	29	175	-101	86	143	-86	-188
-30	102	78	-99	-229	-76	93	78	-308	-158	9	161	-90	79	125	-79	-210
-40	90	42	-87	-221	-67	43	69	-293	-140	-9	143	-76	70	101	-70	-229
-50	76	10	-73	-207	-57	-1	58	-268	-117	-22	120	-58	59	74	-59	-247
-60	59	-12	-57	-187	-44	-26	45	-234	-91	-25	93	-33	46	41	-46	-268
-70	40	0	-39	-166	-30	0	31	-193	-62	0	64	4	31	0	-31	-316
-80	20	0	-20	-174	-15	0	16	-202	-32	0	32	4	16	0	-16	-331
-90	0	0	0	-176	0	0	0	-206	0	0	0	4	0	0	0	-337

given in Table 2, we find the maximum deficit occurred 30,000 years ago at the winter solstice. The change is about 10% of the present value near the south pole. The maximum excess occurred 10,000 years ago at the summer solstice. The change is about 10% of the present value near the north pole.

To compare the insolation from one geological time to another, it may suffice to calculate cumulative insolation for the astronomical summer "half" year and winter "half" year. However, the durations of the summer "half" year T_s and the winter "half" year T_w are dependent on the eccentricity and the longitude of perihelion. The difference between T_s and T_w is mathematically expressed as

$$T_s - T_w = \frac{4T\,e\,\sin\Pi}{\pi}, \qquad (11)$$

where T is the duration of the tropical year.

Figure 4 shows the results of (11) for the past half million years and 25,000 y in the future. At present, the difference between summer and winter half years is 7.6 days. The maximum difference over the period of these calculations was about 22 days; it occurred 200,000 y ago.

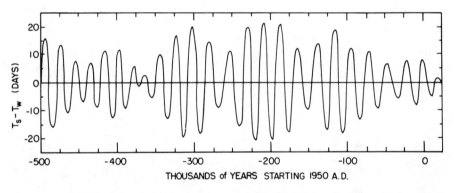

Fig. 4 The variation in the difference between the duration of astronomical summer T_s and that for the winter T_w as a function of time for the period from 500,000 y before 1950 A.D. to 25,000 y in the future.

Since the average insolation over unequal lengths of time is not comparable from one geological time to another, Milankovitch divided the year into two equal halves such that the heat received during any day of the summer half (caloric summer) is greater than that of any day of the winter half (caloric winter). The calculations of the average insolation for the caloric summer half Q_s and the winter half Q_w are shown in Table 4 in units of Wm^{-2}.

The amount by which the cumulative insolation at any time deviates from the present values can be expressed in terms of corresponding deviations in the earth's orbital parameters:

$$\Delta Q_s = p_s \Delta\varepsilon - m\Delta(e \sin\Pi)$$
$$\Delta Q_w = p_w \Delta\varepsilon + m\Delta(e \sin\Pi)$$
(12)

for the northern hemisphere and

$$\Delta Q_s = p_s \Delta\varepsilon + m\Delta(e \sin\Pi)$$
$$\Delta Q_w = p_w \Delta\varepsilon - m\Delta(e \sin\Pi)$$
(13)

for the southern hemisphere, where p_s and p_w are respectively the summer and winter insolation for unit degree change in obliquity and where

$$m = \frac{4}{\pi^2} \frac{S \cos\phi}{(1 - e^2)^{1/2}} \qquad (14)$$

Milankovitch calculated p_s and p_w by differentiating the equations for mean daily insolation for the astronomical summer and winter with respect to obliquity ε and by using the finite-difference approximation.

The results of (12) and (13) are shown in Figure 5 in units of Wm^{-2}. The upper half of the figure shows the insolation variations for the northern hemisphere caloric winter and the southern hemisphere caloric summer; the lower half shows the variations for northern hemisphere caloric summer and southern hemisphere caloric winter. The numbers between the two halves refer to the time in thousands of years. The figure shows the insolation variations from 115,000 y ago to 25,000 y in the future. Over this period the maximum departure from the present value occurred 106,000 y ago. This change is approximately 6% of the present value given in Table 4. Comparing the mean deviations at 30,000 and 10,000 y ago with the corresponding deviations in the daily insolation in Table 3, we find that the magnitudes of the mean deviations are significantly reduced, especially at higher latitudes, by the averaging process.

The significance of these results is their role in determining the efficacy of radiation variations in forcing the Pleistocene climatic fluctuations, or in other words, in testing the validity of Milankovitch's hypothesis of the ice ages. The Milankovitch hypothesis assumes that the glaciations corresponded to periods during which the high northern latitudes received a minimum summertime solar radiation. Such minima occur when the obliquity is small, when the eccentricity is large, and when the northern summer solstice occurs when the earth is at aphelion. As a test of this hypothesis, Sellers (1970) used his zonally averaged climate model based on the energy balance of the earth-atmosphere system (Sellers, 1969) to compute the north-south distribution of the earth's surface temperature under contrasting conditions of insolation when the obliquity was

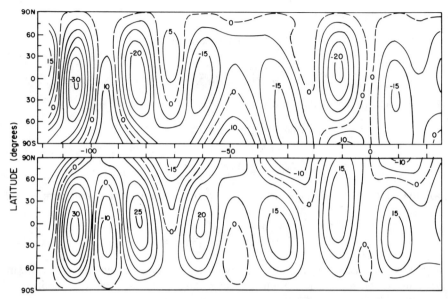

Fig. 5 The variation in ΔQ (deviation of the mean insolation from its present value) as a function of latitude and time in the units of Wm^{-2}. The upper half of the figure shows the variations for the northern caloric winter and southern caloric summer. The lower half shows the variations for the northern caloric summer and southern caloric winter. The abscissa shows the time (numbers between the two halves) in thousands of years. The variations shown in the figure are for the period from 115,000 y before 1950 A.D. to 25,000 y in the future.

very high and when it was very low. Sellers concluded that the difference in surface temperature between these two extreme cases was too small to support the Milankovitch hypothesis. A similar study was made by Saltzman and Vernekar (1971b). They chose two extreme conditions in the summer half-year insolation at 60°N over the past 50,000 y. The summer insolation was at its maximum 10,000 y ago and at a minimum 25,000 y ago; the difference between them is about 7%. Using their climatic model (Saltzman and Vernekar, 1971a), they computed surface temperature, the surface zonal wind, and the difference between evaporation and precipitation as a response to the insolation prevailing 25,000 and 10,000 y ago. The changes in all three variables between the two extreme cases were smaller than the minimum changes postulated by Simpson (1940) as necessary for the initiation of an ice age. On the basis of this it appears doubtful that the summer minimum insolation 25,000 y ago was a triggering mechanism for the development of the Wisconsin glacial stage which attained its maximum around 18,000 y ago.

However, these studies used the mean insolation for caloric summer and winter and kept the surface albedo constant. Recently, Held and Suarez (1974) and

TABLE 4

Mean intensity of solar radiation (Wm^{-2}) at the top of the atmosphere for the caloric winter and summer of 1950 A.D.

Latitude (deg,)	Northern caloric winter and southern caloric summer	Northern caloric summer and southern caloric winter
90	0	345
80	10	346
70	38	355
60	91	381
50	158	410
40	224	432
30	287	443
20	342	443
10	388	430
0	424	406
-10	448	370
-20	460	325
-30	459	271
-40	445	210
-50	422	146
-60	390	83
-70	362	31
-80	349	7
-90	345	0

Suarez and Held (1975) have shown that a change in insolation due either to a change in the solar constant or to a change in obliquity may lead to significant change in the ice cover through the albedo feedback mechanism, if the surface albedo is allowed to change. Saltzman and Vernekar (1975) used their climate model to show that the solution for the surface temperature, calculated as a response to the surface albedo which prevailed 18,000 y ago, is in good agreement with observations. These studies and the significant difference between the results in Table 3 and in Figure 5, especially at high latitudes, where the initial size of the ice cap might change through the albedo feedback mechanism, suggest that one should reexamine the efficacy of insolation variation in initiating the climatic changes of the geological past.

ACKNOWLEDGMENTS

I am indebted to Mr. H-D. Chang for his aid in making the calculations and to Mrs. S. Epstein for typing the manuscript. This work has been supported by the Atmospheric Science Section of the National Science Foundation under Grant ATM 70-00181-A04.

REFERENCES

Berger, A.L., 1976a, *Celestial Mech.*, in press.

―――――――, 1976b, to be published.

―――――――, 1976c, *Quaternary Res.*, submitted.

Brouwer, D., and van Woerkom, A.J., 1950, *Astron. Pap. Am. Ephemeris Part 2, 13*, 83.

Harzer, P., 1895, *Die Säkularen Veranderunger der Bahnen der Grossen Planeten*, Leipzig Preisschrift.

Held, I.M., and Suarez, M.J., 1974, *Tellus 26*, 613.

Michkovitch, V.V., 1931, *Glas. Srp. Kralyereske Akad. 143*, First Class No. 70, Belgrade.

Milankovitch, M., 1930, in *Handbuch der Klimatologie*, Verlage Brontrager, Berlin.

Newcomb, S., 1895, *Suppl. to American Ephemeris and Nautical Almanac for 1897*, Government Printing Office, Washington, D.C.

―――――――, 1905, *Astr. Pap. 8*.

Saltzman, B., and Vernekar, A.D., 1971a, *J. Geophys. Res. 76*, 1498.

――――――― and ―――――――, 1971b, *ibid. 76*, 4195.

――――――― and ―――――――, 1975, *Quaternary Res. 5*, 307.

Sellers, W.D., 1969, *J. Appl. Meteorol. 8*, 392.

―――――――, 1970, *ibid. 9*, 960.

Sharaf, S.G., and Budnikova, N.A., 1967, *Trudy Inst. Teor. Astronomii Akademiya Nauk SSSR 11*, 231. (English transl. National Technical Information Service, Springfield, Va.)

――――――― and ―――――――, 1969, *ibid. 14*, 48.

Simpson, G.C., 1940, *Linn. Soc. London, Proc. 152*, 190.

Stockwell, J.N., 1875, *Smithsonian Contribution to Knowledge No. 232*.

Suarez, M.J., and Held, I.M., 1975, *Proc. WMO/IAMAP Symposium on Long-Term Climatic Fluctuations, Norwich* (18-23 August 1975).

Vernekar, A.D., 1972, *Meteorol. Monog. 12*.

IV. THE SOLAR SPECTRUM

In attempting to define the solar spectrum in absolute units, we are dissecting the solar constant through the application of techniques of absolute spectrophotometry. In this chapter we describe measurements of the entire solar spectrum, the techniques used in these projects, and the intercomparison of different measurements. At first glance, the apparent disagreements and active controversies in this particular field of measurement physics may be dismaying. But the situation is not really so surprising. The measurement problem is a very difficult one, and the source of the radiation varies differently over the spectrum. Nevertheless, we do have some measures of absolute fluxes over the whole spectrum. It is only recently that the placement of permanent space platforms permitting the use of recoverable spectrophotometers has become a possibility; we can thus foresee the measurement of the entire solar spectrum to whatever precision we can afford in terms of both time and money. As in the case of the solar constant measurements, the opportunity exists to complete our record of the relative variation of the solar spectrum in the future.

Two basic difficulties in the measurement of the solar spectrum are the availability of adequate absolute standards and the elimination of systematic errors in the various types of experiments used to measure the radiant flux. In addition to the individual review papers covering each spectral range, attention should be given to the general comments of R. P. Madden and R.F. Donnelly at the end of the chapter. Madden discusses the current situation in the availability of standards and their improvement through the NBS research program. From his experience in ionospheric research, Donnelly speaks directly to the problem of comparing measurements made using different techniques and states views that should be very helpful to those physicists who will undertake to measure the solar spectrum during solar cycle 21. His principal points are that different measurements of short-wavelength solar fluxes can be compared only if they were made at the same level of solar activity and only if a common data base has been obtained through proper correction for instrumental spectral resolution, stray light due to scattering or overlapping orders, and sensitivity variation with wavelength.

Despite the measurement problems, it seems clear that the variability of solar radiation originating in the solar photosphere is hardly detectable in existing measurements; but outside the visible solar spectrum in both very short and very long wavelengths, where the radiation is emitted by the outer solar layers, i.e., the chromosphere and the corona, fluctuations due to solar active regions become very evident. Although these outer spectral bands contain only a small amount of the sun's total radiant energy, the solar spectrum below 3000 Å interacts strongly with the earth's atmosphere and is thereby crucial to our understanding of the chemistry and physics of the outer atmospheric envelope of the earth. If we are to understand the variation in this envelope, we must verify the indications of intrinsic variability in the solar spectrum at short wavelengths and then determine the combined role of both solar radiation and particles in establishing or changing the physical and chemical state of the upper atmosphere at any particular instant in time.

The variability we do detect in the solar spectral output occurs as a direct result of variations in physical processes in the solar atmosphere proper. Since the sun's radiation is hardly affected by its passage through interplanetary space to the earth, the analysis of the solar spectrum from measurements at 1 AU is the most direct method that we have for deducing properties of the solar atmosphere. Avrett (Chapter IV) and Linsky (Chapter VII) describe the current analytic approach for interpreting measurements of the solar spectrum in terms of the temperature and density structure in the solar atmosphere. Accurate values for the radiative flux emitted in both spectral lines and continua are absolutely necessary for realistic estimates of density and temperature throughout the solar atmosphere; therefore, much of our quantitative knowledge of the structure of the visible atmosphere of the sun now rests squarely on the flux values referred to in this chapter. Improvement in this quantitative picture clearly depends on improvement in our ability to measure diagnostically important spectral regions on an absolute scale that can be reliably established for decades.

Oran R. White

THE SOLAR SPECTRUM
ABOVE 1 mm

F.I. Shimabukuro
Electronics Research Laboratory
The Aerospace Corporation
Los Angeles, California

It has long been known that the sun, as a blackbody source, emits electromagnetic waves in the radio spectrum. However, because of a lack of technical development in the field of sensitive radio receivers and highly directional antennas, and unawareness that the sun emits much more strongly than a 6000K blackbody source at longer radio wavelengths, little early research was devoted to trying to detect radio emissions from the sun. Attention was first called to solar radio emission by radio amateurs who, during solar maximum in 1936-37, reported the existence of a curious hiss in their receivers on wavelengths of about 10 m. This noise occurred only in the daytime and was associated with periods of solar activity. For the typical receivers and antennas of that day, the reported solar emission enhancement at 10 m was about 10^4 that of a 6000K blackbody source (Appleton and Hey, 1946). During World War II, with more sensitive receivers in operation, many further reports of this type of noise reception were made. The war effort spawned the development of sensitive radar receivers with high-gain antennas, and in 1942 Hey was able to identify noise bursts in 4-6 m-band radar equipment as emanating from the sun (Hey, 1946). In that same year Southworth, using microwave radar equipment, observed a small but measurable amount of radiation from the sun at three wavelengths between 1 and 10 cm (Southworth, 1946). Independently, Reber (1944) reported detection of solar radiation at 1.87 m.

With the impetus of these pioneering efforts and the rapid development of microwave technology, research activity in the new science of radio astronomy increased greatly during the postwar period, notably in England and Australia. It was soon established from observations that in the 1 to 5 m range the solar intensity was much higher than that derived from blackbody radiation of the photosphere. Furthermore, a striking feature emerged: the solar radiation showed great variability both over long periods and over a few seconds (Appleton and Hey, 1946; Pawsey, Payne-Scott, and McCready, 1946; McCready, Pawsey, and Payne-Scott, 1947). The early investigations delineated three main components of the solar radiation at radio wavelengths: (1) the quiet-sun radiation, (2) the slowly varying component, and (3) transitory radiation, commonly known as bursts.

These will be discussed in more detail in the sections below. Although the solar radio observations are undertaken primarily to unravel the structure and physics of the solar atmosphere, they also give measures of the solar irradiance at the earth; and it is logical to discuss the incident flux density as arising from the three types of variation in the solar atmosphere.

I. THE QUIET-SUN FLUX

The quiet-sun emission is defined as all radiation from the solar atmosphere exclusive of contributions from all discrete sources. It is the emission level below which the solar flux never falls throughout the solar cycle. This level is most nearly approached during the sunspot minimum when there are no active centers of emission on the sun. The quiet-sun radiation is thermal in nature, and the radio brightness is given by the Rayleigh-Jeans approximation to the Planck radiation law

$$B = \frac{2k\, T_b}{\lambda^2} \text{ Wm}^{-2} \text{ Hz}^{-1} \text{ ster}^{-1} \qquad (1)$$

where T_b is the brightness temperature of the sun at wavelength λ and k is Boltzmann's constant. The flux density is then

$$F = \int B d\Omega \text{ Wm}^{-2} \text{ Hz}^{-1} \qquad (2)$$

where Ω is the solid angle subtended by the radio disk. The early observations revealed that the sun had an apparent brightness temperature, which was a function of wavelength, of about 10^4 K at the short (centimeter) wavelengths to 10^6 K at the meter wavelengths. The temperature behavior was consistent with thermal emission from free electrons in the solar atmosphere and was predicted by Ginzburg (1946) and Martyn (1946, 1948) from the temperature and electron density distribution information available at that time. The corona, they predicted, would be optically thick at the meter wavelengths, and the emission would correspond to a source of a million degrees; while at the centimeter wavelengths the intensity would drop to a value (corresponding to the temperature of the chromosphere) of the order of 10^4 K. These ideas were basically correct and have been further refined by subsequent studies.

The current model of the quiet solar atmosphere that gives rise to radio emission is shown in Figure 1. In this model the Harvard-Smithsonian Reference Atmosphere chromosphere (Gingerich *et al.*, 1971) is smoothly connected to a transition region whose base is at a temperature of 50,000K at a height of ~2000 km (Athay, 1966). The corona is Newkirk's (1967) model. This atmosphere model is consistent with measured central disk temperatures and depicts the structure of the quiet solar atmosphere that gives rise to the radio flux density at the earth.

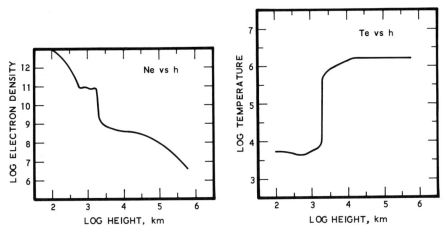

Fig. 1 Electron and temperature distribution of the quiet sun.

A. METHOD OF MEASUREMENT

Radiotelescopes are instruments that directly measure flux density. Typically, measurements are made in the following way: the intensity received from the sun is compared with a blackbody calibration source. The absolute flux is derived from measured or calculated antenna characteristics after correction for atmospheric effects. However, absolute flux measurements are difficult to make in practice; the principal sources of error are the uncertainty in antenna characteristics and the fact that the sun and the calibration signal are not coupled in exactly the same way into the input antenna terminals. In the radio region, an absolute flux determination to 5% is a very good measurement; however, the errors are usually 10% or more.

If the antenna beamwidth is greater than that of the radio disk, the total solar flux is measured directly although the contributions of active regions are included in such a measurement. This situation is encountered at the meter wavelengths where mechanical constraints make high-resolution pencil-beam antennas unfeasible. In the presence of active centers, the quiet-sun emission can be determined by statistical analysis. This is done by plotting daily flux values as a function of projected sunspot areas. Under the assumption that a linear relationship exists between sunspot area and the active emission, the quiet-sun flux is estimated by extrapolating the regression line to zero sunspot area (Pawsey and Yabsley, 1949; Denisse, 1950). The difficulty with this technique is that active radio centers exist even in the apparent absence of sunspots, and at the meter wavelengths the variable emission can be large compared to the basic quiet component. Another method, restricted to periods when the sun is very quiet, is to find the lowest level of the flux density. It is possible to obtain high resolution at meter wavelengths by using interferometric techniques, but the inherently high secondary sidelobes make this type of flux measurement inaccurate.

At the centimeter and millimeter wavelengths, beamwidths are sufficiently small to distinguish the quiet from the active regions, and the quiet flux density is easier to determine, although there are two added problems. First, the solid beam efficiency of the antenna, which is a difficult quantity to measure or calculate, must be known. The solid beam efficiency is the ratio of the energy received by the antenna main lobe and sidelobes out to the solid angle subtended by the sun and the energy received over 4π sr. Second, when high-resolution measurements are made, the brightness distribution must be determined and integrated over the radio disk to get the total flux density.

B. ORIGIN OF THE QUIET RADIO EMISSION

As mentioned before, the quiet-sun radiation is thermal in nature and is the result of free-free emission. A plot of the brightness temperature at various heights for the centimeter and millimeter wavelengths for the model of Figure 1 is shown in Figure 2. At the meter wavelengths the corona is the source of emission, but the

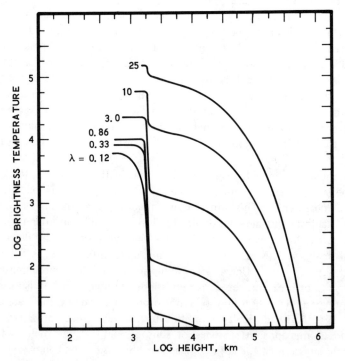

Fig. 2 Plot of brightness temperature of the quiet sun as a function of height, at various wavelengths, in cm, for the model of Figure 1.

index of refraction decreases with height and becomes zero for various wavelengths at different heights. A refraction index of zero is called the turning or critical point. At about 100 MHz the corona is optically thin above the critical point, and the source of the emission can be determined by tracing the radio ray trajectories through the corona, as was done by Jaeger and Westfold (1949); this is shown in Figure 3. In addition, the apparent brightness temperature is less than the kinetic temperature of the coronal gas and decreases with decreasing frequency. Scattering effects of the irregular corona also have to be taken into account (Aubier, Leblanc, and Boischot, 1971).

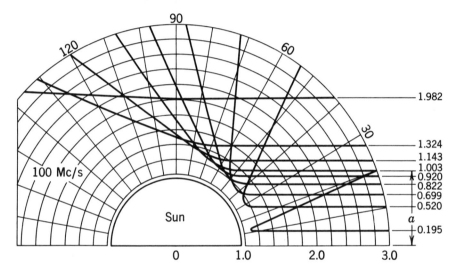

Fig. 3 Ray trajectories in the solar corona for 100 MHz. The radial scale is in units of 705,000 km (= photospheric radius + 7650 km). (After Jaeger and Wesfold, 1949).

C. MEASUREMENTS OF THE QUIET-SUN FLUX

In the past there have been several studies of the quiet-sun emission (e.g., Allen, 1957; Smerd, 1964). Shimabukuro and Stacey (1968) tabulated the quiet central disk temperatures at the centimeter and millimeter wavelengths, and Linsky (1973) recalibrated the quiet-sun temperatures in the millimeter-wave spectrum by using the moon as an absolute radiometric standard. The measurements are found in the above two papers.

The measurements at the centimeter and millimeter wavelengths are generally made with high-resolution radiotelescopes, and the total flux density is obtained by integrating the brightness temperature over the radio disk. Calculations using a model such as that shown in Figure 1 give an overestimate of the limb brightening when compared with the measurements. The interpretation is that the inhomogeneities in the solar atmosphere, such as the spicules, cause limb-darkening at the short centimeter and millimeter wavelengths. According to reported observations, there is little limb brightening at the short centimeter and millimeter wavelengths.

Figure 4 summarizes the measured flux density of the quiet sun. At the longer meter wavelengths, measurements by O'Brien (1953), Leblanc and LeSqueren (1969), Sheridan (1970), and Aubier et al. (1971) have been incorporated.

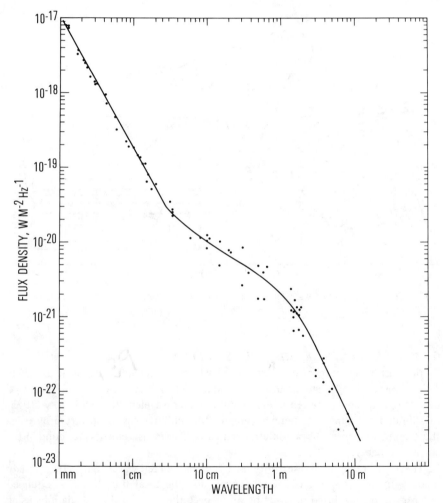

Fig. 4 Summary of measured values of the total quiet-sun flux at the earth, as a function of wavelength.

II. THE SLOWLY VARYING COMPONENT

A. ORIGIN OF THE EMISSION

Covington (1947) made observations at 10.7 cm during the partial solar eclipse of 1946 and reported that there was a marked decrease in the solar flux as the moon occulted a large sunspot group near the center of the disk. Since then, he and other workers have found that in addition to the basic quiet component there is a variable emission, called the slowly varying or S-component, whose intensity is closely correlated with sunspot areas (Covington, 1948; Lehany and Yabsley, 1949; Pawsey and Yabsley, 1949). The slowly varying component is thought to be of thermal origin in localized areas of high electron density and magnetic fields with coronal temperatures that are slightly higher than average (coronal condensations). These areas are characterized by the presence of sunspots and plages. At the meter wavelengths the S-component is believed to be thermal emission from these coronal condensations (Waldmeir and Muller, 1950; Lehany and Yabsley, 1949; Piddington and Minnett, 1951; Leblanc, 1970; Couturier and Leblanc, 1970; Sheridan, 1970). However, if the scattering effects are included in addition to refraction, calculations show that the density enhancements may lead to lower rather than higher emission, as actually measured, and the S-component at the meter wavelengths may be of nonthermal origin (Riddle, 1974). Covington (1949) and Piddington and Minnett (1951) found that the S-component was partially circularly polarized at 10 cm, and further observations revealed that circularly polarized emission was a characteristic feature in the centimeter and millimeter wavelength region (Tanaka and Kakinuma, 1956; 1958a; 1958b; 1960; Kundu, 1959b; Korolkov, Soboleva, and Gelfreikh, 1960; Tanaka, 1961; Kakinuma and Swarup, 1962; Feix, 1969; Edelson, Mayfield, and Shimabukuro, 1971). The source of the polarized emission is located above sunspots and it has been suggested that the emission mechanism is resonance absorption (Kakinuma and Swarup, 1962; Zheleznyakov, 1962) and magnetoionic emission (Shimabukuro et al., 1973). At 20 cm, and presumably at longer wavelengths, there is practically no polarization (Christianson and Mathewson, 1959).

B. MEASUREMENTS OF THE S-COMPONENT FLUX

Figure 5 shows plots of the daily flux values as a function of sunspot number for several frequencies (Smerd, 1964). The correlation between the flux and sunspot number is quite good at the centimeter and decimeter wavelengths while at the longer wavelengths the correlation is rather poor. Christianson and Mathewson (1959) found that at 20 cm, and presumably at longer wavelengths, the slowly varying component is more closely correlated with plage areas. A statistically derived spectrum of the S-component flux density per unit sunspot number is shown in Figure 6. (Smerd, 1964). It has been found that the flux for a strong source peaks in the vicinity of 6-8 cm and that the degree of polarization is highest at the short centimeter wavelengths (Tanaka and Kakinuma, 1958b; Kundu, 1959;

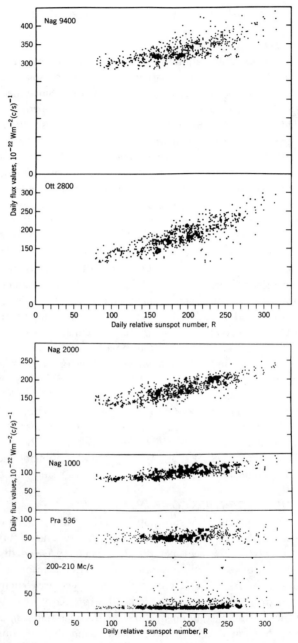

Fig. 5 Diagrams of correlation between daily flux densities and relative sunspot number at different frequencies. (After Smerd, 1964.)

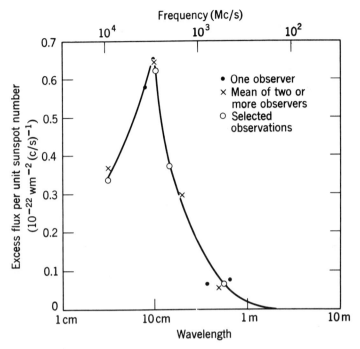

Fig. 6 Statistical spectrum of S-component flux density obtained during the IGY period. (After Smerd, 1964.)

Kakinuma and Swarup, 1962; Swarup et al., 1963). A representative example of a slowly varying component flux density spectrum near its maximum is plotted in Figure 7. Because the slowly varying component is variable, the spectrum depends on the number and intensity of the sources and their location on the disk. An estimate of the flux for a particular day can be obtained by correlation with any of several solar activity indices, such as the sunspot number. The S-component is most prominent at the centimeter wavelengths where the flux is partially circularly polarized and can exceed the basic quiet component when the sun is very active. At the meter wavelengths, S-component measurements are scarce because of the inherent difficulty of the measurement. The absolute flux is low, the sky background brightness temperature is relatively high, and antennas in the wavelength range are generally low gain. Moutot and Boischot (1961) observed the S-component at 1.76 m to be as high as 15-20% of the quiet-sun flux.

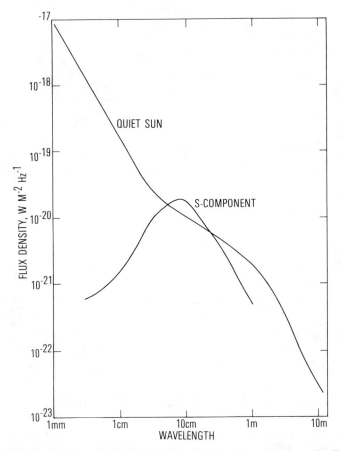

Fig. 7 Plot of slowly varying component flux density with strong sources on the disk.

III. RADIO BURSTS

A. CHARACTERISTICS AND ORIGINS OF THE RADIO BURST FLUX

Like the quiet-sun flux the S-component is thermal in nature and exhibits a slow continuous fluctuation. Its intensity does not exceed the quiet solar flux except on very active days and usually only at the centimeter wavelengths. However, there is a highly transitory component, resulting from an explosive release of a substantial amount of energy in active regions of the solar atmosphere, whose emission runs through the gamut of the radio spectrum. The frequency of occurrence of this component, which is called radio bursts, waxes and wanes with the sunspot cycle. Except for the gradual type observed at the centimeter wavelengths, these bursts are nonthermal in origin. There is a bewildering array of observed burst types. Intensities range from barely detectable levels to outputs that are orders of magnitude greater than that of the quiet sun (at a particular wavelength) on a time scale from less than a second to days. There are many excellent reviews on the observations and theory of bursts, including those by Kundu (1963, 1965), Fokker (1963), Wild (1964), Wild, Smerd, and Weiss (1963), Maxwell (1965), Takakura (1966, 1967), and Wild and Smerd (1972). A brief summary of the characteristics of bursts is shown in Table 1. The summary is by no means a complete description; a fuller discussion of the topic can be found in the above reviews. The idea to be conveyed is that as a great amount of energy is suddenly released in the solar atmosphere, the plasma is heated, particles are accelerated to high energies, plasma and acoustic waves are excited, and many emission mechanisms are possible in the radio region, which encompasses more than four decades of wavelengths.

TABLE 1

Summary of burst types.

WAVELENGTH RANGE	TYPE	CIRCULAR POLARIZATION	DURATION	APPARENT ANGULAR SIZE ARC MIN	APPARENT T_B LARGE EVENT K	EMISSION MECHANISM
m	I	STRONG	BURST: ~1s STORM: hrs TO days	1-6 5-10	$10^{10}-10^{12}$	PLASMA
	II	NONE	5-30 min	6-12		PLASMA
	III	NONE OR PARTIAL	BURST: ~10s GROUP: ~1 min	3-12		PLASMA
	IVmA	WEAK	10 min-2 hr	6-12		SYNCHROTRON
	IVmB	STRONG	FEW HOURS	3-6		PLASMA SYNCHROTRON (?)
	V	PARTIAL	~1 min	3-12		PLASMA
DECIMETER	IVdm	PARTIAL OR STRONG	5 min-2 hr	3-5	$<10^9$	SYNCHROTRON
cm, mm	GRADUAL	NONE OR PARTIAL	10-100 min	1-4	10^6	THERMAL
	IMPULSIVE	PARTIAL	1-5 min	cm, mm <2.5 dm <5	10^7-10^8	SYNCHROTRON
	IVμ	PARTIAL	5 min-1/2 hr, cm, mm 5 min-2h dm	2-4	10^9	SYNCHROTRON

B. MEASUREMENTS OF BURST FLUX

Measured flux densities of bursts are shown in Figure 8. The upper curve is the envelope of single-frequency burst maxima observed in the period 1954 to 1972 (Badillo and Salcedo, 1969; Castelli, Barron, and Aarons, 1973). Any individual burst has its own characteristic spectrum; one is shown at maximum emission for the large event of 4 August 1972 (Castelli et al., 1973). A typical gradual burst,

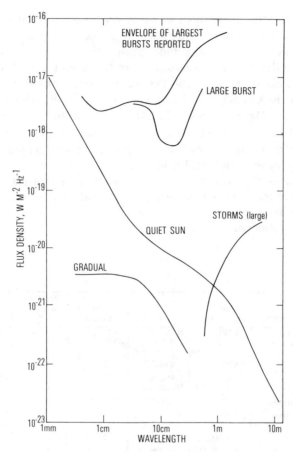

Fig. 8 Flux density spectra of bursts.

observed on 29 March 1969 (Shimabukuro, 1972), is also shown. The flux spectrum of large storms is taken from Wild, Smerd, and Weiss (1963). Radio bursts, in general, are unpredictable, although their probability of occurrence increases markedly if there are especially active regions on the solar disk. The flux spectrum is different from burst to burst, and the time duration and emission are variable. In order to assess the impact of radio bursts on the solar flux incident at the earth, it is necessary to know the intensity of the bursts, their frequency of occurrence, and the time of duration. The very large bursts, like the one on 4 August 1972 and those that produce the maximum envelope, are associated with highly energetic flares, with proton production and with possible white light emission. These proton flares are quite rare; they occur perhaps several times a year near the sunspot maximum and are characterized by broadband centimeter bursts (Kundu and

Haddock, 1960). The larger bursts are longer lived and are usually associated with the larger flares (importance 1b and greater), and their frequency of occurrence can be estimated from the statistics of these larger flares (e.g., see Smith and Smith, 1963). The burst data for the large proton flare of 4 August 1972 is shown in Table 2. It may be seen that while the burst is in progress the solar flux is greatly increased although the impact of the burst flux is significantly diluted when averaged over a longer time period. Maxwell, Howard, and Garmire (1960) made a statistical study of bursts near the sunspot maximum. Their results are summarized in Table 3. These results are for sunspot maximum; the frequency of occurrence decreases at the quieter periods in the solar cycle. Das Gupta and Sarkar (1971) analyzed 3221 simple events observed at the Sagamore Hill Solar Radio Observatory during a period of nearly four years covering the maximum phase of solar cycle 20. They presented relative frequency histograms for the duration, rise time, peak flux density, event energy, and rise rate for bursts at 8.80 GHz, 4.995 GHz, 1.415 GHz, and 606 MHz. Wefer (1973) made a statistical study of 5480 solar bursts at 960 MHz, 2.7 GHz, and 10.7 GHz during the period 1964-1970, which covers the period of sunspot minimum and maximum. Burst data from the Sagamore Hill Solar Radio Observatory for the period 1966-1974 is shown in Table 4 (Castelli, *private communication*). An event is defined as a burst observed on several frequencies simultaneously. From Table 4 it appears that bursts can occur quite often, especially during the solar maximum. However, large bursts are relatively rare, and the variation of frequency of occurrence of bursts with their intensity has been empirically determined to be proportional to I^{-x}, where I is the intensity and the exponent x is in the range 1.5 to 1.8 (Akabane, 1956; Kakinuma, Yamashita, and

TABLE 2

Radio burst data associated with large proton event of 4 August 1972. (After Castelli *et al.*, 1973.)

FREQUENCY MHz	DURATION	FLUX DENSITY		MEAN BURST FLUX QUIET SUN FLUX
		PEAK	MEAN	
8800	26.4	36500	26600	83
4995	26.8	19500	14100	71
2695	26.3	7500	4700	43
1415	28.7	6700	2800	80
606	28.2	~60000	--	--
245	181.0	UNCAL	--	--

TABLE 3

Summary of burst statistics, 1 July 1957 to 30 June 1958. (After Maxwell et al., 1960.)

INTENSITY LEVELS OF INCIDENT RADIATION			
	APPROXIMATE INTENSITY RANGES, Watts m^{-2} $(cps)^{-1}$ x 10^{-22}		
FREQUENCY, Mc/s	INTENSITY 1 (faint)	INTENSITY 2 (moderate)	INTENSITY 3 (strong)
125	5-40	40-200	>200
200	10-60	60-250	>250
425	20-50	50-200	>200
550	20-50	50-200	>200

TOTAL DURATION OF ALL TYPES OF SOLAR BURSTS					
		TOTAL TIMES AT			
TYPE OF ACTIVITY	INTENSITY	125 Mc/s, hr	200 Mc/s, hr	425 Mc/s, hr	550 Mc/s, hr
NOISE STORMS	1	371	225	2.3	1.5
(type I)	2	80	47	0.4	0.3
	3	83	50	0.5	0.3
	TOTAL	534	322	3.2	2.1
SLOW-DRIFT BURSTS	1	0	0	0	0
(type II)	2	0.1	0.1	0	0
	3	1.8	0.7	0.1	0.1
	TOTAL	1.9	0.8	0.1	0.1
FAST-DRIFT BURSTS	1	4.5	2.9	0.4	0.3
(type III)	2	2.4	1.7	0.3	0.2
	3	3.0	1.1	0.3	0.2
	TOTAL	9.9	5.7	1.0	0.7
CONTINUUM	1	5.0	12.0	1.8	3.5
(type IV)	2	3.5	5.3	3.2	5.7
	3	2.8	3.7	11.5	11.3
	TOTAL	11.3	21.0	16.5	20.5
UNCLASSIFIED	1	0.7	0.2	0	0
	2	0.2	0.1	0	0
	3	0.3	0.1	0	0
	TOTAL	1.2	0.4	0	0
ALL ACTIVITY	1	381	240	4.5	5.3
	2	86	54	3.9	6.2
	3	91	56	12.4	11.9
	TOTAL	558	350	20.8	23.4

TOTAL OBSERVING TIME: 4010 hr

PERCENTAGE OF OBSERVING TIME OCCUPIED BY EACH TYPE OF SOLAR BURST				
	PERCENTAGE TIME AT			
TYPE OF ACTIVITY	125 Mc/s	200 Mc/s	425 Mc/s	550 Mc/s
NOISE STORMS (type I)	13.32	8.01	0.08	0.05
SLOW-DRIFT BURSTS (type II)	0.05	0.02	<0.01	<0.01
FAST-DRIFT BURSTS (type III)	0.25	0.14	0.02	0.02
CONTINUUM (type IV)	0.28	0.52	0.41	0.51
UNCLASSIFIED	0.03	0.01	<0.01	<0.01
TOTAL PERCENT	13.93	8.70	0.51	0.58

TABLE 4

Summary of burst events at Sagamore Hill Solar Radio Observatory 1966-1974. (After Castelli, private communication.)

YEAR	EVENTS, SAGAMORE HILL ONLY	ESTIMATED NUMBER OF EVENTS FOR 24 hr OBSERVATION PER DAY	PERCENT OF TOTAL
1966	217	434	5.2
1967	472	944	11.3
1968	655	1310	15.6
1969	827	1654	19.8
1970	632	1264	15.0
1971	346	692	8.2
1972	473	946	11.2
1973	351	702	8.4
1974	224	448	5.3
TOTAL	4197	8394	100.0

Enome, 1969; Kundu, 1959a; 1965). As an illustration, according to data taken during the period 1964-1970, large bursts (flux $> 500 \times 10^{-22}$ W m^{-2} Hz^{-1}) occurred at a rate of .0005 h^{-1} and .0002 h^{-1} at 10 GHz and 2.7 GHz, respectively, during the sunspot minimum, and increased to a rate of .0052 h^{-1} and .004 h^{-1} at sunspot maximum (Wefer, 1973). The results of these statistical studies show that, although burst fluxes can greatly increase the incident solar flux at certain wavelengths at the earth, they do not have a significant effect on the total solar insolation at radio wavelengths when averaged over long periods of time.

IV. CONCLUSIONS

At radio wavelengths the quiet sun emits as a thermal radiator that has an apparent brightness temperature of about 6000K at 1 mm to more than 10^6 K at the meter wavelengths. In addition to the basic quiet-sun flux, there are two variable components associated with active centers of emission on the sun. There is the slowly varying or S-component, which is thermal in origin and is emitted by localized areas of high electron density and magnetic fields. The S-component flux depends on the number and intensity of active regions on the sun and rarely exceeds the quiet flux, and then only at the centimeter wavelengths when there are strong sources on the disk. This usually occurs during the period of sunspot maximum. The S-component changes very gradually, evolving with the active regions on the disk; the flux can be estimated with the use of various activity indices such as the sunspot number or plage areas. There is also a transitory component, called bursts. These, except for the gradual type observed at centimeter wavelengths, are nonthermal in origin. During a burst the flux can greatly exceed the quiet component, but the long-term averaged contribution is small in comparison to the quiet-sun output. Because the sun-to-earth distance changes cyclically during the year, there is a yearly periodic modulation on the solar flux that has a peak-to-peak variation of about 6%. The radio emission of the sun has been studied for more than 20 years, most extensively at 10.7 cm. Since 1964, the quiet 10.7 cm flux has been constant, although there is an indication that it was perhaps a few percent lower in 1954 (Covington, *private communication*). From the 10.7 cm observations it is inferred that quiet solar flux at radio wavelengths has not undergone any significant change since radio measurements were initiated.

The energy output of the sun at the radio wavelengths is insignificant when compared to the output at other regions of the solar spectrum and has little direct physical impact on solar terrestrial relationships. The radio emissions are important because they serve as activity indices measurable continuously from ground-based observatories and as diagnostic tools in identifying the solar sources of the higher energy emissions, such as the UV, X-ray, and particle fluxes important in the earth's upper atmosphere.

REFERENCES

Akabane, K., 1956, *Publ. Astr. Soc. Japan 8*, 173.
Allen, C.W., 1957, *Radio Astronomy, I.A.U. Symposium No. 4*, ed. H.C. Van DeHulst, Cambridge University Press, London, 153.
Appleton, E. V., and Hey, J.S., 1946, *Phil. Mag. 37*, 73.
Athay, R.G., 1966, *Ap. J. 145*, 784.
Aubier, M., Leblanc, Y., and Boischot, A., 1971, *Astr. and Ap. 12*, 435.
Badillo, V.L., and Salcedo, J.E., 1969, *Nature 224*, 503.
Castelli, J.P., Barron, W.R., and Aarons, J., 1973, *AFCRL Report TR-73-0086*.
Christianson, W.N., and Mathewson, D.S., 1959, *Paris Symposium on Radio Astronomy*, ed. R.N. Bracewell, Stanford University Press, Stanford, 108.
Couturier, P., and Leblanc, Y., 1970, *Astr. and Ap. 7*, 254.
Covington, A.E., 1947, *Nature 159*, 405.
_____, 1948, *Proc. IRE 36*, 454.
_____, 1949, *ibid. 37*, 407.
Das Gupta, M.K., and Sarkar, S.K., 1971, *J.R.A.S. Canada 65*, 284.
Denisse, J.F., 1950, *Ann. d'Ap. 13*, 181.
Edelson, S., Mayfield, E.B., and Shimabukuro, F.I., 1971, *Nature Phys. Sci. 232*, 82.
Feix, G., 1969, *Ap. J. 157*, 903.
Fokker, A.D., 1963, *Space Sci. Rev. 2*, 70.
Gingerich, O.; Noyes, R.W.; Kalkofen, W.; and Cuny, Y., 1971, *Solar Phys. 18*, 347.

Ginzburg, V.L., 1946, *C.R. Acad. Sci. U.R.S.S.* 52, 487.
Hey, J.S., 1946, *Nature* 157, 47.
Jaeger, J.C., and Westfold, K.C., 1949, *Australian J. Sci. Res.* A2, 322.
Kakinuma, T., and Swarup, G., 1962, *Ap. J.* 136, 975.
―――――――, Yamashita, T., and Enome, S., 1969, *Proc. Res. Inst. Atmos. Nagoya Univ. Japan* 16, 127.
Korolkov, D.V., Soboleva, N.S., and Gelfreikh, G.B., 1960, *Proc. Pulkovo Obs.* 21, 81.
Kundu, M.R., 1959a, *Ann. d'Ap.* 22, 1.
―――――――, 1959b, *Paris Symposium on Radio Astronomy*, ed. R.N. Bracewell, Stanford University Press, Stanford, 222.
―――――――, 1963, *Space Sci. Rev.* 2, 438.
―――――――, 1965, *Solar Radio Astronomy*, Interscience, New York.
―――――――, and Haddock, F.T., 1960, *Nature* 186, 610.
Leblanc, Y., 1970, *Astr. and Ap.* 4, 315.
―――――――, and LeSqueren, A.M., 1969, *ibid.* 1, 239.
Lehany, F.J., and Yabsley, D.E., 1949, *Australian J. Sci. Res.* A2, 48.
Linsky, J.L., 1973, *Solar Phys.* 28, 409.
Martyn, D.F., 1946, *Nature* 157, 632.
―――――――, 1948, *Proc. Roy. Soc. London* A 193, 44.
Maxwell, A., 1965, *Solar Spectrum*, ed. C. de Jager, D. Reidel, Dordrecht, 342.
―――――――, Howard, W.E., III, and Garmire, G., 1960, *J. Geophys. Res.* 65, 3581.
McCready, L.L., Pawsey, J.L., and Payne-Scott, R., 1947, *Proc. Roy. Soc. London* A 190, 357.
Moutot, M., and Boischot, A., 1961, *Ann. d'Ap.* 24, 181.
Newkirk, G., 1967, *Annu. Rev. Astr. and Ap.* 5, 213.
O'Brien, P.A., 1953, *M.N.R.A.S.* 113, 597.
Pawsey, J.L., and Yabsley, D.E., 1949, *Australian J. Sci. Res.* A2, 198.
―――――――, Payne-Scott, R., and McCready, L.L., 1946, *Nature* 157, 158.
Piddington, J.H., and Minnett, H.C., 1951, *Australian J. Sci. Res.* A4, 131.
Reber, G., 1944, *Ap. J.* 100, 279.
Riddle, A.C., 1974, *Solar Phys.* 36, 375.
Sheridan, K.V., 1970, *Proc. Astr. Soc. Australia* 1, 304.
Shimabukuro, F.I., 1972, *Solar Phys.* 23, 169.
―――――――, and Stacey, J.M., 1968, *Ap. J.* 152, 777.
―――――――; Chapman, G.A.; Mayfield, E.B.; and Edelson, S., 1973, *Solar Phys.* 30, 163.
Smerd, S.F., 1964, *Ann. IGY* 34, 323.
Smith, H.J., and Smith, E.V.P., 1963, *Solar Flares*, Macmillan Co., New York, 34.
Southworth, G.C., 1946, *J. Franklin Inst.* 239, 285.
Swarup, G.; Kakinuma, T.; Covington, A.E.; Harvey, G.A.; Mullaly, R.F.; and Rome, J., 1963, *Ap. J.* 137, 1251.
Takakura, T., 1966, *Space Sci. Rev.* 5, 80.
―――――――, 1967, *Solar Phys.* 1, 304.

Tanaka, H., 1961, *Proc. Res. Inst. Atmos. Nagoya Univ. Japan 8*, 51.
────────── and Kakinuma, T., 1956, *ibid. 4*, 78.
────────── and ──────────, 1958a, *ibid. 5*, 81.
────────── and ──────────, 1958b, *Rept. Ionosphere Res. Japan 12*, 273.
────────── and ──────────, 1960, Nahoya University *Proc. Res. Inst. Atmos. 7*, 79.
Waldmeir, M., and Müller, H., 1950, *Z. Ap. 27*, 58.
Wefer, F.L., 1973, *Penn. St. Univ. Sci. Rept. No. 026*.
Wild, J.P., 1962, *J. Phys. Soc. Japan 17*, A2, 249.
──────────, 1964, *AAS-NAS Symposium on Physics of Solar Flares*, U.S. Government Printing Office, Washington, D.C., 161.
────────── and Smerd, S.F., 1972, *Annu. Rev. Astr. and Ap. 10*, 159.
──────────, ──────────, and Weiss, A.A., 1963, *ibid. 1*, 291.
Zheleznyakov, V.V., 1962, *Soviet Astr. AJ6*, 3.

Comment

The 10.7 cm radio flux has been extensively used as an index of solar activity. Though this index has been the best one available over the last several solar cycles, recent work has demonstrated that it cannot be used to estimate the solar EUV or X-ray flux accurately, to better than 10% at all times. Direct measurements of the solar UV, EUV, and X-ray fluxes are now becoming available from the Atmospheric Explorer and SOLRAD-HI satellites. These direct measurements should be used in preference to flux estimates derived from the 10.7 cm radio flux values, but the 10.7 cm radio measurements should be maintained both for continuity of the established record and for an indicator of solar activity in the low corona. Real-time monitoring of the solar X-ray flux by a stationary satellite is now being conducted on a routine basis by the Space Environment Laboratory in Boulder.

Richard F. Donnelly
Space Environment Laboratory
National Oceanic and Atmospheric Administration

THE SOLAR SPECTRUM
BETWEEN 10 AND 1000 μm

William G. Mankin
National Center for Atmospheric Research*

Solar radiation with wavelengths of 10 to 1000 μm arises primarily in the upper photosphere and lower chromosphere where the solar temperature falls to its minimum value and begins its rise to higher chromospheric values. In this chapter, we review measurements of the radiation in this spectral range, collectively called the "far infrared" (FIR).

From a physical point of view, the far infrared is the simplest part of the solar spectrum. The radiation is almost exclusively continuous; it is formed in local thermodynamic equilibrium by well-understood mechanisms. The source of the opacity is known, as is its value as a function of wavelength. It arises in a region of the solar atmosphere which is particularly homogeneous and where nonthermal processes are comparatively small.

For the observer, on the other hand, this region of the spectrum presents some formidable problems. These arise chiefly from the fact that throughout most of the far infrared the terrestrial atmosphere is totally opaque, owing to the intense absorption of the pure rotation band of water vapor. In addition, the comparative weakness of the emission—the solar spectral intensity at 100 μm is 8×10^7 times less than at 0.5 μm—and the insensitivity of FIR radiation detectors make accurate measurements difficult. Finally, the problems of diffraction, increasingly severe at longer wavelengths, make large telescopes necessary for observation of fine detail. The difficulties are such that the first measurement of the solar FIR radiance was not made until 1958.

Although the FIR radiation is a valuable diagnostic for the solar atmosphere, it has scant influence on the earth. The integrated FIR solar irradiance at the earth amounts to only 0.057% of the total solar irradiance. Some other spectral regions where the energy is small achieve an importance because they provide the main input to certain regions of the atmosphere; not so the far infrared. Its contribution to heating the troposphere and stratosphere is swamped by the thousand-times-more-abundant terrestrial radiation.

---

*The National Center for Atmospheric Research is sponsored by the National Science Foundation.

The intensity of radiation emerging from the top of the solar atmosphere at wavelength λ in the direction specified by the polar angles θ, ϕ may be calculated from the equation of radiative transfer as

$$I(\lambda,\theta,\phi) = \int_{-\infty}^{0} S_\lambda(x) \exp\left[-\int_{x}^{0} k_\lambda \rho dx'\right] k_\lambda \rho dx \quad , \tag{1}$$

where S_λ is the source function, k_λ the mass absorption coefficient, and ρ the density; the integral is along the line of sight. Three assumptions, generally valid for the far infrared region, allow a simplification of this expression.

The source of FIR opacity in the solar atmosphere is almost exclusively free-free transitions of negative hydrogen ions and, to a lesser extent, negative helium ions. In such transitions a free electron accelerated in the field of a neutral hydrogen or helium atom absorbs or emits radiation. The free-free absorption coefficient of negative hydrogen has been computed by numerous authors (Chandrasekhar and Breen, 1946; Geltman, 1956; Ohmura and Ohmura, 1960; John, 1964, 1966) and is found to be proportional to the electron pressure and to the square of the wavelength for wavelengths greater than a few microns. There have been suggestions (Gay, 1970b) of additional opacities, but there appears to be no convincing evidence of any FIR continuous opacity that does not follow the λ^2 law. We shall assume, except where noted below, that the law holds.

The second assumption is that the radiation is emitted in local thermodynamic equilibrium (LTE); this simplifies the results by allowing replacement of the microphysical source function by the Planck function at the local electron temperature. Justification of LTE demands consideration of the atomic processes. Pagel (1957) showed that as long as electron collisions are frequent enough to maintain a Maxwellian velocity distribution, the assumption of LTE is justified for the free-free emission. Mankin (1969) showed that for conditions near the temperature minimum the rate of collisions is at least a thousand times that required to maintain the Maxwellian distribution. Therefore the source function has its LTE value to a high degree of accuracy, although the opacity may differ owing to non-LTE effects on the electron pressure.

The final assumption we make is that the relevant layers of the solar atmosphere are spherically symmetric. This is more difficult to justify and probably is not satisfied in the highest regions we consider. In the upper photosphere and lowest chromosphere, we deal with probably the most homogeneous region of the solar atmosphere. In the photosphere, granules driven by convection exhibit temperature fluctuations on the order of 100 K over distances of a few hundred kilometers. Higher, where the granules become optically thin, radiative coupling should reduce the fluctuations and make the atmosphere more uniform. Above the temperature minimum there must be nonradiative energy input, so we may again expect inhomogeneities to be enhanced. Spectroheliograms in the K_2 line of Ca II, which arises in this region, show granular-type irregularities. Such inhomogeneities are the likely cause of the absence of conspicuous submillimeter limb brightening (Lindsey and Hudson, 1976). Thus, while this assumption breaks down at the highest regions we consider, it should hold at least approximately in the lower regions. With the caveat that limb-darkening measurements at the longest wavelength may be discrepant, we shall use the assumption of spherical symmetry.

Using these assumptions, Equation (1) yields

$$I(\lambda,\mu) = \int_0^\infty B(\lambda,T_e) \exp\left(-\frac{\lambda^2}{\mu}\tau_0\right) \frac{\lambda^2}{\mu} d\tau_0 \quad , \tag{2}$$

where $\tau_0 = \tau_\lambda / \lambda^2$ with $\tau_\lambda = \int_z^\infty k_\lambda \rho\, dz'$, the optical depth. $B(\lambda, T_e)$ is the Planck function at the local electron temperature T_e; μ is the cosine of the angle between the direction of emission and the local vertical to the solar sphere.

Because the Planck function varies so rapidly with wavelength, and in the far infrared is nearly linear with temperature (Rayleigh-Jeans approximation), we find it convenient to use the brightness temperature $\Theta(\lambda, \mu)$ instead of the radiance. The brightness temperature of a surface is the temperature of a blackbody having the same radiance as the surface.

We write $B(\lambda,T)$ as $c_1 T [1 - \phi(c_2/\lambda T)] / (c_2 \lambda^4)$ where c_1 and c_2 are the first and second radiation constants and $\phi(x) = (e^x - 1 - x) / (e^x - 1)$ is a small correction for departures from the Rayleigh-Jeans law. Then Equation 2 yields

$$\begin{aligned}\Theta(\lambda,\mu) = &\frac{\lambda^2}{\mu} \int_0^\infty T_e(\tau_0) \exp\left(-\frac{\lambda^2}{\mu}\tau_0\right) d\tau_0 \\ &+ \frac{\lambda^2}{\mu} \int_0^\infty T_e(\tau_0) \frac{\phi(c_2/\lambda\Theta) - \phi(c_2/\lambda T_e)}{1 - \phi(c_2/\lambda\Theta)} \exp\left(-\frac{\lambda^2}{\mu}\tau_0\right) d\tau_0 \quad .\end{aligned} \tag{3}$$

The second integral is very small; for a reasonable model of the solar atmosphere, it amounts to less than 0.1 K. Therefore we may say that the brightness temperature depends only on the variable λ^2/μ, and from the spectrum of the disk center we may construct the spectrum of the mean disk, or we may use limb-darkening measurements at one wavelength to give the central intensity at a longer wavelength. This fact has been exploited to extend ground-based measurements to wavelengths not accessible from the ground.

There have been no measurements of solar absorption or emission lines in the 10-1000 μm region. There are two possible sources of such lines: pure rotation bands of molecules near the temperature minimum and dielectronic recombination lines in the corona. Calculations indicate that each of these might be barely observable with high-resolution instruments. In any event they can have no significant effect on the total irradiance, and we shall ignore them.

I. MEASUREMENTS OF THE SOLAR CONTINUUM RADIANCE

Measurements of the solar far infrared radiance may be made from the ground in the 8-12 μm atmospheric window. If observations are made from a very dry high-altitude site, measurements may also be taken in the rather poor atmospheric windows between the 15 μm CO_2 bands and the water vapor rotation spectrum, and also at a few windows in the submillimeter region. Otherwise, observations must be made from an aircraft, balloon, rocket, or satellite to get above the water vapor absorption in the troposphere.

Table 1 lists the absolute measurements of the solar intensity reviewed here. The symbol C or M indicates whether the measurement refers to the radiance at the center of the solar disk or the mean radiance of the disk. The two measures are related through the limb darkening, and knowledge of the complete spectrum of one implies the other, given spherical symmetry. The errors indicated are those quoted by the original investigator. Generally, the systematic errors are larger than the random errors, so that estimation of the errors depends upon the experimenter's judgment. Most of the error estimates quoted seem credible, although some are perhaps rather optimistic. I comment below on each of the measurements.

Table 1

Observations of Solar Far Infrared Continuum Intensity

λ (μm)	M or C	Θ (K)	$\Delta\Theta$ (K)	Reference
8.5	C	5220	80	Kondratyev et al. (1965)
8.6	C	5160	40	Saiedy (1960)
9.0	C	5200	70	Kondratyev et al. (1965)
10.0	C	5200	70	Kondratyev et al. (1965)
10.5	C	5210	70	Kondratyev et al. (1965)
11.1	C	5200	70	Kondratyev et al. (1965)
11.1	C	5036	30	Saiedy and Goody (1959)
11.5	C	5160	70	Kondratyev et al. (1965)
12.0	C	5110	75	Kondratyev et al. (1965)
12.0	C	5050	80	Saiedy (1960)
12.5	C	5100	75	Kondratyev et al. (1965)
13.0	C	4740	90	Kondratyev et al. (1965)
13.5	M	4850	100	Baluteau (1971)
17.0	M	4815	120	Baluteau (1971)
19.8	M	4600	200	Baluteau (1971)
20.0	M	4820	370	Beer (1966)
20.1	C	4586	100	Koutchmy, Peyturaux (1970)

λ (μm)	M or C	Θ (K)	ΔΘ (K)	Reference
23.5	M	4350	275	Baluteau (1971)
24.4	C	4460	360	Koutchmy, Peyturaux (1970)
40.0	M	4570	300	Beer (1966)
55.0	M	5460	590	Gay et al. (1968)
60.0	C	4400	300	Clark et al. (1971)
70.0	M	5580	360	Gay et al. (1968)
100.	M	4600	290	Gay et al. (1968)
119.	M	4350	500	Stettler et al. (1972)
150.	M	4700	250	Stettler et al. (1972)
150.	C	4400	300	Clark et al. (1971)
160.	C	4200	350	Eddy et al. (1973)
200.	M	4090	200	Stettler et al. (1975)
200.	M	4670	590	Gay et al. (1966)
222.	M	4950	250	Stettler et al. (1972)
240.	C	4510	200	Eddy et al. (1973)
240.	M	4260	100	Stettler et al. (1975)
275.	C	4370	260	Eddy et al. (1969)
290.	M	4575	150	Stettler et al. (1975)
300.	C	4585	200	Eddy et al. (1973)
300.	M	4550	250	Stettler et al. (1972)
350.	C	4730	150	Eddy et al. (1973)
350.	M	4570	100	Stettler et al. (1975)
350.	M	4540	360	Gezari et al. (1973)
400.	C	4960	150	Eddy et al. (1973)
400.	M	4700	500	Stettler et al. (1972)
450.	C	5210	250	Eddy et al. (1973)
450.	M	4970	220	Stettler et al. (1975)
450.	M	4800	380	Gezari et al. (1973)
500.	C	5420	450	Eddy et al. (1973)
600.	C	5850	500	Eddy et al. (1973)
600.	M	5170	500	Stettler et al. (1975)
700.	C	6250	800	Eddy et al. (1973)
740.	M	5200	1000	Fedoseev et al. (1967)
870.	M	5350	800	Fedoseev et al. (1967)
1000.	C	5900	500	Low and Davidson (1965)
1000.	M	5900	410	Gezari et al. (1973)
1060.	M	5750	600	Fedoseev et al. (1967)

A. GROUND-BASED MEASUREMENTS BELOW 25 μm

The earliest, and still the most accurate, of the measurements are those by Saiedy and Goody (1959) at 11.1 μm and by Saiedy (1960) at 8.63 and 12.02 μm. Saiedy also analyzed drift curves to give limb-darkening parameters. At these wavelengths, the terrestrial atmosphere is comparatively transparent and the correction for atmospheric extinction is small.

By using a grating spectrometer to isolate a narrow spectral region, Saiedy and Goody were able to apply the Lambert-Beer law to the extinction and extrapolate reliably to zero airmass by Bouguer's method.* They attenuated the solar signal with accurately calibrated rotating sectors to bring it to approximate equality with the signal from a carefully constructed blackbody, reducing the requirements on linearity of the detector. Their maximum atmospheric transmission varied from 0.78 to 0.93 on different days. The maximum spread of their intensity measurements was only 4%. By very careful calibration and attention to sources of systematic error, they reduced the error in their quoted result to a credible 0.6%. This is an impressive achievement, and marks their result of a solar central brightness temperature of 5036 ± 30 K at 11.1 μm as the best reference point in the FIR spectrum.

Using the same techniques, Saiedy obtained values of 5160 ± 40 K at 8.63 μm and 5050 ± 80 K at 12.2 μm. His measurements of the limb darkening will be discussed in the next section.

A group from Leningrad State University (Kondratyev et al., 1965) used a technique similar to that of Saiedy and Goody to obtain the 3-13 μm solar radiance from sites above 3000 m on Mt. Elbrus. They used a prism spectrometer which gave lower spectral resolution than used by Saiedy and Goody; if the bandpass includes spectral lines, this can introduce error in the Bouguer extrapolation. Their results, which show the brightness temperature declining from 5300 K at 8 μm to 5100 K at 12.5 μm, average about 75 K higher than those of Saiedy and Goody; this is comparable to the estimated errors in each.

Koutchmy and Peyturaux (1968, 1970) have extended ground-based measurements, similar to those by Saiedy and Goody, to wavelengths as long as 24.4 μm. In the 17-24 μm region the absorption by water vapor becomes important and accurate measurements become much more difficult because of the large atmospheric corrections involved. Koutchmy and Peyturaux observed from a site at a height of 1600 m in the Pyrenees to obtain low absorption. In the case of absorption by water vapor, the variability of the humidity during the observations complicates the Bouguer method of extrapolation to zero extinction. Koutchmy and Peyturaux developed a method of correlating the observed signal

---

* The law of exponential extinction of radiation is known variously by the names of Lambert, Beer, and Bouguer. We shall use "Lambert-Beer law" for the extinction formula and "Bouguer's method" for the application of it to extrapolation of observed intensity of radiation to its value outside the atmosphere.

from the sun with measurements of the total water quantity to solve simultaneously for the solar radiance and the atmospheric extinction. Even so, the extinction correction is large and introduces considerable uncertainty in the results at the longest wavelengths.

In the 8-13 µm region, their results are almost identical with those of Saiedy and Goody. At 20.15 µm they report a central brightness temperature of 4580 ± 100 K and at 24.4 µm, 4460 ± 360 K. These numbers seem somewhat low when compared with the measurements of limb darkening at 11 µm, although the correct value probably lies within the quoted error limits.

B. MEASUREMENTS FROM BALLOONS AND AIRCRAFT

The first balloon-borne observations were by Beer (1966), covering the region from 10 to 70 µm. His results, which presumably relate to the full sun rather than the disk center, were not absolutely calibrated. He deduces a $\lambda^{-4.075 \pm .012}$ dependence for the radiance from 20 to 40 µm. Using the Saiedy and Goody result at 11.1 µm to calibrate his spectrum, he obtains brightness temperatures of 4820 ± 370 K at 20 µm and 4570 ± 300 K at 40 µm. Because of the calibration difficulties, his results are less reliable than others, but they do overlap the measurements of Koutchmy and Peyturaux at 20-25 µm and are valuable in ruling out large deviations from a brightness temperature of 4500 K at 40 µm.

A group from Meudon Observatory has observed the spectrum from 50-200 µm on a series of balloon flights, the best of which was on 16 January 1968 (Gay et al., 1968; Gay, 1970a). They employed two calibration methods to obtain an absolute intensity scale. The first, conducted on the ground before flight, consisted of observing a blackbody cooled by liquid nitrogen; the second involved observing the sky away from the sun during the flight. Both methods rely on the interior of the apparatus approximating a blackbody at a known temperature. The authors did not consider this a possible source of error. There were also some difficulties with the effective field of view when viewing the sky. The method of calibrating on a cold blackbody gives a small signal of the sign opposite that from the sun, and is not as reliable as using a high temperature blackbody. The two calibration methods gave answers differing by 200-700 K. The quoted estimates of systematic error are 300-500 K. Their results for the brightness temperature, between 5000 and 5500 K in the 50-100 µm region, are 500-1000 K above the values obtained by Koutchmy and Peyturaux at shorter wavelengths. In a separate paper, Gay (1970b) argues that this implies an additional source of opacity in the 30-100 µm region. The hypothesized absorber would have to absorb approximately 1000 times more strongly at 60 µm than the free-free absorption. There is no convincing candidate for this absorber. Furthermore, observations by Clark et al. (1971), although not very well calibrated themselves, do not support the high radiances in the 50-100 µm region.

Baluteau (1971) also made measurements of the mean disk intensity with a balloon-borne telescope and spectrometer. His measurements, made with a grating spectrometer, covered the 12 - 24 µm region. The apparatus included a blackbody for absolute calibration, but an alignment problem during inflight calibration limited the calibration to a relative one. The values were converted to an absolute scale by normalizing to the results of Saiedy (1960) and Kondratyev *et al.* (1965) near 12 µm. At the longer wavelengths, systematic errors of up to 3-5% are possible due to uncertainties in emissivities and reflectivities. The results, showing mean brightness temperatures declining from 4850 ± 100 K at 13.5 µm to 4350 ± 275 K at 24 µm, agree well with the results by Koutchmy and Peyturaux (1970), although Baluteau's have somewhat larger uncertainties.

While there is a need for additional measurements in the 50-100 µm region, the most difficult FIR region to observe, the evidence of Gay *et al.* does not seem strong enough to assume a brightness temperature above 4500 K in this region.

The observation by Clark *et al.* (1971) was basically very similar to the one by Gay *et al.* (1968). Their report is very sparse on details and gives neither estimates of their errors nor sufficient information to allow the reader to make his own estimates. Their results average about 4400 K between 60 and 160 µm and show variations with wavelength of about ± 300 K. Other than to say that they do not support the high temperatures reported by Gay *et al.*, we cannot draw very firm conclusions.

In two flights using a lamellar grating interferometer, a Swiss group (Stettler *et al.*, 1972; Stettler *et al.*, 1975) measured the solar brightness temperature, averaged over the central 29 arcmin, from 100 to 600 µm. The results of the second flight are more accurate because of a number of improvements. They used a blackbody as a radiometric reference (constant but not calibrated) and performed an absolute calibration on the ground before the flight. A carefully constructed high-temperature blackbody was used for the primary calibration. They paid careful attention to the sources of radiometric errors. No correction was made for absorption by water vapor lines in the calibration; thus their answers are too large at frequencies of strong absorption, but should be rather reliable in the windows. Their first flight gave temperatures of 4550 to 4450 ± 250 K from 150 to 300 µm, whereas the results from the second flight ranged from 4100 ± 200 K at 200 µm to 5170 ± 500 K at 600 µm. These results agree well with the earlier aircraft measurements by Eddy, Léna, and MacQueen (1969).

Eddy *et al.* used a Michelson interferometer on board the NASA Convair 990 aircraft to obtain spectra of the center of the disk from 238-312 µm. An aircraft observation has two significant disadvantages compared to a balloon observation. Because of the lower altitude, the atmospheric extinction correction is larger. Provided that the aircraft is above the tropopause, however, the transparency between strong water lines is high, at least for wavelengths longer than 200 µm. By using high-resolution spectroscopy, the water quantity may be determined and correction made for the residual absorption. The error in the solar radiance from atmospheric extinction is probably less than 1%. The second disadvantage is the necessity of a window. Eddy et al. reported an average brightness temperature of 4370 ± 260 K over the range 238-312 μm, with no spectral variation greater than the errors. They recognized the necessity of using two reference blackbodies for absolute calibration and realistically assessed the sources of error.

Their work was extended in wavelength range at higher resolution in a series of flights during October 1971 on the NCAR Sabreliner (Eddy et al., 1973). These results showed brightness temperatures near 4200 ± 350 K at 160 μm, rising to 4600 ± 200 K at 300 μm, 4950 ± 150 K at 400 μm, and continuing to rise to 6000 ± 600 K at 650 μm. They paid careful attention to radiometric problems, using inflight calibration with two blackbodies and making the signal and calibration optical paths as nearly identical as possible, even putting a cooled window in the calibration beam to match the aircraft window. The emissivity of the blackbody had been determined to be greater than 0.99 at 350 μm (Eddy et al., 1970), but at the longer wavelengths there remained some question about its value. If the emissivity were actually lower than 0.99, the calculated solar radiance would be greater than its true value; thus their values for the brightness temperature at wavelengths greater than 400 μm should be considered upper limits; at shorter wavelengths they should be very accurate.

C. GROUND-BASED OBSERVATIONS BEYOND 300 μm

At the longest FIR wavelengths, there are window regions of sufficiently small absorption that observations may be made from dry mountaintop observatories. Gezari, Joyce, and Simon (1973) exploited the windows at 345, 450, and 1000 μm to measure the solar radiance with a filter radiometer on a 61 cm telescope on Mauna Kea. They measured brightness temperatures of 4540 ± 360 K, 4800 ± 380 K, and 5900 ± 410 K in their rather broad spectral bands. Their results fit well with other FIR measurements despite two weaknesses in their method. They avoided the blackbody question by measuring the ratio of solar to lunar radiance and multiplying the ratio by the calculated lunar brightness temperature (Linsky, 1972). This procedure depends upon photometric measurements at other wavelengths and a model of the lunar thermal properties to calibrate the solar measurement. Because the moon is comparatively cool, any error in its temperature is magnified more than 10 times in the solar temperature. Because the full moon and the sun must be observed at widely separated times, atmospheric conditions may change drastically. When the spectral bandwidth is so large that the absorption coefficient varies greatly within the band, extrapolation from the observed signal to zero airmass by Bouguer's method is unreliable. In view of these difficulties it is perhaps surprising that the measurements of Gezari et al. agree as well as they do with other observations. They do provide useful points for mapping the rising chromospheric temperature, but further and more detailed measurements are needed in this wavelength region.

Low and Davidson (1965) measured the solar brightness temperature at 1000 µm as 5900 ± 800 K using a filter radiometer. Their measurement was made almost incidentally to measuring the lunar radiance, and in effect uses the moon as a calibration standard. It therefore suffers from the same shortcomings as the work of Gezari et al. mentioned above. In addition, very little information is given from which one can evaluate the systematic errors.

At the longest FIR wavelengths, radio astronomy techniques may be used. Fedoseev et al. (1967) used radio telescopes from a high-altitude site in the Soviet Union to measure the solar brightness temperature at 740, 870, and 1060 µm. The atmospheric transmission was on the order of 50%. They used the emission from the atmosphere in both a vertical and a horizontal path to calibrate the radiometer and to determine the extinction. This procedure introduces fairly large errors; the reported errors are largely systematic. They report the brightness temperature rising from 5200 ± 1000 K at 740 µm to 5750 ± 600 K at 1060 µm.

II. CENTER-TO-LIMB MEASUREMENTS

Several measurements have been made of the solar FIR limb darkening, particularly at the shorter wavelengths. These do not give the solar spectrum directly, but they do relate the mean and central radiances and predict the ratios of intensities at different wavelengths. They are particularly useful for filling in the regions where direct spectral measurements have not been made (e.g., in the 15 µm CO_2 bands) and for comparing measurements made at different wavelengths. For example, the radiance measurements at 11 µm (Saiedy and Goody, 1959; Koutchmy and Peyturaux, 1970) and the limb-darkening measurements (Saiedy, 1960; Léna, 1969; Mankin and Strong, 1969) indicate a 20.15 µm brightness temperature near 4800 K, while Koutchmy and Peyturaux (1970) measure 4590 ± 100 K.

The validity of the limb-darkening interpretation depends critically upon two assumptions: (1) that the opacity is due solely to free-free transitions of H^- and He^-, or at least that the opacity varies as λ^2; (2) that the solar atmosphere is spherically symmetric at the level where $\tau_\lambda \sim \mu^{-1}$. The first assumption has been questioned by Gay (1970b), but is generally accepted (see Léna, 1970, for further discussion). It implies an effective wavelength $\lambda_{eff} = \lambda \, \mu^{-1/2}$ where $\mu = (1-(r/r_\odot)^2)^{1/2}$. The second assumption holds accurately enough in the photosphere for proper interpretation of visible limb-darkening curves; it is almost surely valid up to the height of the temperature minimum. However, at the longest FIR wavelengths, its validity appears to be questionable. Models of the temperature distribution based on disk center spectra predict fairly large limb brightening for $\lambda > 300$ µm. This has not been observed, although there is evidence, discussed below, for some submillimeter limb brightening. The absence of conspicuous limb brightening more likely invalidates the spherical symmetry assumption than casts doubts on our understanding of the vertical temperature profile.

Limb measurements, although very simple in principle, are difficult in practice because of the necessity of measuring intensity with high precision very close to the limb, since the pertinent parameter, $\mu = (1-(r/r_\odot)^2)^{1/2}$, varies rapidly near the limb. The proximity of the limb makes observation of very small limb darkening extremely difficult; even small side lobes of the instrument angular response function must be well known. The difficulties are compounded by increased diffraction as the wavelength is increased. For example, to observe at $\mu = 0.11$, corresponding to an effective increase of a factor of 3 in wavelength, one must observe within 6 arcsec of the limb. At a wavelength of 100 μm this corresponds to the diffraction limit of the Hale 200-inch telescope! At wavelengths greater than 25 μm, observations require instruments on balloons or aircraft. One experiment (Noyes et al., 1968) used occultation by the moon at eclipse to provide high spatial resolution. Limb-darkening results are summarized in Table 2.

Table 2

Infrared Limb Darkening Measurements

Wavelength (μm)	$I(\mu)/I(\mu=1)$				Reference
	$\mu = 0.6$	$\mu = 0.4$	$\mu = 0.3$	$\mu = 0.2$	
8.6	0.977 ± 0.002	0.955 ± 0.003	–	–	Saiedy (1960)
11.1	0.986 ± 0.0015	0.968 ± 0.003	–	–	"
12.0	0.984 ± 0.0016	0.966 ± 0.002	–	–	"
12.9	0.986 ± 0.002	0.969 ± 0.0025	–	–	"
10.4	0.986 ± 0.002	0.961 ± 0.003	0.940 ± 0.004	0.921 ± 0.004	Lena (1968,1970)
17.9	0.977 ± 0.004	0.962 ± 0.005	0.958 ± 0.005	0.952 ± 0.006	"
20.4	–	0.985 ± 0.006	0.971 ± 0.007	–	"
24.2	–	1.000*	0.995*	0.983*	"
11.1	0.982 ± 0.005	0.982 ± 0.008	0.968 ± 0.011	–	Mankin and Strong (1969)
18	0.985 ± 0.005	0.981 ± 0.010	0.978 ± 0.017	–	
31	0.980 ± 0.005	0.985 ± 0.014	1.000 ± 0.016	–	"
52	1.000 ± 0.005	0.986 ± 0.012	0.999 ± 0.05	–	"
83	0.93 ± 0.05	–	–	–	"
115	0.97 ± 0.04	–	–	–	"

*May be systematically low by up to 0.02

Saiedy (1960) reported limb-darkening observations in the range from $\mu = 1.0$ to $\mu = 0.4$ for wavelengths of 8.3-13 µm, with uncertainties in the intensity ratios of 0.003. His curves were mutually consistent, showing less limb darkening at the longer wavelengths. His measurements at 11.1 µm and 12.95 µm, together with a 5036 K brightness temperature at 11.1 µm, imply a brightness temperature of 4850 K at 20.15 µm. His measurements, being drift scans, referred to equatorial regions.

Léna (1968, 1970) used the 150 cm McMath telescope at Kitt Peak to measure limb darkening at 10.4, 17.9, 20.4, and 24.2 µm. His results were extended to within 2 arcsec of the limb by deconvolving the effects of diffraction and seeing. At 10.4 µm his results agree well with Saiedy's over their common range, but Léna's go to smaller values of μ. They indicate a brightness temperature of 4780 K at 20.15 µm. At 10.4 µm the limb darkening amounts to 7% at $\mu = 0.2$. The values at 20.4 µm and 24.2 µm, although less precise, are about 4% and 2% respectively, this last figure implying decreasing solar central brightness temperature out to 55 µm.

Noyes, Beckers, and Low (1968) measured limb darkening at 24 µm by chopping the beam between two positions separated by 18 arcsec, and scanning the two positions across the sun in a north-south direction. A slope in the resulting curve indicates darkening or brightening. They obtained a darkening $(1/I)\, dI/d\mu = +0.16 \pm 0.2$ at $\mu = 0.28$ (there is a sign error in their paper), about twice the value obtained by Léna, and indications that darkening persists to $\mu = 0.14$. At the 1966 eclipse in Peru they attempted to extend the measurements (at 22.5 µm) even closer to the limb and observed no brightening even in the last 10 arcsec, indicating that the minimum brightness temperature occurs at wavelengths greater than 60 µm.

To make limb-darkening measurements at still longer wavelengths, Mankin and Strong (1969) flew a 40 cm telescope and a filter radiometer on a balloon. One filter was an interference filter at 11.1 µm for comparison with Saiedy (1960); the others were broadband filters with effective wavelengths of 18, 31, 52, 85, and 115 µm. The first four filters comprised one channel of the radiometer; the last two, a second channel. The results at 11 µm agreed very well with those of Saiedy (1960) and of Léna (1970). The 18 µm filter showed less darkening than Lena's results, but the filter had large bandwidth with significant response out to 25 µm where the limb darkening is much less. When the effect of this is considered, the results of Mankin and Léna are consistent. The 31 µm filter showed about 2% limb darkening at $\mu = 0.4$, while the 52 µm curve was almost flat. The second channel showed some limb darkening at 85 µm but little at 115 µm; these results, however, are much less reliable than the results for the first channel and do not permit us to draw any firm conclusions.

The 10-50 µm limb-darkening curves taken together indicate that the solar brightness temperature continues to decrease, although at a decreasing rate, to at least 60 µm and is almost flat at 80 µm, and that the brightness temperature declines from 5050 K at 11 μm to about 4800 K at 20 μm and 4500 K at 80 μm.

At wavelengths of 350-1000 μm it again is possible to observe from the ground. The sky transparency is not as severe a problem in attempting to measure relative intensity as in attempting absolute measurements, although transparency fluctuations make photometry difficult. Diffraction makes it impossible to measure with precision close to the limb, even with large telescopes, but the rapid rise of central brightness temperature beyond the minimum indicates that limb brightening should occur for fairly large values of μ . There have been several attempts to observe this limb brightening. All the observations indicate that the brightening, if it occurs at all, is much less than predicted.

Noyes et al. (1968) made drift scans at an effective wavelength of 1.2 mm, using a 150 cm telescope. The resolution was not sufficient to resolve a brightening at the edge of the disk directly, but they did infer a small brightening from comparison of solar and lunar drift scans. They calculated 1.14 ± 0.03 for the ratio of mean to central brightness temperatures, but were not able to evaluate the effects of systematic errors.

Lindsey and Hudson (1976) have used a 350-1000 μm two-beam filter radiometer to measure the gradient of intensity across the sun in the submillimeter region. The technique is similar to that used by Noyes et al. at 24 μm, although at the longer wavelengths the resolution becomes poorer (2 arcmin full width at half maximum) and the required beam separation is greater. Thus it is not possible to observe as close to the limb. Evidence obtained over the central 80% of the disk showed 1.3% brightening at μ = 0.6, only about one third the predicted value. This brightening was considerably less than the value reported by Noyes et al. at 1.2 mm.

Beckman and Clark (1972), using a 105 cm telescope at Pic du Midi, have made maps of the sun at 400, 800, and 1200 μm. The resolving power of the telescope was 1.6 arcmin at 400 μm and correspondingly greater at longer wavelengths. Their radial scans gave no clear indication of limb darkening or brightening. It is not possible to draw conclusions other than that the limb brightening, if present, is smaller than that predicted from the spectrum. Kundu (1971), on the other hand, has observed limb darkening at 1.2 mm using the 10 m telescope at Kitt Peak. Although the 40 arcsec spatial resolution of this telescope is better than others used for submillimeter observations, the side lobes are important (Ade et al., 1974) and incomplete compensation for them may cause spurious apparent limb darkening. See also Kundu and Liu (1975) on this point.

The absence of conspicuous limb brightening at submillimeter wavelengths, in view of the well-established rise in brightness temperature with wavelength, indicates that the assumption of spherical symmetry is not valid in regions well above the temperature minimum. Since fairly large amounts of nonradiative energy must be deposited here to maintain the temperature above its radiative equilibrium value, it is perhaps not surprising that the simple plane-parallel atmosphere model is inadequate. Simon and Zirin (1969), Beckman et al. (1973), and Lindsey and Hudson (1976) discuss models of the chromosphere in which limb brightening is suppressed by the small-scale structure. Above the temperature minimum, the central brightness temperature provides a lower limit for the mean brightness temperature, while the value calculated on a spherically symmetric model provides an upper limit. At the longer wavelengths it appears that the true mean brightness temperature approaches the lower limit rather than the upper limit.

III. STRUCTURE AND VARIABILITY

The astrophysics group at Queen Mary College has made ground-based observations of the enhanced submillimeter emission from active regions (Ade et al., 1971; Beckman and Clark, 1972). They have produced maps of the sun at wavelengths of 400, 800, and 1200 μm with resolution of a few arcmin. The regions of enhanced emission are well correlated with calcium II plage areas, which is not surprising since K_{12} arises in the same height region. The enhancement in brightness temperature is greater at longer wavelengths, amounting to 200 K at 400 μm, 300 K at 800 μm, and 350 K at 1200 μm. The infrared brightenings of plage regions extend to shorter wavelengths as well (Hudson, 1975); at the shorter wavelengths where radiation arises in the photosphere, sunspots become evident by reduced radiance (Saiedy, 1960; Turon and Lena, 1970).

Hudson and Lindsey (1974) have directly observed time variation in the 20 μm radiance with an effective aperture area of 33 arcsec. The fluctuations over periods of around 300 s, corresponding to the well known "300-s oscillation" in the velocity field, amounted to 3 K rms. They have also observed oscillation at other wavelengths.

No flare-associated variations in FIR emission have been detected (Hudson, 1975). Dall 'Oglio et al. (1974) have reported linear polarization of whole sun radiation amounting to 6-10% in the 300-600 μm region. They speculated that this polarization might result from synchrotron radiation from relativistic electrons produced by flares.

IV. SUMMARY AND CONCLUSIONS

Considering all the various measurements of the solar spectrum and limb darkening discussed above, we have taken the values shown in Table 3 for the brightness temperature at the center of the disk, along with the estimated uncertainties given, as representing the likeliest FIR spectrum of the sun. The values agree generally with the various measurements within the stated errors, with two exceptions. We have raised the value at 20.15 μm by about 100 K above the

Table 3

Wavelength (μm)	Brightness Temperature			Spectral Irradiance	
	Central	Mean	$\pm \Delta\theta$	(watts m^{-1} μm^{-1})	$\pm \Delta S_\lambda$
10	5100	4949	93	2.40(−1)	4.7(−3)
15	4841	4740	155	4.76(−3)	1.6(−3)
20	4709	4648	186	1.51(−2)	6.1(−4)
30	4610	4570	223	3.01(−3)	1.5(−4)
50	4530	4498	255	3.92(−4)	2.2(−5)
75	4478	4450	249	7.75(−5)	4.4(−6)
100	4440	4423	231	2.45(−5)	1.3(−6)
150	4394	4442	204	4.89(−6)	2.2(−7)
200	4417	4541	202	1.58(−6)	6.8(−8)
300	4600	4820	341	3.34(−7)	2.2(−8)
500	5100	5301	470	4.76(−8)	4.1(−9)
750	5612	5700	491	1.01(−8)	8.6(−10)
1000	6000	6000	321	3.37(−9)	1.8(−10)

result of Koutchmy and Peyturaux (1970), based largely on the limb-darkening measurements in the 8-13 μm region. Secondly, the large brightness temperatures from 50 to 100 μm reported by Gay et al. (1968) cannot be supported without the assumption of a large additional source of opacity there; rather than hypothesize such an opacity source, we prefer to let the adopted spectrum fall below the error limits given by Gay et al. The adopted spectrum agrees well with the predictions of empirical models. See the paper by Avrett in this volume for a discussion of models and the influence of infrared measurements on their construction.

Figure 1 shows the curve of adopted central brightness temperature along with the reported measurements. From this spectrum we have calculated the mean solar brightness temperature under the assumption of spherical symmetry. For wavelengths less than 250 μm we have adopted this value in calculating the spectral irradiance. At longer wavelengths, we have allowed for the increasing "roughness" of the chromosphere by taking $T_m(\lambda) = \alpha_\lambda T_c(\lambda) + (1-\alpha_\lambda) T_s(\lambda)$, where T_m is the adopted mean solar brightness temperature, T_c is the central brightness temperature, and T_s is the mean brightness temperature in a spherical model. The parameter α_λ varies linearly with wavelength from zero at 250 μm to unity at 1000 μm.

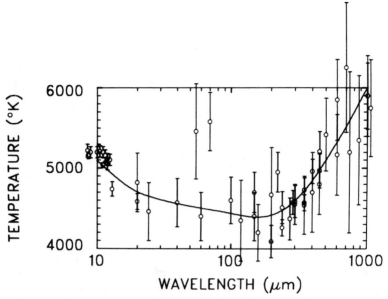

Fig. 1 The adopted value for the brightness temperature of the center of the solar disk, with various measurements of the brightness temperature. For references to the observed points, see Table I.

From the mean brightness temperature, we calculate the spectral irradiance at 1 AU from Planck's law, assuming $\Omega_\odot = 6.8 \times 10^{-5}$ sr. The adopted value for the spectral irradiance is shown in Figure 2, and values are given in Table 3 along with estimated uncertainties. The value for the integrated spectral irradiance from 10 μm to 1000 μm is 0.802 ± 0.026 W/m^2, amounting to about 0.057% of the total solar irradiance.

A convenient rule of thumb for the FIR spectral irradiance at 1 AU is $s_\lambda = 10^4/(4\lambda^4)$, where s_λ is the spectral irradiance in W m^{-2}-μm^{-1} and λ is the wavelength in μm. This results in errors scarcely larger than the uncertainties of the measurement from 10 μm to 500 μm.

It seems likely that there is some variability in the FIR irradiance with the solar cycle owing to enhanced submillimeter emission from active regions, but this will not be appreciable. Even if we assume that in an active region the brightness temperature increases by an amount numerically equal to the wavelength (e.g., $\Delta T_\lambda = 450$ K at $\lambda = 450$ μm) and that these active regions cover 10% of the sun, then the increase in integrated irradiance is only 2.7×10^{-4} W/m^2, an increase of approximately 0.03%. At the extreme limit, the 1000 μm irradiance would only be increased by 1.6%.

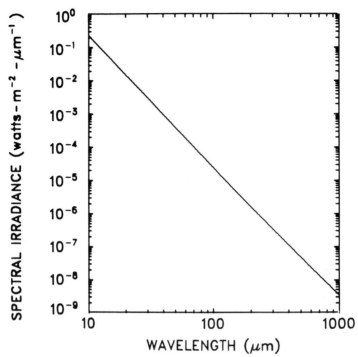

Fig. 2 Estimated far infrared solar spectral irradiance at 1 AU based on observation.

ACKNOWLEDGMENTS

I wish to express my appreciation to Dr. Hugh Hudson for his valuable suggestions on various stages of the manuscript, to the members of the conference working group on solar irradiance for their comments, and to the editors and editorial staff for their cheerful and unobtrusive help at all times.

REFERENCES

Ade, P.A.R., Beckman, J.E., and Clark, C.D., 1971, *Nature Phys. Sci. 231*, 55.
_____, Rather, J.D.G., and Clegg, P.E., 1974, *Ap. J. 187*, 389.
Baluteau, J.P., 1971, *Astr. and Ap. 14*, 428.
Beckman, J.E., and Clark, C.D., 1972, *Solar Phys. 29*, 25.
_____, _____, and Ross, J., 1973, *ibid. 31*, 319.
Beer, R., 1966, *Nature 209*, 1226.
Chandrasekhar, S., and Breen, F., 1946, *Ap. J. 194*, 430.

Clark, T.A., Courts, G.R., and Jennings, R.E., 1971, *Phil. Trans. Roy. Soc. London A 270*, 55.
Dall'Oglio, G.; Fonti, S.; Guidi, I.; Melchiorri, B.; Melchiorri, F.; Natale, V.; Lombardini, P.; and Trivero, P., 1974, *Infrared Phys. 14*, 323.
Eddy, J.A., Léna, P.J., and MacQueen, R.M., 1969, *Solar Phys. 10*, 330.
―――――――; Lee, R.H.; Lena, P.J.; and MacQueen, R.M., 1970, *Appl. Optics 9*, 439.
―――――――; ―――――――; MacQueen, R.M.; and Mankin, W.G., 1973, *Bull. AAS 5*, 271.
Fedoseev, L.I., Lubyako, L.V., and Kukin, L.M., 1967, *Astr. Zh. 44*, 1191 (English translation in *Soviet Astr. 11*, 953).
Gay, J., 1970a, *Astr. and Ap. 6*, 327.
―――――――, 1970b, *ibid. 7*, 24.
―――――――; Lequeux, J.; Verdet, J.P.; Turon-Lacarrieu, P.; Bardet, M.; Roucher, J.; and Zeau, Y., 1968, *Ap. Letters 2*, 169.
Geltman, S., 1956, *Ap. J. 141*, 376.
Gezari, D.Y., Joyce, R.R., and Simon, M., 1973, *Astr. and Ap. 26*, 409.
Hudson, H.S., 1975, *Solar Phys. 45*, 69.
――――――― and Lindsey, C.A., 1974, *Ap. J. (Letters) 187*, L35.
John, T.L., 1964, *M.N.R.A.S. 128*, 93.
―――――――, 1966, *ibid. 131*, 315.
Kondratyev, K.Y.; Andreev, S.D.; Badinov, I.Y.; Grishechkin, V.S.; and Popova, L.V., 1965, *Appl. Optics 4*, 1069.
Koutchmy, S., and Peyturaux, R., 1968, *C.R. Acad. Sci. Paris 267B*, 905.
――――――― and ―――――――, 1970, *Astr. and Ap. 5*, 470.
Kundu, M.R., 1971, *Solar Phys. 21*, 130.
――――――― and Liu, S.-Y., 1975, *ibid. 44*, 361.
Léna, P.J., 1968, *ibid. 3*, 28.
―――――――, 1970, *Astr. and Ap. 4*, 202.
Lindsey, C.A., and Hudson, H.S., 1976, *Ap. J. 204*, 753.
Linsky, J.L., 1972, *Ap. J. Suppl. 25*, No. 216.
Low, F.J., and Davidson, A.W., 1965, *Ap. J. 142*, 1278.
Mankin, W.G., 1969, Ph.D. thesis, Johns Hopkins University.
――――――― and Strong, J., 1969, *Bull. AAS 1*, 200.
Noyes, R.W., Beckers, J.M., and Low, F.J., 1968, *Solar Phys. 3*, 36.
Ohmura, T., and Ohmura, H., 1960, *Ap. J. 131*, 8.
Pagel, B.E.J., 1957, *ibid. 125*, 298.
Saiedy, F., 1960, *M.N.R.A.S. 121*, 482.
――――――― and Goody, R.M., 1959, *ibid. 119*, 213.
Simon, M., and Zirin, H., 1969, *Solar Phys. 9*, 317.
Stettler, P., Kneubühl, F.K., and Müller, E.A., 1972, *Astr. and Ap. 20*, 309.
―――――――; Rast, J.; Kneubühl, F.K.; and Muller, E.A., 1975, *Solar Phys. 40*, 337.
Turon, P., and Léna, P.J., 1970, *ibid. 14*, 112.

THE SOLAR SPECTRUM
BETWEEN .3 and 10 μm

A. Keith Pierce
Richard G. Allen
Kitt Peak National Observatory*
Tucson, Arizona

One of the most difficult problems of observational solar astronomy has been the determination of the solar constant and the sun's spectral energy distribution. The difficulty is directly traceable to the physical problem of measuring radiant energy on an absolute basis and relating it to the thermodynamic temperature scale.

For many years the field of radiometry has been given little attention by the scientific community. Efforts have generally been channeled into more exciting projects, and the highly accurate seven-decimal figures of modern science appear elsewhere. In contrast, when one compares reasonably well-calibrated measures of the solar radiation obtained by different experimenters, one still may find differences as large as 20%. However, a few new techniques have recently appeared in the field of radiometry, and the situation will eventually improve.

The solar irradiance at a particular wavelength is defined as the radiant power from the sun, per unit area and per unit wavelength interval, at the sun's mean distance. As shown by Figure 1, the solar irradiance is approximately equivalent to that of a blackbody at 5700 K. However, when examined in detail, the actual distribution is found to be dissected by numerous Fraunhofer lines of solar origin. The true solar spectrum bears very little resemblance to the smooth curve in Figure 1 and can only be shown properly on a greatly expanded scale. In Figure 2 we show a very small portion of this spectrum with high resolution.

In addition, when viewed from the ground the solar spectrum exhibits many absorption bands from the terrestrial atmospheric gases, such as water vapor, carbon dioxide, oxygen, and methane, among others. As might be expected, the strengths of these bands vary greatly with the altitude of the sun and the atmospheric conditions above the observer.

Very high resolution atlases of the solar irradiance are being prepared but do not currently exist in published form. Mapping has been carried out for the region from

---

* *Operated by the Association of Universities for Research in Astronomy, Inc., under contract with the National Science Foundation.*

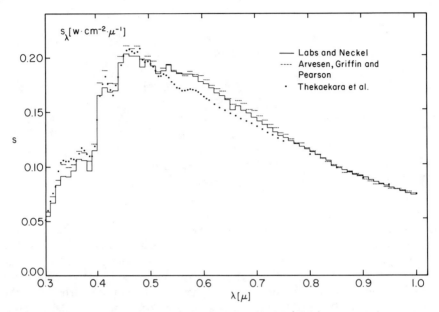

Fig. 1 The solar irradiance as determined by Labs and Neckel (1968) from a discussion of all the data up to 1967. The dashed curve represents the observations of Arvesen, Griffin and Pearson (1969). The dots are from the work of several observers as given by Thekaekara (1974).

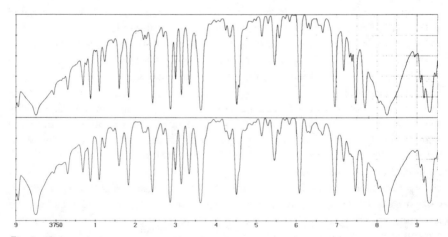

Fig. 2 High resolution tracings λλ4665-4676 Å of the solar spectrum. The upper tracing refers to disc center; the lower to a point close to the limb at solar radius 98% of R.

3200 to 6500 Å by Slaughter at the Kitt Peak National Observatory. The region from 4000 to 7000 Å has been mapped by Beckers at the Sacramento Peak Observatory and is now being prepared for publication.

Although detailed atlases of the solar irradiance are not yet available, a number of atlases of the solar spectrum at disk center do exist. These are: Minnaert, Mulders, and Houtgast (1940); Mohler *et al* (1950); Bruckner (1960); Delbouille and Roland (1963); Hall (1970); and Delbouille, Roland, and Neven (1973). None of these atlases are on an absolute scale. In fact, all of them have been constructed piecemeal from separate sections of the solar spectrum that have each been normalized to a local continuum. The rectification of each section and the factors needed to put them on an absolute energy scale have been determined for several of the atlases by Houtgast (1965, 1968), Houtgast and Namba (1968), and Labs and Neckel (1968, 1970).

In principle, the comparison of the sun with an absolute standard is straightforward. In practice, care must be taken because of the great disparity between the intensity of the standard and the intensity of the sun. A determination of the solar radiation outside the earth's atmosphere from observations made within the atmospheric envelope requires an extrapolation to zero air mass. In most observations, a telescope is used with a monochromator that isolates a narrow passband in the solar spectrum. Measurements are made by comparing the intensity of a small region at the center of the solar disk to the intensity of a standard source of radiation. The conversion of intensity measures at disk center to irradiance measures for the entire solar disk then requires a transformation through solar limb-darkening curves. The observational techniques that were used in a number of specific solar radiation experiments are discussed below. The results of these experiments are presented in tabular form. A brief review is given of the evidence for possible variations with time.

I. CALIBRATION OF TELESCOPE, SPECTROMETER, AND RECEIVERS

If one can build a detector with a neutral or blackbody response to energy at each wavelength, then it is possible to use a classical procedure developed by Abbot and Fowle (1908) to determine the spectral energy distribution of sunlight. In their experiment these investigators constructed two similar spectrometers that could be used either singly or in series. As measured at the exit slit, the sun's energy at wavelength λ was $K \cdot I(\lambda) \cdot T_1(\lambda)$ where K was a scale factor, $I(\lambda)$ was the sun's intensity at λ, and $T_1(\lambda)$ was the transmission of spectrometer 1. When spectrometers 1 and 2 were used in series, the measured output was $K \cdot I(\lambda) \cdot T_1(\lambda) \cdot T_2(\lambda)$. The ratio of these two measures gave $T_2(\lambda)$. Thus, the transmission of one or the other of the spectrometers was evaluated as a function of wavelength from a set of measurements taken by using the spectrometers both singly and in series. The wavelength dependence of the transmission of the telescope (or coelostat) was also determined, and one of the calibrated spectrometers was then used to determine the *relative energy* at each wavelength. Finally, the derived curve was integrated, and the sum equated to a measure of the total energy received from the sun in order to place the curve on an absolute scale.

Because of the fact that the transmission of monochromators and telescopes and the sensitivity of detectors vary with time (and temperature, humidity, and the like), subsequent workers in the field of absolute solar photometry have preferred to calibrate their total systems against a standard radiation source such as a blackbody, a synchrotron (which is just beginning to be used as a reference source in the ultraviolet), or both. Laboratory blackbodies at the melting point of gold, T_{Au} = 1337.58 K, or tungsten sources from 2700 to 2900 K have spectral energy distributions that differ greatly from that of the sun, which has an effective temperature of 5700 K. A list of possible intensity ratios encountered in solar irradiance work is shown in Table 1. The figures reveal the basic problem that is involved in accurately measuring the radiation from the sun. However, the very large ratios can be bridged through use of filters and modern, highly linear detectors and amplifiers.

For practical reasons, a tungsten lamp is often used as an intermediate standard between the blackbody or synchrotron and the sun. There are several potential pitfalls in calibrating and using such comparison lamps. Particular attention must be paid to the variation in brightness with position on the lamp filament, the constancy and accuracy of the filament current, and the gradual aging of the lamp. An improper treatment of any of these can result in large systematic errors.

Table 1

Ratio of emission of the sun, T = 5700 K, to a gold point blackbody, T = 1337 K and sun to a tungsten blackbody T = 2700 K.

λ	Sun/Gold	Sun/Tungsten
0.3 μm	8.4×10^{11}	11500
0.4	8.8×10^{8}	1100
0.6	9.3×10^{5}	109
1.0	4100	17.9
2.0	85.3	5.3
5.0	11.6	2.9
10.0	6.7	2.45

The observations of the sun and the comparison source should be made as differentially as possible because any change in the optical path from one source to the other is a potential source of error. In addition, in most experiments there are a number of optical and electrical components that have to be calibrated as part of the observations. Such calibrations usually involve measuring mirror reflectivities and filter transmissions at each observational wavelength.

Absolute solar intensities are usually referenced directly to sources that have been calibrated either on the International Practical Temperature Scale of 1948 (IPTS-48) or the International Practical Temperature Scale of 1968 (IPTS-68). The details of the IPTS-68 can be found in Barber (1969). A similar description of the older IPTS-48 is given in Stimson (1949).

The highest temperature primary fixed point in both the IPTS-48 and the IPTS-68 is the freezing point of gold. All temperatures for blackbodies hotter than the gold point (T_{Au}) are defined photometrically by the following equation:

$$\frac{I_\lambda(T)}{I_\lambda(T_{Au})} = \frac{e^{c_2/\lambda T_{Au}} - 1}{e^{c_2/\lambda T} - 1} \cdot \quad (1)$$

The gold point T_{Au} in the IPTS-68 is 1337.58 K, and the second radiation constant c_2 is 1.4388 cm K. Solar data based on the older system can be revised, when necessary, by using the formulas that were given by Labs and Neckel (1970).

II. CORRECTION TO ZERO AIR MASS

Definitive observations have not yet been made from space; hence the observations made from the ground or from aircraft must still be corrected for the absorption and scattering by aerosols and molecules. (Saturated band structures cannot be corrected, and it is necessary to interpolate the observations over the wavelength interval affected by the bands.) Bouguer's extinction formula for plane-parallel layers is

$$I(\lambda) = I_0(\lambda) e^{-n(\lambda) \sec z} \quad (2)$$

or the linear form

$$\ln I(\lambda) - \ln I_0(\lambda) = -n(\lambda) \sec z$$
$$= -n(\lambda) \cdot \text{Air Mass} , \quad (3)$$

where $I_0(\lambda)$ is the intensity of radiation outside the earth's atmosphere, $I(\lambda)$ is the measured intensity, $n(\lambda)$ is the extinction coefficient (proportional to λ^{-4} for excellent atmospheric conditions), and z is the zenith distance of the sun.

This formula can be generalized to include the effects of refraction and curvature of the layers. The general case has been numerically evaluated by Link and Neuzil (1969). A publication of Kasten (1964) also gives the air mass for various altitudes of the observer above sea level and for different solar zenith distances.

The intensity $I_0(\lambda)$ may be determined by measuring $I(\lambda)$ for a set of diverse z-values with a sufficiently large range in sec z and then extrapolating the linear form, Equation (3), to zero air mass. During the measuring period the atmospheric conditions must remain constant. An example of this procedure is shown in Figure 3.

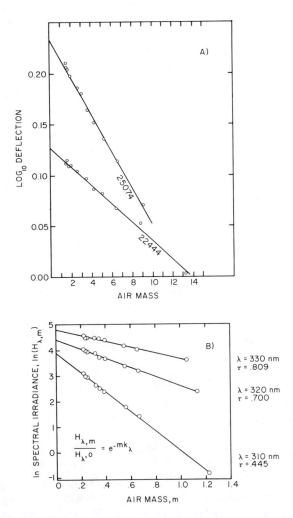

Fig. 3 Bouguer extrapolations to zero air mass. Figure A) from the work of Pierce (1954); figure B) from Arvesen, Griffin and Pearson (1969) as obtained from an aircraft.

It is sometimes assumed that there is a great advantage in work at high altitudes, where the absolute air mass is small. As pointed out by Labs and Neckel (1973), there is a great gain to be had when measuring total irradiance. However, for measurements of spectral irradiance, in which observations are made in transparent regions of the atmospheric spectrum where the residual extinction is nearly that of a Rayleigh atmosphere, the gain in going to high altitudes is very small. When the sky conditions are favorable, the intensities that one obtains by extrapolating Equation (3) to zero air mass have a probable error of less than 0.3%.

III. THE RATIO OF MEAN TO CENTRAL INTENSITY

As a practical matter, because of the small angular diameter of the sun (1/108 rad), observations are often made through a telescope that forms an image of the sun at the slit of the spectrometer and transforms the sunlight to a much larger solid angle. As a result, the standard lamp or blackbody ideally is compared with continuum radiation from the center of the sun's disk, which differs from the mean intensity of the total disk because of solar limb darkening. Limb darkening in the continuum spectrum results from the temperature gradient in the solar atmosphere and is a function of wavelength. The two curves shown in Figure 4 graphically illustrate how the intensity I_λ varies across a solar diameter at two different wavelengths, $\lambda = 6791$ Å and $\lambda = 3389$ Å. The ratio of the mean intensity F_λ to the central intensity $I_\lambda(o)$ is

$$\frac{F_\lambda}{I_\lambda(o)} = \frac{\int_0^1 \frac{I_\lambda(\rho)}{I_\lambda(o)} 2\pi\rho d\rho}{\int_0^1 2\pi\rho d\rho} = \int \frac{I_\lambda(\rho^2)}{I_\lambda(o)} d\rho^2 \quad (4)$$

with $\rho = r/R = \sin \theta$ = distance from the center of the disk in units of the disk radius R. The ratio $F_\lambda/I_\lambda(o)$ is about 0.64 at $\lambda = 0.30$ μm, 0.750 at $\lambda = 0.5$ μm, 0.884 at $\lambda = 1.0$ μm, and 0.980 at $\lambda = 10$ μm. The estimated error in the transfer from central intensity to mean intensity (for the continuum) is 0.5% (see Labs and Neckel, 1968, 1973).

The problem of correcting for the variation of the Fraunhofer lines is much more difficult and uncertain. Figure 2 shows the spectrum at 4670 Å both at disk center and very near the limb at radius 0.98. The changes in going from disk center to the limb are easily discerned; at the limb the lines are broader and shallower. The difference between the integrated light spectrum and the spectrum at disk center is much less than the difference between center and limb, as is shown. Let $\eta_{\Delta\lambda}$(center) be the fraction of light absorbed by the lines in the interval $\Delta\lambda$ at the center of the solar image and $\eta_{\Delta\lambda}$ (disk) represent the same quantity for integrated

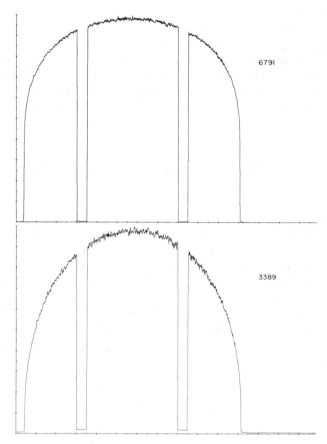

Fig. 4 Draft curves across the solar diameter at λ6791 and λ3389 Å.

light from the whole disk. By using the method of Labs and Neckel (1968, 1973), we obtain the following modification to Equation (4):

$$\left[\frac{F}{I}\right]_{line} = \left[\frac{F}{I}\right]_{cont.} \left[\frac{1 - \eta(disc)}{1 - \eta(center)}\right]_{\Delta\lambda} , \qquad (5)$$

where F (line) is the flux in the spectrum including Fraunhofer lines in the interval $\Delta\lambda$ and I (line) is the corresponding intensity in the line spectrum at disk center. We can write

$$\frac{\eta(disc)}{\eta(center)} = 1 + \varepsilon_\lambda , \qquad (6)$$

and Equation (5) becomes

$$\left[\frac{F}{I}\right]_{\text{line}} = \left[\frac{F}{I}\right]_{\text{cont.}} \left[1 - \frac{\eta(\text{center})}{1 - \eta(\text{center})} \cdot \epsilon_\lambda\right]_{\Delta\lambda} \quad (7).$$

Precision determinations of η (center) are rather easily obtained from the high-resolution atlases, but until the tracings in integrated light become available, η (disk) can only be estimated from past studies of the center-to-limb variation of selected lines. For disk center see the recent work on line blocking in the solar spectrum by Ardeberg and Virdefors (1975) for the range from 4300 to 6200 Å and by Wohl (1975) for the region from 0.3 to 2.5 μm. The line-blocking coefficient η (center) is about 1/2 at 3200 Å, 1/3 at 4000 Å, 1/10 at 5000 Å, and 1/30 at 6000 Å. However, as is apparent from Figure 2, it is highly variable with wavelength.

Observations of weak and moderately strong lines show a slight increase in strength at the limb; the strongest lines show a decrease. In their work, Labs and Neckel (1968) calculated the mean intensity from the approximate expression

$$F_{\text{line}} = \left[\frac{F}{I}\right]_{\text{cont.}} \left[1 - 0.03\, \eta(\text{center})\right] \cdot I_{\text{line}} \quad (8)$$

with the η (center) values given above, the correction factor in the bracket becomes 0.985 at 3200 Å, 0.990 at 4000 Å, 0.997 at 5000 Å, and 0.999 at 6000 Å.

V. EXPERIMENTAL TECHNIQUES AND OBSERVATIONS

Many of the older determinations of the solar spectral energy distribution are described in several detailed reviews that summarize the available observational material of the last 75 years. They are Labs and Neckel (1968, 1973), Neckel and Labs (1973), Vernazza, Avrett, and Loeser (1976), Makarova and Kharitonov (1972), and Thekaekara (1965).

A. THE OBSERVATIONS BY LABS AND NECKEL

A comprehensive set of solar absolute intensity measures (obtained at the Jungfraujoch Scientific Station in Switzerland, altitude 3.6 km) in the spectral range from 3288 to 12,480 Å was published and discussed by Labs and Neckel (1962, 1963, 1967, 1968, 1970, and 1973). The instrumentation is shown in Figure 5b. An image of the sun was formed by a 25-cm-aperture Cassegrain telescope stopped down to an 8 by 8 cm opening located to one side of the secondary. The field of view of the telescope was restricted to 30 arcsec near the disk center by a 0.7-mm circular aperture in the focal plane. The radiation was then relayed by a glass achromatic lens onto the entrance slit of a grating double monochromator attached to the telescope. The position of the relay lens was adjusted to produce a 5:1

reduction in image scale between the focal plane of the telescope and the 10 μm entrance slit of the monochromator. The radiation was dispersed by a pair of Czerny-Turner spectrometers with 52 by 52 mm Bausch and Lomb gratings ruled at 600 lines per millimeter, blazed at 5000 Å, and used in first order. The gratings were both mounted on the same mechanical axis of rotation; radiation initially dispersed in one spectrometer passed through an intermediate slit (0.7 mm wide) to be further dispersed in the other. Wavelengths shorter than 3300 Å were blocked by the glass achromatic lens between the telescope and the monochromator. The theoretical limit for the resolution of the double monochromator was nearly achieved. With an exit slit width of 1.4 mm, the spectral bandpass was 20.0 Å wide and almost perfectly rectangular for each of the 132 measurements over the spectral range from 4010 to 6569 Å. The dispersed radiation from the double monochromator was detected with either an EMI 6256 or an RCA 1P21 photomultiplier.

During the observations the Cassegrain telescope was switched back and forth between the sun and the comparison lamp at regular intervals of 15 minutes or less. Neutral density filters mounted in front of the photomultiplier had to be used during observations of the sun to cut down the intensity difference between the sun and the lamp. The filter transmissions were recalibrated every day using the sun as a source. The intensity of the sunlight for such measurements was reduced by an additional neutral density filter located just behind the focal plane of the Cassegrain telescope.

The lamp measurements used for comparison were taken by placing the filament of a tungsten ribbon lamp at the exact focus of a collimating mirror aimed at the Cassegrain telescope. Except for the neutral-density filters used with the sun, the subsequent optical path through the equipment was the same for both the sun and the comparison lamp. The two tungsten ribbon lamps that served alternately as the comparison source were calibrated against a 2500 K blackbody at the Happel Laboratory of the Heidelberg Observatory both before and after the observing run. Comparisons of the Heidelberg radiation scale with that of other standard laboratories show a good agreement (see, for example, Hayes, Oke, and Schild, 1970).

The reflectivity of the collimating mirror was calibrated as a function of wavelength on several days during the observations. An auxiliary tungsten ribbon lamp mounted on the back of the telescope was arranged in such a way that the ratio of the intensity measured from an image formed by the collimation mirror to the intensity measured directly gave the reflectivity of the mirror.

Atmospheric corrections for the solar intensity measurements were made on each day of observation by determining extinction coefficients at four preselected wavelengths separated from each other by about 120 Å. The extinction coefficients computed at these wavelengths were then used to extrapolate the solar intensity measurements to zero air mass.

For the spectral interval from 6389 to 12,480 Å the instrument was slightly

5a

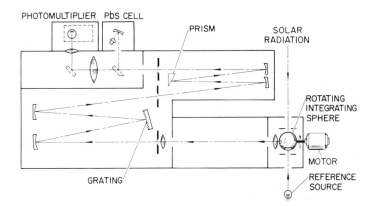

Fig. 5a   Schematic of the Cary spectroradiometer used by Arvesen, Griffin and Pearson (1969). The use of an integrating sphere for alternate viewing of reference and sun is illustrated.

5b

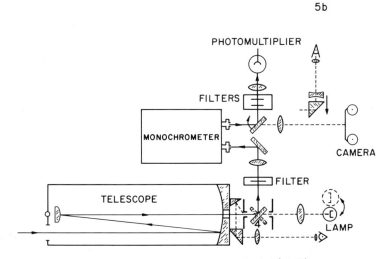

Fig. 5b   Schematic of the optical system used by Labs and Neckel (1962).

modified. The diffraction gratings were replaced with similar gratings blazed at 7500 Å. Again, they were used in the first order, with the higher orders blocked by filters. The detector was changed to a 12-stage infrared photomultiplier cooled with dry ice. The observation procedure and the reduction scheme for the raw data were unchanged from the previous observing run.

Labs and Neckel (1967) published a complete revision of their original absolute intensity measurements from 4010 to 12,480 Å. Included as part of this revision were new measurements of the absolute solar intensity in 44 passbands from 3287.9 to 4127.1 Å. A further detailed description of the experiments has been given by Labs and Neckel (1973).

B. THE ARC, NASA 711 GALILEO EXPERIMENT

Arvesen, Griffin, and Pearson (1969) of the (NASA) Ames Research Center were one of three groups that flew equipment on a research aircraft outfitted to measure the solar constant and the solar spectral irradiance. The observing platform was a Convair CV-990 four-engine jet belonging to NASA, which provided ample equipment payload, flight duration, and stability. For their experiment, sunlight was brought into the cabin through either of two windows (2.5 cm thick and approximately 30 cm in diameter) located at 14° and 65° elevation. In this way, the necessary observations to extrapolate to zero air mass could be obtained. Sunlight was directed to the spectrometer by an aluminized flat 40 cm in diameter. Eleven flights were made at pressure altitudes between 11.6 and 12.5 km.

A modified Cary spectrometer (see Figure 5a) was used, forming a prism-grating double monochromator with a resolution power of 0.1 nm in the ultraviolet and 0.3 nm in the near infrared. Stray light was less than 10^{-6} over most of the range. The slit of the spectrometer faced a rotating integrating sphere 5 cm in diameter that chopped the radiation at 30 Hz. This technique integrated sunlight and reference source, eliminated the geometrical differences between them, and gave the irradiance.

The transmission of the quartz windows was obtained by the use of integrating spheres and two different types of spectrometers. The estimated accuracy was ± 0.4% in the transmission. The reflectivity of the aluminum flat was determined for one angle of incidence by a multiple reflection technique and extended to other angles by the Fresnel equations. The estimated accuracy was ± 0.5%.

The weakest part of the experiment, as in nearly all determinations, seems to lie in the standard lamp used to calibrate the system. In this work, the standard of spectral irradiance was a 1000 W coiled tungsten filament enclosed in a quartz envelope that contained a small amount of iodine. This lamp was used without auxiliary optics at a distance of 50 cm (see Figure 5a). The lamp, calibrated to a blackbody at a known temperature, was from the Eppley Laboratory (Newport, Rhode Island), which traced its standard to the work of Stair, Schneider, and Jackson (1963). The estimated error from this source is ± 3%. Recent work (see

Kostkowski, 1974) suggests changes in the 1963 spectral irradiance scale, which bring the results of Arvesen, Griffin, and Pearson into closer agreement with the irradiance curve of Labs and Neckel (see Figure 1).

C. THE NASA/GSFC 711 GALILEO EXPERIMENT, AUGUST 1967

The work of the following experimenters is to be found in Thekaekara (1970) and also in Thekaekara, Kruger, and Duncan (1969). The observations were made concurrently on board the NASA Convair jet mentioned above. Here we report on the results from instruments operated by teams from Goddard Space Flight Center (GSFC).

McIntosh and Park operated a Zeiss double prism monochromator fed by an integrating sphere 7.5 cm in diameter pointed directly at the sun through the aircraft window or directly at the standard lamp. Forty scans, each of five minutes duration, were obtained in the interval from 0.3 to 1.6 μm; resolution is not indicated by the investigators. The monochromator was calibrated with the use of a 1000 W, NBS-type standard lamp. Because of drift (probably from heating) it was impossible to place the observations on an absolute scale except by normalizing to the solar constant. However, the relative distribution was thought to be satisfactory. Two humps (not previously observed) appear in this data at 0.9 and 1.3 μm. Therefore, we question this set of observations. The estimate of rms error is 5.5%.

Thekaekara, Stair, and Winker operated a Perkin-Elmer model 112 monochromator on board the aircraft. Rather than using an integrating sphere, they chose to allow sunlight to fall on a diffusing mirror (aluminized ground glass) in order to increase the energy (10 times) in the 2 to 4 μm region of the spectrum. Calibration was performed with a quartz-iodine standard lamp. The transmission of the sapphire window (10 cm in diameter) and the reflection of the aluminum mirror, both experimentally determined, were compared to the theoretical values. Close agreement was found, and the theoretical values were used in the reductions. Extrapolation to zero air mass was carried out for a number of wavelengths. From these a table was prepared that gave the percentage zenith atmospheric transmittance to the aircraft altitude. A computer program was used to reduce the data, apply the corrections, and convert to absolute units for approximately 50 scans of the spectrum. Resolution varied from 2.7 Å at 3000 Å to 440 Å at 4 μm. Data was reported in 100 Å bandwidths in the interval from 3000 to 6300 Å and at 1000 Å for the thermocouple range to the red. An estimated accuracy of ± 5% is reported.

Stair, Webb, and Lester made spectral measurements from 0.3 to 1.1 μm by using an Eppley Mark V filter radiometer with 22 interference filters that provided approximately 0.01 μm bandpass in the ultraviolet and wider bandpass in the visible and near infrared. An integrating sphere received light directly from the sun or standard. Because of the breadth and the shape of the bands transmitted by the filters it is difficult to compare these results with higher resolution data. The investigators estimate an uncertainty of 5%.

Ward, Thekaekara, and Rogers made observations with a solar polarizing interferometer (manufactured by Block Engineering, Cambridge, Mass.) for the range from 0.3 to 2.5 μm and with a Michelson interferometer in the interval from 2.6 to 15 μm. This appears to be the first application of this important technique to solar intensity measurements. The Soleil prism interferometer develops its path difference, in polarized light, by linear displacement of one element of the prism with respect to its compensating prism. An interferogram was obtained in 0.8 s. For the interval 0.7 to 2.2 μm, 120 wavelengths were placed on a relative scale through calibration with a standard lamp. Absolute values were obtained by fitting to the red end of the curve obtained with the Perkin-Elmer monochromator.

In the companion experiment an interferometer of the Michelson type was used to make direct observations of the sun through an Irtran-4 window mounted in the aircraft port. A ramp function generator drove the cube beamsplitter with a voice coil. The output appeared in the form of an interferogram that was later processed to obtain the spectrum. A problem, well-known to the infrared observer, was the radiation loss from the detector to the sky surrounding the sun. To take this into account, a series of calibration measurements was made with the instrument pointed to a 1200 K blackbody while the remainder of the field of view remained the same as in the solar observation but with a field temperature of 37°C, equal to that of the bolometer. Their results were presented in graphical form, and we estimate that their resolution, which they did not state, was about 0.1 μm.

The results from these experimenters were augmented by an extensive set of filter radiometer measurements obtained by an Eppley-Jet Propulsion Laboratory team under A.J. Drummond (Thekaekara and Drummond, 1971; Thekaekara, 1974). Measurements were made on several series of jet aircraft and X-15 rocket flights. The instruments were Eppley thermopiles with a series of filters that gave irradiance values over 20 wavelength bands. The spectral range of the filters extended from 0.234 to 2 μm. The X-15 measurements provided the only aircraft data from completely outside the ozonosphere.

D. THE OBSERVATIONS BY SAIEDY AND GOODY

Saiedy and Goody (1959) and Saiedy (1960) published a calibrated set of solar absolute intensity measurements at 8.63, 11.10, and 12.02 μm. A 30-cm siderostat and a parabolic mirror of similar size were used to form a 17-mm image of the sun on the entrance slit of a grating double monochromator with a spectral bandpass of 310 Å (for the observations at 11.10 μm). Scattered radiation was found to be less than 0.05%. The infrared radiation was detected with a Golay cell that was coupled to a phase-sensitive rectifier and a Honeywell-Brown recorder. A chopping wheel mounted in front of the monochromator modulated the incoming radiation at 11 Hz. For calibration an extra parabolic mirror was used to send radiation to the siderostat from a carefully regulated blackbody at 1300 K. The temperature of the blackbody was measured with a platinum and rhodium thermocouple which had been calibrated at the National Physical Laboratory to an accuracy of ±2°C. The reflectivity of the collimating mirror and the angular dependence of the reflectivity of the siderostat were both accurately determined. In order to make the measurement, as far as possible, a null one, the large intrinsic intensity difference between the sun and the blackbody was reduced by a sector wheel that chopped with a much higher frequency than the chopper for the Golay cell, attenuating the solar beam by the required amount. The intensity measures from each set of daily observations were extrapolated to zero air mass in order to find the true intensity outside the atmosphere, and an additional small correction in the intensity values was made for the effects of solar limb darkening.

E. NEAR INFRARED OBSERVATIONS FROM MT. ELBRUS

As a byproduct of an investigation from Mt. Elbrus of the infrared absorption of the earth's atmosphere, Kondratyev et al. (1965) obtained measurements of the solar flux at wavelengths from 3 to 13 µm. We have few details of their work except that spectral observations were carried out with an NaCl prism spectrometer fed by a Cassegrain telescope that was directed to the central part of the sun's disk. Each solar record was succeeded by a recording of a reference lamp to establish the stability of the system. Absolute calibration was obtained by a conic blackbody operated at 712 and 802 K. The errors range from 2 to 3.5%.

F. THE BALLOON EXPERIMENT BY MURCRAY, MURCRAY, AND WILLIAMS

The spectral radiance of the sun from 4 to 5 µm was observed from a balloon at an altitude of 31 km in an experiment by Murcray, Murcray, and Williams (1964). Their spectrometer was a single-pass, Littrow-type prism monochromator. Scattered light was blocked by a long-pass filter that was opaque short of 3.4 µm. A telescope system focused an image of the sun on the spectrometer slit; the length equaled 18 arcmin. For calibration the observers constructed a tungsten foil blackbody enclosure, heated electrically to $T \sim 2{,}500°C$ in a partial vacuum. Temperatures were determined to ± 10° by an optical pyrometer. The estimated error is 2.5% in the final determination.

G. PEYTURAUX'S OBSERVATIONS FROM MT. LOUIS

A number of disk-center solar intensity measures in the spectral range from 4477 to 8636 Å were published by Peyturaux (1968). These observations were taken at an altitude of 1.6 km on Mt. Louis in the Pyrenees. A 14-cm siderostat and two optical flats were used to direct radiation from the sun to a parabolic primary mirror with a 3-m focal length. The radiation was then brought to a focus on the entrance slit of a prism monochromator. The siderostat could also be positioned to send radiation to the monochromator from a blackbody operated at 2,600 K. The temperature of the blackbody was monitored with an optical pyrometer. The detector was either a seven-stage Lallemand tube or a lead sulfide cell. Observations were taken by alternately observing the sun and the blackbody. The signals from the two sources were accurately balanced with optical filters and a calibrated electrical attenuator. Atmospheric extinction was determined at each observational wavelength with an auxiliary telescope and spectrophotometer. Polarization effects were estimated to have less than 1% effect on the final measures. The error estimate may be too low, however, because the computation neglected a 45° mirror over the blackbody.

Koutchmy and Peyturaux (1970) presented additional measures in several passbands from 3.5 to 25 μm. These observations were made with a Schwartz cell that was operated at the ambient air temperature. The signal was chopped at 20 Hz to reduce the thermal background. The auxiliary telescope was used to monitor variations in the water vapor content of the atmosphere during the course of the observations.

V. DISCUSSION OF THE OBSERVATIONS: VARIATIONS IN THE SOLAR SPECTRAL RADIATION

Observations of solar radiation made at wavelength λ with well-calibrated but different instruments differ because of the Fraunhofer line spectrum. Each spectrometer has its own instrumental contour or passband or both that is determined by the instrument and its configuration. Thus, an accurate determination of the passband and its profile is a necessary part of the reductions. Because of the Fraunhofer spectrum, intercomparison between observers is only possible if all the observations are reduced to identical passbands.

The sun's spectral radiation as obtained by several observers is shown in Table 2. In the first column we give the central wavelength of each interval, with bandwidth 0.01 or 0.1 μm. Three succeeding columns give the mean value of the spectral irradiance, $\bar{s}_{\Delta\lambda}$, for the solar absorption-line spectrum in units of mW cm^{-2} μm^{-1}. The values of Labs and Neckel (1970, 1973) are those they derived from their observations in the interval from 0.33 to 1.25 μm. Values quoted by Labs and Neckel between 0.3 and 0.33 μm were taken from the summary by Tousey (1963). Beyond 1.25 μm they used the relative observations by Pierce (1954) normalized to a least-squares model fitting their observations at shorter wavelengths. The quantities listed under Arveson, Griffin, and Pearson (1969) were calculated from the tabular values they gave for each angstrom. The values listed for Thekaekara, Kruger, and Duncan (1969) were generated in part from measurements obtained on the Galileo 711 flight described here and in part from later flights. The spectral curves for the two monochromators, a filter radiometer and an interferometer, were not identical but lay within the error limits assigned to each experiment. The NASA 711 experimenters also had on board four total irradiance instruments that gave a solar constant value of

Table 2

Mean values of the solar irradiance, $\bar{s}_{\Delta\lambda}$, $mW \cdot cm^{-2} \cdot \mu m^{-1}$

Mean intensity at disc center (line spectrum), $\bar{I}_{\Delta\lambda}(o)$, $KW \cdot cm^{-2} \cdot sterad^{-1} \cdot \mu m^{-1}$

Limb darkening, $F_{\Delta\lambda}/\bar{I}_{\Delta\lambda}(o)$

Intensity at disc center between spectrum lines, $I'_{\lambda}(o)$, $KW \cdot cm^{-2} \cdot sterad^{-1} \cdot \mu m^{-1}$

$\bar{\lambda}$ μm / $\bar{s}_{\Delta\lambda}$	Labs Neckel	Arvesen Griffin Pearson	Thekaekara et al.	$\bar{I}_{\Delta\lambda}(o)$	$\dfrac{F_{\Delta\lambda}}{\bar{I}_{\Delta\lambda}(o)}$	λ μm	$I'_{\lambda}(o)$
Bandwidth = 0.01 µ							
0.305	53.7	59.2	60.3	1.24	0.636	0.30	2.13
0.315	65.8	72.9	76.4	1.47	0.657	0.31	2.58
0.325	81.8	87.8	97.5	1.78	0.674	0.32	2.80
0.335	90.0	102	108	1.93	0.685	0.33	2.93
0.345	89.4	99.7	107	1.89	0.694	0.34	3.02
0.355	94.9	102	108	1.99	0.703	0.35	3.10
0.365	105	115	113	2.24	0.691	0.36	3.16
0.375	104	113	116	2.25	0.680	0.37	3.60
0.385	94.5	106	110	2.02	0.687	0.38	4.20
0.395	113	121	119	2.39	0.698	0.39	4.50
0.405	163	177	164	3.38	0.710	0.40	4.56
0.415	170	188	177	3.48	0.718		
0.425	166	177	169	3.36	0.726	0.42	4.60
0.435	167	176	166	3.35	0.734		
0.445	193	201	192	3.81	0.744	0.44	4.53
0.455	201	211	206	3.92	0.753		
0.465	199	208	205	3.85	0.759	0.46	4.38
0.475	199	211	204	3.82	0.765		
0.485	189	199	198	3.61	0.769	0.48	4.22
0.495	196	202	196	3.71	0.775		
0.505	190	197	192	3.59	0.779	0.50	4.06
0.515	183	187	183	3.44	0.784		
0.525	186	191	185	3.47	0.788	0.52	3.90
0.535	192	193	182	3.56	0.793		
0.545	186	188	175	3.42	0.798	0.54	3.75
0.555	184	185	172	3.38	0.802		
0.565	183	184	171	3.34	0.805	0.56	3.59
0.575	183	188	172	3.34	0.809		
0.585	181	184	171	3.28	0.812	0.58	3.43
0.595	176	184	168	3.18	0.815		
0.605	174	180	165	3.14	0.817	0.60	3.27
0.615	171	174	162	3.06	0.820		
0.625	166	172	159	2.96	0.823	0.62	3.11
0.635	164	169	156	2.92	0.825		
0.645	160	162	153	2.84	0.828	0.64	2.96
0.655	152	158	150	2.69	0.830		
0.665	156	161	147	2.74	0.834	0.66	2.81
0.675	152	158	144	2.67	0.836		
0.685	149	153	141	2.61	0.838	0.68	2.67
0.695	145	152	139	2.54	0.841		

IV SOLAR SPECTRUM

Table 2 (continued)

λ μm / $\bar{s}_{\Delta\lambda}$	Labs Neckel	Arvesen Griffin Pearson	Thekaekara et al.	$\bar{I}_{\Delta\lambda}(o)$	$\dfrac{\bar{F}_{\Delta\lambda}}{\bar{I}_{\Delta\lambda}(o)}$	λ μm	$I'_\lambda(o)$
0.705	142	146	136	2.47	0.843	0.70	2.53
0.715	138	142	133	2.41	0.845		
0.725	136	139	130	2.36	0.847	0.72	2.42
0.735	132	136	128	2.28	0.849		
0.745	128	131	125	2.22	0.851	0.74	2.30
0.755	126	130	122	2.18	0.853		
0.765	124	126	120	2.13	0.855	0.76	2.18
0.775	121	121	117	2.07	0.857		
0.785	118	122	115	2.03	0.859	0.78	2.07
0.795	116	117	112	1.98	0.861		
0.805	114	114	110	1.94	0.863	0.80	1.97
0.815	110	110	107	1.88	0.864		
0.825	108	107	105	1.83	0.865		
0.835	105	104	102	1.78	0.867		
0.845	101	102	100	1.70	0.868		
0.855	98.6	95.1	97.9	1.67	0.869	0.85	1.74
0.865	96.8	94.5	95.8	1.64	0.870		
0.875	94.7	95.2	93.7	1.60	0.872		
0.885	92.4	92.8	91.7	1.56	0.873		
0.895	92.0	91.2	90.0	1.55	0.874		
0.905	89.8	87.9	88.6	1.51	0.875	0.90	1.38
0.915	87.4	86.5	87.5	1.47	0.876		
0.925	85.7	83.6	86.4	1.44	0.877		
0.935	84.1	83.4	85.3	1.41	0.878		
0.945	82.3	81.3	84.2	1.38	0.879		
0.955	80.6	79.6	82.9	1.35	0.880		
0.965	78.9	78.9	81.2	1.32	0.881		
0.975	77.3	77.9	79.4	1.29	0.882		
0.985	75.6	76.9	77.6	1.26	0.883		
0.995	73.9	75.1	75.8	1.23	0.884		
Bandwidth = 0.1 μ							
1.05	66.1	67.4	66.8	1.10	0.887	1.00	1.23
1.15	54.0	55.1	53.5	0.890	0.894	1.10	0.990
1.25	44.7	45.6	43.8	0.732	0.901	1.20	0.812
1.35	37.4	38.4	35.8	0.610	0.907	1.30	0.674
1.45	31.9	32.3	31.2	0.516	0.913	1.40	0.566
1.55	27.3	27.5	26.7	0.441	0.918	1.50	0.482
1.65	22.8	23.7	22.3	0.368	0.923	1.60	0.408
1.75	18.6	19.2	18.0	0.298	0.928	1.70	0.333
1.85	15.3	15.2	14.2	0.243	0.931	1.80	0.270
1.95	12.7	13.1	11.4	0.200	0.933	1.90	0.222
2.05	10.6	10.66	9.65	0.167	0.935	2.00	0.184
2.15	8.95	8.44	8.45	0.141	0.936	2.10	0.154
2.25	7.59	7.22	7.40	0.119	0.938	2.20	0.129
2.35	6.48	6.04	6.55	0.101	0.940	2.30	0.110
2.45	5.58	5.30	5.85	0.087	0.941	2.40	0.094
2.55	4.83		5.15	0.075	0.943	2.50	0.081
2.65	4.19		4.55	0.065	0.945	2.60	0.070
2.75	3.65		4.10	0.057	0.946	2.70	0.061
2.85	3.20		3.70	0.050	0.948	2.80	0.053
2.95	2.81		3.30	0.044	0.950	2.90	0.046

1351 W m^{-2}. The weighted average of the spectral instruments was scaled down by 0.1% to make the integrated area under the spectral curve agree with the solar constant value (see Thekaekara, Kruger, and Duncan, 1969). A revised table published in Thekaekara (1970), with additions from other sources, covers the range from 0.115 to 1000 µm. For wavelengths greater than 0.7 µm, the older values were retained because Drummond's integral data from cutoff filters were identical to the GSFC data. In the wavelength range from 0.3 to 0.7 µm, the results from Drummond's bandpass interference filters were in fairly close agreement although there were a few differences. Therefore, a weighted average of the GSFC data and Drummond data was taken. The maximum changes from the GSFC curve are +2.3% at 0.34 µm, −0.7% at 0.45 µm, and +1.6% at 0.63 µm. The percentages vary between zero and these maxima at wavelengths between 0.3 and 0.7 µm. With these changes the integral under the spectral curve was raised from the GSFC value, 1351 W m^{-2}, to the revised solar constant, 1353 W m^{-2}.

The difference between the three determinations is an indication of our present knowledge of the sun's radiation. The work of Labs and Neckel is the most thorough and detailed. They report possible systematic errors of 2% below 0.4 µm, 1.8% at 0.4 µm, 0.7% at 1.0 µm, and somewhat less at longer wavelengths. The results of Arvesen, Griffin, and Pearson suffer from an uncertain standard. The results given by Thekaekara, Kruger, and Duncan were adopted as the NASA Space Vehicles Design Criteria (Anonymous, 1971) and the engineering standard of the American Society of Testing and Materials (Anonymous, 1974).

Column 5 of Table 2 gives the mean radiance at disk center averaged over the interval $\Delta\lambda$. In column 6 we show the ratio of the flux to the intensity at disk center for the line spectrum (table, Labs and Neckel, 1973). The intensity of the continuum spectrum, derived from the "windows" in the spectral radiation from disk center, is shown in column 8 for the wavelengths given in column 7.

Table 3 summarizes the results of several observers who have made observations from 2.5 to 10 µm. Terrestrial water vapor, methane, and carbon dioxide bands obscure the spectrum in several intervals (see Goldberg, 1954). However, the spectrum is largely devoid of solar lines and the continuum spectrum is easily interpolated over these gaps. It is difficult to assess the accuracy of the results in this region. An intercomparison between the observers suggests errors of 5 to 15%; thus, although the region contributes only 3.6% of the total solar radiation, further observations are desirable.

The values of Tables 2 and 3 give the sun's radiation at a mean distance of 1 AU. The radius vector r, the sun-to-earth distance, varies throughout the year and is tabulated in the American Ephemeris and Nautical Almanac. It is 1.00000 on April 3 and October 1. The minimum distance of 0.98327 occurs on January 2 and the maximum distance of 1.01673 occurs on July 2. Thus there is a 6.6% variation in the solar flux reaching the earth throughout the year.

Figure 6 shows the intensity of the line spectrum at disk center integrated over

Table 3

Solar spectral irradiance s_λ in $mW \cdot cm^{-2} \cdot \mu m^{-1}$

μ	KABGP	T	MMW	S	KP
2.5		5.5			
2.6		4.8			
2.7		4.3			
2.8		3.9			
2.9		3.5			
3.0	2.4	3.1			
3.1		2.6			
3.2		2.3			
3.3		1.9			
3.4		1.7			
3.5		1.5			
3.6	1.2	1.4			
3.7		1.2			
3.8		1.1			1.1
3.9		1.03			
4.0	0.82	0.95	0.82		
4.06					
4.1		0.87	0.75		
4.2		0.78	0.69		
4.3		0.71	0.62		
4.4		0.65	0.56		
4.5	0.55	0.59	0.53		
4.6		0.54	0.47		
4.7		0.49	0.43		
4.8		0.45	0.41		
4.9		0.41	0.39		0.34
5.0	0.37	0.38	0.34		0.33
5.5	0.26				
6.0	0.18	0.18			
6.5	0.133				
7.0	0.100	0.099			
7.5	0.078				
8.0	0.057	0.059			
8.5	0.046				
8.63				0.044	
9.0	0.038	0.037			
10.0	0.024	0.025			

KABGP: Kondratyev, Andreev, Badinov, Grishechkin, Popova (1965)
T   : Thekaekara (1970)
MMW : Murcray, Murcray, Williams (1964)
S   : Saiedy (1960)
KP : Koutchmy, Peyturaux (1970)

0.01 μm intervals. The continuum spectrum between the absorption lines that was derived by Labs and Neckel (1973) is shown, together with the observational results by Peyturaux (1968) and Sitnik (1965). The latter are in good agreement but differ from those given by Labs and Neckel. The cause of the discrepancy is not known.

In the interval 0.3 to 10 μm the solar flux ranges over several orders of magnitude. In plots of intensity or log intensity versus wavelength it is difficult

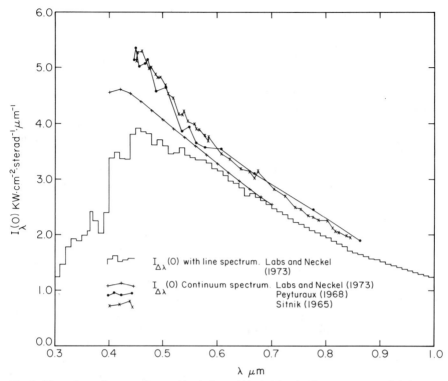

Fig. 6 The solar radiance as observed by Labs and Neckel for the line spectrum and their derived continuum spectrum. The continuum observation of Peyturaux (1968) and Sitnik (1965) are plotted on the same scale.

to illustrate the small differences between solar radiation and an equivalent blackbody. Therefore, it is often more convenient to plot solar brightness temperatures, which can be calculated from

$$T_b(\text{disc}) = \frac{14388}{\lambda \ln[(0.80978/\lambda^5 S_\lambda) + 1]} \quad (9a)$$

or

$$T_b(\text{center}) = \frac{14388}{\lambda \ln[(11909/\lambda^5 I_\lambda) + 1]} \quad (9b)$$

With the constants given, the units of the variables are λ in μm, I_λ in W cm^{-2} sr^{-1} μm^{-1} and s_λ in W cm^{-2} μm^{-1}. From Wien's law we can relate the proportional change in flux to the proportional change in temperature:

$$\frac{dF_\lambda}{F_\lambda} = \frac{c_2}{\lambda T} \cdot \frac{dT}{T} \quad . \quad (10)$$

Values of the brightness temperature for the total disk have been given by Vernazza et al. (1976) and are plotted in Figure 7. The effect of strong line-absorption in lowering the effective temperature in the interval from 0.3 to 0.4 μm is shown. Note also at 1.6 μm the increased radiation that escapes from deeper solar layers through a "window," produced by the H⁻ opacity minimum. In the far red, with increasing absorption by H⁻ the brightness temperature decreases to a minimum of about 4400 K near $\lambda = 200$ μm.

Spectroheliograms taken in the strongest Fraunhofer lines, such as hydrogen alpha and beta, calcium H and K, and the strong ultraviolet iron lines, show bright and dark areas. The bright areas are plages. They vary in intensity and area with the magnetic field structures in the sunspot zones and hence with the well-known 11-y period of sunspots. We estimate that the total solar energy output would change from this cause alone by less than 10^{-3} %.

Pettit (1932) observed over a period of seven years the ratio of the ultraviolet solar radiation at 0.32 μm transmitted by a silver film to the green radiation at 0.50 μm transmitted by a gold film. The monthly average for the ratio varied from 0.95 to 1.57 during the period and, except for one year (1929), closely followed the number of sunspot groups. Although 0.32 μm has strong bands of ozone that are known to vary, Pettit was able to demonstrate that terrestrial ozone was probably not the cause of the observed variation. Unfortunately it is not now possible to convert this variation to a numerical value of the change in the solar radiation or even to be sure of its solar origin because we are not entirely convinced of Pettit's demonstration. We thus conclude that there is as yet little or no observational evidence for any measurable change in the spectral radiation with time.

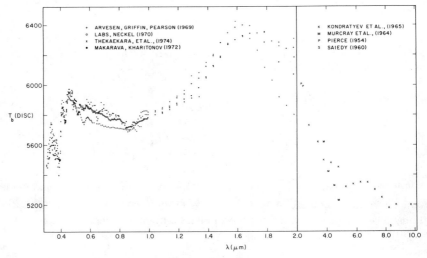

Fig. 7 Solar brightness temperatures for the whole disc.

ACKNOWLEDGMENTS

We are greatly indebted to Dietrich Labs, Heinz Neckel, and Mathew Thekaekara for their personal help, interest, and attention to this manuscript. The remaining errors are ours.

REFERENCES

Abbot, C. G., and Fowle, F. E., 1908, *Ann. Smithsonian Ap. Obs. 2*.

Anon., 1971, *Solar Electromagnetic Radiation, NASA Space Vehicles Design Criteria, NASA-SP 8005*, Washington, D.C.

Anon., 1974, "Standard Specification for Solar Constant and Airmass Zero Solar Spectral Irradiance," *1974 Annual Book of ASTM Standards*, ASTM Standard, E4773a, Philadelphia, Part 41, 609-615.

Ardeberg, A., and Virdefors, B., 1975, *Astr. and Ap. 45*, 19.

Arvesen, J.C., Griffin, N.R., Jr., and Pearson, B.D., Jr., 1969, *Appl. Optics 8*, 2215.

Barber, C.R., 1969, *Nature 222*, 929.

Bruckner, G., 1960, *Photometrischer Atlas des Nahen Ultravioletten Sonnenspektrums*, Gottingen.

Delbouille, L., and Roland G., 1963, *Photometric Atlas of the Solar Spectrum from $\lambda 7498$ to $\lambda 12016$*, Liège.

_____, _____, and Neven, L., 1973, *Photometric Atlas of the Solar Spectrum from $\lambda 3000$ to $\lambda 10000$*, Liège.

Goldberg, L., 1954, *The Solar System*, Vol. 2, ed. G.P. Kuiper, University of Chicago Press, Chicago, 434.

Hall, D.B., 1970, *An Atlas of Infrared Spectra of the Solar Photosphere and of Sunspot Umbrae*, Kitt Peak National Observatory.

Hayes, D.S., Oke, J.B., and Schild, R.E., 1970, *Ap. J. 162*, 361.

Houtgast, J., 1965, *Proc. Kon. Akad. Wetensch. Ser. B68*, 306.

_____, 1968, *Solar Phys. 3*, 47.

_____ and Namba, O., 1968, *B.A.N. 20*, 87.

Kasten, F., 1964, *U.S. Army Material Command, Cold Regions Res. and Eng. Lab. Rept. 136*.

Kondratyev, K.Y.; Andreev, S.D.; Badinov, I.Y.; Grishechkin, V.S.; and Popova, L.V., 1965, *Appl. Optics 4*, 1069.

Kostkowski, H.J., 1974, *Opt. Rad. News N.B.S.*, No. 3.

Koutchmy, S., and Peyturaux, R., 1970, *Astr. and Ap. 5*, 470.

Labs, D., and Neckel, H., 1962, *Z. Ap. 55*, 269.

―――――― and ――――――, 1963, *ibid. 57*, 283.

―――――― and ――――――, 1967, *ibid. 65*, 133.

―――――― and ――――――, 1968, *ibid. 69*, 1.

―――――― and ――――――, 1970, *Solar Phys. 15*, 79.

―――――― and ――――――, 1973, *Proceedings of Symposium on Solar Radiation*, Smithsonian Institution, Washington, D.C., 269.

Link, F., and Neuzil, L., 1969, *Tables of Light Trajectories in the Terrestrial Atmosphere*, Hermann, Paris.

Makarova, Y.A., and Kharitonov, A.V., 1972, Nauka Publishing House, Moscow (Translation: NASA TT F-803), 288.

Minnaert, M., Mulders, G.F.W., and Houtgast, J., 1940, *Photometric Atlas of the Solar Spectrum*, Utrecht.

Mohler, O.C.; Pierce, A.K.; McMath, R.R.; and Goldberg, L., 1950, *Photometric Atlas of the Near Infrared Solar Spectrum 8465 to 25242 Å*, University of Michigan Press, Ann Arbor.

Murcray, F.H., Murcray, D.G., and Williams, W.J., 1964, *Appl. Optics 3*, 1373.

Neckel, H., and Labs, D., 1973, *Proceedings of Symposium on Solar Radiation*, Smithsonian Institution, Washington, D.C., 326.

Pettit, E., 1932, *Ap. J. 75*, 185.

Peyturaux, R., 1952, *Ann. d'Ap. 15*, 302.

――――――, 1968, *ibid. 31*, 227.

Pierce, A.K., 1954, *Ap. J. 119*, 312.

Saiedy, F., 1960, *M.N.R.A.S. 121*, 483.

―――――― and Goody, R.M., 1959, *ibid. 119*, 213.

Sitnik, G.F., 1965, *Astr. Zh. 42*, 59.

Stair, R., Schneider, W.E., and Jackson, J.K., 1963, *Appl. Optics 2*, 1151.

Stimson, H.F., 1949, *J. Res. N.B.S. 42*, 209.

Thekaekara, M.P., 1965, *NASA Tech. Report SP-74*.

――――――, 1970, *NASA Tech. Report TR R-351*.

――――――, 1974, *Appl. Optics 13*, 518.

―――――― and Drummond, A.J., 1971, *Nature Phys. Sci. 229*, 6.

――――――, Kruger, R., and Duncan, C.H., 1969, *Appl. Optics 8*, 1713.

Tousey, R., 1963, *Space Sci. Rev. 2*, 3.

Vernazza, J.E., Averett, E.H., and Loeser, R., 1976, *Ap. J. Suppl. Ser. 30*, No. 1, 1.

Wöhl, H., 1975, *Astr. and Ap. 40*, 343.

THE SOLAR SPECTRUM BETWEEN 1200 AND 3000 Å

Donald F. Heath
and
Matthew P. Thekaekara
NASA/Goddard Space Flight Center

The solar flux emitted in the wavelength region from 1200 to 3000 Å is of great interest to both solar and atmospheric physicists. In the visible and infrared, the solar spectrum is essentially a continuum; however, absorption lines superimposed upon the continuum become increasingly more pronounced at shorter wavelengths in the near ultraviolet near 3000 Å. Continuing to still shorter wavelengths, a very steep decrease is observed in the continuum flux associated with the Al I ionization edge and its associated continuum at shorter wavelengths. Similar ionization edges with their associated continua are observed for H, Mg, Si, Fe, and C. These elements represent the principal sources of solar opacity in the wavelength region from 1200 to 3000 Å. From 3000 to 2100 Å the absorption lines superimposed on the continuum absorb increasingly more energy. Further into the ultraviolet, past the Al I ionization edge, the importance of emission lines increases rapidly and the absorption lines disappear near 1500 Å. For wavelengths shorter than 1400 Å, the chromospheric and coronal emission lines begin to dominate the emission in the continuum.

The source of the solar continuum radiation changes from the photosphere to the chromosphere as the wavelength decreases from 3000 to 1200 Å. In passing through this transition region, the brightness temperature of the solar continuum goes through a minimum between 1500 and 1800 Å. The increase in the absorption cross-sections with decreasing wavelengths means that the effective emitting height of the solar continuum moves upwards in the solar atmosphere. These effects are shown in Figure 1 (Vernazza, Avrett, and Loeser, 1976). The shaded band represents the brightness temperature T_b of the continuum at the center of the solar disk. The solid line gives the height h in the solar atmosphere at which the solar spectral flux has an attenuation of $1/e$ (or has an optical depth τ_λ equal to one). The height is measured from the level where τ_λ is 1 for $\lambda = 5000$ Å. The value of T_b is from observations and that of h from the solar model computations by Vernazza, Avrett, and Loeser (1976).

In the terrestrial atmosphere, solar radiation from 1250 to 2000 Å is absorbed in the lower thermosphere and mesosphere by O_2, which yields atomic oxygen through dissociation. The longer wavelengths (from 2000 to 3000 Å) are absorbed in the upper and lower stratosphere. This radiation is responsible for the photochemistry of stratospheric ozone. At the long-wavelength end the absorption cutoff beyond which solar flux will not reach the ground is strongly dependent upon the total ozone column amount, which is highly variable with season and geographical location. Figure 2 is an approximate representation of the absorption of normally incident solar flux by the earth's atmosphere as a function of wavelength (Friedman, 1960). The y-axis gives the altitude above sea level at which solar flux is reduced to 1/e of the extraterrestrial value.

Thermal gradients, which vary with altitude, geographical location, season, and solar flux, represent a major forcing term in the circulation of the stratosphere. Investigations of atmospheric phenomena on synoptic and climatologic time scales can serve as useful indicators for the behavior of the sun as an ultraviolet variable star. The variability of the sun below 1200 Å is well established not only from direct observations of the sun, but also from in-situ measurements of atmospheric constituents and parameters and satellite drag effects. The problem of determining the temporal behavior of the solar flux in the region from 1200 to 3000 Å is much more difficult, since the magnitude of the variability is less than in the EUV, and satellites cannot operate in the region of the atmosphere bounded by the lower thermosphere (below 120 km) and the stratosphere.

In this region, only remote atmospheric sounding techniques are possible. Since

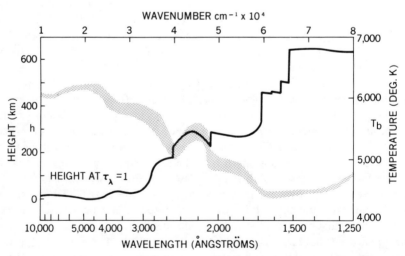

Fig. 1 Brightness temperature T_b (legend on right, shaded area) and height in the solar atmosphere for optical depth $\tau_\lambda = 1$ above $\tau_{5000} = 1$ (legend on left, continuous line). From Vernazza, Avrett, and Loeser (1976).

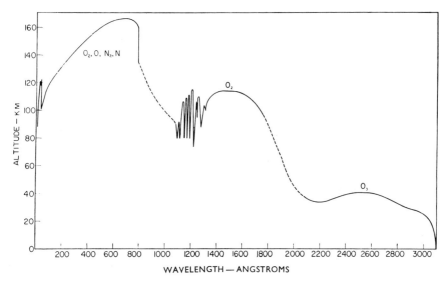

Fig. 2 Descriptive representation of the altitude above sea level in the earth's atmosphere of unit optical depth for solar spectral irradiance with zero zenith angle.

the dynamics of the earth's upper atmosphere plays an important role in atmospheric structure and composition, it is quite difficult in this region to separate dynamic effects from those that can be attributed to the intrinsic variability of the sun.

I. SOLAR SPECTRAL IRRADIANCE

Measurements of the solar spectral energy distribution are usually given in terms of either the spectral radiance at the center of the solar disk (in units of ergs s^{-1} cm^{-2} nm^{-1} sr^{-1}) or of the spectral irradiance of the whole sun at 1 AU. Because of limb darkening (or brightening, at short wavelengths) and increased brightening in active regions, the radiance is not uniform at all points on the disk. The relation between the average spectral radiance of the disk, L_λ, and that at the disk center, L_{λ_c}, is

$$L_\lambda = L_{\lambda_c} \int_0^{\pi/2} R_{\lambda\theta} \sin\theta \cos\theta \, d\theta \quad , \tag{1}$$

where θ is the angle subtended at the center of the sun by two solar radii, one passing through the disk center and the other passing through any other surface element on the sun, and $R_{\lambda\theta}$ is the limb-darkening (or brightening) function, namely, the ratio of radiance of a surface element at angle θ to the radiance of the disk center at wavelength λ. To the extent that L_{λ_c} and $R_{\lambda\theta}$ are known, L_λ can be evaluated.

The spectral irradiance E_λ of the sun at 1 AU is

$$E_\lambda = 2\pi(r^2/R^2) L_\lambda \int_0^{\pi/2} \sin\theta \cos\theta \, d\theta$$

$$= \pi(r^2/R^2) L_\lambda = 6.7997 \times 10^{-5} L_\lambda \quad , \tag{2}$$

where r is the solar radius and R is 1 AU.

The total irradiance E, or the solar constant, is

$$E = \int_0^\infty E_\lambda \, d\lambda \quad . \tag{3}$$

The type of solar observation (solar spectral irradiance or radiance) is usually determined by the nature of the scientific investigation. In general, measurements of the solar spectral energy distribution are of irradiance or radiance according as the investigations are for terrestrial or solar phenomena, respectively. Because measurements of the solar spectral energy distribution from 1200 to 3000 Å must be made from altitudes where the effects of atmospheric attenuation are negligible, it is easier to make measurements of irradiance, since telescope systems with high pointing accuracy are not required.

A separate topic, which cannot be discussed here in detail but which is obviously of great importance, is the question of instrument calibration techniques.

Basically, there are two ways to calibrate. One is to measure the wavelength dependence of the instrument transfer function and detector efficiency; the other is to illuminate the instrument with standards of spectral radiance or irradiance. In the latter technique, it is important that the spectral energy distribution of the radiation standard and of the sun not be too dissimilar over the bandpass of the instrument.

Conversion of measurements of the central disk solar spectral radiance to irradiance for the region from 1200 to 3000 Å is complicated because the character of the solar spectrum from longer to shorter wavelengths changes from line absorption to line emission and from limb darkening to limb brightening. In addition, the limb darkening function not only varies with wavelength, but also is dependent on each individual spectral line. This problem has been discussed in great detail by Vernazza, Avrett, and Loeser (1976) and Avrett in this chapter.

Table I gives a list of the major sources of data on solar spectral irradiance of the sun. It is not an exhaustive list, and the information about type of measurement and method of calibration is necessarily very brief. The authors and the references are given in the first column and the dates on which the measurements were made

Table 1

Major Measurements of the Solar Spectral Radiance and Irradiance in the Wavelength Range 1200 to 3000 Å.

Author Reference	Date	Wavelength Range Å	Observing Platform	Type of Measurement	Method of Calibration
Detwiler et al. (1961)	4/19/60	850-2600	Rocket	Photographic Recording	Mt. Lemmon Measurements
Carver et al. (1972)	1967	1430-1470 1580-1640	WRESAT-1 Spacecraft	Photo Ion Chambers	Spectral Response of the System
Dupree and Reeves (1971)	Oct. 26 and 27, 1967	300-1400	OSO IV Spacecraft	Photomultiplier	Normalization to Rocket Flights
Parkinson and Reeves (1969)	9/24/68	1400-1875	Rocket	EMR 641 G Photomultiplier	Reeder Thermopile and Total Irr. Std.
Heath (1973)	April 70 to March 72	2550-3400	Nimbus 4 Spacecraft	EMR 541-N Photomultiplier	NBS CsTe Std.
Broadfoot (1972)	6/15/70	2100-3400	Rocket	EMR 541-F Photomultiplier	Reeder Thermopile and Total Irr. Std.
Simon (1975)	9/23/72 and 5/16/73	1960-2300 2840-3540	Balloon	EMR 542 Photomultiplier	Reeder Thermopile and Deuterium Lamp
Rottman (1974)	12/13/72 and 8/30/73	1150-1850	Rocket	EMR Photomultiplier	NBS Calibrated Detector
Nishi (1973)	2/19/73	1400-2000	Rocket	Hamamatsu R-600 Photomultiplier	Spectral Response of the System
Brueckner et al. (1976)	9/4/73	1750-2100	Rocket	Photographic Recording	Deuterium Lamp and Hydrogen Arc
Heroux and Swirbalus (1976)	11/2/73 and 4/23/74	1230-1940	Rocket	EMR 641 G Photomultiplier	NBS Calibrated Detector
Samain and Simon (1976)	4/17/73	1510-1090	Rocket	Photographic Recording	Spectral Response of the System

in the next column. The listing is in the order of these dates.

The values of solar spectral irradiance from eight different sources are given in Table II. The wavelength range for the table is 1000 to 3000 Å, though not all of the measurements cover the full range. The unit of irradiance is erg sec^{-1} cm^{-2} per 10 Å bandwidth (or W m^{-2} μm^{-1}), except for the last column, which is in units of 10^{10} photon cm^{-2} sec^{-1} nm^{-1}. The sources for these data are as follows: column 3, Donnelly and Pope (1973); column 4, Heroux and Swirbalus (1976); column 5, Samain and Simon (1976) and Broadfoot (1972); column 6, Detwiler et al. (1961); column 7, Thekaekara (1974); column 8, Brueckner et al. (1976); columns 9 and 10, Rottman (1974) data from 13 December 1972 and 30 August 1973, respectively; and column 11, Donnelly and Pope (1973). The entries in columns 3 and 11 are equivalent and are the only ones that cover the full range from 1000 to 3000 Å and have 10 Å resolution. They are taken from a detailed NOAA Technical Report which presents the solar flux in the wavelength range from 1 to 3000 Å for a moderate level of solar activity (10.7 cm radio flux = 150 × 10^{-22} W m^{-2} Hz^{-1} at 1 AU). The report does not present new measurements but gives a critical evaluation of all previous measurements made from balloons, rockets, and satellites. The major sources of data for the wavelength range from 1000 to 3000 Å are Hinteregger (1970); Timothy et al. (1972); Vidal-Madjar, Blamont, and Phissamay (1973); Rottman (private communication); Dupree and Reeves (1971); Detwiler et al. (1961); Prag and Morse (1970);

Table 2

Solar Spectral Irradiance in the Wavelength Range 1000 to 3000 Å.

Sources: D. & P., Donnelly and Pope (1973); H. & S., Heroux and Swirbalus (1976); S. & S./ A. L. B., Samain and Simon (1976) for 1510 to 2090 Å and A. L. Broadfoot (1972) for 2100 to 2990 Å; D. et al., Detwiler et al. (1961); M. P. T., M. P. Thekaekara (1974); B. et al., Brueckner et al. (1976); G. J. R. (1), G. J. Rottman (1974) 12/13/1972; G. J. R. (2), G. J. Rottman (1974) 8/30/1973; D. & P., Donnelly and Pope (1973) in units of 10^{10} photons s^{-1} cm^{-2} nm^{-1}.

Wavelength Range		D. & P.	H. & S.	S. & S./ A.L.B.	D. et al.	M.P.T.	B. et al.	G.J.R. (1)	G.J.R. (2)	D. & P. Photons
From	To									
1000	1010	0.0030								0.015
1010	1020	0.0030								0.015
1020	1030	0.0709								0.3656
1030	1040	0.0805								0.4172
1040	1050	0.0028			0.002					0.015
1050	1060	0.0024								0.013
1060	1070	0.0025								0.013
1070	1080	0.0024								0.013
1080	1090	0.0132								0.072
1090	1100	0.0029			0.012					0.016
1100	1110	0.0027								0.015
1110	1120	0.0024								0.0135
1120	1130	0.0114								0.065
1130	1140	0.0018								0.01
1140	1150	0.0016			0.016	0.007				0.0092
1150	1160	0.0018								0.0105
1160	1170	0.0020						0.0163	0.0155	0.012
1170	1180	0.0394						0.0565	0.0415	0.234
1180	1190	0.0027						0.0139	0.0128	0.016
1190	1200	0.0028			1.14	0.9		0.0343	0.0247	0.017
1200	1210	0.0646						0.102	0.0940	0.392
1210	1220	5.102						5.18	3.70	31.012
1220	1230	0.0026						0.0339	0.0264	0.016
1230	1240	0.0106	0.026					0.0247	0.0202	0.066
1240	1250	0.0071	0.0150		0.03	0.007		0.0179	0.0137	0.045
1250	1260	0.0026	0.0187					0.0184	0.0166	0.016
1260	1270	0.0130	0.0187					0.0229	0.0197	0.083
1270	1280	0.004	0.0123					0.0151	0.0110	0.026
1280	1290	0.0029	0.0091					0.0116	0.0117	0.019
1290	1300	0.008	0.0115		0.036	0.007		0.0112	0.0123	0.052
1300	1310	0.068	0.069					0.111	0.102	0.44
1310	1320	0.024	0.0131					0.0198	0.0179	0.16
1320	1330	0.019	0.0135					0.0146	0.0124	0.13
1330	1340	0.106	0.049					0.125	0.111	0.71
1340	1350	0.015	0.0142		0.052			0.0127	0.0130	0.10
1350	1360	0.039	0.024					0.0312	0.0291	0.26
1360	1370	0.024	0.0184					0.0202	0.0195	0.17
1370	1380	0.023	0.0186					0.0192	0.0194	0.16
1380	1390	0.025	0.0182					0.0195	0.0198	0.17
1390	1400	0.069	0.043		0.052	0.030		0.0549	0.0587	0.48

Table 2 (continued)

Wavelength Range		D. & P.	H. & S.	S. & S.	D. et al.	M.P.T.	B. et al.	G.J.R. (1)	G.J.R. (2)	D. & P. Photons
From	To									
1400	1410	0.060	0.044					0.0463	0.0483	0.43
1410	1420	0.034	0.028					0.0286	0.0292	0.24
1420	1430	0.036	0.028					0.0297	0.0337	0.26
1430	1440	0.043	0.035					0.0361	0.0375	0.31
1440	1450	0.040	0.034		0.10			0.0342	0.0389	0.29
1450	1460	0.044	0.036					0.0387	0.0435	0.32
1460	1470	0.061	0.045					0.0516	0.0524	0.45
1470	1480	0.067	0.056					0.0624	0.0674	0.50
1480	1490	0.067	0.054					0.0666	0.0700	0.50
1490	1500	0.072	0.053		0.19	0.070		0.0630	0.0622	0.54
1500	1510	0.080	0.060					0.0692	0.0717	0.61
1510	1520	0.084	0.066	0.0694				0.0727	0.0775	0.64
1520	1530	0.112	0.078	0.0786				0.0964	0.0991	0.86
1530	1540	0.13	0.088	0.0813				0.109	0.108	0.98
1540	1550	0.201	0.109	0.124	0.34			0.175	0.175	1.55
1550	1560	0.17	0.139	0.108				0.161	0.155	1.31
1560	1570	0.17	0.116	0.107				0.150	0.154	1.3
1570	1580	0.15	0.104	0.106				0.127	0.127	1.2
1580	1590	0.14	0.103	0.106				0.119	0.120	1.1
1590	1600	0.15	0.091	0.108	0.64	0.230		0.122	0.118	1.2
1600	1610	0.17	0.109	0.119				0.140	0.123	1.4
1610	1620	0.18	0.124	0.137				0.145	0.144	1.5
1620	1630	0.22	0.156	0.150				0.174	0.169	1.8
1630	1640	0.27	0.179	0.167				0.200	0.175	2.2
1640	1650	0.29	0.182	0.192	1.0			0.215	0.207	2.4
1650	1660	0.42	0.33	0.278				0.323	0.328	3.5
1660	1670	0.35	0.25	0.201				0.250	0.236	2.9
1670	1680	0.40	0.31	0.226				0.291	0.277	3.4
1680	1690	0.47	0.35	0.260				0.358	0.342	4.0
1690	1700	0.62	0.45	0.349	1.62	0.630		0.456	0.438	5.3
1700	1710	0.73	0.55	0.411				0.535	0.512	6.2
1710	1720	0.73	0.55	0.428				0.543	0.519	6.3
1720	1730	0.79	0.58	0.495				0.571	0.543	6.8
1730	1740	0.76	0.56	0.503				0.555	0.539	6.7
1740	1750	0.90	0.62	0.625	2.4		0.697	0.659	0.627	7.9
1750	1760	1.02	0.70	0.853			0.859	0.732	0.753	9.0
1760	1770	1.1	0.73	0.958			0.934	0.788	0.747	9.8
1770	1780	1.2	0.85	1.29			1.15	0.873	0.887	11
1780	1790	1.3	0.93	1.47			1.28	0.990	0.993	12
1790	1800	1.3	0.95	1.43	3.8	1.250	1.27	0.968	0.977	12
1800	1810	1.4	1.24	1.69			1.56	1.14	1.16	13
1810	1820	1.6	1.56	2.19			1.77	1.31	1.33	15
1820	1830	1.5	1.51	2.02			1.86	1.23	1.25	14
1830	1840	1.5	1.50	2.13			1.99	1.08	1.26	14
1840	1850	1.4	1.31	1.79	5.6		1.69	0.875	0.958	13

Table 2 (continued)

Wavelength Range From	Wavelength Range To	D. & P.	H. & S.	S. & S.	D. et al.	M.P.T.	B. et al	G.J.R. (1)	G.J.R. (2)	D. & P. Photons
1850	1860	2.0	1.31	2.04			1.89			19
1860	1870	2.2	1.60	2.53			2.20			21
1870	1880	2.5	1.88	2.81			2.46			23
1880	1890	2.7	1.83	2.95			2.60			25
1890	1900	2.9	2.1	3.08	8.2	2.710	2.82			28
1900	1910	3.1	2.3	3.06			3.07			30
1910	1920	3.4	2.4	3.45			3.25			33
1920	1930	3.7	2.7	3.59			3.54			36
1930	1940	4.0	2.1	2.61			2.69			39
1940	1950	4.4		4.56	11		4.43			43
1950	1960	4.8		4.34			4.32			47
1960	1970	5.1		4.91			4.84			51
1970	1980	5.6		4.90			4.92			56
1980	1990	6.0		4.93			4.98			60
1990	2000	6.6		5.50	14	10.7	5.46			66
2000	2010	7.1		6.19			5.92			72
2010	2020	7.8		6.21			6.66			79
2020	2030	8.4		6.30			6.91			85
2030	2040	9.1		7.45			7.18			93
2040	2050	9.8		8.70	18		8.96			100
2050	2060	10.7		8.89			9.18			110
2060	2070	11.5		9.28			9.80			120
2070	2080	12.		10.8			11.5			130
2080	2090	18.		12.1			13.9			190
2090	2100	26.		A.L.B	29	22.9	20.8			280
2100	2110	37.2		39.4						394
2110	2120	46.3		44.9						494
2120	2130	39.7		41.2						425
2130	2140	43.1		50.7						463
2140	2150	55.2		53.2	48					597
2150	2160	48.3		47.3						524
2160	2170	45.3		43.0						494
2170	2180	48.9		57.5						536
2180	2190	58.9		58.2						648
2190	2200	63.3		66.7	62	57.5				700
2200	2210	61.1		52.1						679
2210	2220	52.		59.3						580
2220	2230	66.		74.8						740
2230	2240	85.5		78.4						963
2240	2250	74.7		73.7	70					844
2250	2260	66.5		60.9						755
2260	2270	49.6		45.1						566
2270	2280	51.8		63.6						594
2280	2290	66.7		62.8						767
2290	2300	58.8		58.6	72	66.7				680

Table 2 (continued)

Wavelength Range From	Wavelength Range To	D. & P.	H. & S.	A.L.B.	D. et al.	M.P.T.	B. et al.	G.J.R. (1)	G.J.R. (2)	D. & P. Photons
2300	2310	68.3		65.3						793
2310	2320	60.5		68.1						706
2320	2330	66.4		60.4						777
2330	2340	54.9		50.7						646
2340	2350	48.1		58.3	64	59.3				568
2350	2360	67.3		62.1						798
2360	2370	59.2		62.1						705
2370	2380	62.1		52.4						743
2380	2390	50.5		58.2						606
2390	2400	54.9		51.1	68	63.0				662
2400	2410	50.7		49.9						614
2410	2420	64.		79.3						778
2420	2430	86.		82.9						1050
2430	2440	77.3		78.2						949
2440	2450	73.3		63.6	78	72.3				902
2450	2460	60.1		58.9						743
2460	2470	61.1		66.3						758
2470	2480	66.9		61.3						834
2480	2490	52.5		55.7						657
2490	2500	70.2		74.8	76	70.4				882
2500	2510	68.5		61.8						865
2510	2520	53.3		50.2						676
2520	2530	51.2		56.9						651
2530	2540	64.7		68.3						826
2540	2550	70.7		82.0	112	104.0				906
2550	2560	99.8		101.6						1280
2560	2570	120.		141.1						1550
2570	2580	144.		145.3						1870
2580	2590	148.		138.5						1930
2590	2600	115.		106.4	140	130.				1500
2600	2610	107.		96						1400
2610	2620	108.		128.						1420
2620	2630	123.		119.						1630
2630	2640	190.		270.						2530
2640	2650	279.		267.		185				3720
2650	2660	292.		288.						3910
2660	2670	269.		274.						3610
2670	2680	278.		278.						3750
2680	2690	267.		261.						3620
2690	2700	264		288.		232				3580
2700	2710	302.		283.						4110
2710	2720	234.		201.						3210
2720	2730	228.		247.						3140
2730	2740	204.		158.						2800
2740	2750	142.		164.		204				1970

Table 2 (continued)

Wavelength Range		D. & P.	H. & S.	A.L.B.	D. et al.	M.P.T.	B. et al.	G.J.R. (1)	G.J.R. (2)	D. & P. Photons
From	To									
2750	2760	212.		245.						2950
2760	2770	267.		273.						3710
2770	2780	244.		202.						3410
2780	2790	169.		131.						2370
2790	2800	89.8		87.1		222				1260
2800	2810	121.		175.						1710
2810	2820	244.		289.						3460
2820	2830	320.		336.						4550
2830	2840	340.		318.						4850
2840	2850	241.		155.		315				3450
2850	2860	183.		295.						2640
2860	2870	359.		380.						5190
2870	2880	358.		301.						5180
2880	2890	353.		429.						5130
2890	2900	509.		598.		482				7410
2900	2910	623.		611.						9110
2910	2920	592.		550.						8690
2920	2930	520.		562.						7670
2930	2940	554.		525.						8190
2940	2950	521.		531.		584				7730
2950	2960	585.		593.						8700
2960	2970	499.		451.						7460
2970	2980	567.		524.						8500
2980	2990	442.		497.						6650
2990	3000	496.		434.		514				7490

Parkinson and Reeves (1969); Widing, Purcell, and Sandelin (1970); Carver et al. (1972); and Broadfoot (1972). Two of these data sets, those of Broadfoot and Detwiler et al., are listed separately in columns 5 and 6, respectively. There are several emission lines below 1560 Å. Donnelly and Pope list 20 of these single lines or groups of lines. In Table II, the energies of these lines have been added to that of the continuum in the corresponding 10 Å range. The strongest of these is Lyman-alpha at 1215.7 Å, with an energy flux (according to Donnelly and Pope) of 5.1 erg s^{-1} cm^{-2}, which is nearly twice the total solar irradiance below this wavelength.

In column 4 the values of spectral irradiance obtained by Heroux and Swirbalus (1976) from a rocket flight of 2 November 1973 are presented. The authors also made a later flight on 23 April 1974, and values from the second flight are shown graphically but not listed in tabular form in their publication. The 23 April values were 5 to 10% lower than those from 2 November, a difference they attribute to a change in spectrometer efficiency, not to a decrease in solar flux. Unlike the other entries in Table II, these values have not been adjusted to 1 AU. Since the sun-earth distance on 2 November was 0.8% less than 1 AU., the values of column

4 should be adjusted downward by 1.6%. The wavelength range for these measurements is 1230 to 1940 Å.

In column 5 the values are from two sources. For the range from 1510 to 2090 Å, the data are from a preprint of Samain and Simon (1976), values they obtained with a rocketborne spectrograph flown on 17 April 1973. Direct measurements over another part of the spectrum, 2105 to 3005 Å, are from Broadfoot (1972), in the same column. For Broadfoot's data and for the next two columns, the entries fall midway between the lines of the other columns, since the measurements published in the literature apply to bands centered at integral multiples of 10 Å. Thus, for example, the first entry in column 5 for 2100 to 2110 Å is 39.4, which is the average irradiance in the range 2105 to 2115 Å. The values listed were derived by converting Broadfoot's photon flux data into ergs and adding 3.18% to adjust to 1 AU.

The data in column 6 are from rocket flights of a much earlier period, made under the auspices of the group at the Naval Research Laboratory. The measurements were on a relative scale, and conversion to absolute units was made by comparing the results in the range beyond 3000 Å with the data obtained by Dunkelman and Scolnik (1959) from Mt. Lemmon and scale adjustments of these data made by Johnson (1954). The values are averages over 50 Å bandwidths.

In column 7 a similar set of values derived by Thekaekara (1973, 1974) and averaged over wide wavelength bands is given. These values are from the first portion of Thekaekara's table, which extends from 1150 Å to 1000 μm. The direct measurements of the Goddard Space Flight Center group from the Convair 990 research aircraft (Thekaekara, Kruger, and Duncan, 1969) cover the wavelength range from 3000 Å to 15 μm. The extension of the results to wavelengths below 3000 Å was based on the results of Heath (1969), Detwiler et al. (1961), and Parkinson and Reeves (1969), with adjustments in the scale based on the absolute measurements beyond 3000 Å made from the Convair 990 by the Goddard group.

Data over a more limited wavelength range, from 1740 to 2105 Å, are given in column 8. They are from a preprint of Brueckner et al. (1976) and were obtained from a rocket flown on 4 September 1973. Data were also obtained on this flight for the lower wavelength range, down to 1175 Å, but they have not yet been published. The measurements were made with a spectrometer pointed at a 60 × 60 arcsec area of the quiet sun. The authors give four sets of data:(1) plage, sun center; (2) quiet sun, sun center; (3) quiet sun, average disk intensity; and (4) quiet sun flux at 1 AU. The fourth set of values is quoted here. Brueckner's listing is for 5 Å averages and pairs of values have been summed to give irradiance over 10 Å bandwidths.

Two sets of values obtained by Rottman (1974) from rocket flights of 13 December 1972 and 30 August 1973, respectively, are given in columns 9 and 10. The wavelength range is 1160 to 1850 Å.

The final column, as stated earlier, is the solar flux in units of 10^{10} photon cm^{-2} s^{-1} nm^{-1}; the values are from Donnelly and Pope.

Most of these data sets are also shown graphically in Figures 3 and 4 for wavelength ranges from 1300 to 2300 Å and from 2000 to 3000 Å, respectively. Data not included in these figures are those of Detwiler et al., which are significantly different from all later measurements, and the data from Rottman's second rocket flight, which are very close to those from the first flight. Three other sets of data are also shown graphically that could not be entered conveniently in Table II because the wavelength intervals are not multiples of 10 Å. They are from Ackerman (1971), Simon (1975), and Heath (1973). Ackerman's listing is based on a review of several earlier measurements and covers the whole wavelength range of these figures. Simon's data were obtained from balloon measurements and extrapolated to zero air mass. They cover the wavelength ranges from 1961 to 2299 Å and from 2857 to 3525 Å. Heath lists solar irradiance at 12 wavelengths between 2557 and 3399 Å. The measurements were made from the Nimbus 4 spacecraft with a double-grating monochromator that had a 10 Å spectral bandpass and a triangular slit function.

The results presented in Table II and Figures 3 and 4 fail to show the fine

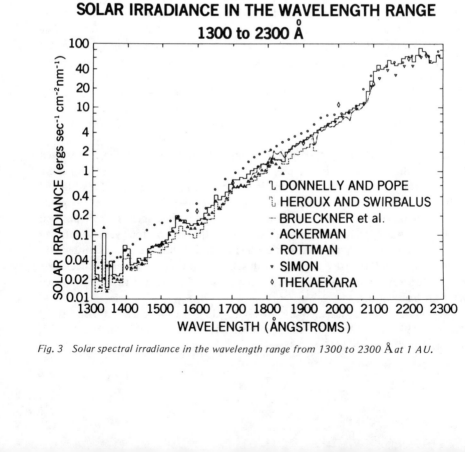

Fig. 3 Solar spectral irradiance in the wavelength range from 1300 to 2300 Å at 1 AU.

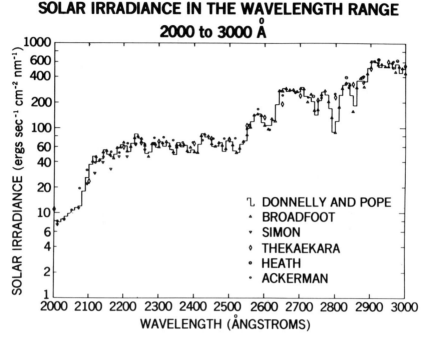

Fig. 4 Solar spectral irradiance in the wavelength range from 2000 to 3000 Å at 1 AU.

structure that exists in the solar spectrum. Many of the observers obtained spectra with considerably higher resolution than the 10 Å bandwidth adopted for these figures. Figure 5, from Broadfoot (1972), gives an example of high-resolution spectra. Broadfoot also gives an extended table of irradiance values at 1 Å intervals and includes with his spectral curve a comparison spectrum based on earlier data from the Naval Research Laboratory. These data are not reproduced here. The line structure of Broadfoot's spectra is essentially the same as that of the earlier Naval Research Laboratory spectra, but the flux values are lower. The maximum differences between the two sets of data when the ratio curve is smeared with a triangular function of 25 Å halfwidth are about 30% at 2500 Å and 40% at 3200 Å. Without the smearing, the differences are greater. These differences—for example, a factor of 2 at 3160 Å—should be attributed to measurement problems rather than to intrinsic variations in solar output.

Digitized versions of high-resolution spectra are available in two major reports. A NASA report prepared at the Jet Propulsion Laboratory by Brinkman, Green, and Barth (1966) gives in digital form the densitometer tracings of the Naval Research Laboratory spectra. The regions covered are from 880 to 1550 Å and from 1760 to 2990 Å. The tables give the irradiance at intervals of 0.1 Å and

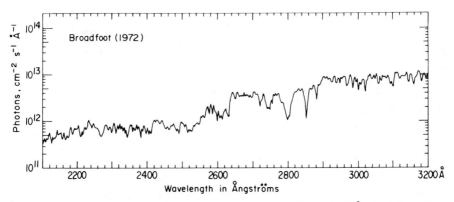

Fig. 5 Solar spectral irradiance in the wavelength range from 2100 to 3200 Å. From Broadfoot (1972).

0.2 Å and integrated values of the irradiance over intervals of 1, 10, and 50 Å. A National Center for Atmospheric Research report published by Furukawa, Haagenson, and Scharberg (1967) gave a new estimate of the absolute solar flux from the digital data of Brinkman, Green, and Barth, using unpublished Naval Research Laboratory spectra, Soviet data, and ground-based measurements. The wavelength range is from 2080 to 3600 Å, with over 2400 data points for the range from 2080 to 3050 Å and about 10,000 data points for the range from 3050 to 3600 Å.

A major problem with all measurements of spectral radiance or irradiance is the absolute radiometric accuracy. Tungsten coiled coil lamps, referred to as quartz-iodine lamps (1000W), are available from the National Bureau of Standards (NBS) and many commercial suppliers as standards of spectral irradiance. They cover the range from 2500 to 25,000 Å. Such lamps have been issued since 1964 (Stair, Schneider, and Jackson, 1963) by NBS, but since 1967 doubts have been raised as to their absolute accuracy. As a result of more recent research conducted at NBS, new spectral irradiance standards became available in 1973. A comparison of the scales of 1964 and 1973 shows that in the wavelength range between 2500 and 3000 Å the new scale is lower by percentages varying between 5 and 10%. The uncertainties in other sources and in the detectors used below 2500 Å are somewhat greater. With possible inaccuracies in the so-called standards, the absolute accuracy of the transfer in calibration to an experiment on rocket, balloon, or spacecraft decreases. Hence, it is not surprising that measurements of the sun made by different observers do not yield identical values. As an example of the variant readings, the data from six different measurements are shown in Figure 6, which is taken from Donnelly and Pope (1973). The wavelength range, from 1400 to 1750 Å, is one-third of that covered in Figure 3, and the irradiance scale is expanded. Over most of the range, the data of Widing, Purcell, and Sandlin (1970) have values about three times greater than those of Parkinson and Reeves (1969), and at 1600 Å the former is 5.2 times the latter. The brightness temperature that best fits the data is 4700 K for Widing, Purcell, and Sandlin and 4400 K for Parkinson and Reeves. The ratio of solar irradiances at these two temperatures decreases gradually from 4.44 at 1400 Å to 3.30 at 1750 Å.

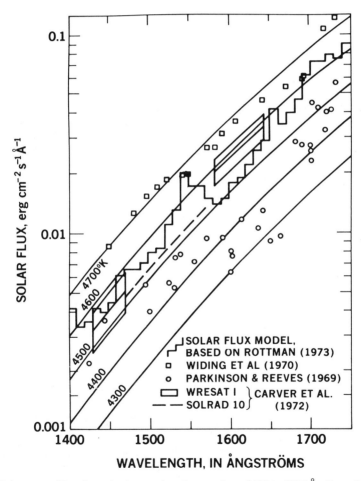

Fig. 6 Solar spectral irradiance in the wavelength range from 1400 to 1750 Å. From Donnelly and Pope (1973).

II. SOLAR FLUX VARIABILITY

The effects of solar variability on the density of the thermosphere are well established from satellite drag measurements (King-Hele and Quinn, 1965) for sunspot maximum and sunspot minimum and for the 27-d solar rotational period (Jacchia, 1963). The principal region of the solar flux that produces this effect originates below 1200 Å. Direct measurements of the variability associated with the 27-d solar rotational period of the solar EUV flux below 1200 Å have been reported by numerous authors (e.g., Hall and Hinteregger, 1970). The magnitude of the variability of the solar EUV flux below 1200 Å between solar minimum and solar maximum, however, has not been measured directly.

Above 1200 Å, the 27-d variability in the UV flux has been measured from satellites over extended periods by Vidal-Madjar, Blamont, and Phissamay (1973) at Lyman-alpha and by Heath (1973) at Lyman-alpha and at 1750 Å. Other satellite measurements of a 27-d variability of the solar flux coming from the region of the solar temperature minimum have been made by Prag and Morse (1970) and Hinteregger (*private communication*). The observations by Prag and Morse are not consistent with those by Heath and Hinteregger. An example of the wavelength dependence of the 27-d variability observed with three channels of an experiment on Nimbus 3 (Heath, 1973) is shown in Figure 7 for a period of very high solar activity in May 1969.

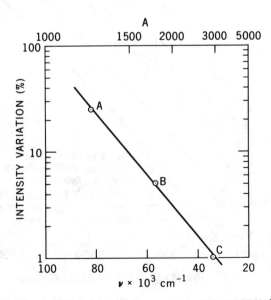

Fig. 7 Variation in UV solar spectral irradiance per solar rotation, $100 \times (E_{max} - E_{min})/E_{min}$, observed with three sensors on Nimbus 3 near solar maximum.

Since the apparent 27-d variability of the incident solar flux that originates between the transition region and the photosphere is small, it may be difficult to observe direct effects in the stratosphere and mesosphere. While the principal source of the 27-d variability of the shortest wavelength (channel A in Figure 7) is Lyman-alpha, that centered about 1750 Å has been shown by Brueckner et al. (1976) to be quantitatively related (within 40%) to an enhanced continuum and line emission in active regions.

III. THE 11-Y CYCLE

The sunspot cycle, a nominal period of 11-y, is defined in terms of a modulation in the number of sunspots and sunspot groups with time. This cycle is characterized by the Wolf or Zurich sunspot number, R, which is a function of the total number of spots and the number of spot groups.

An 11-y solar variability is of considerable interest in the area of sun-weather relations, since many correlations with atmospheric phenomena have been reported (e.g., see the recent review by King, 1975). Recently Smith and Gottlieb (1974) have concluded that there is no variation longward of 1500 Å over the 11-y sunspot cycle.

Observations by Heath along with those obtained from a balloon flight by Brewer and Wilson (1965), have indicated possible evidence for an 11-y solar-cycle variability in the ultraviolet solar flux. Some of these observations are shown in Figure 8 as deviations from an arbitrary model of solar spectral irradiance. The observations by Heath (1973, 1976) are based on a combination of rocket and satellite measurements that began in August 1966 and extended through May 1976. The open data points corresponding to the years 1966, 1969, and 1970 represent broad-band photometric observations by the Monitor of Ultraviolet Solar Energy Experiment (MUSE) from a rocket flight in August 1966 and from the satellites Nimbus 3 and 4, which were launched in April 1969 and April 1970, respectively. These experiments were calibrated against a standard CsTe vacuum photodiode which had been calibrated by NBS. The solid circle is from the measurements reported by Ackerman (1973) and the open triangle represents the balloon observations by Brewer and Wilson (1965). The crosses were obtained with a double-monochromator experiment, which was flown on Nimbus 4 in April 1970 and on Explorer 55 in November 1975. The latter represents the flight of a backscattered ultraviolet (BUV) experiment flight model from the Nimbus program, a unit identical to the one launched in April 1970. Both double monochromators observed the sun with a 10 Å spectral bandpass at 12 discrete wavelengths from 2550 to 3400 Å, and both were calibrated against 1000W, quartz-iodine tungsten lamp standards of spectral irradiance.

Envelopes of the data obtained near solar maximum and solar minimum are

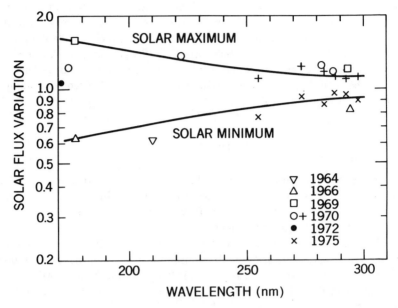

Fig. 8 Variations in solar spectral irradiance apparently related to the 11-y sunspot cycle, based on observations from 1964 to 1975.

indicated by the two solid lines in Figure 8. These observations indicate that the solar flux at 1750 Å is about a factor of 2.5 greater at solar maximum than at solar minimum, and at 3000 Å the effect is about 18%. These changes over the solar cycle correspond to increases in the equivalent brightness temperature of the sun of about 240 K at 1800 Å and 120 K at 3000 Å.

The variability that has been observed to be associated with the 11-y sunspot cycle is about a factor of 20 greater than that observed over the 27-d solar rotational period. An interesting question then remains as to whether the variation in UV flux through the maxima and minima of solar cycles is also characteristic of a similar change in the total solar irradiance at all wavelengths or whether this UV flux emitted principally by the solar chromosphere and enhanced at solar maximum is compensated by a corresponding decrease in the longer-wavelength flux that originates lower in the solar photosphere. Such a spectral redistribution with the solar cycle could then result in the apparent constancy of the "solar constant" with time.

REFERENCES

Ackerman, M., 1971, *Mesospheric Models and Related Experiments*, ed. G. Fiocco, D. Reidel, Dordrecht, 149.

_____, and Simon, P., 1973, *Solar Phys. 30*, 345.

Anon., 1971, *Solar Electromagnetic Radiation, NASA Space Vehicles Design Criteria*, NASA SP 8005.

Brewer, A.W., and Wilson, A.W., 1965, *Quart. J.Roy. Meteorol. Soc. 91*, 452.

Brinkman, R.T., Green, A.S., and Barth, C.A., 1966, *JPL Tech. Rept. No. 32-951*.

Broadfoot, A.L., 1972, *Ap. J. 173*, 681.

Brueckner, G.E., and Moe, O.K., 1971, *Space Research XII*, Akademie-Verlag, Berlin, 1595.

_____ ; Bartoe, J-D. F.; Moe, O.K.; and Van Hoosier, M.E., 1976, *Ap. J.*, preprint.

Carver, J.H.; Horton, B.H.; Lockey, G.W.A.; and Rofe, B., 1972, *Solar Phys. 27*, 347.

Detwiler, C.R.; Garrett, D.L.; Purcell, J.D.; and Tousey, R., 1961, *Ann. Geophys. 17*, 263.

Donnelly, R.F., and Pope, J.H., 1973, *NOAA Technical Report ERL 276-SEL 25*, U.S. Government Printing Office, Washington, D.C.

Dunkelman, L., and Scolnik, R., 1959, *J. Opt. Soc. Am. 49*, 356.

Dupree, A.K., and Reeves, E.M., 1971, *Ap. J. 165*, 599.

Friedman, H., 1960, *Physics of the Upper Atmosphere*, ed. J.A. Ratcliffe, Academic Press, New York, 134.

Furukawa, P.M., Haagensen, P.L., and Scharberg, M.J., 1967, *NCAR Technical Notes*, NCAR-TN-26.

Hall, L.A., and Hinteregger, H.E., 1970, *J. Geophys. Res. 74*, 4181.

Heath, D.F., 1969, *J. Atmos. Sci. 26*, 1157.

_____ , 1973, *J. Geophys. Res. 78*, 2779.

_____ , 1976, *International Symposium on Solar-Terrestrial Physics*, Boulder, Colo., 7-18 June (abstract).

Heroux, L., and Swirbalus, R.A., 1976, *J. Geophys. Res. 81*, 436.

Hinteregger, H.E., 1970, *Ann. Geophys. 26*, 547.

Jacchia, L.G., 1963, *Rev. Mod. Phys. 35*, 973.

Johnson, F.S., 1954, *J. Meteorol. 11*, 431.

King, J.W., 1975, *Astr. and Aeronaut. 13*, 10.

King-Hele, D.G., and Quinn, E., 1965, *J. Atmos. Terr. Phys. 27*, 197.

Nishi, K., 1975, *Solar Phys. 42*, 37.

Parkinson, J.H., and Reeves, E.M., 1969, *ibid. 10*, 342.

Prag, A.B., and Morse, F.A., 1970, *J. Geophys. Res. 75*, 4613.

Rottman, G.J., 1974, *A.G.U. Meeting, San Francisco*, Dec.

Samain, D., and Simon, P.C., 1976, *Solar Phys.*, preprint.

Simon, P., 1975, *Bull. Acad. Roy. Belgique 61*, 399.

Smith, E.V.P., and Gottlieb, D.M., 1974, *Space Sci. Rev. 16*, 771.

Stair, R., Schneider, W.E., and Jackson, J.K., 1963, *Appl. Optics 2*, 1151.

Thekaekara, M.P., 1973, *Solar Energy 14*, 109.

——————, 1974, *Appl. Optics 13*, 518.

——————, Kruger, R., and Duncan, C.H., 1969, *ibid. 8*, 1713.

Timothy, A.F.; Timothy, J.G.; Willmore, A.P.; and Wagner, J.H., 1972, *J. Atmos. Terr. Phys. 34*, 969.

Tousey, R., 1963, *Space Sci. Rev. 2*, 3.

Vernazza, J.E., Avrett, E.H., and Loeser, R., 1976, *Ap. J. Suppl. 60*, 1.

Vidal-Madjar, A., Blamont, J.E., and Phissamay, B., 1973, *J. Geophys. Res. 78*, 1115.

Widing, K.G., Purcell, J.D., and Sandlin, G.D., 1970, *Solar Phys. 12*, 52.

Comment

Additional solar flux values between 1500 and 2100 Å from work by Samain and Simon (1976) might have been included in this review paper. This recent work also provides data for the center-to-limb variation in this wavelength band. Such measurements are very useful in estimating the solar irradiance values from intensities measured at the center of the solar disk.

The authors' conclusion concerning the 11-y cycle variability of the solar irradiance between 1700 and 2000 Å must be considered very carefully. The number of reliable measurements available at this time does not seem large enough to definitely establish the trend reported in this review. Further measurements are needed to give the complete picture of the variation in this wavelength band. Figure 8 of this paper has to be considered as a trend. A factor of 2 variation in the irradiance at 2000 Å between solar minimum and maximum seems too high.

P.C. Simon
Institut d'Aeronomie
Spatiale de Belgique

Reference

Samain, D., and Simon, P.C., 1976, *Solar Phys. 49*, 1, 33.

THE SOLAR SPECTRUM AT LYMAN-ALPHA 1216Å

Alfred Vidal-Madjar
Laboratoire de Physique Stellaire et Planétaire
Verrières-le-Buisson

The Lyman-alpha line of hydrogen was first discovered in the laboratory in 1914 by Theodore Lyman. This line corresponds to the $1s^2 S$-$2p^2 P^o$ resonance transition at 1215.668Å and 1215.674Å. The first attempt to detect that emission line from the sun was made by Friedman, Lichtman, and Byram (1952), who found a total irradiance value between 1 and 10 ergs cm^{-2} s^{-1}. Rense (1953) was the first to give an absolute identification of the solar Lyman-alpha emission line; he observed it with a narrow-band spectrograph of about 1Å resolution. Hydrogen Lyman-alpha is a unique, very bright chromospheric emission line of the solar spectrum; its flux is equivalent to the total flux emitted by the sun from 0 to 1500Å. The emission plays a number of important roles in the solar system. It is responsible for the D-region ionization in the upper atmosphere under quiet conditions, as first suggested by Nicolet (1945); it interacts with planetary and cometary outer envelopes; and it controls the dynamics of hydrogen atoms entering the solar system from nearby interstellar space (the irradiance from the center of the line is large enough to compensate, through radiation pressure, for the solar attraction).

After the first detection, several attempts were made during the 1950s to define the Lyman-alpha emission line from the sun more precisely. The line profile, as observed by Purcell and Tousey (1960), is of the order of 1Å wide with a large reversal at the center; the two peaks are separated by about 0.4Å. Purcell and Tousey's spectral resolution (0.03Å) was good enough to show very clearly a central narrow absorption feature due to the atomic hydrogen in the earth's atmosphere.

The distribution of the Lyman-alpha emission over the solar disk was also observed. First, Mercure et al. (1956) and then Purcell, Packer, and Tousey (1959), the latter with 30 arcsec resolution, demonstrated very clearly that the Lyman-alpha emission was well correlated with the Ca K emission in bright areas of the sun.

During the late 1950s, several observations were made of the solar Lyman-alpha irradiance; these are shown in Table 1. After various evaluations using different detection methods, the estimated values of the irradiance stabilized at approximately 0.006 W m^{-2} as measured by the ion-chamber detector.

Table 1

Lyman-α Measurements Taken From Freidman (1960) p. 167, From 1949 to 1958

References	Launch Time	Intensity Above Atmosphere (ergs cm^{-2}sec^{-1})	Instrumentation	Band Width (Å)
a	1000 MST, 9/29/49	1-10	Photon counter	1100-1350
b	1101 MST, 2/17/50	0.4	Thermoluminescent phosphor	1050-1240
c	0659 MST, 4/30/52	0.15	Photon counter	1180-1300
c	0644 MST, 5/5/52	0.10	Photon counter	1180-1300
d	1238 MST, 12/12/52	1.5	Spectrograph	
e	0830 MST, 2/2/54	0.6	Spectrograph	
e	0830 MST, 2/24/55	0.6	Spectrograph	
f	1550 MST, 10/18/55	5.7 (-1, +3)	Ion chamber	1065-1350
f	1715 MST, 10/21/55	4.0 (± 0.8)	Ion chamber	1065-1350
f	0830 MST, 11/4/55	9.0 (± 3)	Ion chamber	1065-1350
g	1915 UT, 7/20/56	6.1 ± 0.5	Ion chamber	1065-1350
g	2113 UT, 7/25/56	6.7 ± 0.3	Ion chamber	1065-1350
h	1600 CST, 7/20/57	6.1 ± 0.3	Ion chamber	1065-1350
h	1208 CST, 3/23/58	6.3 ± 0.3	Ion chamber	1065-1350

Table 1 (continued)

a   H. Friedman, S. W. Lichtman, and E. T. Byram, Phys. Rev. 83, 1025 (1952).

b   R. Tousey, K. Watanabe, and J. D. Purcell, Phys. Rev. 83, 165 (1951).

c   E. T. Byram, T. A. Chubb, H. Friedman, and N. Gailar, Phys. Rev. 91, 1278 (1953).

d   W. A. Rense, Phys. Rec. 91, 200 (1953).

e   F. S. Johnson, H. H. Malitson, J. D. Purcell, and R. Tousey, Astrophys. J. 127, 80 (1958).

f   E. T. Byram, T. A. Chubb, H. Friedman, and J. E. Kupperian, Jr. Astrophys. J. 121, 480 (1956).

g   T. A. Chubb, H. Friedman, R. W. Kreplin, and J. E. Kupperian, Jr. J. Geophys. Res. 62, 389 (1957).

h   E. T. Byram, T. A. Chubb, H. Friedman, and J. E. Kupperian, Jr., private communication; corrected for lines other than Lyman-α, included in band width.

In summary, by 1960 the solar Lyman-alpha irradiance was approximately known; but clearly, according to line profile and distribution studies, some variation was to be expected.

I. STUDY OF THE VARIATION OF THE LYMAN-ALPHA IRRADIANCE

The problem is to find whether the variations observed are due to the absolute calibration of any instrument or to a real change in solar emission. Some of the first results obtained during the late 1950s and early 1960s are presented in Table 2.

Table 2

Other Observations of the Lyman Alpha Irradiance Prior to 1965

Author	Time	ergs* $cm^{-2}s^{-1}$	10^{11} ph $cm^{-2}s^{-1}$	Instrumentation
Miller et al. (1956)	Dec 13, 1955	3.0	1.8	Spectrograph
Aboud et al. (1959)	Aug 6, 1957	3.43	2.1	Spectrograph
Purcell and Tousey (1960)	Jul 21, 1959	6.0	3.7	Ion Chamber
Hinteregger (1960)	Jan 19, 1960	6.0	3.7	Ion Chamber
Detwiler et al. (1961)	Apr 19, 1960	6.0	3.7	Ion Chamber
Efremov et al. (1962)	Aug 20, 1960	5.5	3.4	Photoelectric
Hinteregger (1961)	Aug 23, 1960	3.3	2.0	Monochromator
Hall et al. (1963)	Aug 23, 1961	5.1	3.1	Monochromator
Hall et al. (1965)	Dec 12, 1963	4.5	2.7	Monochromator

*or in 10^{-3} $W.m^{-2}$ unit

From this table, except for the August 23, 1960 measurement, the Lyman-alpha irradiance appears to decrease from 0.006 to 0.0045 Wm^{-2} between 1960 (solar maximum) and 1965 (solar minimum). But such a decrease is very hard to prove, since this variation is of the same order as the calibration uncertainty. Weeks

(1967) made a more precise study with a larger number of measurements made with the same observational technique (ion chambers) in order to minimize the possible systematic errors. Figure 1 (from Weeks, 1967) shows the different measurements made from 1955 to 1965 along with a statistical study of the results. Weeks concluded that at the maximum of the solar cycle the Lyman-alpha irradiance was equal to 6.1 ±0.45 ergs cm^{-2} s^{-1}, and for the minimum of solar activity it was equal to 4.3 ±0.35 ergs $c_\odot^{-2}$ s^{-1}. From this first attempt, the variation of the solar Lyman-alpha irradiance appears to be of the order of 40% over the 11 y solar cycle, or almost a factor-of-two increase from solar minimum to solar maximum.

To obtain information concerning the variation of irradiance with the 27 d solar rotation, an orbiting instrument was needed. The first result was presented by Kreplin, Chubb, and Friedman (1962) using SR-1 satellite data obtained in July 1960. They observed less than 18% variation in the Lyman-alpha irradiance. Fossi, Poletto, and Tagliaferri (1968), using SOLRAD 8 data from March to May 1966, obtained absolute fluxes through comparison with rocket data (Weeks, 1967). A flux variation of up to 15% was clearly visible during that period and it related to the 27 d solar rotation through motion of active regions across the solar disk.

In a similar manner, Hall and Hinteregger (1970) analyzed OSO-3 results to find that in May 1967 a variation of up to 30% occurred in the Lyman-alpha irradiance and that this variation was due to the solar rotation. Prag and Morse (1970), using OV1-15 atmospheric density satellite data from a broadband detector (1150-1600Å), observed a 60% variation in solar irradiance during one

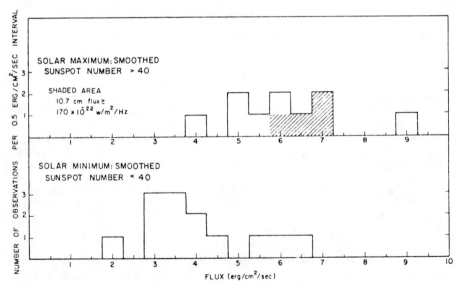

Fig. 1 Rocket measurements of Lyman-alpha irradiance by ion chambers for solar maximum and solar minimum (from Weeks, 1967).

solar rotation in July-August 1968, but they noted that the length of the survey was too short to produce a precise result.

More recently, from Nimbus 3 and 4 data over a much longer period of observation (April 1969 to March 1972), Heath (1973) observed Lyman-alpha in a broad band (1150-1600Å) and deduced a 30% variation in Lyman-alpha irradiance related to the solar rotation. Woodgate et al. (1973), using OSO-6 data over a six-month period from August 1969 to February 1970, also observed a 30% variation with solar rotation. Furthermore, they found a good correlation between the solar Lyman-alpha irradiance and the Zurich sunspot number, R_Z, and the solar 10.7 cm flux, F10.7. The regression relations deduced from these data were also presented.

Finally, I surveyed the longest record of Lyman-alpha irradiance (Vidal-Madjar, 1975), four years of data (January 1969 to December 1972) from the OSO-5 satellite. I found that no more than 30% variation in the Lyman-alpha irradiance can be related to the 27 d solar rotation. Good correlations with the Zurich sunspot number R_Z, the F10.7 solar flux, and the calcium plage index, Ca II, were also found. Here again, regression relations were presented. Table 3 summarizes the different results concerning the variation of Lyman-alpha irradiance with the rotation of bright chromospheric features.

Table 3

Lyman Alpha Irradiance Variations Observed From In-Orbit Spacecrafts

Author	Year	Spacecraft	Ly-α Irradiance Variation	Correlation Coefficient		
				R_z	$F_{10.7}$	CaII
Kreplin et al. (1962)	1960	SR-1	<18%			
Fossi et al. (1968)	1966	Solrad-8	15%		0.7	0.77
Hall et al. (1970)	1967	OSO-3	30%			
Prag et al. (1970)	1968	OV1-15	60%			
Heath (1973)	1969–1972	Nimbus 3 & 4	30%	0.52	0.51	0.66
Vidal-Madjar (1975)	1969	OSO-5	30%	0.75	0.68	0.54
Woodgate et al. (1973)	1969	OSO-6	30%	0.66	0.66	
Vidal-Madjar (1975)	1970	OSO-5	28%	0.60	0.64	0.65
Vidal-Madjar (1975)	1971	OSO-5	24%	0.46	0.73	0.65

At the peak of the solar cycle, the 60% variation observed by Prag and Morse (1970) seems to be overestimated, and this underlines the difficulty of such long-term variation studies using instruments that suffer aging at the same time. In many cases, the aging was very severe, and the deductions were difficult. Nevertheless, in some cases the instrument aging was slow enough compared to the 27 d rotation period to give confidence in the variation data. Although my studies indicate that Lyman-alpha irradiance should not vary more than 30% at solar maximum, even this point might be questioned; according to Brueckner (1975), two instruments (SOLRAD and OSO-5) measuring the same solar irradiance with the same bandpass simultaneously may differ by up to 40%. Under such conditions any observed variation with an amplitude smaller than 40% may be doubted and no conclusion should be drawn.

The advantages of orbital instruments are thus balanced by the uncertainties of their sensitivity once in operation. The need for rocket flights to calibrate the space measurements is obvious. Table 4 presents more recent measurements of Lyman-alpha irradiance for purposes of comparison.

From this table, it is apparent that, for rocket flights also, measuring the solar Lyman-alpha irradiance with identical observational techniques at the same time (solar eclipse March 7, 1970) does not necessarily produce compatible results. To attempt to verify the previous variability results and to estimate the variation over the 11 y solar cycle, let us compare the satellite data with rocket flight measurements during that period. For this comparison we may use the OSO-5 data (Vidal-Madjar, 1975), which represents the longest survey with the least amount of instrument aging (10% degradation per year). Table 5 presents all OSO-5 measurements that may be compared to rocket flight observations.

Comparison of Tables 4 and 5 shows that OSO-5 data are consistent with four different rocket measurements to an accuracy of better than 10%. This agreement remains very good even after four years of OSO-5 operation. This result should make us confident that the 30% variation observed with the 27 d cycle is a realistic figure.

The OSO-5 data also show, as already suggested by Tousey (1963), Fossi, Polett, and Tagliaferri (1968), and Prinz (1974), that the variation of Lyman-alpha irradiance does not merely reflect a constant "quiet-sun" irradiance over which there is superimposed a 27 d variation due to active areas. The quiet-sun irradiance also varies with the 11 y solar cycle. Using empirical relations, I showed quantitatively that from year to year the quiet-sun Lyman-alpha irradiance F_q may change in the following manner (Vidal-Madjar, 1975):

$$F_q = 0.63 \times 10^{-2} \times \overline{F}_{10.7} + 1.49 \quad , \tag{1}$$

where F_q is in units of 10^{11} photons cm^{-2} s^{-1} and $\overline{F}10.7$ represents the yearly average of the 10.7 cm solar flux in units of 10^{22} W m^{-2} Hz^{-1}. This relationship

Table 4

The Lyman Alpha Irradiance Observed From Rocket Flights or Spacecrafts From 1966 to 1975

Author	Time	Ly-α Irradiance in erg cm^{-2}s^{-1}	10^{11} ph cm^{-2}s^{-1}	Remarks S = Spacecraft R = Rocket
Fossi et al. (1969)	Mar 20, 1966	4.2	2.6	S
Bruner and Parker (1969)	Apr 14, 1966	5.2	3.2	R ion chamber
Hinteregger (1970)	Mar 11, 1967	5.3	3.2	S
Bruner and Rense (1969)	Oct 19, 1967	5.3	3.2	S private comm
Dupree and Reeves (1971)	Oct 26, 1967	3.6	2.2	S $\sim$ quiet sun from 1x1' area
Dupree et al. (1973)	Aug 14, 1969	5.6	3.4	S $\sim$ quiet sun from 35x35" area
Woodgate et al. (1973)	Oct 2, 1969	4.9	3.0	S
Dickinson (1972)	Mar 7, 1970	5.3	3.2	R ion chamber
Smith (1972)	Mar 7, 1970	3.1	1.9	R ion chamber
Higgins (1965)	Nov 9, 1971	5.78	3.53	R
Ackerman and Simon (1973)	Feb 28, 1972	3.4	2.1	Two rockets
Prinz (1974)	Jul 10, 1972	5.31	3.25	R
Heroux et al. (1974)	Aug 23, 1972	4.72	2.89	R
Rottman (1974)	Dec 13, 1972	5.18	3.17	R
Timothy (1976)	Jun 1, 1973	5.33	3.26	S
Rottman (1974)	Aug 30, 1973	3.70	2.26	R
Bruner (1976)	Jul 28, 1975	4.04	2.47	R

*or in 10^{-3} W.m^{-2} unit

Table 5

Comparison of OSO-5 Data to Rocket Flight Observations

Time	Lyman Alpha Irradiance From OSO-5 in ergs cm^{-2} s^{-1} *	10^{11} ph cm^{-2} s^{-1}	Comparison With Rocket Data
Mar 7, 1970	5.57	3.41	-6%
Nov 9, 1971	~4.6	~2.8	+26%
Feb 28, 1972	~4.9	~3.0	-30%
Jul 10, 1972	4.92	3.01	+8%
Aug 23, 1972	4.89	2.99	-3%
Dec 13, 1972	4.71	2.88	+10%

*or in 10^{-3} W.m^{-2} unit

shows that from solar minimum to maximum, the quiet-sun Lyman-alpha irradiance may vary by close to a factor of two. This result is in good agreement with the first evaluation made by Weeks (1967). The OSO-5 data thus seem to represent quite well the short-and long-period variations of the solar Lyman-alpha irradiance, and the deduced empirical laws give probably a good representation of the Lyman-alpha irradiance for any given day.

In periods of high solar activity, the irradiance $F_{L\alpha}$ may be represented by

$$F_{L\alpha} = 0.77 \times 10^{-2} \times \bar{R}_Z + 0.38 \times 10^{-2} \times R_Z + 2.10 \quad (\pm 0.16) \quad , \quad (2)$$

where R_Z is the Zurich sunspot number and $\bar{R}_Z$ is the yearly average. On the other hand, during conditions of low solar activity, the Lyman-alpha irradiance is better represented by

$$F_{L\alpha} = 0.63 \times 10^{-2} \times \bar{F}_{10.7} + 0.54 \times 10^{-2} \times F_{10.7} + 1.49 \quad (\pm 0.17) \quad (3)$$

Nevertheless, these relations need to be confirmed or modified through observations made at a solar minimum. Such observations were made from August 1974 to August 1975 when OSO-5 was reactivated, and are now under study (Vidal-Madjar and Phissamay, 1977).

II. VARIATION OF THE SOLAR LYMAN-ALPHA IRRADIANCE DURING FLARES

The first attempt to observe a flare in Lyman-alpha was made by Byram et al. (1956) during a rocket flight. They observed no Lyman-alpha enhancement corresponding to a small class 1 flare but found positive evidence of X-ray emission. Once again, orbital observations were necessary to make a detailed study. In 1960, Kreplin et al. (1962), using an ion chamber on board the SR-1 satellite, showed that for a class 2 flare the corresponding Lyman-alpha variation is less than 11%. Yefremov et al. (1963) observed no detectable variation of Lyman-alpha irradiance from Spaceship II during the flare that took place at 1540 on August 19, 1960. With the SOLRAD 8 spacecraft Fossi, Polett, and Tagliaferri (1969) observed a 20% increase of the Lyman-alpha irradiance during a flare on March 20, 1966, a much larger value than previously reported. These observations demonstrated that Lyman-alpha enhancement during flares is small and may reach 20% at most. This result is fully confirmed and quantified by the large number of solar flare observations made by Hall (1971) with the OSO-3 satellite. He found that for a subflare no Lyman-alpha enhancement may be expected; for a class 1 flare, 2% enhancement is observed; and for a class 2 flare, 8% enhancement. Hall gave the following relation for the Lyman-alpha irradiance enhancement E, with A representing the corrected area of the flare over the solar disk:

$$E = (0.3 \pm 0.1) \times A^{3/2} \quad . \tag{4}$$

He also found that the average rise time of these Lyman-alpha flares was 2.4 ±1.6 minutes, the average decay time 4.4 ±3.0 minutes, and the time lag with the Hα maximum 3.4 ±0.9 minutes. More recently Heath (1973) observed a 16% Lyman-alpha enhancement on April 21, 1969 from the Nimbus spacecraft during a class 3B flare. This result is in excellent agreement with Hall's prediction of 18% for that flare.

In summary, although some experimenters (Martini et al., 1974) observed strong decreases (not related to instrumental difficulties) in the solar Lyman-alpha irradiance during flares, it seems quite well established that Lyman-alpha irradiance enhancements associated with solar flares are negligible except for very unusual flares; it is reasonable to conclude that ionospheric events are certainly more related to X-ray than to Lyman-alpha events.

III. ABSOLUTE CALIBRATION DIFFICULTIES

The existence of relative variations of the solar Lyman-alpha irradiance has been established, at least for short periods of time like the 27 d rotation or flare events.

But whether the absolute value of the irradiance is a realistic value is a remaining question.

On one hand, great effort has been devoted to developing precise standards: sources such as hydrogen arcs, stabilized uranium-getter hydrogen and deuterium lamps, synchrotron radiation, storage ring synchrotron radiation; or detectors such as ion chambers, thermopiles, tungsten photocathodes, and NBS photodiodes (Canfield, Johnston, and Madden, 1973). On the other hand, several independent arguments suggest values two times greater for irradiance from the sun.

First Lejeune and Petit (1969), then Wesley, Swartz, and Nisbet (1973) compared electron temperatures observed in the earth's atmosphere by an incoherent scatter technique with computed electron temperatures from the published EUV solar irradiance values. They found that values twice as high for the solar EUV irradiance were necessary to make the observations compatible with the theoretical calculations. Roble and Dickinson (1973), comparing observed thermospheric temperatures with model temperatures, found that the Hinteregger (1970) solar EUV irradiance was low by a factor of two. Keller and Thomas (1973) used Lyman-alpha data from comet Bennett to deduce absolute values of the Lyman-alpha solar irradiance at the center of the line, independent of instrumental calibration (comet morphology is controlled by radiation pressure). They found that twice the Lyman-alpha irradiance is necessary to explain the observed cometary isophotes when compared with OSO-5 data. Each one of these points may be individually criticized on the theoretical model used or on possibly overlooked mechanisms; but it is surprising that all of them reached a similar conclusion—that the irradiance must be doubled to fit theoretical models.

Possibly, all measured fluxes were underestimated due to instrumental difficulties (Meier and Mange, 1973). However, according to Heroux, Cohen, and Higgins (1974), who simultaneously measured the solar EUV irradiance and electron densities between 110 and 300 km, both types of observations were compatible (30% difference). Heroux, Cohen, and Higgins (1975a) reevaluated the electron densities and found even better agreement with the same EUV fluxes, showing that absolute values of the EUV fluxes are known with a ±30% accuracy. The debate between two opposing groups ended recently in the literature (Nisbet, 1975; Heroux, Cohen, and Higgins, 1975b), showing that probably some agreement has been reached between solar flux measurements and atmospheric observations. Finally, as mentioned by Roble (1976), the solar EUV irradiance is probably doubled from the minimum to the maximum of a solar cycle in agreement with the observed variation of the Lyman-alpha irradiance during the 11 y solar cycle.

In conclusion, it seems reasonable to say that the absolute value of the solar Lyman-alpha irradiance is known within the ±30% error bars quoted by Heroux et al. (1974) or as it showed up through the comparisons made in Table 5.

IV. OTHER STUDIES RELATED TO THE SOLAR LYMAN-ALPHA LINE

A. OBSERVATION OF THE LINE PROFILE

As already mentioned, Purcell and Tousey (1960) observed almost all the main characteristics of the solar Lyman-alpha line, and its general shape has been explained theoretically (Thomas, 1957; Jefferies and Thomas, 1959; 1960; Avrett, 1968; Cuny, 1968; Vernazza, 1972).

Bruner and Parker (1969) and Bruner and Rense (1969) obtained a very high-quality (low-noise) profile with 0.03Å resolution clearly showing the details of the Lyman line profile. Charra (1973), using data from the D2A satellite with 3 arcmin resolution, found almost no reversal of the Lyman-alpha profile over an active region. Bruner et al. (1973) reached 20 arcsec spatial resolution and demonstrated clearly that the Lyman-alpha line profile was quite variable over the solar disk and that the central reversal almost disappeared over some active areas. In 1975 during two Salyut 4 missions, cosmonauts also observed many peculiar Lyman-alpha profiles with 0.4Å resolution over faculae, prominences, and even flares (Severny, 1975).

More recently, two instruments on board the OSO-8 satellite observed the Lyman-alpha line profile over the solar disk with increased spectral and spatial resolution. Typical Lyman-alpha line profiles (University of Colorado instrument, Bruner, *private communication*, presented in Figure 2; and LPSP instrument, Bonnet and colleagues, *private communication*, presented in Figure 3) show again the general characteristics of the line. Due to the very high quality of the profile, two new peculiar details were observed. The first was a systematic red shift between the solar line and the geocoronal absorption core. This may be indicative of falling material on the sun (Bruner, *private communication*). The second was an apparent incompatibility between the two edges of the solar reversal near the central geocoronal absorption core. To make them fit one must either build up a discontinuity or suppose another line in the center of the solar line (if no geocoronal absorption core was present) or a change in the curvature at the center of the line (concavity parameter < 0), as already observed in another manner by Vidal-Madjar, Blamont, and Phissamay (1973). Interestingly, on almost all the Lyman-alpha profiles obtained either with the CU instrument or the LPSP instrument on OSO-8 (see Figure 3 showing Lyman-alpha as well as Lyman-beta profiles from the LPSP instrument) this feature appears.

B. DISTRIBUTION OF THE LYMAN-ALPHA EMISSION OVER THE SOLAR DISK

The first Lyman-alpha solar disk images were made by Mercure et al. (1956). Later, Purcell et al. (1959) clearly correlated Lyman-alpha enhancement with Ca K emissions, but they did not reach better than 30 arcsec resolution. Sloan (1968) observed the distribution of Lyman-alpha emission over a quiet area of the sun

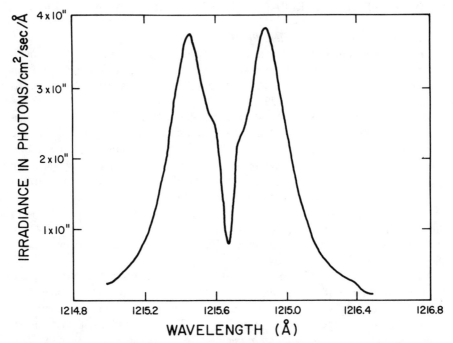

Fig. 2 The solar Lyman-alpha line profile over a quiet area as seen by the OSO-8 satellite (University of Colorado instrument, Bruner, 1976), on 21 July 1975, with the 900 arcsec slit. The total intensity was normalized to the 28 July 1975 value measured during a calibration rocket flight: 2.47×10^{11} ph cm^{-2} s^{-1}.

with 2 arcsec resolution. He observed structures as small as 2 arcsec and calculated a ratio of 1.7 from bright to dark areas.

More recently Prinz (1973, 1974) obtained extraordinarily good images of the whole solar disk with 3 arcsec spatial resolution. A number of features were observed and identified for the first time at Lyman-alpha, particularly; a factor of 10 enhancement was observed over active regions, which showed the role of these areas in the 27 d fluctuation.

The OSO-8 instruments now in operation are able to obtain either spectroheliograms of the whole sun with 20 arcsec resolution or images of smaller areas with up to 1 arcsec spatial resolution (internal raster of the LPSP instrument). The spectroheliograms can be completed on any specific point of the line profile (with 0.02Å resolution); such observations have never been made before. They show (Fig. 4) that when the solar image is obtained on the blue peak (CU instrument, Bruner, *private communication*) dark areas appear very clearly near the poles, related to the permanent coronal holes present there. This particular type of

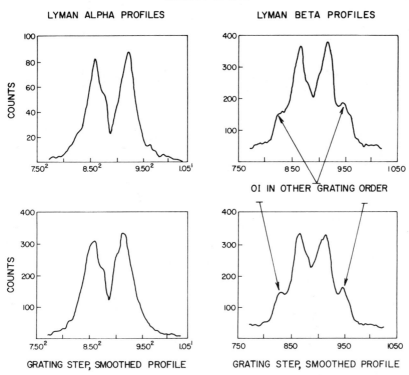

Fig. 3 The geocoronal absorption is visible on Lyman-alpha and Lyman-beta profiles. An apparent discontinuity seems to exist on the solar line profiles between the two edges of the geocoronal feature. From the LPSP instrument on board the OSO-8 spacecraft (Bonnet et.al., 1976).

observation gives the solar irradiance in any point of the Lyman-alpha line, spectra information which is sometimes necessary for interpreting Lyman-alpha observations made in the solar system (see paragraphs D and E).

Finally, simultaneous measurements in Ca K and Lyman-alpha (see Fig. 5) made by the OSO-8 LPSP instrument (Bonnet, *private communication*) with high spatial resolution (1 arcsec) over an active area showed clearly that the correlation between the two emissions over plage areas or sunspots is often very poor. This may explain why no really good correlation between Lyman-alpha irradiance and solar activity indices was ever found (Table 4).

C. INTERACTION WITH THE UPPER ATMOSPHERE OF THE EARTH

As already shown, the Lyman-alpha radiation is first absorbed by atomic hydrogen

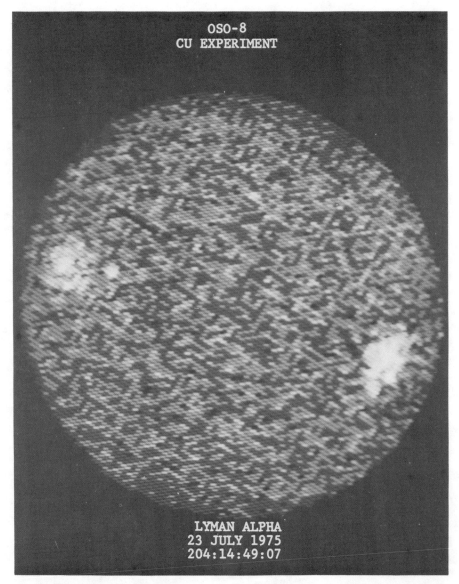

Fig. 4 Spectroheliogram over the blue peak of the Lyman-alpha line as observed by the OSO-8 spacecraft (University of Colorado instrument, Bruner, 1976).

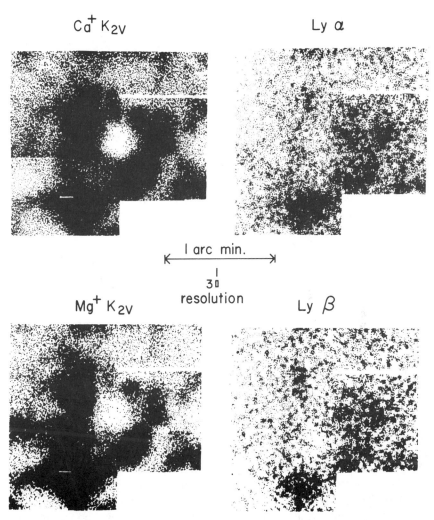

Fig. 5 The comparison of the two upper spectroheliograms in the Ca^+ and Lyman-alpha line shows that the Lyman-alpha emission is correlated neither with the calcium plages nor with the sunspot area. Bright areas are dark (negative pictures) and the sunspot is the white circle in the middle of the dark plage structure (Bonnet et al., 1976).

at high altitudes, producing the narrow geocoronal core observed on any high-spectral resolution profile. The Lyman-alpha radiation then is absorbed by molecular oxygen around 80 km. From 80 km to 110 km, the Lyman-alpha radiation ionizes nitric oxide, producing the well-known ionospheric D layer under quiet solar conditions. Lastovicka (1973) found that even under medium solar activity conditions (F10.7 $\sim$ 150) the ionization of the D layer seems more likely to be controlled by the 27 d Lyman-alpha irradiance variation than by the solar X-ray irradiance fluctuation. This ionization of the D layer is possible only because Lyman-alpha enters deep enough into the earth's atmosphere owing to a coincidence that places the Lyman-alpha wavelength at the bottom of one of the sharpest and deepest transmission windows of O_2, which is the principal absorber in this wavelength range (Lee, 1955; Watanabe, 1958; Ogawa, 1968).

The O_2 absorption cross section at Lyman-alpha measured under laboratory conditions is equal to 1.0×10^{-20} cm^{-2}. But several observers (Smith and Miller, 1974; Thrane and Johannessen, 1975) who tried to deduce O_2 number density in the earth's atmosphere from Lyman-alpha extinction curves found systematic differences with other observational techniques. They conclude that a 0.8×10^{-20} cm^{-2} cross section is more adequate according to their data. The disagreement may arise from the laboratory measurements being made at pressures and temperatures higher than in the upper atmosphere. Weeks (1975) sought another source of error in the determination of O_2 number density by investigating whether Lyman-alpha extinction could be due to other absorbers. He found that only nitric oxide may play such a role, and only then in the upper layers from 105 to 115 km. Recently, the OSO-8 LPSP instrument was able to scan very quickly across the Lyman-alpha line (1.28 s). During sunrises and sunsets the Lyman-alpha extinction was very clearly observed, and Bonnet (*private communication*) found that the cross section of O_2 varies quickly across the line, but not in agreement with the laboratory measurements (Ogawa, 1968). This again may be owing either to different physical conditions in the upper atmosphere or to more than one absorber being responsible for the extinction.

D. THE LYMAN-ALPHA IRRADIANCE FROM THE CENTER OF THE LINE

The observation of the Lyman-alpha irradiance from the center of the line is important since it concerns the interpretation of many Lyman-alpha observations made in the solar system, such as the observation of the planetary exospheres, the study of the outer hydrogen envelopes of comets, and the measurement of the local interstellar medium density, temperature, and relative velocity.

All these observations are related to the study of the origin and evolution of the solar system. Also, knowing the variation of the center of the line irradiance is important to the interpretation of any time-dependent phenomenon.

Figure 6, from Thomas (1973), summarizes the role of the Lyman-alpha line in the solar system through radiation pressure or through scattering of hot or cold

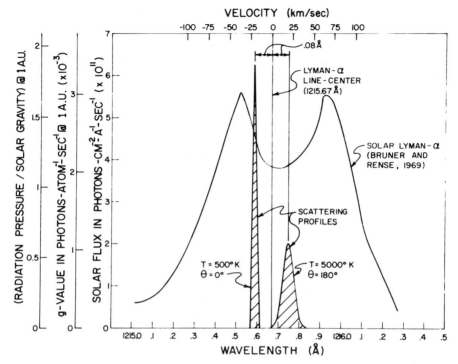

Fig. 6 Taken from Thomas (1973), this figure shows the different interaction of the solar Lyman-alpha line with hydrogen atoms in the solar system. The scattering profiles (shaded areas) show what part of the solar irradiance will contribute to the excitation of the resonance of hydrogen atoms assuming a temperature (500 or 5000 K) and a relative velocity (±25 km s^{-1}) for the cloud. Also in the left scales is shown the solar irradiance per Å along with the corresponding excitation rate of hydrogen atoms at 1 AU and the radiation pressure-solar gravity balance. Surprisingly, the central irradiance of the line is very close to balancing the solar gravity exactly.

hydrogen atoms. Such observations of the central Lyman-alpha irradiance are difficult because most instruments have generally low spectral resolution over the whole solar disk. These observations have been made by using the resonance scattering of hydrogen atoms either inside a cell in an instrument on board a spacecraft or in a natural hydrogen cell, the earth's geocorona.

Several authors used the second method, after removing the effect of the atmospheric variations, to evaluate the center of the Lyman-alpha line irradiance and its variation. Meier (1969) and Sheffer (1971) observed a 30% variation of the geocoronal emission and explained that this number represented a lower limit of the center of the Lyman-alpha line irradiance.

Then Meier and Mange (1973), after correcting OGO-4 data for the atmospheric perturbation, found a 50% variation of the central line irradiance. More recently, Thomas and Anderson (1976), using the same observing method, found an excellent correlation between the central irradiance and the Zurich sunspot number R_Z (from June 1969 to June 1970, when R_Z remained near 100). They found a linear relation between the two parameters:

$$F_0 = 2.25 + (7.38 \times 10^{-3}) R_Z \quad (\pm 0.25) \; , \tag{5}$$

where F_0 is the central irradiance of the Lyman-alpha line in 10^{11} ph cm^{-2} s^{-1} Å^{-1} unit.

On the other hand, Vidal-Madjar (1975) made the same study using a hydrogen resonance cell on board the OSO-5 spacecraft. The empirical relation he found over a four-year period for an average R_Z value of 100 agrees with Thomas and Anderson's to better than 10%. A very strong correlation was also found between the total Lyman-alpha irradiance and the central irradiance of the form:

$$F_0 = 0.54 \times (F_{L\alpha})^{1.53} \quad (\pm .33) \; . \tag{6}$$

This relation, along with relations (2) and (3), gives a good prediction of the central Lyman-alpha irradiance at all solar activity levels. This prediction is in agreement with the measurements of the total Lyman-alpha irradiance since both were made with the same instrument on OSO-5. From relation (6) we find that the 30% variation found for the total irradiance related to the 27 d rotation corresponds to a 46% variation of the central line irradiance. This result is in good agreement with the previous evaluations.

New instruments on OSO-8 provide the opportunity to make the same observation with a completely different method. The Lyman-alpha irradiance of the sun can be measured with both the CU and LPSP instruments on any point of the line with a 0.02Å resolution. This type of observation will then give the evolution of the whole line profile with solar activity changes.

E. THE PROBLEM OF DEUTERIUM IN THE SOLAR SYSTEM

The study of deuterium abundance in the solar system is very exciting since its abundance is directly related to that system's origin and formation. In particular, observations of deuterium in comets (as attempted by the Copernicus satellite for Comet Kohoutek) may be a clear indication of whether their origin is solar, interstellar, or related directly to the birth of the solar system.

Again, the solar irradiance on the blue wing of the Lyman-alpha line (at 0.33Å from the center) is important in interpreting any observation of the Lyman-alpha deuterium resonance scattering in the solar system. The only measurements have been made by Charra (1973) and Vidal-Madjar (1975) using a deuterium resonance cell on the D2A and OSO-5 spacecraft, respectively. They both found that for a 30% variation of the total irradiance there is a corresponding 40% variation of the blue-wing irradiance at 0.33Å from the center. Empirical relations representing these variations are also presented by Vidal-Madjar (1975).

V. CONCLUSION

The solar Lyman-alpha irradiance over the whole line and at some precise point of the line is now well observed. Empirical relations do exist to estimate the Lyman-alpha irradiance for any given day when no observation is available. The absolute accuracy of the irradiance evaluation seems to be smaller than ±30%. The relative accuracy of the observed variations is less than ±5%. Also it was shown that from minimum to maximum solar activity conditions, Lyman-alpha irradiance increases by a factor of 2, essentially owing to the change over the solar cycle of the quiet-sun irradiance. Nevertheless, precision in determining the absolute values is necessary to interpret properly many observations related directly or theoretically to the solar Lyman-alpha irradiance.

Two approaches to these observations are necessary: long-term variation studies with orbital instruments, possibly with inflight absolute calibrations; and rocket experiments to improve the absolute calibration and correct orbital measurements.

Direct observations are always preferable to evaluations through an empirical relation. Table 6 lists the past, present, and possible future Lyman-alpha irradiance observations available from orbiting spacecraft.

Table 6

Observation of the Lyman Alpha Irradiance by Orbiting Spacecrafts Since 1965

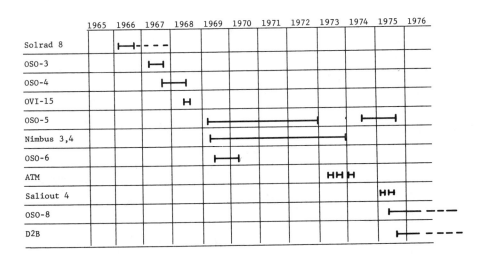

As shown in Table 6, observations from three spacecraft may be interesting for a future study, namely OSO-5 (if reactivated); OSO-8 (in operation now); and D2B, a French satellite launched in September 1975, which also measures the Lyman-alpha irradiance of the sun. Interestingly, the absolute calibration for this experiment was made in an independent manner using the synchrotron storage ring of Orsay (France) as a standard source, as well as NBS standard detectors (Delaboudinière and Millier, 1974).

Finally, we should mention that the observation of the central irradiance is always hidden by the earth's geocoronal absorption. For a future study a high-altitude platform (above 10,000 km) may be necessary to overcome the geocoronal absorption effect.

ACKNOWLEDGMENTS

I would like to thank all the teams which are making the OSO-8 spacecraft operational, at the National Aeronautics and Space Administration, the University of Colorado, the Centre National d'Etudes Spatiales, and Laboratoire de Physique Stellaire et Planétaire (LPSP). I also wish to thank Dr. C. Barth for providing the facilities necessary to operate the OSO-8 LPSP instrument and Drs. O.R. White and E.C. Bruner from the CU team for their help in preparing this review and for allowing me to use some of their unpublished data.

K. Murray and R. Jouchoux of the Laboratory of Atmospheric and Space Physics (University of Colorado) assisted in the preparation of this manuscript.

REFERENCES

Aboud, A., Begring, W.E., and Rense, W.A., 1959, *Ap. J. 130*, 381.
Ackerman, M., and Simon, P., 1973, *Solar Phys. 30*, 345.
Aiken, A.C., Kane, J.A., and Troim, J., 1964, *J. Geophys. Res. 69*, 4621.
Avrett, E.H., 1968, *Resonance Lines in Astrophysics* NCAR, Boulder, Colo., 27.
Brueckner, G., 1975, *The Solar Constant and the Earth's Atmosphere* (Proceedings of the Workshop, Session A), 26.
Bruner, E.C., Jr., and Parker, R.W., 1969, *J. Geophys. Res. 74*, 107.
────── and Rense, W.A., 1969, *Ap. J. 157*, 417.
──────; Parker, R.W.; Chipman, E.; and Stevens, R., 1973, *Ap. J. (Letters) 182*, L33.
Byram, E.T.; Chubb, T.A.; Friedman, H.; and Gailar, N., 1953, *Phys. Rev. 91*, 1278.
──────; ──────; ──────; and Kupperian, J.E., Jr., 1956, *Ap. J. 124*, 480.
──────; ──────; ──────; ──────; and Kreplin, R.W., 1958, *ibid. 128*, 738.
Canfield, L.R., Johnston, R.G., and Madden, R.P., 1973, *Appl. Optics 12*, 1611.
Carver, J.H.; Mitchell, P.; Murray, E.L.; and Hunt, B.G., 1964, *J. Geophys. Res. 69*, 3755.
Charra, J., 1973, Ph.D. thesis, University of Paris VI.
Chubb, T.A.; Friedman, H.; Kreplin, R.W.; and Kupperian, J.E., Jr., 1957, *J. Geophys. Res. 62*, 389.

Cuny, Y., 1968, *Solar Phys.* **3**, 204.
Delaboudinière, J.P., and Miller, F., 1974, *Calibration of the Experiment D2B with the Storage Ring of Orsay (France)* (Third Workshop on the VUV Radiometric Calibration of Space Experiment, NCAR, Boulder, Colo., 19-20 September 1974, sponsored by NBS).
Detwiler, C.R.; Garrett, D.L.; Purcell, J.D.; and Tousey, R., 1961, *Ann. Geophys.* **17**, Fascicule 3, 263.
Dickinson, P.H.G., 1972, *J. Atmos. Terr. Phys.* **34**, 621.
Dupree, A.K., and Reeves, E.M., 1971, *Ap. J.* **165**, 599.
_____; Huber, M.C.E.; Noyes, R.W.; Parkinson, W.H.; Reeves, E.M.; and Withbroe, G.L., 1973, *ibid.* **182**, 321.
Efremov, A.I.; Podmoshenskii, A.L.; Efimov, O.N.; and Lebedev, A.A., 1962, *Planet. Space Sci.* **9**, 969.
Fossi, M.B., Poletto, G., and Tagliaferri, G.L., 1968, *Mem. Soc. Astr. Italy* **39**, 201.
_____, _____, and _____, 1969, *Solar Phys.* **10**, 196.
Friedman, H., 1960, *Physics of the Upper Atmosphere*, ed. J.A. Ratcliffe, Academic Press, New York.
_____, Lichtman, S.W., and Byram, E.T., 1952, *Phys. Rev.* **83**, 1025.
Hall, L.A., 1971, *Solar Phys.* **21**, 167.
_____ and Hinteregger, H.E., 1970, *J. Geophys. Res.* **75**, 6959.
_____, Damon, K.R., and Hinteregger, H.E., 1963, *Space Res.* **3**, 745.
_____, Schweizer, W., and Hinteregger, H.E., 1965, *J. Geophys. Res.* **70**, 2241.
Heath, D.F., 1973, *ibid.* **78**, 2779.
Heroux, L., Cohen, M., and Higgins, J.E., 1974, *ibid.* **79**, 5237.
_____, _____, and _____, 1975a, *ibid.* **80**, 4732.
_____, _____, and _____, 1975b, *ibid.* **80**, 4771.
Higgins, J.E., 1976, *ibid.* **81**, 1301.
Hinteregger, H.E., 1960, *Ap. J.* **132**, 801.
_____, 1961, *J. Geophys. Res.* **66**, 2367.
_____, 1970, *Ann. Geophys.* **26**, Fascicule 2, 547.
Jeffries, J.T., and Thomas, R.N., 1959, *Ap. J.* **129**, 401.
_____ and _____, 1960, *ibid.* **131**, 695.
Johnson, F.S.; Malitson, H.H.; Purcell, J.D.; and Tousey, R., 1958, *ibid.* **127**, 80.
Keller, H.U., and Thomas, G.E., 1973, *Ap. J. (Letters)* **186**, L87.
Kreplin, R.W., Chubb, T.A., and Friedman, H., 1962, *J. Geophys. Res.* **67**, 2231.
Lastovicka, J., 1973, *J. Atmos. Terr. Phys.* **35**, 815.
Lee, P., 1955, *J. Opt. Soc. Am.* **45**, 703.
Lejeune, G., and Petit, M., 1969, *Planet. Space Sci.* **17**, 1763.
Martini, L.; Bishoff, K.G.; Pfau, G.; Shtark, B.; and Ulrich, E., 1974, *Kosmicheskie Issledovanya* **12**, 736.
McNutt, D.P., 1966, *private communication to Weeks (1967)*.
Meier, R.R., 1969, *J. Geophys. Res.* **74**, 6487.
_____ and Mange, P., 1973, *Planet. Space Sci.* **21**, 309.
Mercure, R.; Miller, S.C., Jr.; Rense, W.A.; and Stuart, F., 1956, *J. Geophys. Res.* **61**, 571.
Miller, S.C., Jr., Mercure, R., and Rense, W.A., 1956, *Ap. J.* **124**, 580.
Nicolet, M., 1945, *Mem. Inst. Roy. Meteorol. Belgique* **19**.
Nisbet, J.S., 1975, *J. Geophys. Res.* **80**, 4770.
Ogawa, M., 1968, *ibid.* **73**, 6759.
Prag, A.B., and Morse, F.A., 1970, *ibid.* **75**, 4613.
Prinz, D.K., 1973, *Solar Phys.* **28**, 35.
_____, 1974, *Ap. J.* **187**, 369.

Purcell, J.D., and Tousey, R., 1960, *J. Geophys. Res. 65*, 370.
─────────, Packer, D.M., and Tousey, R., 1959, *Nature 184*, 8.
Rense, W.A., 1953, *Phys. Rev. 91*, 299.
Roble, R.G., 1976, *J. Geophys. Res. 81*, 265.
───────── and Dickinson, R.E., 1973, *ibid. 78*, 249.
Rottman, G.J., 1974, paper presented at AGU Meeting, San Francisco, 13 December 1974.
Severny, 1975, Rapport Commission 44 de U.A.I.
Sheffer, E, K., 1971, *Kosmicheskie Issledovanya*, 9 435.
Sloan, W.A., 1968, *Solar Phys. 4*, 196.
Smith, L.G., 1972, *J. Atmos. Terr. Phys. 34*, 601.
───────── and Miller, K.L., 1974, *J. Geophys. Res. 79*, 1965.
─────────; Accardo, C.A.; Weeks, L.H.; and McKinnon, P.J., 1965, *J. Atmos. Terr. Phys. 27*,
Thomas, G.E., 1973, paper 73-547, presented at AIAA/AGU Space Science Conference on the Exploration of the Outer Solar System, Denver, Colo.,
───────── and Anderson, D.E., Jr., 1976, *Planet. Space Sci.*, in press.
Thomas, R.N., 1957, *Ap. J. 125*, 260.
Thrane, E.V., and Johannessen, A., 1975, *J. Atmos. Terr. Phys. 37*, 655.
Timothy, J.G., 1976, this volume.
Tousey, R., 1963, *Space Sci. Rev. 2*, 3.
─────────, 1964, *13th International Astronomy Congress*, ed. N. Boneff and I. Hersey, Springer-Verlag, Vienna, 166.
─────────, Watanabe, K., and Purcell, J.D., 1951, *Phys. Rev. 83*, 165.
Vernazza, J.E., 1972, Ph.D. thesis, Harvard University.
Vidal-Madjar, A., 1975, *Solar Phys. 40*, 69.
───────── and Phissamay, B., 1976, to be published.
─────────, Blamont, J.E., and Phissamay, B., 1973, *J. Geophys. Res. 78*, 1115.
Watanabe, K., 1958, *Advance Geophysics*, ed. H.E. Landsberg and J. Van Mieghem, Academic Press, New York, 183.
Weeks, L.H., 1967, *Ap. J. 147*, 1203.
─────────, 1975, *J. Geophys. Res. 80*, 3655.
Wesley, E., Swartz, and Nisbet, S.J., 1973, *ibid. 78*, 5640.
Woodgate, B.E.; Knight, D.E.; Uribe, R.; Sheather, P.; Bowles, J.; and Nettleship, R., 1973, *Proc. Roy. Soc. London A. 332*, 291.
Yefremov, A.I.; Podmoshensky, A.L.; Yefimov, O.N.; and Lebedev, A.A., 1963, *Space Res. 3*, 843.

Comment

As a contribution to high-resolution measurements of solar lines near Lyman-beta, we report recent results obtained for the solar C III line at 977 Å.

We have measured the line profile of the 977 Å line of C III with good wavelength resolution from spectra obtained with the Naval Research Laboratory slit spectrograph on Skylab. Previously the flux of the 977 Å line had been measured by Detwiler *et al.* (1961), giving the value 0.05 erg cm^{-2} s^{-1} at a distance of 1 AU from the sun. This measurement was an intensity integrated over the entire solar disk. The specific intensity of the line at the center of the disk has recently been given as 900 erg cm^{-2} sr^{-1} by Dupree et al. (1973) from OSO-6 data. This corresponds to a flux of 0.06 erg cm^{-2} s^{-1} at 1 AU, assuming the intensity distribution to be uniform across the solar disk. Even if the brightening at the limb and contributions from active regions are taken into account, the two measurements must be considered in agreement within their error limits.

The Skylab instrument measured the spectral intensity between 970 Å and 4000 Å averaged over a 2 arcsec by 60 arcsec area on the sun. This implies that the profiles and intensities are at least partially averaged over the inhomogeneous intensity structures on the sun. Because of the low reflectivity of the spectrograph at wavelengths below 1170 Å, very long exposure times were needed to register the line at 977 Å. For the quiet sun at the disk a 44-min exposure was required even when we used the most sensitive UV emulsion (Eastman Kodak 101) available to us. Exposure times of 10 minutes or more had to be used in active regions. As a result, very few profiles of the 977 Å line were observed.

The results of the observations are shown in Figure 1, which gives the profile of the 977 Å line for a quiet area at the center of the solar disk and for two active regions located at the solar limb. In the case of the active regions, the slit was placed tangential to the limb at a position 2-4 arcsec above the limb, where the emission in the transition-zone lines reaches its maximum.

The profiles from the quiet and active regions are drawn on the same intensity scale. Because absolute calibration of the instrument is not available below 1240 Å, the absolute intensity scale in Figure 1 was obtained by normalizing the total intensity of the line at sun center to the intensity value given by Dupree et al. (1973).

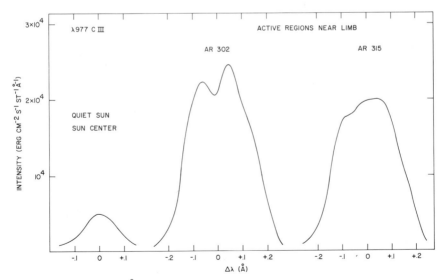

Fig. 1 Profile of the 977 Å line of C III.

The observed profile of the 977 Å line for the quiet region can be approximated well by a doppler profile of width $\Delta\lambda_D = 100$ Å in the core of the line. Correcting for the instrument profile would reduce the doppler width to 0.07 Å. In the wings ($\Delta\lambda > 1.2\Delta\lambda_D$) the observed intensities are higher than those of a doppler profile, the deviation being particularly pronounced in the blue wing.

In the active regions the line profile does not have a gaussian shape typical of optically thin, doppler-broadened lines, and it is wider than expected. It should be noted that Figure 1 probably does not give the correct impression of the relative contribution from active regions on the disk to the flux in the 977 Å line. One would expect the intensity measured at the limb to be considerably higher than if the active region were located on the disk. If the limb brightening of the 977 Å line is similar to that of other transition-zone lines and if the limb brightening of active regions is similar to that of quiet regions, the intensity of the 977 Å line in active regions on the disk may be lower by a factor of about 0.2 than the intensities displayed in Figure 1.

Finally we will mention an observation of the 977 Å line in an active region during a flare. The line profile, which has not been included in Figure 1, shows the 977 Å line to be asymmetrical with a full width of 0.8 Å.

O. Kjeldseth Moe
K.R. Nicholas
J-D. F. Bartoe
U.S. Naval Research Laboratory

References

Detwiler, C.R.; Garrett, D.L.; Purcell, J.D.; and Tousey, R., 1961, *Ann. de Geophys.* 17, 9.
Dupree, A.D.; Huber, M.C.E.; Noyes, R.W.; Parkinson, W.H.; Reeves, E.M.; and Withbroe, G.L., 1973, *Ap. J. 182*, 321.

THE SOLAR SPECTRUM BETWEEN 300 AND 1200 Å

J. Gethyn Timothy
Center for Astrophysics
Harvard College Observatory and
Smithsonian Astrophysical Observatory
Cambridge, Massachusetts

The solar extreme ultraviolet (EUV) radiation at wavelengths below 1200 Å is the dominant source of energy for heating and ionization in the terrestrial upper atmosphere at altitudes above 90 km. The photoionization limits for the major neutral constituents of the terrestrial atmosphere, O_2, O, and N_2, occur at wavelengths of 1027, 911, and 796 Å, respectively. Radiation at wavelengths between 911 and 1200 Å is accordingly absorbed principally by molecular oxygen, with the range from 911 to 1027 Å contributing a large fraction of the available energy for ionization in the E-region at altitudes between about 100 and 150 km. Radiation at wavelengths between 911 and 796 Å is absorbed by both O_2 and N_2, while below 796 Å absorption by O dominates. The wavelength range from 911 to about 300 Å accordingly provides the bulk of the available energy for ionization of the three major neutral species in the F-region of the ionosphere at altitudes between about 150 and 400 km. A knowledge of the solar spectral irradiance at EUV wavelengths is thus of critical importance in any analysis of the photochemistry and energy balance of the ionosphere and the thermosphere. Furthermore, since the EUV spectral region contains a wealth of emission lines and continua formed at temperatures in the range from 8×10^3 to 4×10^6 K, spectrophotometric observations at these wavelengths can provide unique quantitative information on the state of the plasma in the chromospheric, transition-region, and coronal layers of the outer solar atmosphere.

This paper will review the current state of knowledge on the solar spectral irradiance at EUV wavelengths between 300 and 1200 Å. Techniques available for photometric measurements at EUV wavelengths will be reviewed briefly and the observational data will be presented in detail, together with estimates of errors. Finally, current ideas about the effects of specific solar features on the variability of the spectral irradiance will be discussed, and suggestions for future measurements will be outlined.

I. THE NATURE OF THE SOLAR EUV SPECTRUM

The solar EUV spectrum in the wavelength range from 300 to 1200 Å contains emission lines and continua produced by neutral atoms and ionized species up to and including Fe XVI. Detailed reviews of the form of the solar EUV and soft X-ray spectrum have recently been presented by Fawcett (1974) and Walker (1972). Examples of spectra from 300 to 1350 Å obtained by the Harvard College Observatory instrument on the Skylab Apollo Telescope Mount (ATM) (Reeves, Huber, and Timothy, 1976) are shown in Figure 1. It can be seen that the wavelength range covered in this review is effectively bounded by the strong Lyman-alpha emission lines of neutral hydrogen at 1215.7 Å and of ionized helium at 303.8 Å. Differences in the shape of the spectrum for radiation emitted from active regions, average quiet regions on the solar disk, and coronal holes can be seen clearly in these data.

The concentration of emission from lines of high excitation potential at coronal altitudes and particularly in the dense plasma in active regions and bright points can be seen in the images of the sun (Figure 2) taken at wavelengths corresponding to emission lines characteristic of chromospheric, transition-region, and coronal

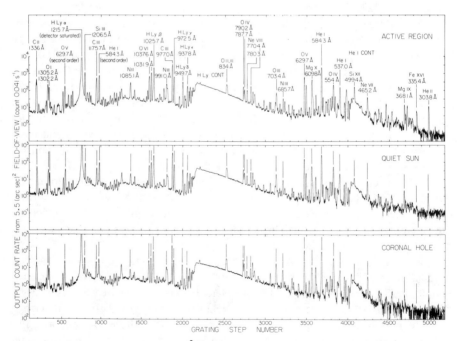

Fig. 1 EUV spectra for a 25 arcsec² field of view located in an active region, at the quiet-sun center, and in a coronal hole.

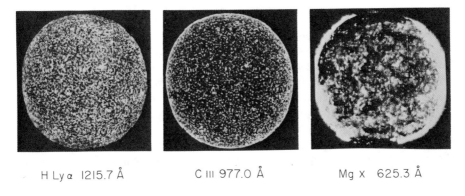

H Lyα 1215.7 Å   C III 977.0 Å   Mg X 625.3 Å

SOLAR DISK   28 JANUARY 1974

Fig. 2   EUV images of the solar disk at wavelengths corresponding to chromospheric (H Ly α, 1215.7 Å), transition-region (C III, 977.0 Å), and coronal (Mg X, 625.3 Å) emission lines.

temperatures. The highly structured appearance of the solar disk, particularly in coronal lines of high excitation potential, indicates that the spectral irradiance is highly dependent on the appearance and disappearance of specific features. Furthermore, because the emission lines and continua at EUV wavelengths are formed over such a wide temperature range, dramatically different variabilities in the irradiance are observed over narrow wavelength intervals.

A detailed discussion of plasma physics is outside the scope of this review. However, it should be noted that the intensity of a particular emission line or continuum will be directly controlled by the radiative, conductive, and mechanical energy-exchange processes within the solar atmosphere. Consequently, a final understanding of the variability of the EUV spectral irradiance throughout the solar cycle will be achieved only through a detailed analysis of the behavior of the solar plasma.

II. MEASUREMENT TECHNIQUES

Photometric measurements at EUV wavelengths are particularly difficult to implement since no rugged window materials are available for use between about 1050 Å (the lower wavelength limit for transmission by LiF) and about 50 Å (the upper wavelength limit for transmission by thin plastic membranes). Furthermore, the reflectance of normal-incidence optical systems falls rapidly below 500 Å, and below 300 Å grazing-incidence optics must be employed. Consequently, photometric calibration at EUV wavelengths in the laboratory is a time-consuming and difficult task, and EUV instruments with unprotected optical surfaces and either open-structure photoelectric detection systems or Schumann photographic emulsions are highly vulnerable to contamination or damage.

A summary of the photometric standards available for use at visible and EUV wavelengths is given in Figure 3 (Timothy and Lapson, 1974). It should be noted that, first, the use of an absolute standard detector is the only practical method for routine EUV calibration in most laboratories, and, second, *none* of the absolute standard light sources or detectors has so far been used in space for photometric calibration at EUV wavelengths.

As the direct result of these difficulties, essentially all the available data on the EUV spectral irradiance have been obtained by means of instruments flown on sounding rockets; for such flights, the instrument may be calibrated in the laboratory shortly before and also, in many cases, immediately after the flight. Great care must be taken in analyses of these data to ensure that several possible sources of error are eliminated. First, the total flight time for a rocket is very short, usually less than ten minutes. Consequently, the instrumentation is required to pump down from ambient atmospheric pressure to a safe operating pressure in a time of approximately 100 s. It is accordingly critical to determine that the pressure in the instrument is sufficiently low to prevent either residual absorption in the observing path or instabilities in the performance of any open-structure photoelectric detection system. Typically, the pressure must be below 10^{-4} Torr during the observing sequence. Second, there will still be significant attenuation by the terrestrial atmosphere at the peak altitudes for most sounding rockets ($\lesssim$ 300 km), particularly at EUV wavelengths below about 700 Å. Corrections

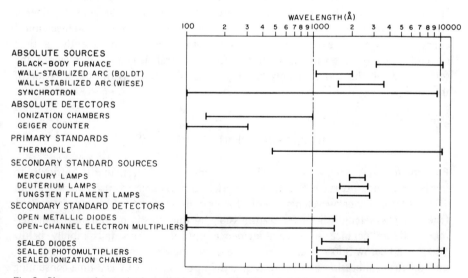

Fig. 3 Photometric standards available for use at EUV and visible wavelengths.

must be made for this absorption in order to determine the incident spectral irradiance. These corrections require a good knowledge of the atmospheric conditions at the time of the flight.

The data from instruments on orbiting satellites are free from ambient pressure and residual absorption problems, but in many cases the instruments have been plagued by major sensitivity changes during their orbital lifetimes. Consequently, most of the accurate spectral irradiance data from satellite instruments have been obtained during short periods of time immediately following launch, or, alternatively, when it has been possible to make a cross-calibration against a sounding-rocket instrument.

The procedures adopted for the photometric calibration of EUV instruments in the laboratory have been described by a number of investigators (Samson, 1967; Huber, Reeves, and Timothy, 1974; Heroux, Cohen, and Malinovsky, 1972; Higgins, 1976). All authors define the basic problems in a similar manner, with a consensus that the accuracy of the laboratory standards is of the order of $\pm$ 15% or better, and that the calculated rms error on the published flight data is typically of the order of $\pm$ 30% or better. Nonsystematic errors in the flight data are, of course, extremely difficult to quantify.

As discussed in the following sections, the apparent differences in the measured values of the EUV spectral irradiance considerably exceed the calculated errors. However, the recent intercomparison of laboratory standards through the National Bureau of Standards (NBS) and the evident care taken by investigators in applying the admittedly tedious calibration techniques should eliminate the possibility that systematic errors in the standards are a cause of major discrepancies in the measured data.

III. OBSERVATIONAL DATA

The first spectrophotometric measurements of the solar spectral irradiance at EUV wavelengths between 300 and 1200 Å were obtained by Hinteregger and colleagues at Air Force Cambridge Research Laboratories (AFCRL)*. The earliest results (Hall, Damon, and Hinteregger, 1963), were later reviewed and corrected (Hinteregger, 1970) in a spectrum that was stated to be typical of "medium" solar activity and nonflaring conditions. However, studies of the photochemistry and energy balance of the terrestrial upper atmosphere have led a number of investigators (Roble and Dickinson, 1973; Swartz and Nisbet, 1973; Roble, 1976) to the conclusion that the EUV irradiance values of Hinteregger are too low by a factor of two for a moderate level of solar activity. This fundamental question

---

*Now Air Force Geophysics Laboratories (AFGL).

concerning the magnitude of the solar EUV spectral irradiance has not so far been resolved.

Timothy et al. (1972) were able to obtain agreement within the experimental errors between the measured electron density in the F1-region of the ionosphere and the computed electron density derived from the simultaneously measured EUV irradiance. However, their values for the EUV irradiance were a factor of two greater than those of Hinteregger. Additional measurements are available both from the AFCRL group (Heroux, Cohen, and Malinovsky, 1972; Heroux, Cohen, and Higgins, 1974; Higgins, 1976), whose results form the cornerstone of the observational data, and from investigators at other institutions in the United States and in Europe (Timothy and Timothy, 1970; Woodgate et al., 1973; Timothy and Reeves, 1976; Schmidtke, 1976). Many other studies of either the solar EUV irradiance or of the brightness of specific solar features at EUV wavelengths have been reported in the literature (e.g., Dupree and Reeves, 1971; Reeves and Parkinson, 1970; Chapman and Neupert, 1973; Huber et al., 1973). However, it should be noted that these investigators have chosen to relate their measurements to the spectral irradiance data of one or more of the authors previously listed, regardless of whether or not an independent photometric calibration was undertaken. Although many excellent photographic studies of the solar EUV spectrum have been published, (e.g., Freeman and Jones, 1970; Behring, Cohen, and Feldman, 1972; Tousey et al., 1973), the photometric calibrations of these data are either inadequate or nonexistent. Accordingly, this review will be confined to a discussion of measurements made with photoelectric detection systems.

The published values for the EUV spectral irradiances determined by instruments having both good spectral resolution ($\Delta\lambda \lesssim 10$ Å) and a well-defined photometric calibration are presented in Table 1. The irradiances are tabulated as photon fluxes rather than in units of energy, since these values are directly related to the instrument responses. Systematic differences in the measurements for broad wavelength bands are accordingly more easily seen with the data in this format. The relation of the times of these ten measurements to the phase of the solar cycle is shown in Figure 4. For ease of comparison, the summary values of the spectral irradiance adopted by Donnelly and Pope (1973) for the National Oceanic and Atmospheric Administration (NOAA) model are also included in the table (column II). It has become customary to employ the 2800 MHz (10.7 cm) radio noise flux (Covington, 1969) as a basic index of the level of solar activity for intercomparison of the EUV spectral irradiances. Accordingly, both the 2800 MHz index (F10.7) for the date of the observation and the smoothed 2800 MHz index ($\overline{F}10.7$) for the 90-day period centered about the date of the observation have been included at the top of the table. The validity of relating these indices to the EUV spectral irradiances will, however, be discussed in the following section of this paper.

It should be noted that there is a major problem in correlating all the available

Table 1
Solar EUV Spectral Irradiance Measurements

Reference			1	2	3	4	5	6	7	8	9	10	11
Date			11 Mar. 1967	3 Apr. 1969	4 Apr. 1969	2 Oct. 1969	9 Nov. 1971	23 Aug. 1972	19 Jan. 1973	1 June 1973	9 Aug. 1973	10 Dec. 1973	–
$F_{10.7}$ (×10^{-22} Wm^{-2}Hz^{-1})			141.6	189.8	177.3	143.2	101.6	122.9	94.8	85.2	87.1	72.3	–
$\bar{F}_{10.7}$ (×10^{-22} Wm^{-2}Hz^{-1})			144.4	158.6	158.6	148.3	112.9	123.3	98.3	93.5	93.0	80.5	–
Wavelength Range (Å)	Wavelength (Å)	Ion	Solar Irradiance (×10^9 photons cm^{-2} s^{-1})										
1220-1210													0.12
	1215.7	H Ly α	~300^a			300	353.21	289		326			310
1210-1200													0.22
	1206.5	Si III	3.6				3.38	2.93					3.7
1200-1190													0.17
1190-1180													0.16
1180-1170													0.14
	1175	C III Gr	2.2				2.10	1.89^a					2.2
1170-1160													0.12
1160-1150													0.105
1150-1140													0.092
1140-1130													0.10
1130-1120													0.15
	1128.3	Si IV	0.27^b										0.28
	1122.5	Si IV	0.22^b										0.22
1120-1110													0.135
1110-1100													0.15
1100-1090													0.16
1090-1080													0.11
	1085	N II Gr	0.59				0.49	0.40^a					0.61
1080-1070													0.13
1070-1060													0.13
1060-1050				0.550									0.13
1050-1040				5.700									0.15
1040-1030													0.099^a
	1037.6	O VI, (C II)	1.7^c			1.480	1.32	1.03	4.489	2.9^a	2.5^a	1.8^a	1.7
	1031.9	O VI	2.3			2.950	2.26	2.26					2.4
1030-1020													0.13^b
	1310-1027	Unresolved	3.7										
	1310-1027	Integral	317.52										
	1220-1027	Unresolved					3.04	297.51					
	1220-1027	Integral					365.8	4.95	3.983				
	1025.7	H Ly β	3.5	5.100, 5.14^b			3.24						3.5
1020-1010				} 1.380									0.15
1010-1000													0.15
1000-990													0.16
	991.6	N III					0.50	0.39					0.61
	990	N III Gr	0.60				0.23	0.17	0.723				0.089
990-980				1.100									0.044
980-970													4.5
	977.0	C III	4.4	6.158^c			4.62	4.60	5.206	5.6^a	6.1^a	3.6^a	0.82
	972.5	H Ly γ	0.80	0.800^d			0.31	0.36	0.766				0.034
970-960				} 0.164									0.029
960-950				1.099									0.033
950-940													0.035
	949.7	H Ly δ	0.39				0.40	0.37	0.411				0.093
	944.5	S VI	0.10				0.08	0.08	0.137				0.052
940-930				0.925									0.22
	937.8	H Ly ε	0.22				0.20	0.16					0.134
	933.4	S VI	0.13				0.13	0.16					0.132
	930.7	H Ly ζ	0.13				0.22	0.18					0.117
930-920													0.132
	926.2	H Ly η	0.13	} 4.278			0.10	0.08					0.23
920-910													
	1027-910	Unresolved	1.21				0.92		2.181				
	1027-910	Integral	11.61				10.95	9.62	13.407				
910-900				3.237									1.92
	904	C II Gr	0.13										1.31
900-890													
	911-890	H cont.	3.2				2.29	2.6	4.162				1.07
890-880				2.52^e									0.93
880-870				9.191									0.78
870-860													
	890-860	H cont.	2.7				2.35	2.4	3.559				0.42^c
860-850													0.67^d
850-840				} 5.714									0.40
840-830				0.71^f									0.53
	835-834	O II, III Gr	0.52				0.72	0.66^a	0.602				
	860-830	H cont.	1.5				1.33	2.0	1.930				
830-820													0.33
820-810													0.27
810-800													0.25^e
	911-800	Unresolved	0.10	3.986			0.20		0.420				
	911-800	Integral	8.93				7.88		11.634				
	830-800	H cont.	0.78				0.99	1.0	0.961				
800-790													0.15^f
	790.2-790.1	O IV	0.26				0.36	0.34					0.27
790-780											0.544		0.099^g
	787.7	O IV	0.13				0.22	0.20					0.133
	786.5	S V	0.08	} 5.449	0.20^a		0.12	0.13					0.082
	780.3	Ne VIII	0.12				0.25	0.19					0.125
780-770													0.117
	770.4	Ne VIII	0.23		0.34^b		0.29	0.29	0.265				0.24
	800-770	H cont.	0.39				0.38	1.32	0.623				
770-760													0.077
	765.1-764.6	N III, IV	0.18				0.18	0.20					0.185
	760.4-760	O V	0.08				0.10	0.19^a	0.223				0.077
760-750				} 3.592									0.08
750-740													0.049
	770-740	H cont.	0.19				0.33	0.53	0.366				
740-730													0.041
730-720													0.066
720-710				} 2.793									0.076
	740-710	H cont.	0.09				~0	<0.15					
710-700													0.028
	703	O III Gr	0.23				0.32	0.13^a	0.272				0.23
700-690													0.049
690-680				} 2.578									0.076
	710-680	H cont.	0.04				~0.20	<0.28					
680-670													0.044
670-660													0.047
660-650				} 1.564									0.049
650-640													0.036
640-630													0.032
	800-631	Unresolved	0.39				0.27		0.918				
	800-631	Integral	2.41				3.02		3.211				

Table 1 (continued)

Wavelength Range (Å)	Wavelength (Å)	Ion	1	2	3	4	5	6	7	8	9	10	11
630-620	680-631			3.188		~0	~0						0.035
	629.7	O V	0.92				1.41	1.60	1.913				0.93
	625.3-624.9	Mg X	0.25	0.308[c]			0.32	0.44	0.484	0.26[a]	0.16[a]	0.14[a]	0.26
620-610													0.04
610-600													0.031
600-590	609.8	Mg X	0.5	1.078[g]	0.623[c]		0.62	0.83	0.939				0.52
	599.6	O III	0.08				0.14	0.16					0.033
590-580													0.082
	584.3	He I	0.89	1.67[h]		2.4	1.28	1.47	1.610				0.029
580-570				2.401									0.92
570-560													0.038
560-550													0.036
	554	O IV Gr	0.31				0.77	0.68[a]	0.737	0.61[a]	0.61[a]	0.45[a]	0.031
550-540				1.708									0.32
540-530													0.025
	537.0	He I				3.2							0.035
530-520													0.021
	521.0-520.7	Si XII	0.19				0.28	0.29	0.325				0.20
520-510													0.023
510-500													0.020[h]
	508	O III Gr	0.08	3.732			0.35						0.08
	631-505	Unresolved					5.17	5.65					
	631-505	Integral							1.160				
	504→	He I cont	0.50[d]										
500-490													0.31[i]
	499.4-499.3	Si XII	0.38	2.978			0.36	0.61	0.384				0.40
490-480													0.15
480-470													0.096
470-460													0.070
	465.2	Ne VII	0.16				0.26	0.26	0.225				0.16
	630-460	Unresolved	0.44						1.021				
	630-460	Integral	4.70						8.798				
460-450													0.069
450-440				3.168									0.067
440-430													0.066
430-420													0.086
	505-425	Unresolved		1.901			0.83						
	505-425	Integral					1.45	2.07					
420-410													
	417.2	Fe XV	0.10				0.10	0.18					0.10
410-400													0.082
400-390				2.205									0.060
390-380													0.058
380-370													0.057
	460-370	Unresolved	0.53						0.689				
	460-370	Integral	0.63						0.689				
	425-370	Unresolved					0.66						
	425-370	Integral					0.76	0.18					
370-360													0.20
	368.1	Mg IX	0.56	4.985			0.70	0.62	0.807				0.57
	365		0.17										
	360.8-360.7	Fe XVI	0.36				0.22	0.31	0.330				0.39
360-350													0.25
350-340													0.24
340-330													0.25
	335.4	Fe XVI	0.72	2.834			0.52	0.83	0.677				0.79
330-320													0.23
320-310													0.33
310-300													0.18
	303.8	He II	5.4	9.04[i](9.3)[i,j]		7.9	7.44	7.2	8.749				9.0
	370-300	Unresolved		2.809[k]			2.41		3.614[a]				
	370-300	Integral					12.13	12.4	14.177[b]				
300-290													0.25
290-280													0.16

Footnotes:
Gr (Group) means that there are several emission lines from that ion near the wavelength given.
Cont. means continuum emission.

Reference:
1. Hall and Hinteregger, 1970. AFCRL, Bedford, Mass. NASA OSO 3 satellite.
   (a) measured in non-linear region
   (b) blend with C I
   (c) blend with C II
   (d) integral for total continuum (at the head: 0.04 × 10⁹ ph cm⁻² sec⁻¹ Å⁻¹)
2. Timothy, Timothy, Willmore and Wager, 1972. University College, London, England. British sounding rocket.
   (a) all data are from scanning spectrometer, unless otherwise noted
   (b) scanning spectrometer and fixed polychromator
   (c) 975 Å
   (d) level of Hall and Hinteregger subtracted from total flux
   (e) 885-865 Å
   (f) fixed polychromator
   (g) excluding 584 Å
   (h) fixed polychromator
   (i) fixed polychromator
   (j) from OSO-4 monochromator at 550 km altitude (Timothy and Timothy, 1970)
   (k) excluding 304 Å
3. Heroux, Cohen and Malinovsky, 1972. AFCRL, Bedford, Mass. AFCRL rocket.
   (a) blended with fourth order Fe XII; 5% correction applied for atmospheric absorption
   (b) 5% correction applied for atmospheric absorption
   (c) possibly blended with O IV; 15% correction applied for atmospheric absorption
4. Woodgate, Knight, Uribe, Sheather, Bowles and Nettleship, 1973. University College, London, England. NASA OSO 4 satellite.
5. Higgins, 1976. AFCRL, Bedford, Mass. AFCRL sounding rocket.
6. Heroux, Cohen and Higgins, 1974. AFCRL, Bedford, Mass. AFCRL sounding rocket.
   (a) from averaged weighted energy levels of the multiplet. Integrated intensity for the multiplet is given.
7. Schmidtke, 1976. Institute for Space Physics, Fraunhofer Society, Freiburg, Federal Republic of Germany. German/NASA AEROS-A satellite.
   (a) includes unresolved from 300-280 Å
   (b) includes unresolved from 300-280 Å
8. Timothy and Reeves, 1976. Harvard College Observatory, Cambridge, Mass. NASA Skylab-Apollo Telescope Mount (ATM).
   (a) Quiet sun irradiances—no allowance for effects of active regions.
9. Timothy and Reeves, 1976. Harvard College Observatory, Cambridge, Mass. NASA Skylab calibration rocket.
   (a) Quiet sun irradiances—no allowance for effects of active regions.
10. As for 9.
11. Donnelly and Pope, 1976. NOAA model for solar EUV spectral irradiance.
    (a) 1040-1027 Å
    (b) 1027-1020 Å
    (c) 860-853 Å
    (d) 853-840 Å
    (e) 810-796 Å
    (f) 796-787 Å
    (g) 787-780 Å
    (h) 510-504 Å
    (i) 504-490 Å

data on the solar EUV spectrum because of the divergence of interests between solar physicists and aeronomers. The primary objective of the solar physicists has been to increase the spatial and spectral resolution of the EUV instrumentation to optimize the observation of the fine structure of the solar disk. The application of these high-spatial-resolution data to the calculation of the total irradiance is

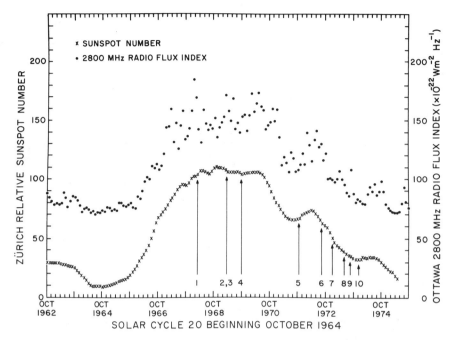

Fig. 4 Relationship of the EUV spectral irradiance data of Table 1 to the phase of the solar cycle. Numbered measures are keyed to Table 1.

an extremely difficult or, in many cases, impossible task. Much of the solar data taken with the more sophisticated and hence, in many cases, best calibrated instruments is thus only of limited interest to the problem of the total irradiance. The measurements of the EUV spectral irradiance taken for the purposes of aeronomy have, of course, concentrated on obtaining the maximum photometric accuracy. However, many of these instruments have been flown on spacecraft as part of a larger complement of instruments for in-situ plasma measurements. Consequently, the size and complexity of the instruments has been severely limited, and the spectral resolution has often proved insufficient for an accurate determination of the total irradiance in a particular spectral feature.

For ease of comparison of those measurements that cover the complete spectral range, the data of Table 1 have been grouped into broad wavelength bands and listed in Table 2. The wavelength bands are defined by the ionization limits for O_2, O, and N_2 and hence are of direct aeronomic interest. In both tables, the values for the Lyman-alpha (1215.7 Å) emission line of hydrogen have been included to provide continuity with the review of Vidal-Madjar in this volume and the values of the He II (303.8 Å) emission line to provide continuity with the review of Manson, also in this volume.

Table 2
Solar EUV Spectral Irradiances in Wavelength Bands of Aeronomic Interest

Reference	1	2	5	6	7	11
Date	11 March 1967	4 April 1969	9 November 1971	23 August 1972	19 January 1973	--
$F_{10.7} (\times 10^{-22} \text{Wm}^{-2}\text{Hz}^{-1})$	141.6	189.8	101.6	122.9	94.8	--
$\bar{F}_{10.7} (\times 10^{-22} \text{Wm}^{-2}\text{Hz}^{-1})$	144.4	158.6	112.9	123.3	98.3	--
Wavelength range (Å)	\multicolumn{6}{c}{Solar irradiance ($\times 10^9$ photons cm^{-2} s^{-1})}
1220-1027	314.58[a]	-	365.80	297.51	-	323.68
1027-911	11.61	20.84	10.95	9.62	13.41	11.70
911-800	8.93	22.84[a]	7.88	8.66	11.63	8.88
800-630	2.41	15.98[b]	3.02	3.95	3.21	2.46
630-300	12.54	40.89[c]	19.51	20.30	23.66[a]	17.97
1027-300	35.49	100.55	41.36	42.53	51.91	41.01

Footnotes:

Reference

1. Hall and Hinteregger, 1970. ARCRL, Bedford, Mass. OSO 3 satellite.
   a. Includes background from 1310 to 1220 Å (∼1.0).
2. Timothy, Timothy, Willmore and Wager, 1972. University College, London, England. British sounding rocket.
   a. 911 to 790 Å
   b. 790 to 640 Å
   c. 640 to 300 Å
5. Higgins, 1976. AFCRL, Bedford, Mass. AFCRL sounding rocket.
6. Heroux, Cohen and Higgins, 1974. AFCRL, Bedford, Mass. AFCRL sounding rocket.
7. Schmidtke, 1976. Institute for Space Physics, Fraunhofer Society, Freiburg, Federal Republic of Germany. German/NASA AEROS-A satellite.
   a. Includes background from 300 to 280 Å (∼1.0).
11. Donnelly and Pope, 1976. NOAA model for solar EUV spectral irradiance.

Because of problems of instrument degradation in orbit, it has so far not been possible to determine unambiguously the variability of the EUV spectral irradiance over the 11-y solar cycle. However, several instruments have demonstrated sufficient stability of response to measure the spectral irradiance over times longer than the 28-d solar rotation period. Measurements have been taken at a number of EUV wavelengths, both during the period of maximum solar activity (Hall and Hinteregger, 1970), as shown in Figure 5, and near solar minimum (Hinteregger, 1976), shown in Figure 6. Measurements of the total irradiance of the He II and Si XI emission lines at 304 Å over a period of almost two years from 1967 to 1969 have been reported by Timothy and Timothy (1970). Data for the period from June to December 1968 are shown together with the F10.7 values in Figure 7.

Additional data on the solar irradiance at wavelengths between 1027 and 150

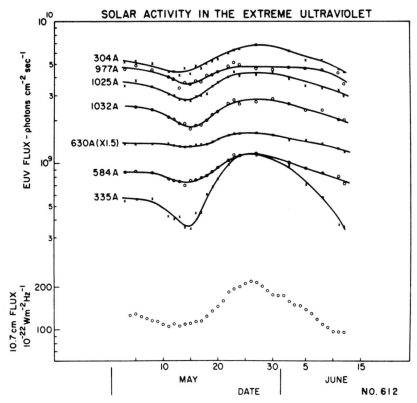

Fig. 5 Variations of EUV irradiances observed in May and June 1967 (Hall and Hinteregger, 1970).

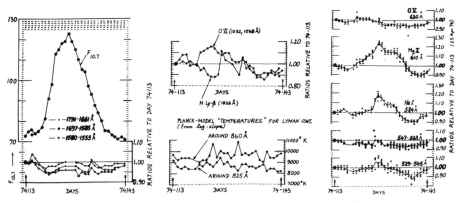

Fig. 6 Variations of EUV irradiances observed in April and May 1974 (Hinteregger, 1976).

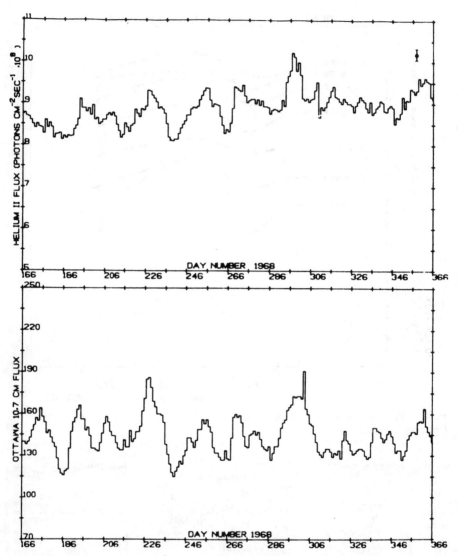

Fig. 7 Variations of the 304 Å irradiance (He II and Si XI emission lines) during the period from June to December 1968 (Timothy and Timothy, 1970).

Å have been recorded by instruments on the German/NASA AEROS-A and -B satellites, and these results are now becoming available in the literature (Schmidtke, 1976). The variations observed in the wavelength range from 911 to 800 Å for the first six months of 1973 are shown in Figure 8. The results of all these measure-

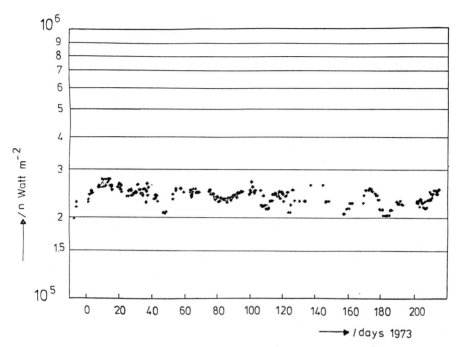

Fig. 8 Variations of the irradiance in the wavelength range 911 to 800 Å during the first six months of 1973 (Schmidtke, 1976).

ments are in general agreement throughout the solar cycle, variations in the irradiance of lines and continua characteristic of chromospheric and transition-region temperatures being typically less than about ± 15% over a 28-d period. The coronal lines, having excitation temperatures in excess of 10^6 K, are, however, much more sensitive to changes in solar activity with variations as large as a factor of three being observed during a single rotation. The effects of dynamic events such as solar flares, which can significantly increase the irradiance at EUV wavelengths on a time scale of a few minutes, have specifically not been included in this analysis, but will be noted briefly in the following section.

Evidence for a possible long-term variation in the EUV irradiance associated with the 11-y solar cycle is now beginning to emerge. A general increase in the early EUV irradiances with increasing solar activity was observed in the early AFCRL sounding rocket data for the period from 1961 to 1968 (Hall *et al.*, 1969). The steady decline in the EUV irradiances observed by the AEROS-A instrument (see, for example, Figure 8) tracks the general reduction in the level of solar activity (Figure 4) and can apparently be separated from the effects of detector degradation (Schmidtke, G., *private communication*). A similar reduction in the quiet sun irradiances at specific EUV wavelengths (see Table 1) was observed by the instruments on the ATM and the ATM calibration rockets (Timothy and Reeves, 1976). Again, these variations track the reduction in the general level of solar activity. The ATM observations also indicate changes in the structure of the quiet solar atmosphere (Timothy, 1975), supporting the view that the observed variations are solar rather than instrumental in origin. The data are, however, insufficient to determine the magnitude of the variation over the 11-y solar cycle.

IV. DISCUSSION

The differences in the values of the irradiances listed in Tables 1 and 2 are clearly greater than would be expected from a combination of the experimental errors and the observed variations during one or more solar rotations. Since the 2800 MHz radio noise index ($F_{10.7}$) is routinely employed as an indicator of the level of solar activity, the measured irradiances for the 1027 to 300 Å band are plotted against the daily ($F_{10.7}$) and the smoothed ($\overline{F}_{10.7}$) indices in Figure 9. Further, in order to show more clearly any systematic variations as a function of wavelength, the measured irradiances for the chromospheric emission lines C III, 977.0 Å; He I, 584.3 Å; and He II, 303.8 Å have been plotted against these same indices (Figure 10). It is important at this point to discuss the validity of employing the radio noise indices as measures of solar activity. Several authors reporting long-term measurements of EUV irradiances (Timothy and Timothy, 1970; Chapman and Neupert, 1973; Woodgate et al., 1973; Hinteregger, 1976) have pointed out the problems inherent in attempting correlations with the radio flux. The source of the radio emission in the solar atmosphere lies at lower altitudes in the quiet sun than in active regions. In the quiet sun the 2800 MHz flux originates at transition-region altitudes having temperatures of the order of 8×10^5 K, while in active regions having high plasma densities, the emission originates at higher altitudes, in the corona where the temperatures are 10^6 K or greater (Vernazza, J.E., *private communication*). Consequently, the irradiances of coronal EUV emission lines can be expected to track the radio flux more closely than will the irradiances of chromospheric or transition-region lines. However, the coronal EUV emission will have a much stronger dependence on the density of the active-region plasma than will the radio emission. Accordingly, there can be many different sets of conditions on the solar disk that will yield the same value for the radio index but that will produce drastically different EUV irradiances. Nevertheless, at this time, no superior ground-based index of the general level of solar activity has been successfully employed. Typically, F10.7 values of the order of 90, 150, and 220 respectively are taken to define quiet, medium, and active solar conditions. It can be seen from Figure 4 that, on this basis, solar cycle 20, which began in October 1964, was not a particularly active cycle; F10.7 peaked at about

180 to 190.

The irradiance values plotted in Figures 9 and 10 appear to cluster in two groups. The AFCRL data show generally consistent values for the EUV irradiance throughout the solar cycle. However, the results of other investigators tend to lie above the AFCRL values and, when combined with the AFCRL data, indicate a dependence of the EUV irradiance on the general level of solar activity. The Hinteregger (1970) irradiance values lie below the majority of the data, particularly at wavelengths below 600 Å. This fact has been noted in a number of recent publications (Higgins, 1976; Hinteregger, 1976), which state that the irradiance

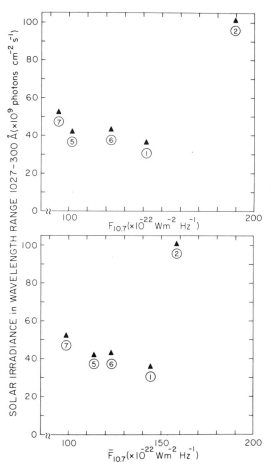

Fig. 9   Variations of the measured irradiance in the wavelength range from 1027 to 300 Å as functions of the daily (F10.7) and smoothed ($\bar{F}$10.7) 2800 MHz radio noise indices. Data from Table 2.

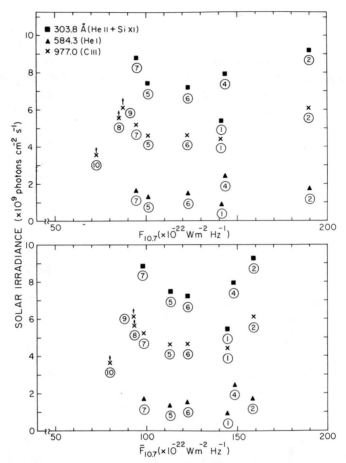

Fig. 10   Variations of the measured irradiances at 977, 584, and 304 Å as functions of the daily (F10.7) and smoothed ($\bar{F}$10.7) 2800 MHz radio noise indices. Data from Table 1.

in the wavelength range from 630 to 300 Å should be increased by about 15 to 30 % because of a recently discovered error in the calibration of the laboratory standard. Furthermore, the corrected values are essentially identical to those reported by Hinteregger (1976) for very quiet solar conditions (F10.7 ≲ 70).

The irradiance values of Timothy et al. (1972) lie well above the general level. A re-analysis indicates that the value for the total irradiance probably needs some additional correction for the effects of overlapping spectral orders, particularly in the 600 Å region. The magnitude of this correction is estimated to be of the order of 10%. It should, however, be noted that these data were taken during an active period (F10.7 = 190).

Making these corrections and replotting the irradiances as shown in Figures 11 and 12, there is an indication from the complete data set of a general increase in the irradiance with solar activity as suggested by the early AFCRL rocket data (Hall et al., 1969). The magnitude of this variation could be as large as a factor of two for the irradiance in the range 1027 to 300 Å during solar cycle 20. However, there is a large scatter in the data and the later AFCRL results are self-consistent throughout the solar cycle. This argues against a major change in the mean EUV irradiance over the solar cycle even though the AFCRL data were always recorded under the quietest available conditions (Heroux, L., *private communication*). Clearly, taking into account both the calculated experimental errors and the observed variations during a solar rotation, additional measurements will be required to resolve this fundamental question.

A significant amount of information on the emission of EUV radiation from specific solar features is now becoming available, particularly from the high-spatial resolution Skylab ATM observations. The state of knowledge of the effects of different features on the EUV irradiance is as follows:

Quiet Sun. The quiet sun is characterized by the chromospheric network, which has an almost constant structure when observed at EUV wavelengths

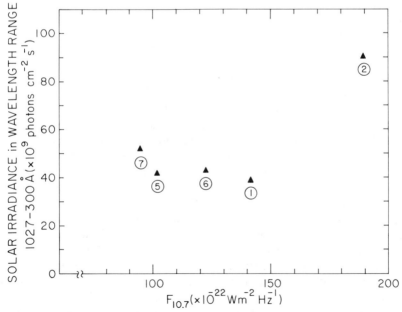

Fig. 11 Variation of the measured irradiance in the wavelength range from 1027 to 300 Å as a function of the daily (F10.7) 2800 MHz radio noise index. Data from Table 2 with corrected values for references 1 and 2.

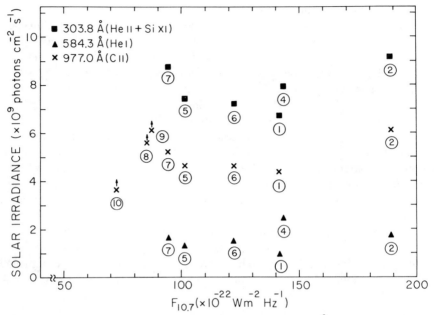

Fig. 12 Variations of the measured irradiances at 977, 584, and 304 Å as a function of the daily (F10.7) 2800 MHz radio noise index. Data from Table 1 with corrected values for reference 1.

characteristic of temperatures in the range 10^4 to about 1×10^6 K (see Figure 2). At coronal altitudes (temperatures $>1.5 \times 10^6$ K) the network is no longer visible and the majority of the radiation originates in active regions and large-scale magnetic loop structures. Some evidence suggests that there may be long-term changes in the EUV emission from the chromospheric network (Timothy, 1975). Since the network contributes more than 60% of the radiation from the quiet sun in chromospheric and transition-region emission lines (Reeves, 1976), these changes could have a significant effect on the EUV irradiance during the solar cycle.

Active Regions. The emission at all EUV wavelengths is enhanced in the dense plasma of active regions. The observed ratios of the average active-region and quiet-sun intensities for specific EUV emission lines are shown in Figure 13 (Noyes, Withbroe, and Kirshner, 1970). These enhancements are sufficient to explain the observed variabilities of the EUV irradiance during the solar rotation period, but are insufficient to account for a change of a factor of two in the irradiance over the solar cycle.

Flares. Dynamic events such as flares can cause enhancements of a factor of two or greater in chromospheric and transition-region line irradiances and of more than an order of magnitude in coronal emission-line irradiances, in a time scale of

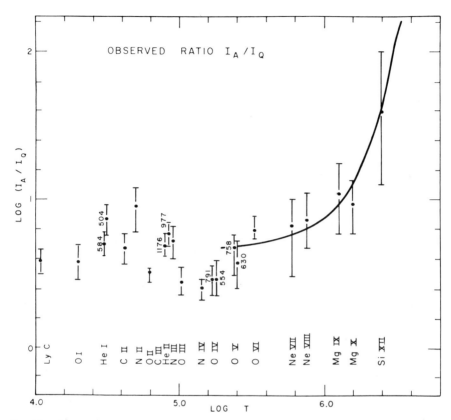

Fig. 13 Mean enhancement of EUV emission from active regions (I_A/I_Q) where I_A = average intensity per unit area in active region and I_Q = average intensity per unit area in quiet regions of the solar disk (Noyes, Withbroe, and Kirshner, 1970).

a few minutes. These effects are not directly relevant to this discussion of the average EUV irradiance. However, detailed reviews are available in the literature (e.g., Donnelly, 1976).

Coronal Holes. These are regions of very low intensity at coronal wavelengths which have lifetimes of several solar rotations. Examples of coronal holes can be seen near the poles in the Mg X (625.3 Å) image in Figure 2. These regions have a negligible effect on the irradiance of chromospheric and transition-region lines but contribute to the variability of coronal lines.

Bright Points. Bright-point sources of EUV and X-ray radiation have been observed to be uniformly distributed across the solar disk (see Figure 2). The integrated effect of these bright points is not significant in terms of the total irradiance at EUV wavelengths even for coronal emission lines. However, the possibility that a variation in the number of bright points in the chromospheric network may cause a long-term change in the quiet-sun irradiance during the solar cycle needs investigation.

VI. CONCLUSIONS

1. The Hinteregger (1970) values for the EUV irradiance need to be revised upward by approximately 15-30% in the wavelength range from 630-300 Å. Furthermore, on the basis of the most recent data (Hinteregger, 1976) it appears that these values are more appropriate for times of extremely low solar activity (F10.7 < 70). Since the NOAA model spectrum (Donnelly and Pope, 1973) is primarily based on the Hinteregger (1970) data, it should also be revised accordingly.

2. No definitive statement can be made at this time concerning the long-term variability of the EUV irradiance over the 11-y solar cycle. It is possible, on the basis of the available data, that a variation of up to a factor of two in the mean EUV irradiance may have occurred during solar cycle 20. However, the scatter of the available data argues against this hypothesis, and there is consequently a critical need for further measurements, particularly during times of high solar activity, to resolve this fundamental question.

3. The variability of the EUV irradiance during the solar rotation period arises principally from the appearance and disappearance of active regions on the solar disk. No systematic change in the magnitude of the variability has been observed from the maximum to minimum phases of the solar cycle. Variations of up to $\pm$ 15% can be expected in chromospheric and transition-region lines, while variations of up to a factor of three may be observed in coronal emission lines. These variations are, however, insufficient to account for a change of a factor of two in the EUV irradiance over the solar cycle.

4. The 2800 MHz radio noise index is a highly unreliable indicator of the magnitude of the EUV irradiance and should be replaced by direct measurements at EUV wavelengths on a routine basis.

5. A systematic program of EUV spectrophotometry should be initiated as soon as possible and should be continued throughout the period of maximum solar activity at the end of this decade. The instrumentation for these measurements should be capable of determining both the solar irradiance and the average brightness of the quiet solar atmosphere in order to determine the magnitude of possible long-term changes not associated with centers of activity. The wavelength resolution of these measurements should be sufficient to clearly resolve important spectral features; the optimum resolution should be of the order of 2 to 5 Å.

ACKNOWLEDGMENTS

The assistance of a large number of people with the preparation of this review is gratefully acknowledged. In particular, I thank R.W. Noyes, E.M. Reeves, J.E. Vernazza, and G.L. Withbroe of Harvard College Observatory; L. Heroux, J.E. Higgins, H.E. Hinteregger, and J. Manson of Air Force Cambridge Research Laboratories; and G. Schmidtke of the Fraunhofer Society for many informative discussions and for making available currently unpublished data. The efforts of Mrs. Linda S. Bayliss at Harvard College Observatory in tabulating the irradiance and solar activity data are also gratefully acknowledged. The solar activity data in Figure 4 and in Tables 1 and 2 are published from NOAA Solar-Geophysical Data bulletins by kind permission of the editor, J. Virginia Lincoln. This work was supported by NASA under Contract NAS5-3949.

REFERENCES

Behring, W.E., Cohen, L., and Feldman, U., 1972, *Ap. J. 175*, 493.

Chapman, R.D., and Neupert, W.M., 1973, *Goddard Space Flight Center Report X-680-73-304*.

Covington, A.E., 1969, *J. R.A.S. Canada 63*, 125.

Donnelly, R.F., 1976, *J. Geophys. Res.*, in press.

──────────── and Pope, J.H., 1973, *NOAA Technical Report ERL 276-SEL 25*, U.S. Government Printing Office, Washington, D.C.

Dupree, A.K., and Reeves, E.M., 1971, *Ap. J. 165*, 599.

Fawcett, B.C., 1974, *Adv. Atom. and Mol. Phys. 10*, 223.

Freeman, F.F., and Jones, B.B., 1970, *Solar Phys. 15*, 288.

Hall, L.A., and Hinteregger, H.E., 1970, *J. Geophys. Res. 75*, 6959.

──────────── , Damon, K.R., and Hinteregger, H.E., 1963, *Space Research III*, ed. W. Priester, John Wiley & Sons, Inc., New York, 745.

──────────── ; Higgins, J.E.; Chagnon, C.W.; and Hinteregger, H.E., 1969, *J. Geophys. Res. 74*, 4181.

Heroux, L., Cohen, M., and Higgins, J.E., 1974, *J. Geophys. Res. 79*, 5237.

──────────── , ──────────── , and Malinovsky, M., 1972, *Solar Phys. 23*, 369.

Higgins, J.E., 1976, *J. Geophys. Res. 81*, 1301.

Hinteregger, H.E., 1970, *Ann. Geophys. 26*, 547.

──────────── , 1976, *J. Atmos. Terr. Phys.*, in press.

Huber, M.C.E., Reeves, E.M., and Timothy, J.G., 1974, *Space Optics*, National Academy of Sciences, Washington, D.C., 33.

_____; Dupree, A.K.; Goldberg, L.; Noyes, R.W.; Parkinson, W.H.; Reeves, E.M.; and Withbroe, G.L., 1973, *Ap. J. 183*, 291.

Manson, J.E., 1976, this volume.

Noyes, R.W., Withbroe, G.L., and Kirshner, R.P., 1970, *Solar Phys. 11*, 388.

Reeves, E.M., 1976, *Solar Phys.*, in press.

_____ and Parkinson, W.H., 1970, *Ap. J. Suppl. 21*, 181.

_____, Huber, M.C.E., and Timothy, J.G., 1976, *Appl. Optics*, 1976.

Roble, R.G., 1976, *J. Geophys. Res. 81*, 265.

_____ and Dickinson, R.E., 1973, *J. Geophys. Res. 78*, 249.

Samson, J.A.R., 1967, *Techniques of Vacuum Ultraviolet Spectroscopy*, John Wiley & Sons, New York, 263.

Schmidtke, G., 1976, *J. Geophys. Res. Letters*, in press.

Swartz, W.E., and Nisbet, J.G., 1973, *J. Geophys. Res. 78*, 5641.

Timothy, A.F., and Timothy, J.G., 1970, *J. Geophys. Res. 75*, 6950.

_____ ; _____ ; Willmore, A.P.; and Wager, J.H., 1972, *J. Atmos. Terr. Phys. 34*, 969.

Timothy, J.G., 1975, *Bull. AAS 7*, 407 (abstract).

_____ and Lapson, L.B., 1974, *Appl. Optics 13*, 1417.

_____ and Reeves, E.M., 1976, *Proceedings AIAA/AGU Conference on Scientific Experiments of Skylab*, Huntsville, Ala., Oct.-Nov. 1974, in press.

Tousey, R.; Bartoe, J.-D.F.; Bohlin, J.D.; Brueckner, G.E.; Purcell, J.D.; Scherrer, V.E.; Sheeley, N.R., Jr.; Schumaker, R.J.; and Vanhoosier, M.E., 1973, *Solar Phys. 33*, 265.

Vidal-Madjar, A., 1976, this volume.

Walker, A.B.C., 1972, *Space Sci. Rev. 13*, 672.

Woodgate, B.E.; Knight, D.E.; Uribe, R.; Sheather, P.; Bowles, J.; and Nettleship, R., 1973, *Proc. Roy. Soc. London A 332*, 291.

Comment

With regard to the availability of line profile data, colleagues measuring Lyman-alpha or He 584 emission in the solar system are anxious to know not only the integrated solar irradiance in these lines, but also the value of the spectral irradiance at the line center. Early measurements of both the Lyman-alpha and Lyman-beta line profiles from OSO-8 are presented in my review of the Lyman-alpha flux measurements in this book.

With the launch of the French D2B satellite in September 1975, we now have a new source of solar irradiance data in the ranges from 300 Å to Lyman-alpha. This is a low-spectral-resolution experiment with a 10 Å bandwidth and the capability of building solar images with a 1 arcmin spatial resolution. These data should be available on a daily basis, and will be presented in *Solar Geophysical Data* starting in April 1976. J.P. Delaboudinière is in charge of the D2B spectrometer, and he is also working with G. Schmidtke in the COSPAR review of the measurements of the solar radiation between 1000 and 2000 Å.

Alfred Vidal-Madjar
Laboratoire de Physique Stellaire et Planétaire

THE SOLAR SPECTRUM
BETWEEN 10 AND 300 Å

James E. Manson
Air Force Geophysics Laboratory
Bedford, Massachusetts

The solar electromagnetic spectrum between 10 and 300 Å has been the subject of several reviews in recent years. Walker (1972, 1975), Culhane and Acton (1974), Parkinson (1975a), and Doschek (1975) have discussed the solar and atomic physics to be studied by measurements in this spectral region, and have clearly indicated that those interested in the effects of this radiation on the atmosphere of the earth can no longer expect to measure full-disk solar radiation by any of the instruments intended for solar physics research within the 10-300 Å region. The monitoring instruments on the Naval Research Laboratory (NRL) series of SOLRAD satellites, particularly those launched during 1976, will continue to provide very useful broad-band measurements, but they are designed to provide good time resolution, not spectral resolution.

The purpose of this review is to examine those full-disk measurements already made that satisfy the requirements of a quantitative measure of the solar spectral irradiance between 10 and 300 Å. The measurements will all be presented as the irradiance at one astronomical unit (AU) from the sun, a measure variously referred to as the photon flux distribution, the intensity at earth distance, or the solar flux on top of the atmosphere. Since each photon within this range of wavelengths is capable of causing primary and secondary ionization of the atmospheric constituents, the units of irradiance will be energy per unit area per unit time, within the given wavelength interval.

The discussion will be concerned mainly with the total irradiance within wavelength bands that are broad relative to both the resolution of the measuring instruments and the natural width of the discrete emission lines containing most of the radiant energy within this region of the spectrum. In order to avoid the pitfalls of a priori assumptions concerning the shape of the spectral distribution, I have chosen relatively high-resolution spectrometric measurements as the primary data base, numerically degrading the resolution for use in the present study. I will show that a thorough spectrometric analysis of a few carefully prepared rocket-probe spectrometers can be used to extend the application and improve the understanding of the much larger collection of data acquired by satellite instruments having lower resolution and sensitivity and less secure intensity calibration.

I. THE 10–31 Å REGION

This region has been studied by the use of Bragg spectrometers, generally using a potassium acid phthalate (KAP) crystal, which places an upper limit on the observed wavelengths approximately at the N VII line at 24.78 Å. Two intensity-calibrated, grazing-incidence grating spectrometers have also been used to record photographically some of the lines in this short wavelength region, as well as in the longer wavelengths. The results of the experiments chosen to represent this region are given in Table 1. The line intensities reported by each experimenter have been summed over 2-Å intervals, and I have added any continuum measurements given by each experimenter. The columns of Table 1 are headed by the experimental measurement date, the first author of the report in which the data are communicated, and the value of the solar radio flux at 10.7 cm reported by the Algonquin Observatory at Ottawa for that day, in units of 10^{-22} W m^{-2} Hz^{-1}. A short discussion of each of the ten experiments follows, in the order of the listing in Table 1.

The first results from measuring spectrally resolved solar emission lines in this range were reported by Blake et al. (1965); since these measurements were made

TABLE 1. Solar irradiance at 1 AU between 10 and 31 Å summed in bands. Units 10^{-3} ergs cm^{-2} s^{-1}.

Wave-length Range Å	1963 July 25 Blake F74	1966 May 5 Evans F90	1966 Dec 22 Walker F105 †	1967 Jan 4 Walker F161 †	1969 Nov 20 Freeman F189*	1968 Mar 20 Freeman F130*	1971 Nov 30 Parkinson 3'x3' Active area	1967 Feb 13 Walker Post 3B †	1969 May 18 Neupert Max SN	1969 Feb 27 Neupert Max 2B †
10–12	---	---	0.13	0.40	---	---	0.026	1.90	1.87	141.
12–14	---	0.90	0.78	1.13	---	---	0.118	6.00	6.65	211.
14–16	0.35	2.52	2.32	3.01	8.00	---	0.314	10.41	.11.03	123.
16–18	0.73	2.21	2.53	3.06	16.00	47.00	0.320	9.37	7.24	53.9
18–20	0.72	2.50	2.36	3.37	10.40	26.00	0.134	5.45	---	24.6
20–22	5.00	5.60	2.10	3.35	6.20	22.00	0.119	4.15	---	7.8
22–24	0.48	---	2.34	1.80	3.80	11.00	0.073	4.45	---	16.8
24.78 ‡	0.41	---	0.24	0.24	---	5.00	---	0.38	---	22.5
24–31	---	---	---	---	7.50	---	---	---	---	---
Accuracy quoted?	YES	YES	NO	NO	YES	YES	NO	NO	NO	NO
Value	25–35%	x2–x4	--	--	x2	x2	--	--	--	--
Review** value	same	50%	50%	50%	same	same	20%	50%	x2	x2

* Evidence of enhanced activity or X-ray flare.
† Continuum has been added to line sum by reviewer.
‡ This single line from N VII is the only entry within 24-31Å interval for these experiments.
** See discussion in text.

during very quiet solar conditions the results are listed first in the table. These authors also provided a discussion of the errors of their experiment, a practice unfortunately rare in this region. This resolved-line spectrum showed that the range below 17 Å is dominated by lines from small, high-temperature regions of the solar corona. These lines come from ions such as Fe XVII and O VIII, even during very quiet solar conditions. The range from 17 to 25 Å was found to contain lines emitted from the entire corona, such as those from the ions O VII and N VII.

Under slightly more active solar conditions, Evans and Pounds (1968) found similar results at a higher flux level in their rocket-probe measurement on 5 May 1966. Although these authors directed their analysis to the active regions, their experiment measured the flux from the entire solar disk, and they included a graphical presentation of their total solar irradiance measurements. The values presented in Table 1 for this date were read from that graph. Evans and Pounds' own estimates of maximum error limits for individual line intensities were a factor of 2 below 17 Å and a factor of 4 at 22 Å. They presented a comparison between their total irradiance at 8-20 Å and a recalculation of the ion chamber measurements over this range reported by NRL for the same time interval, which indicated agreement within a few percent. I have therefore suggested an error estimate of 50% for their results, and the accuracy for such integrated flux values may well be much better than that. Evans and Pounds also presented a convincing argument that the true continuum-to-line ratio in this range is no greater than 0.25.

The measurements obtained by Walker and Rugge (1969) from their instrument on the OV1-10 satellite under nonflaring solar conditions on 22 December 1966 are given next. Because of the possibility of detector photocathode variations between the time of laboratory calibration and launch, no estimate of accuracy is given. However, Rugge (*private communication*) has estimated that the errors are of the order of 50%. A similar estimate of errors was given by Rugge for the measurements listed in the next column, headed 4 January 1967, which were reported by Walker, Rugge, and Weiss (1974) using data from the same instrument. This set of data included a firm measurement of continuum and unresolved lines within the 10-20 Å range, in the form of a graphical display. I have taken values from this plot and included them in the data listed in Table 1 for that experiment date.

The next two columns show results reported by Freeman and Jones (1970) from two rocket probes using grazing-incidence grating spectrometers. These instruments were calibrated by means of a reference proportional counter, but the use of film as the detection medium limited accuracy of flux measurement to a factor of 2. Their 1968 instrument was recovered and recalibrated; these authors proposed an unreported X-ray flare to explain the rather high fluxes reported. As will be seen later, the longer wavelength measurements made with this instrument do not display such high flux values relative to other measurements. Their measurements of 20 November 1969 actually were made during a small X-ray flare, as noted by Manson (1972) in discussing the high intensity value of the C VI line reported for that date.

The small active region observed by Parkinson (1975b) in a spatially resolved experiment flown on a rocket probe on 30 November 1971 is included here, even though the area of the solar disk that was viewed was limited to a 3-by-3 arcmin square, because these results are the most complete sample of the emission from an active, nonflaring region. Parkinson has stated that no evidence of continuum radiation was detected. No estimate of accuracy is given, although the comparison of the line intensities from three separate crystal spectrometers in his experiment suggest that the crystal reflectivities were known within 15%. No information on the accuracy of the counter window transmission is given, but the reviewer believes that the University of Leicester group, which prepared this experiment, controls their windows within the 20% error estimate I have assigned these results.

The next three columns relate to measurements of the flaring sun. Those made on 13 February 1967 were reported by Walker and Rugge (1969) and were made by the same satellite instrument that yielded the measurements of 22 December 1966 and 4 January 1967. A class 3B flare occurred at 1749 UT, and these measurements were made between 1810 and 1843 UT. Rugge's estimate of errors is applied to these measurements. He has also stated that they cannot distinguish between true continuum and blended, unresolved lines in what appears as a distinct "hump" in their recorded spectrum, between 10 and 17 Å (Rugge, *private communication*). If we multiply Parkinson's results by a factor of 35, the resulting column closely resembles that of Walker and Rugge for this post-flare observation, which suggests that the "hump" can be explained by a collection of emission lines such as those observed by Parkinson. The lines would be unresolved in Walker and Rugge's instrument as it was operated that day.

The results obtained by Neupert, Swartz, and Kastner (1973) from their instrument aboard OSO-5 on 18 May 1969 and 27 February 1969, during the maximum phase of a class SN flare and a class 2B flare, respectively, are listed in the last two columns. No estimate of absolute error is given, and I have assigned a factor of two for these results. The relative accuracy is probably of more interest for these essentially unrepeatable flare observations, and this is probably well within 50%, since the experiment was calibrated in the laboratory prior to flight, and the authors state that their relative observations of some lines were checked against the reports of other experiments also in orbit during their experiment's life. Unfortunately, no error limits were given for this cross comparison. In order to include the obviously important continuum or unresolved line contribution which is shown for the 27 February 1969 experiment by Neupert, Swartz, and Kastner (1973) in their Figure 3a, I have made bold use of a planimeter and the sensitivity figures they give in their Table I. I assumed that the entire "hump" in their data, between 6 and 16 Å, is indeed real first-order, and have included my estimate of this irradiance contribution in each of the wavelength intervals listed for this date in Table 1. The additional flux amounts to 60% of the values listed, summed over the full 10-24 Å range. The enhancement in the total solar irradiance between 10 and 20 Å indicated for this flare is a factor of 20, if the unresolved lines and continuum are added in this way.

In summary, the region of the solar spectrum between 10 and 31 Å is made up of emission lines, and the moderately active sun is well represented by the measurements listed in Table 1 for 22 December 1966 and 4 January 1967. To assemble even the small collection of measurements shown in Table 1 required somewhat speculative interpretation on the part of the reviewer, and it would be very helpful if the investigators reporting such measurements could be more forthcoming with information on the accuracy of their experiments. More particularly, descriptions of the experiments and information on data handling methods are very important in assessing results years later, when attempts are made to put the essentially unrepeatable solar observations in proper perspective. For instance, in this volume Kreplin *et al.* ascribe the variability of the reported irradiance in the O VIII 18.97 Å line to errors in the crystal reflectivities used in reducing the spectrometer observations. The original reports indicate that the very high results obtained by Argo, Bergey, and Evans (1970) were based on the use of nonscanning, fixed-station crystal monochromators. Because it is very difficult to determine the background in this type of measurement, and because much lower values are obtained when the more complex scanning Bragg spectrometers are used, we ought to consider this source of error as well in determining possible explanations for the wide range of irradiance values reported for the O VIII line at 18.97 Å. The point is that in this case the reports do yield full details from which the reader may make his or her own decisions.

II. THE 31–300 Å REGION

An analysis of the results of the measurements made in this region by the rocket-probe instruments of Air Force Cambridge Research Laboratory (now Air Force Geophysics Laboratory) has been reported recently (Manson, 1976). In addition to those measurements, the results of Chapman and Neupert (1974), Freeman and Jones (1970), and Timothy *et al.* (1972) were chosen to represent the solar irradiance in this region. Following the general procedure used in preparing Table 1, these measurements were summed over 10-Å bands (and one 9-Å band), and the results are arranged in Table 2, with the solar activity increasing from left to right. In this region the authors have all reported some estimate of the accuracy of their irradiance measurements. The consistency of the measurements shown in the region of intense line radiation around 200 Å is quite good; the high results on 3 April 1969 reported by Timothy *et al.* (1972) were probably caused by the inclusion of some scattered and second-order radiation in their fairly wide 6 Å instrumental bandwidth. The values for the Timothy *et al.* experiment were scaled by the

TABLE 2. Solar irradiance at 1 AU between 31 and 300 Å summed in bands. Units 10^{-3} ergs cm^{-2} s^{-1}.

Wavelength Range Å	1965 Nov 3 Manson F81	1967 Jun 11 Chapman * F96	1971 Nov 9 Manson Higgins F102	1972 Aug 23 Heroux F120	1968 Mar 20 Freeman † F130	1967 Aug 8 Manson F143	1969 Apr 4 Heroux F177	1969 Nov 20 Freeman † F189	1969 Apr 3 Timothy F190	1967 May 27 Chapman * F214
31- 40	30	---	50	---	---	---	---	14	---	---
40- 50	18	---	42	---	---	25	---	29	---	---
50- 60	33	---	63	161	---	51	111	36	---	---
60- 70	29	---	72	116	---	44	89	28	---	---
70- 80	27	---	66	98	---	36	71	13	---	---
80- 90	31	---	69	72	---	39	56	---	---	---
90-100	28	---	53	56	---	32	59	---	---	---
100-110	19	---	33	34	---	19	28	---	---	---
110-120	11	---	19	16	---	9	12	---	(73)‡	---
120-130	---	---	10	4	---	---	2	---	(21)	---
130-140	---	---	11	3	---	---	3	---	(18)	---
140-150	---	---	19	14	3	---	18	---	(93)	---
150-160	---	28	20	19	8	---	24	---	45	42
160-170	---	5	32	25	4	---	35	---	54	26
170-180	---	124	159	155	184	---	197	---	227	167
180-190	---	138	138	169	136	---	206	---	220	205
190-200	---	125	109	147	104	---	183	---	275	180
200-210	---	92	---	134	50	---	146	---	174	127
210-220	---	63	---	98	27	---	110	---	139	118
220-230	---	41	---	107	4	---	150	---	150	62
230-240	---	6	55	55	7	---	82	---	101	9
240-250	---	---	47	68	17	---	109	---	97	---
250-260	---	58	104	123	4	---	205	---	148	106
260-270	---	15	34	40	4	---	71	---	97	31
270-280	---	43	39	56	---	---	95	---	69	80
280-290	---	83	57	110	38	---	198	---	167	214
290-300	---	19	16	23	---	---	38	---	34	26

* The sums for these two experiments contain no allowance for continuum or unresolved lines.
† Evidence of enhanced activity or X-ray flare during these experiments.
‡ Values in parentheses are believed to contain scattered background and higher orders.

reviewer from a small graphical display in their report, which seems to show signs of second-order contamination in the 130-, 400-, and 600-Å regions. For the range from 250 to 300 Å they claim an accuracy of 30%, but below 250 Å they report that the error may increase to as much as a factor of 2. The results of Chapman and Neupert (1974) given in the 27 May 1967 column may be underestimated since their fairly large but not complete set of lines was used to represent the total solar irradiance. This point will be discussed in a later section. These authors used a post-launch analysis and comparison with other orbiting experiments to upgrade their intensity accuracy to about 30% at 304 Å.

The short-wavelength results of 3 November 1965 and 8 August 1967 reported by Manson (1967, 1972) had a very high background and below 50 Å the accuracy of these summed values is no better than a factor of 2. Between 50 and 120 Å an accuracy of 25% is applied to the summed values. For the 9 November 1971 measurements, Manson (1976) and Higgins (1976) claim an accuracy of 20%, although the errors may be larger for the sums reported in the weak regions of the spectrum. The measurements reported by Manson (1976) on 9 November for the range between 44 and 60 Å were summed and tabulated especially for comparison with the results from the Naval Research Laboratory satellite SOLRAD 10 on the same date. The agreement of ±10% about the mean of these measurements is remarkable, considering the simple graybody spectrum used in the standardized NRL data reduction method. The NRL results were corrected for the short-wavelength passband of their ion chamber (Horan, *private communication*, 1972).

The data from 23 August 1972 and 4 April 1969 reported by Heroux, Cohen, and Higgins (1974) have accuracies of 50% below 150 Å, exclusive of the very weak region between 120 and 140 Å. Between 150 and 300 Å these measurements were reported to be accurate to 30%. It is noted that Heroux, Cohen, and Higgins (1974) have reevaluated the results of the 4 April 1969 experiment previously reported by Malinovsky and Heroux (1973), specifically giving the results integrated over 10-Å bands, and including unidentified and blended lines in these sums. These new values are quoted in Table 2.

The measurements made on 20 March 1968 and 20 November 1969 reported by Freeman and Jones (1970) are the longer wavelength ranges of the same instruments contributing to Table 1 for these dates. These photographic instruments are seen to yield very good agreement with the photoelectric detector instruments in regions of strong line radiation. There is no reason to expect these measurements to give good results in regions with many small lines that are not resolved, since the photographic technique is not suited to such measurements. The instruments were calibrated accurately enough for the major source of the reported factor-of-2 uncertainty in the intensities to be ascribed primarily to the reproducibility of the photographic technique.

Schmidtke (1976) has presented an important new set of EUV irradiance measurements based on the results of an experiment on the AEROS-A satellite. Although it was not possible to include his results in Table 2, we can compare the two compilations by summing over the interval 150-280 Å. Schmidtke's total of 1.45 ergs cm^{-2} s^{-1} is about 20% higher than that obtained from the 23 August 1972 column of Table 2. His result is about double that listed under the date 11 June 1967, and if we interpolate three missing values into the 9 November 1971 column in Table 2, we obtain a sum which is about 2/3 that of Schmidtke. Since Schmidtke's table is proposed as typical of a moderate level of solar activity (F10.7 equal to 100), these comparisons demonstrate fairly good agreement within the context of experiments having about 30% calibration accuracy. It is to be

III. VARIABILITY UNDER CHANGING SOLAR CONDITIONS

The results given in Table 1 and Table 2 are displayed in Figure 1. The measurements are presented as a differential distribution, so that equal areas represent equal radiant energy per unit area per unit time at 1 AU. The extremes shown for the shaded areas are the limits indicated by the data in Table 2. As is shown by Heroux, Cohen, and Higgins (1974) and Manson (1976), one cannot assume that all of these intervals will vary in even the same direction under varying solar conditions. The range shorter than 31 Å is represented by three sample distributions. The one labeled "normal" is that of 22 December 1966, used to represent the sun under moderate conditions. The data from Table 1 for 13 February 1967 and 27 February 1969 represent the postflare and flaring sun, respectively. The shift of energy to the shorter wavelengths in the flare results is quite striking.

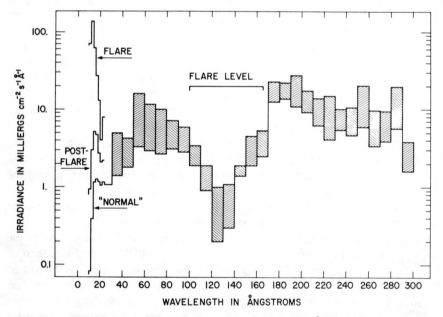

Fig. 1 The solar spectral irradiance at 1 AU between 10 and 300 Å. Three states of solar activity are shown for the region 10-31 Å. The vertical extent of the shaded interval sums between 31 and 300 Å include most of the values in Table 2. The irradiance units are 10^{-3} ergs cm^{-2} s^{-1} Å^{-1}.

In order to include the effect of a 2B solar flare in the range between 100 and 170 Å, indicated by the line labeled "flare level" in the figure, I once again made bold use of published illustrations, this time from the report of Kastner, Neupert, and Swartz (1974). Their Figure 3 shows an enhancement in the 100-170 Å region during a class 2B flare on 25 February 1969. I have used their nonflaring spectrum in the region of 150 Å, and the calculated instrumental full width at half maximum of 0.38 Å to estimate the flare irradiance level shown in Figure 1 for this range. I assume that the general rise in the "background" level that occurred during this event was due to real radiation, and included this in my estimate. The estimate is given only to indicate that changes of this magnitude do take place when the sun flares; much more information is needed before any quantitative limits can be placed on such variation.

Any discussion of the variability under changing solar conditions would be incomplete without reference to the results from the ion chambers used by NRL on the SOLRAD satellites (Kreplin, 1970). Summarizing the experience of about five years of almost continuous monitoring, he reported variations, from solar maximum to minimum in the 8-20 Å band, by a factor of 200, and in the 44-60 Å band by a factor of 20. At solar minimum, the results fluctuated by about a factor of 5 and a factor of 2 for these two bands, respectively. The interpretation of such broad-band ion chamber currents can be sensitive to changes in the shape of the incident irradiance distribution. For instance, the activity-related spectral shift toward shorter wavelengths shown in the 10-20 Å section of Figure 1 could cause an increase in the observed ion current quite a bit larger than the actual enhancement in the total irradiance over this range.

Another large collection of satellite observations is that reported by Chapman and Neupert (1974). They give plots of the daily observations of seven spectrally resolved lines in the solar spectrum between 171 and 335 Å, for 226 d during a 293-d period in 1967. Five of these lines are within the range of this study, and in order to get some idea of the extremes of variability which normally occur as the sun rotates about its own axis, I determined the limits of count-rate that would include the entire set of observations for each of these lines. I have assumed that these observations excluded any intervals of flare activity. The entire range of variation of the Fe IX line at 171 Å was ±33%, and even the very active Fe XV line at 284 Å was restricted to a ±58% range. The variation is stated in this way in order to make a comparison with the absolute accuracy limits discussed above as typical for this region; the limits are quite comparable. The accuracy of most satellite measurements (which generally are not calibrated in this wavelength region) would therefore be improved if we were to select, as an irradiance standard to use for in-flight calibration, the average observed intensity of a single, rather constant line such as the Fe IX line at 171 Å. This suggestion that the sun be its own calibration standard is not meant to discourage continued laboratory calibrations, but to allow the establishment of some limits to the sensitivity shifts that concern all experimenters in dealing with satellite instruments. The solar 304 He II line shows even less variability in the plots given by Chapman and Neupert (1974), and this line is usually used in laboratory calibrations. I believe that the use of these two lines as transfer standards would improve the accuracy of most of the existing satellite and rocket irradiance observations in this range.

IV. CORRELATION OF SOLAR EUV AND DECIMETRIC RADIO FLUX

The early success of the correlation between the "27-d variation" in atmospheric densities and the solar decimetric radio flux has been reviewed by Jacchia (1963). Variations in atmospheric temperature inferred from measurements of satellite orbit variations were shown to be very highly correlated with the 20 cm radio flux (Priester, 1961). Subsequent use of the 10.7 cm radio flux (F10.7) as measured at Ottawa was similarly successful. It was pointed out in most of these early reports that the actual energy input causing these observed effects on the atmosphere must be solar X-ray and EUV radiation. The first quantitative measurements of the solar EUV between 250 and 1300 Å by Hinteregger et al. (1960) demonstrated that there was sufficient energy in this region of the solar spectrum to account for at least a large part of the observed atmospheric effect. With the advent of satellite programs to measure the solar X-ray and EUV flux, all expectations were that a similarly tight correlation between the directly measured X-ray and EUV radiation and F10.7 would be demonstrated. Unfortunately, it has not worked out that way. Not only is there more intrinsic scatter in the direct EUV measurement correlation than in the satellite drag studies, but for some solar rotations the correlation shifts in phase or decreases in amplitude. The expectation is that sufficiently accurate data over a wide spectral range and long time interval would provide a correlation between the integrated daily average EUV and F10.7 as tight as that originally found between F10.7 and the atmospheric temperature. On the other hand, one must not totally ignore the possibility that these preconceptions are in error, and that there exists a relationship, as yet undiscovered, between another energy source and F10.7 that is indicated by just such failures of the simple EUV/F10.7 correlation.

It is well understood that one of the main reasons for the great interest in the correlation between solar EUV and F10.7 is to allow a determination of the EUV input to the atmosphere during times when no direct measurement of solar EUV exists. For this reason, several studies have been carried out to examine this possibility. Early OSO-1 measurements of solar lines between 171 and 400 Å were shown to be highly correlated with F10.7 (Neupert, Behring, and Lindsay, 1964; Neupert, 1967). Figure 2 is reproduced from the latter work and illustrates the wide range of variability in a set of lines from various stages of the ionization of iron. This figure also includes a set of observations of the solar radio noise at various frequencies, illustrating the good correlation between radio noise around

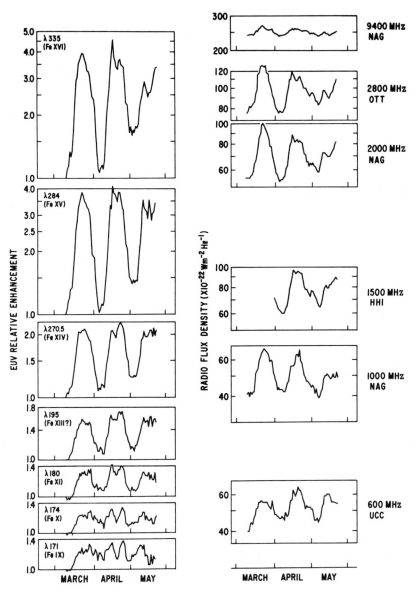

Fig. 2 An example of the observations of the variation of the solar EUV and the solar radio flux. The similarity in the behavior of the lines from highly ionized iron and the radio flux in the 2000 MHz range is evident. (Illustration courtesy of W. M. Neupert.)

2000 MHz and some of the solar EUV lines from high-temperature regions of the solar atmosphere. The purpose of the more recent study of this data and additional similar data from OSO-3, reported by Chapman and Neupert (1974), was to develop a quantitative relationship (linear regression) between solar EUV and F10.7. Timothy and Timothy (1970) measured the solar He II line at 304 Å for the better part of two years and give a linear regression relating the intensity of this line to F10.7. A similar study reported by Woodgate et al. (1973) also includes the relationship of the 304 Å line to F10.7 for six months during 1969-1970. Although this radiation is slightly beyond the range of this review, it is included because it provides an important link with the results reported by Chapman and Neupert (1974). Rugge and Walker (1970) and Rugge, Walker, and Anderson (1970) present straight-line fits to scatter diagrams of soft X-ray versus F10.7. I have derived the regression coefficients for their results from the graphic presentations. In addition to these published results, I have included some new results from our Extreme Ultraviolet Spectrophotometer (EUVS) experiment on the AE-C satellite (Hinteregger, 1976) in Figure 3, and have derived best-fit linear regression coefficients from this data for the three ranges illustrated. This figure illustrates the most clearly correlated data interval observed by the EUVS during the year 1974. One more very small sample of three rocket flight measurements of the range 130-200 Å has been reported by Manson (1976) from measurements in 1969, 1971, and 1972.

It cannot be emphasized too strongly that these statistically demonstrated relationships between solar EUV and F10.7 are indeed only statistical. The fairly tight distribution shown in the 304 Å-versus-F10.7 scatter diagram of Woodgate et al. (1973) probably results from their use of the daily mean of their 304 Å measurements. Presumably the values used by Chapman and Neupert (1974) are also daily averages, since they mention that bursts are removed from their data. It must be noted here that the error limits given by Chapman and Neupert for their linear regression between 140 and 400 Å in the EUV and F10.7 amount to ±50% at an F10.7 value of 150 flux units, if the extremes of the error limits of both of their coefficients are used. These error limits are stated to be statistical probable errors only, with no allowance for systematic error.

The results of the above studies are given in various irradiance units and even in terms of simple recorded counts of the measuring instrument. If the entire sun is viewed and the detectors are linear, the variability information can be extracted by conversion to a dimensionless form. I have expressed the irradiance reported by the experimenter, I, in units of the value of I for F10.7 equal to 100. This was done by adopting the form used by Chapman and Neupert, with a transformation to the independent variable (F10.7 - 100). This results in the following equation:

$$I = A' \left[1 + B'(F10.7 - 100) \right] \quad . \tag{1}$$

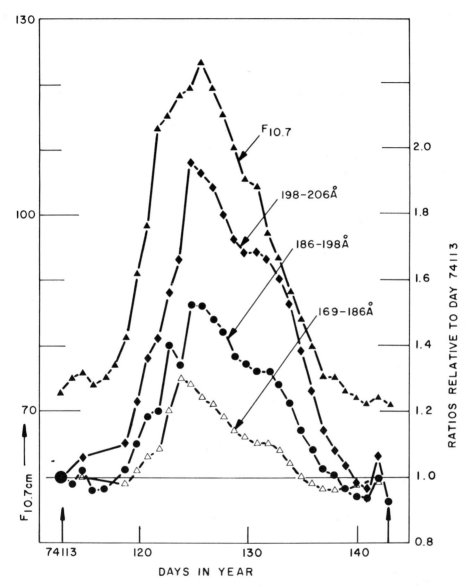

Fig. 3 An example of a solar rotation illustrating good correlation between EUV measurements and F10.7. The EUV data were acquired by the Air Force Geophysics Laboratory experiment on the AE-C satellite.

Essentially, the transformation to (F10.7 - 100) places the normalization constant, A', in the middle of the collection of data points fitted by the straight-line regression. The use of the zero intercept for this purpose, as was done by Chapman and Neupert, introduces a sign change and a discontinuity into the range of the slope constant.

Table 3 lists values of wavelength or wavelength range, and the variability index, B', for six of the studies discussed above. The range of values of B' within any one set of data can be related to the spectroscopy of the solar plasma, which is of no concern to us in this study. What is important for understanding the response of the atmosphere to varying solar conditions is that such a range of variability exists, and that a decent model of the solar X-ray and EUV irradiance must include fairly complicated wavelength-dependent variability indices, which may be correlated with F10.7 as shown here.

The normalized irradiance, I/A', is given as a function of F10.7 in Figure 4 for a sample of the cases presented in Table 3. To make the figure readable, I have plotted the relationship of the averages for the X-ray and the EUVS results, rather than presenting each wavelength independently. The dashed line is a straight line through the point (100,1) and the origin (0,0). It is included for visual reference. Lest the impression be left that there is a monotonic variation of B' with wavelength, it should be noted that the slope of the line corresponding to the 335 Å Fe XVI line is even larger than that shown for the 18.96 Å X-ray line.

TABLE 3. Normalized variability index, B', from irradiance - F10.7 regression.

Source	Wavelength Range (Å)	Normalized Slope, B'		Data interval
Rugge and Walker (1970)	18.96	.0132		
	21.60	.0089		Dec 1966–
Rugge, Walker and	21.80	.0071		Mar 1967
Anderson (1970)	22.10	.0073		
Chapman and Neupert,		OSO-1	OSO-3	
(1974)	171.1	.0025	.0030	OSO-1
	174.5	.0030	.0034	Mar 1962–
	195.1	.0075	.0038	May 1962
	256.3	.0068	.0059	
	284.2	.0211	.0117	OSO-3
	303.8	.0037	.0034	Mar 1967–
	140-400		.0050	Dec 1967
Timothy and Timothy (1970)	303.8	.0017		June 1968– Dec 1969
Woodgate, et al. (1970)	303.8	.0033		Aug 1969– Feb 1970
Hinteregger (1976)	169-186	.0043		
	186-198	.0073		Apr 1974
	198-206	.0109		
Manson (1976)	130-200	.0055		1969 - 1972 (Three rocket flights)

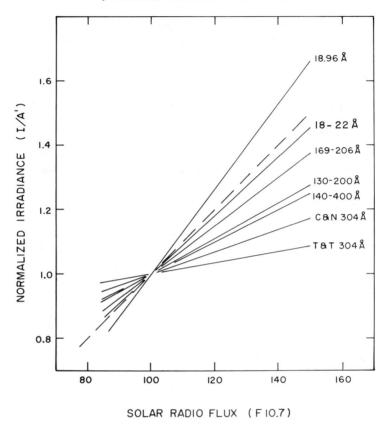

Fig. 4 The normalized irradiance, I/A', as a function of F10.7 for several examples from Table 3. The dashed line is drawn through the origin for visual reference.

V. REPRESENTATIVE SAMPLING OF THE SPECTRUM

The accurate measurement of the entire spectrum has been seen to be a very demanding job. Therefore the possibility of measuring a representative sample of the spectrum has recommended itself to those interested in the effects of solar radiation rather than in the diagnostic study of the spectrum itself. The study by Chapman and Neupert (1974) concerned three such samples. The largest, comprising 81 lines between 148 and 388 Å, was used to represent the entire spectrum. Smaller sets, made up of 21 and 7 lines covering the same wavelength region, were studied for various purposes and were stated to represent a constant fraction of the larger set, independent of the level of solar activity. Since this result is exactly the one sought, I have selected for further study those portions of the sets that lie within the region under study here. The largest such subset, here called S52, is made up of Chapman and Neupert's lines 1 through 52, ranging in wavelength from 148.4 through 296.1 Å. Next in size is S13, made up of the first thirteen lines in Chapman and Neupert's Table 1, ranging from 171.1 through 284.2 Å. The smallest subset, S5, contains the lines with wavelengths 171.1, 175.5, 195.1, 256.3, and 284.2 Å.

The intensities of most of the lines in these sets have been published by Manson (1976), Higgins (1976), Heroux, Cohen, and Higgins (1974), and Malinovsky and Heroux (1973) for the flight dates 9 November 1971, 23 August 1972, and 4 April 1969 used in Table 2. The intensities for the 4 April flight have been corrected and extended and communicated to me by the authors. These corrections are discussed by Heroux, Cohen, and Higgins (1974). The total integrated irradiance values used were those of Table 2, with a few missing values being filled in by extrapolation for the purpose of this examination. The parent sets of Chapman and Neupert (1974) were used for the three dates in 1967 chosen by them to represent their very large body of observations. What I have calculated is the fraction of each total spectrum represented by the subsets S52, S13, and S5.

The results, shown in Table 4, are obviously very encouraging. Most noteworthy, however, is that by using the higher resolution and lower background measurements provided by the three rocket-probe experiments, we can see that the fairly large set S52 contains approximately 50% of the irradiance, rather than most of it. Also remarkable, and in full qualitative agreement with the observations by Chapman and Neupert, is the indication that the smaller subsets contain a fixed proportion of the total irradiance over the range of solar activity represented by this small sample. My own results overlap the region studied by Chapman and Neupert only between 140 and 205 Å, and I have analyzed this data by assigning spectral features of the proper instrumental line profile in sufficient numbers to satisfy the envelope of the observed experimental data points. The result was that 203 such "lines" were required, or an average of about seven features in between each of the members of S52. Similar results were found by Malinovsky and Heroux (1973), as can be seen from their recorded spectra. It is remarkable that the reported sum over S52 of the data reported by Chapman and Neupert should agree as well as it does with the numerically integrated irradiance of these far more complex recorded solar spectra, and one must assume that the features reported as lines by Chapman and Neupert must have included a considerable contribution from neighboring unresolved lines. Considerations such as these are quite important to anyone wishing to make quantitative use of the measurements of irradiance reported by various experimenters. They are vital if the entire spectrum is to be represented by the true measurement of a small set of lines, such as S13 or S5, since the fraction of the total irradiance represented by such a sampling set is very much dependent on the measurement of both the sample and the total population, as can be seen from the results in Table 4.

TABLE 4. The relationship between sets of lines and the total irradiance, between 140 and 300 Å.

Section of the Spectrum	1971 Nov 9 F102	1972 Aug 23 F120	1969 Apr 4 F177	1967 Jun 11 F96	1967 Aug 12 F139	1967 May 27 F214
Total *	1123	1343	1867	---	---	---
S52 *	602	685	997	854	1052	1395
S13 *	413	487	689	532	660	905
S5 *	225	263	394	270	348	486
$\left(\frac{S13}{S52}\right)$	0.69	0.71	0.69	0.62	0.62	0.65
$\left(\frac{S5}{S52}\right)$	0.37	0.38	0.40	0.33	0.32	0.35
$\left(\frac{S52}{Total}\right)$	0.54	0.51	0.53	---	---	---
$\left(\frac{S13}{Total}\right)$	0.37	0.36	0.37	---	---	---
$\left(\frac{S5}{Total}\right)$	0.20	0.20	0.21	---	---	---

* The entries in these lines are in units of 10^{-3} ergs cm^{-2}s^{-1}.

REFERENCES

Argo, H.V., Bergey, J.A., and Evans, W.D., 1970, *Ap. J.* 160, 283.

Blake, R.L.; Chubb, T.A.; Friedman, H.; and Unzicker, A.E., 1965, *ibid.* 142, 1.

Chapman, R.D., and Neupert, W.M., 1974, *J. Geophys. Res.* 79, 4138.

Culhane, J.L., and Acton, L.W., 1974, *Annu. Rev. Astr. and Ap.* 12, 359.

Doschek, G.A., 1975, *Solar Gamma-, X-, and EUV Radiation, IAU Symposium No. 68*, ed. S.R. Kane, D.Reidel, Dordrecht, 165.

Evans, K., and Pounds, K.A., 1968, *Ap. J.* 152, 319.

Freeman, F.F., and Jones, B.B., 1970, *Solar Phys.* 15, 288.

Heroux, L., Cohen, M., and Higgins, J.E., 1974, *J. Geophys. Res.* 79, 5237.

Higgins, J.E., 1976, *ibid.* 81, 1301.

Hinteregger, H.E., 1976, *J. Atmos. Terr. Phys.*, in press.

_____; Damon, K.R.; Heroux, L.; and Hall, L.A., 1960, *Space Res.* 1, 615.

Jacchia, L.G., 1963, *Rev. Mod. Phys.* 35, 973.

Kastner, S.O., Neupert, W.M., and Swartz, M., 1974, *Ap. J.* 191, 261.

Kreplin, R.W., 1970, *Ann. Geophys.* 26, 567.

Malinovsky, M., and Heroux, L., 1973, *Ap. J.* 181, 1009.

Manson, J.E., 1967, *ibid.* 147, 703.

_____, 1972, *Solar Phys.* 27, 107.

_____, 1976, *J. Geophys. Res.* 81, 1629.

Neupert, W.M., 1967, *Solar Phys. 2*, 294.
———, Behring, W.E., and Lindsay, J.C., 1964, *Space Res. 4*, 719.
———, Swartz, M., and Kastner, S.O., 1973, *Solar Phys. 31*, 171.
Parkinson, J.H., 1975a, *Solar Gamma-, X-, and EUV Radiation, IAU Symposium No. 68*, ed. S.R. Kane, D.Reidel, Dordrecht, 45.
———, 1975b, *Solar Phys. 42*, 183.
Priester, W., 1961, *J. Geophys. Res. 66*, 4143.
Rugge, H.R., and Walker, A.B.C., Jr., 1970, *ibid. 15*, 372.
———, ———, and Anderson, M., 1970, *Aerospace Rept. No. TR-0059 (9260-02)-1*, Aerospace Corp., Los Angeles.
Schmidtke, G., 1976, *Geophys. Res. Letters*, submitted for publication.
Timothy, A.F., and Timothy, J.G., 1970, *J. Geophys. Res. 75*, 6950.
———; ———; Willmore, A.P.; and Wager, J.H., 1972, *J. Atmos. Terr. Phys. 34*, 969.
Walker, A.B.C., Jr., 1972, *Space Sci. Rev. 13*, 672.
———, 1975, *Solar Gamma-, X-, and EUV Radiation, IAU Symposium No. 68*, ed. S.R. Kane, D.Reidel, Dordrecht, 73.
——— and Rugge, H.R., 1969, *Solar Flares and Space Research*, ed. C. de Jager and Z. Svestka, North-Holland, Amsterdam, 102.
———, ———, and Weiss, K., 1974, *Ap. J. 192*, 169.
Woodgate, B.E.; Knight, D.E.; Uribe, R.; Sheather, P.; Bowles, J.; and Nettleship, R., 1973, *Proc. Roy. Soc. London A 332*, 291.

COMMENT ON THE SOLAR SPECTRUM BETWEEN 10 AND 300Å

Arthur B.C. Walker, Jr.
Institute for Plasma Research and
Department of Applied Physics
Stanford University

In studying the total irradiance in the solar EUV and soft X-ray spectrum we have three major objectives: to measure the total spectrum with sufficient spectral resolution and precision to be able to specify irradiance in wavelength bands $\Delta\lambda/\lambda \sim 20$ within a few percent; to determine the variation of the irradiance in each wavelength interval so that a "restricted" set of measurements will allow total irradiance to be specified (by "restricted" measurements we mean either a selected set of emission lines or broad-band measurements); and to establish a relationship between the solar soft X-ray irradiance and the easily obtained, universally available solar radio indices.

Manson (this volume) has pointed out the limited observational data base in the wavelength range between 10 and 300Å, and the uncertainties in calibration and interpretation that limit how precisely these objectives can be met. The solar corona, which is responsible for the solar emission between 10 and 300Å, is optically thin; thus in this wavelength range the dependence of the solar irradiance on coronal structure can be separated from its dependence on the atomic parameters that specify the emission function of a Maxwellian gas. This capability of analytically modeling the solar spectrum is of great importance in interpreting the presently available observations in order to obtain the total solar irradiance.

We will comment briefly on the current state of models of the solar spectrum, and on the current status of each of the three major objectives.

I. MODELS OF THE CORONAL SPECTRUM

The irradiance in a coronal emission line (wavelength λ) that is emitted by ionization stage z of element Z can be written (Walker *et al.*, 1974a)

$$I(\lambda,t) = a_H A_Z \int dT_e \, a_{Zz}(T_e) M(T_e,t) \, J_{Zz\lambda}(T_e) \tag{1}$$

where a_H is the number of free electrons per hydrogen atom (essentially constant for the corona), A_Z is the abundance of the element Z, a_{Zz} is the relative population of the parent ion of the line, $M(T_e,t)$ is the coronal emission measure function, and $J_{Zz\lambda}(T_e)$ is the atomic rate constant for line emission (Walker, 1975). We have specifically included the time dependence of the emission measure function and consequently of the irradiance. The slight dependence of a_{Zz} and $J_{Zz\lambda}$ (for some ions) on the coronal density structure is not significant for the present discussion. For the X-ray continuum we have a similar expression,

$$I(\lambda_1,\lambda_2,t) = a_H \sum_Z \sum_z A_Z \int_{\lambda_1}^{\lambda_2} d\lambda \int dT_e\, a_{Zz}(T_e) M(T_e,t)$$

$$\times \left[J_{Zz}^{rad}(T_e,\lambda) + J_{Zz}^{b}(T_e,\lambda) + J_{Zz}^{2\gamma}(T_e,\lambda) \right] \quad (2)$$

where J^{rad}, J^b, and $J^{2\gamma}$ refer to radiative recombination, bremsstrahlung, and two-photon decay of metastable levels.

The population of a particular ionization stage, z, represents a substantial fraction of the ions of a particular element only over a restricted range of temperature (Jordan, 1969, 1970; Landini and Fossi, 1973), consequently the irradiance in a spectral line contains information on the magnitude of the emission measure function M, at a particular temperature, T_{eff}. The temperature T_{eff} is generally defined as the temperature at which the function $a_{Zz}(T_e)J_{Zz\lambda}(T_e)$ has a maximum (Pottasch, 1964). Since the irradiance in the continuum over a given wavelength interval depends on the sum over the population functions of a large number of ions, the continuum irradiance contains much less information on the shape of the emission measure function. If a sufficiently large number of atomic rate constants, J, were known with sufficient precision, and if precise measurements of a carefully selected set of emission lines which are excited over the full range of coronal temperatures were available, it should be possible to determine the emission measure function, and, from this function, to calculate the solar irradiance directly. Walker (1975) and Jordan (1975) have reviewed the status of attempts to develop models of the coronal emission measure function. Considerable progress has been made recently in modeling the spectrum of a low-density Maxwellian gas (Kato, 1975; Mewe, 1972, 1975; Raymond et al., 1975; Shapiro and Moore, 1975; Walker, 1975); however, the emission coefficients for many important emission lines are known only to a factor of 2. Furthermore, great precision in our measurements of J and of the irradiance in the lines selected for analysis will be required (Craig and Brown, 1975) if the function M(T) is to be derived with a high level of confidence. Consequently, the analytical approach to obtaining the total solar irradiance between 10 and 300Å still lacks the necessary precision. However, as we point out in subsequent sections, our knowledge of the analytical nature of the coronal spectrum can have important consequences in improving the precision of new measurements of irradiance, and in relating variations in one portion of the spectrum to variations in another portion of the spectrum.

The usefulness of the analytical approach is suggested by the data presented in Table 1. The correlation of the irradiance in several EUV and soft X-ray emission lines with the solar 2800 MHz flux is shown, along with the effective excitation temperature of each line. The EUV correlations are taken from the comprehensive paper by Chapman and Neupert (1974), and the soft X-ray correlations are taken from Rugge et al. (1970). Equation 1 implies that two lines with the same effective excitation temperature must, independent of the wavelength at which they fall, vary with solar activity in the same way. Table 1 indicates that lines excited between ~ 1.5×10^6 K and 4×10^6 K correlate in the same way with 2800 MHz flux, while lines excited at higher temperatures correlate better with the radio irradiance at higher frequency.

TABLE 1

CORRELATION OF XUV LINE FLUXES AND RADIO FLUXES

Line		Effective Temperature °K	2800 MHz Correlation	Frequency of Max Correlation
Mg XI	9.17A	6.0×10^6 °K	.30	9000 MHz
Fe XVIII	14.27A	5.5×10^6 °K	.52	9000 MHz
FE XVII	15.01A	4.0×10^6 °K	.72	2800 MHz
Ne IX	13.44, 13.55A	3.5×10^6 °K	.78	$\geq$ 2800 MHz
FE XVI	335.4A	3.0×10^6 °K	.79	2000 MHz
O VIII	18.97A	3.0×10^6 °K	.84	2800 MHz
Fe XV	284.2A	2.0×10^6 °K	.80	2000 MHz
O VII	21.6A	2.0×10^6 °K	.82	2000 MHz
Fe XIV	274.8A	1.8×10^6 °K	.80	2000 MHz
Fe XII	195.1A	1.3×10^6 °K	.76	2000 MHz
Fe X	174.5A	1.0×10^6 °K	.60	2000 MHz
Fe IX	171.1A	8.0×10^5 °K	.44	2000 MHz
He II	303.8A	$< 10^5$ °K	.58-.82	2800 MHz

II. ACCURACY OF OBSERVATIONAL DATA

Manson has discussed the problems that are inherent in satellite observations in the wavelength interval from 10 to 300Å. There are two principle problems: the difficulty of absolute laboratory calibrations in this wavelength region and the instability of the windowless photoelectric detectors used in this wavelength region. There are no well-established standard absolute photometric detectors in the wavelength range below 584Å. The NBS standard far ultraviolet detectors, which are windowless photodiodes (Canfield et al., 1973), have been used down to ~ 200Å (Stern, *private communication*); however the efficiency of these devices falls very rapidly below ~ 400Å, and they are not suitable much below 200Å. It has been the practice of experimenters to use thin-window proportional counters as absolute laboratory standards in the wavelength region below ~ 300Å (Caruso and Neupert, 1965; Sampson, 1967; Manson, 1972; Walker et al., 1974a, b). The difficulty of these procedures has resulted in unreliable calibration data for some satellite spectrometers or in the availability of calibration data at only one or two wavelengths. The problem of instability is more severe, since satellite instruments may be subjected to a variety of environments between their final calibration and insertion into orbit. Consequently, as Manson has pointed out, for the wavelength range between ~30 and 300Å all absolute calibration data has been derived from rocket experiments, and satellite data has been used only to establish the variation of the irradiance with solar activity.

The problem of laboratory calibration should be resolvable with sufficient diligence on the part of the experimenters. Two approaches are possible for improving the stability of detectors in orbit. The first approach is to use flow gas detectors in orbit; in fact a number of flow gas systems have been developed and flown on satellites to supply gas for counters with thin windows. These absolute detectors have become the standard for use on rockets to obtain accurate photometric data. The second approach is to use inflight calibration. Walker et al. (1974a) have flown a radioactive calibration source on a satellite spectrometer which made observations below 10Å, and stable radioactive fluorescent sources with sufficient intensity can be constructed out to at least 20Å. As an adjunct to inflight calibration, branching ratio calibration can be most productive and useful. Branching ratio calibration makes use of the fact that the irradiance in two lines having the same upper level can be calculated with a knowledge of atomic transition probabilities alone. Examples of such pairs of lines are the O VIII 1s-3p and 2s-3p lines at 16.006 and 102.43Å, respectively, and the O V 2p-4d and 3p-4d lines at 129.872 and 498.431Å, respectively. While the usefulness of branching ratio calibration is limited by line blending and by the weakness of the non-resonance lines involved, a spectrometer of sufficiently careful design could use this technique to greatly enhance the accuracy of future measurements.

III. REPRESENTATION OF THE SPECTRUM BY A RESTRICTED DATA SET

Despite the usefulness of broad-band observations for the study of solar variability, the sensitivity of broad-band detectors to spectral shape makes them impractical for high-precision irradiance studies. However, if sufficient simultaneous data on spectral shape are available to allow corrections to be made for the effects of spectral variations, broad-band monitors may ultimately be used to obtain irradiance with a high degree of precision. Manson has discussed the excellent correlation between the total flux in the 170-300Å range and the flux in restricted sets of 52, 13, and five lines. The five-line set (Chapman and Neupert, 1974) contains the lines given in Table 2. Since these lines are excited over the full range of coronal temperatures up to 2.0×10^6 K, they should give a good account of the total irradiance in the entire wavelength range from ~20 to 300Å, except for flare conditions. The similarity of behavior of EUV lines and soft X-ray lines is demonstrated by Figure 1 (Rugge et al., 1970), which shows the correlation of the O VII and O VIII lines with the daily 2800 MHz flux (compare the EUV analysis of Chapman and Neupert, 1974). The wavelength interval between 10 and 20Å contains lines which are excited at higher temperatures, such as Fe XVII 15.01Å, and Ne IX 13.45Å, and which do not correlate as well with the daily 2800 MHz flux. Representation of this portion of the spectrum would require the addition of lines of ions such as O VIII, Fe XVII, Ne IX, Ne X, and Mg XI to the data set in Table 2. During flares, the spectrum is very strongly enhanced at wavelengths between 10 and 20Å, chiefly by lines of Fe XVIII-Fe XXIII with transitions of the type 2p-3d and 2p-3s (Doschek, 1975). In addition, lines of Fe XVIII-Fe XXIII of the type 2s-2p greatly enhance the spectrum between 90 and 150Å. In order to represent flare spectra, lines of ions such as Fe XXII, or lines of ions such as Si XIII, Mg XII, and Si XIV, which occur below 10Å, must be added to the data set.

TABLE 2

REPRESENTATIVE SET OF EUV LINES

Wavelength	Ion	T_{eff}
256.3Å	He II	$< 10^5$ °K
171.5Å	Fe IX	8.0×10^5 °K
174.5Å	Fe X	1.0×10^5 °K
195.1Å	Fe XII	1.3×10^5 °K
284.2Å	Fe XV	2.0×10^6 °K

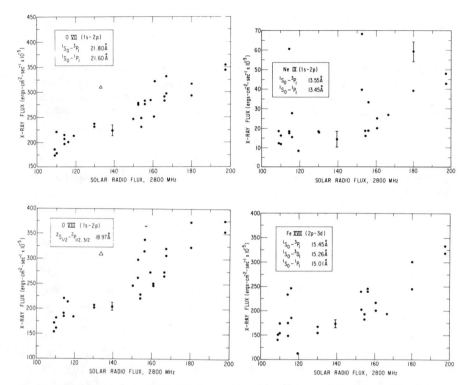

Fig. 1 Correlation between 2800 MHz flux and irradiance in four soft X-ray lines. The triangular points correspond to times during which Hα flares were reported. The correlation is significantly better for the cool O VII and O VIII lines.

We conclude from the results presented by Manson, and from the above discussion, that the monitoring of a representative set of 10-15 lines should allow the level of the total irradiance between 10 and 300Å to be specified with considerable precision, perhaps to an accuracy of a few percent. The selection of the line set must be carried out with considerable care, however, so that the effective excitation temperatures of the lines in the set cover the range of temperatures from $\sim 5 \times 10^5$ K to $\sim 20 \times 10^6$ K. It should then be possible to analytically determine the response of each wavelength interval to changes in the relative intensities of the lines in the set, and to weigh the importance of each of the monitored flux levels to the irradiance in that interval. Before this is possible, however, significantly improved, *simultaneous* measurements of the entire spectrum between 10 and 300Å must be obtained at a number of representative levels of solar activity.

IV. THE CORRELATION OF SOLAR XUV AND RADIO FLUX

Figure 2 (Rugge et al., 1970) shows the correlation of the irradiance in several soft X-ray lines with solar radio irradiance at frequencies between 500 MHz and 9000 MHz. Similar data for EUV lines has been tabulated by Chapman and Neupert (1974), and is summarized in Table 1. There is, however, a caveat that must be considered in interpreting the results of Chapman and Neupert and of Rugge et al. The available solar radio indices generally represent observations taken at a given time of the day and are not usually coincident with the times of observation of the EUV and soft X-ray fluxes. Since it has been well established that many X-ray enhancements do not correspond to recorded optical flares (Datlowe, 1975), and that X-ray and microwave bursts are well correlated (DeFeiter, 1975; Takakura, 1975), it is likely that the analysis of simultaneous observations of soft X-ray and 2800 MHz irradiance will result in improved correlation for the lines with higher T_{eff}.

Manson (cf. his Fig. 3) has clearly demonstrated that it is possible to define a set of relationships between the irradiance in selected wavelength intervals and the solar 2800 MHz flux. The data presented in Figure 2, and by Chapman and Neupert (1974), suggest that the use of radio indices at several frequencies may improve the accuracy of these relationships. Further improvements should result if instantaneous radio indices or true daily averages are used rather than observations obtained at a particular time of day. However, with the presently available data base it is impossible to determine the ultimate precision with which solar EUV and soft X-ray irradiance can be predicted from solar radio indices.

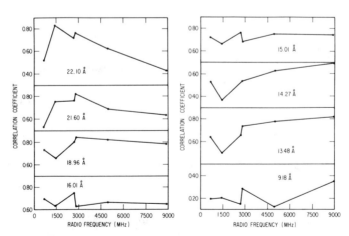

Fig. 2 Correlation between a number of soft X-ray lines and solar radio indices at a number of frequencies. The correlation for the Mg IX line at 9.18Å ($T_{eff} \sim 6 \times 10^6$ K) is noticeably worse than for the cooler lines.

REFERENCES

Canfield, L.R., Johnson, R.G., and Madden, R.P., 1973, *Appl. Optics 12*, 1611.

Caruso, A.J., and Neupert, W.M., 1969, *ibid. 4*, 247.

Chapman, R.D., and Neupert, W.M., 1974, *J. Geophys. Res. 79*, 4138.

Craig, I.J.D., and Brown, J.C., 1975, *Astr. and Ap.*, in press.

Datlowe, D.W., 1975, *Solar Gamma-, X-, and EUV Radiation*, IAU Symposium No. 68, ed. S.R. Kane, D. Reidel, Dordrecht, 191.

DeFeiter, L.D., 1975, *ibid.*, 283.

Doschek, G., 1975, *ibid.*, 165.

Jordan, C., 1969, *M.N.R.A.S. 142*, 499.

—————, 1970, *ibid. 149*, 1.

—————, 1975, *Solar Gamma-, X-, and EUV Radiation*, IAU Symposium No. 68, ed. S.R. Kane, D. Reidel, Dordrecht, 109.

Kato, T., 1975, *Ap. J. Suppl.*, in press.

Landini, M., and Monsignori Fossi, B.C., 1972, *Astr. and Ap. Suppl. 7*, 291.

Manson, J.E., 1972, *Solar Phys. 27*, 107.

—————, 1976, this volume.

Mewe, R., 1972, *Solar Phys. 22*, 459.

—————, 1975, *ibid. 44*, 383.

Pottasch, S.R., 1964, *Space Sci. Rev. 3*, 816.

Raymond, J.C., Cox, D.P., and Smith, B.W., 1975, Space Physics Laboratory, University of Wisconsin, Madison, Preprint No. 12.

Rugge, H.R., Walker, A.B.C., Jr., and Anderson, M., 1970, *Aerospace Report TR-0059 (9260-02)-1*, Aerospace Corp., Los Angeles.

Sampson, J.A.R., 1967, *Techniques of Vacuum Ultraviolet Spectroscopy*, John Wiley & Sons, New York.

Shapiro, P.R., and Moore, R.T., 1975, Smithsonian Center for Astrophysics, Harvard College Observatory Preprint.

Takakura, T., 1975, *Solar Gamma-, X-, and EUV Radiation*, IAU Symposium No. 68, ed. S.R. Kane, D. Reidel, Dordrecht, 299.

Walker, A.B.C. Jr., 1975, *ibid.*, 73.

—————, Rugge, H.R., and Weiss, K. 1974a, *Ap. J. 188*, 423.

—————, —————, and —————, 1974b, *ibid. 192*, 169.

THE SOLAR SPECTRUM BELOW 10 Å

R. W. Kreplin, K. P. Dere, D. M. Horan, and J. F. Meekins

*E. O. Hulburt Center for Space Research,
Naval Research Laboratory,
Washington, D. C.*

Measurement of solar spectral irradiance implies measurement not only of the wavelength distribution of the emitted energy but also of the intensity of that radiation. The objective of a specific study dictates the accuracy with which these measurements are made. The vast majority of published papers deal with solar physics. In these papers the relative spectral distribution is sufficient to derive the temperature of the emitting plasma or the change in emission measure with time. Therefore, over the past decade, solar physicists have developed instruments in which high spectral, spatial, and temporal resolution have been stressed. However, absolute measurement of X-ray flux is also important for determination of such solar parameters as emission measure, coronal element abundances, and line intensities in widely separated parts of the solar spectrum.

Knowledge of the absolute X-ray flux and its spectral distribution is also very important to the ionospheric physicist studying the D-region and, in particular, the disturbances that are produced in that region by X-ray emitting solar flares. For this reason the Naval Research Laboratory (NRL) SOLar RADiation (SOLRAD) program and the Space Environmental Monitoring System aboard the NOAA SMS/GOES weather satellites have been designed to provide accurate absolute measurements of X-ray flux with high time resolution. As a consequence, high spectral and spatial resolution have been sacrificed.

The X-ray spectrum below 10 Å is characterized by its steepness and its extreme variability with solar activity. This variability far exceeds the possible errors in flux measurement. Therefore, average values have little meaning and are not very useful for ionospheric physics. Determination of the absolute flux values below 10 Å requires not only carefully calibrated photometers but continuous measurement with reasonably high time resolution.

In this paper we describe both the instruments currently used aboard satellites and rockets for measuring solar flux and some of the more recent measurements that describe the wavelength distribution, intensity, and variability of the sun's X-ray emission below 10 Å.

I. METHODS OF MEASUREMENT

Instruments thus far developed and flown to measure solar X-ray radiation below 10 Å include band-sensitive ionization chambers and Geiger counters, proportional counter spectrometers, and Bragg crystal spectrometers. At wavelengths less than about 1.2 Å, scintillation counter spectrometers have been used to measure the spectral character and intensity of solar X-ray bursts. Each of these instruments has certain advantages and disadvantages. The Bragg crystal spectrometer is capable of very high spectral resolution but calibration for measurement of absolute intensities is difficult. The proportional counter spectrometer may be calibrated rather easily for intensity measurement, but above 1 Å its wavelength resolution is poor, and it requires one to make assumptions about the spectral distribution of the incident radiation in order to derive intensities in absolute units. The band-sensitive photometers can be calibrated for absolute flux measurement, but the spectral distribution of the incident radiation must be assumed in order to derive flux values from the measured current or counting rate. This requirement is even more severe than with the proportional counter, where a multichannel pulse height analyzer may be used to give some indication of the spectral distribution.

A. BAND-SENSITIVE PHOTOMETERS

The band-sensitive photometers are the simplest of the instruments used in the measurement of solar X-ray emission. The detector is a gas-filled diode whose wavelength sensitivity is determined by the window material and thickness and by the nature, pressure, and depth of the gas filling. The lower wavelength cutoff is determined by the absorption of the X-ray photon by the gas filling, and the upper wavelength cutoff is determined by the transmission of the window material. Because the currents generated within the ionization chamber are very small, sensitive and very stable electrometer amplifiers are required to convert the detected current to a signal suitable to drive a telemetry system.

The range of wavelengths over which the Geiger counter is sensitive is determined by the same factors that govern the range of the ionization chamber photometer. Gain, however, is provided by a discharge or avalanche that takes place when a number of electrons are freed within the volume of the counter by absorption of an X-ray photon. An electric field must be applied that is sufficient to impart to the free electrons an energy equivalent to or larger than the ionization potential of the gas in one mean free path. The discharge is propagated throughout the volume of the counter by emission of UV radiation in the discharge process. The process is stopped by the buildup of space charge around the anode and by a small percentage of quenching agent in the gas filling. The rise time of the Geiger counter pulse is of the order of tens of microseconds and the decay or dead time may be several hundred microseconds. During this period it is not possible to produce a discharge by absorption of an X-ray photon. Consequently, the dynamic range of the Geiger counter is severely limited in comparison with other, nonsaturating types of detectors.

The major source of error in photometer measurements is the necessity of postulating the form of the incident spectrum in order to convert the photometer current, or counting rate in the case of a Geiger counter, to X-ray flux. Another source of error arises when a calculated photometer efficiency is used. This can be reduced by laboratory calibration; and such calibration is routine with the equipment now available. The techniques involved are discussed below.

Calibration of X-Ray Photometers. Meekins et al. (1974) have discussed the problem of absolute calibration of the SOLRAD X-ray ionization chambers. They express the response of the ionization chamber to a photon flux F_o as

$$\bar{I} = \varepsilon_{IC}(\lambda) \, W \, Q(\lambda) \, \frac{F_o \, hc}{\lambda} \, A \tag{1}$$

where $\varepsilon_{IC}(\lambda) = T_W(1-T_G)$ is the quantum efficiency of the ionization chamber, T_W is the transmission of the window, and T_G is the transmission of the gas filling.

$$T = \exp(-\Sigma \mu_m \rho x) \tag{2}$$

where μ_m is the photoelectric mass absorption coefficient and ρx is the area density of the material through which the photons pass.

$$W \cong \frac{e}{3I} \cong e\omega \quad , \tag{3}$$

where e is the electronic charge, I is the ionization potential of the absorbing molecule from the valence shell, ω is the number of ion pairs produced per unit of absorbed energy, and

$$Q(\lambda) = \begin{cases} 1 & \text{for } \lambda > \frac{hc}{I_K} \\ 1 - \frac{\lambda \alpha(\lambda)}{\lambda_K} & \text{for } \lambda \leq \frac{hc}{I_K} \end{cases} \tag{4}$$

That is, $Q(\lambda) = 1$ when the energy of the absorbed photon is less than the ionization potential of a K-shell electron and $Q(\lambda) = 1 - \lambda\alpha(\lambda)/\lambda_K$ when the absorbed photon has energy in excess of that required to ionize by removal of a K-shell electron. The product of the K X-ray fluorescence yield and the probability that the characteristic K X-ray will reach the chamber wall is $\alpha(\lambda)$.

Using these relationships to convert measured current to incident flux requires knowledge only of the ratio

$$R(\lambda) = \frac{\bar{I}}{F_o \, A \, \frac{hc}{\lambda}} = \varepsilon_{IC}(\lambda) \, W \, Q(\lambda) \quad , \tag{5}$$

that is, the ratio of the measured current and the total power incident on the chamber. Before the development of the Meekins et al. (1974) technique for laboratory calibration, the ratio $R(\lambda)$ was calculated from published values of absorption coefficients (McMaster et al., 1969) and from measurements of the energy necessary to produce an ion pair in the gas filling (Whyte, 1963).

Measurement of $R(\lambda)$ requires a source of monochromatic X-ray radiation whose intensity is accurately known as well as a means of accurately measuring the current produced in the ionization chamber. Such a technique has been described by Meekins et al. (1974) and by Unzicker and Donnelly (1974). The monochromatic X-ray beam is produced by an X-ray generator and Bragg crystal spectrometer. The beam is limited to an area smaller than that of the ionization chamber window and is calibrated with a thin-window proportional counter. Ionization chamber current is measured with a Keithley Model 640 Vibrating Capacitor Electrometer calibrated against a standard resistor to give the measurement of the current in the 10^{-14} ampere region to an accuracy of about 2%.

The experimentally determined $\varepsilon_{IC}(\lambda)$ are plotted in Figure 1 for the 1-8 Å and 0.5-3 Å ionization chamber photometers flown aboard SOLRAD 11. The error bars include all sources of error entering these measurements; near the peak efficiency the absolute accuracy of the measurements is ±10%. The dashed lines show the efficiencies calculated on the basis of published values of the mass absorption coefficients for beryllium, the measured surface density of the window, and the gas pressure.

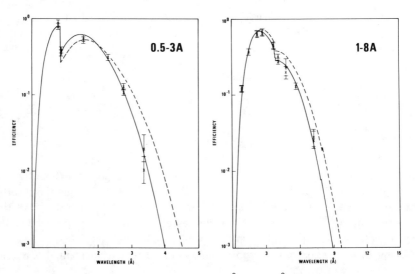

Fig. 1 Efficiency $\varepsilon(\lambda)$ for the SOLRAD-type 0.5-3 Å and 1-8 Å photometers. Solid line represents the best fit to two sets of experimental measurements. Dashed line is the earlier calculated efficiency.

The 0.5-3 Å photometer is shown to be sensitive in the region from 0.3 to 2.8 Å, and the 1-8 Å photometer in the region from 0.5 to 7.2 Å, as determined by reading the wavelengths from the experimental curve that correspond to the efficiencies at 3 and 8 Å from the theoretical curves. Letfus (1972) and Wende (1969) have examined the relationship between the signal-to-energy conversion constant, K_{GB}, and the limits λ_1 and λ_2. Letfus (1972) found that when these limits agree with the actual limits of sensitivity of the photometer, the dependence of K_{GB} on the shape of the spectral distribution, described by its temperature, is at a minimum. According to Letfus (1972) the 1-8 Å photometer wavelength limits should be 0 and 6.5 Å. Wende (1971) found that the OGO-4 1-8 Å ionization chamber, which was similar to those flown on SOLRADs 9, 10, and 11, should have a passband of 0.75 - 7 Å and the 0.5 - 3 Å chamber passband should have been 0.25 - 3 Å. Thus, the actual measurement of $\epsilon(\lambda)$ for these two photometers appears to confirm the conclusions of these studies.

The value of ω used in the computed efficiency was 2.57×10^{10} ion pairs/erg for krypton in the 0.5 -3 Å photometer and 2.38×10^{10} ion pairs/erg for argon in the 1-8 Å photometer. The corresponding values obtained by measurement of $R(\lambda)$ for the SOLRAD 11 ionization chambers ranged from 2.01×10^{10} ion pairs/erg to 2.30×10^{10} ion pairs/erg for the 0.5 - 3 Å photometers and from 1.97×10^{10} ion pairs/erg to 2.37×10^{10} ion pairs/erg for the 1-8 Å photometers. Thus, when compared to the earlier calculations of efficiencies, the results show a shift in the sensitive range to shorter wavelengths. The number of ion pairs formed per erg is also consistently less than was previously assumed. These differences have the effect of increasing the previously published values of X-ray flux measured by the SOLRAD photometers by approximately 100%.

Calculation of X-Ray Flux From Photometer Current: The Graybody Approximation. The graybody approximation permits determination of the energy flux in some wavelength band under the assumption that the solar emission spectrum can be approximated by a blackbody spectral distribution with an emission substantially less than that given by the blackbody formula. That is,

$$E_{GB}(\lambda, T) = (2\pi hc^2 D)\lambda^{-5}\left[\exp(c_2/T\lambda) - 1\right]^{-1} \quad \text{erg cm}^{-2} \text{ sec}^{-1} \text{ Å}^{-1}, \qquad (6)$$

where $C_2 = hc/k$, h is Planck's constant, c is the speed of light, k is Boltzmann's constant, and D is a dilution factor. Because a single temperature will not serve to describe the solar spectrum over a wide range of wavelengths, temperatures of 10^7K and 2×10^6K are used for the 0.5 - 3Å and 1-8 Å regions, respectively. With the use of the appropriate temperature, the conversion constant is calculated from

$$K = \frac{A \int_{\lambda_1}^{\lambda_2} E_{GB}(\lambda) \, d\lambda}{eA\omega \int_0^\infty \varepsilon(\lambda) E_{GB}(\lambda) \, d\lambda} \quad \text{ergs sec}^{-1} \text{ amp}^{-1} \quad , \tag{7}$$

where λ_1 and λ_2 are the wavelength limits of the spectral region over which the emission flux is calculated, e is the electron charge, ω is the number of ion pairs produced in the photometer per unit of absorbed energy, and $\varepsilon(\lambda)$ is the detector efficiency. This conversion constant is then used to derive the emission energy flux from the photometer current by the simple relation

$$F = \frac{K}{A} I \quad \text{erg cm}^{-2} \text{ sec}^{-1} \tag{8}$$

where F is the solar energy flux and I is the detector current. This simple conversion technique has been used consistently over a period of more than 14 y. These measurements are easily converted to flux values on the basis of other solar input spectra.

Single Temperature Spectral Distribution. Since SOLRAD monitoring began, much has been learned about the solar spectrum below 10 Å. The review by Doschek (1972) provides a detailed description of the line and continuum spectrum in this region. Theoretical studies by Tucker and Koren (1971) and by Culhane (1969) have provided a basis for describing the solar X-ray emission spectrum in a way that is consistent with measurements by instruments such as the NRL Bragg crystal spectrometer aboard OSO-6.

Horan (1970) applied the formulas of Culhane to derive the absolute flux values from the SOLRAD ionization chamber data. Dere, Horan, and Kreplin (1974a) discuss this technique as applied to the SOLRAD 9 0.5 - 3 Å and 1-8 Å detectors. They consider the source region to be isothermal and the differential thermal emission spectrum to be a function only of the temperature and the wavelength, i.e.,

$$\frac{d}{dV} E_{Th}(\lambda, T) = f(\lambda, T) n_e^2 \quad \text{erg cm}^{-2} \text{ sec}^{-1} \text{ Å}^{-1} \text{ cm}^{-3} \quad . \tag{9}$$

The ratio of the graybody fluxes in two ionization chambers is written

$$S(T) = \frac{F_{GB}(i)}{F_{GB}(j)} = \frac{K_{GB}(i)\omega_i \int_0^\infty \varepsilon_i(\lambda) f(\lambda, T) \, d\lambda \int n_e^2 \, dV}{K_{GB}(j)\omega_j \int_0^\infty \varepsilon_j(\lambda) f(\lambda, T) \, d\lambda \int n_e^2 \, dV} \quad , \tag{10}$$

where i and j refer to the two detectors and $\int n_e^2 dV$ is the emission measure. We measured $\omega_i \varepsilon_i(\lambda)$ and $\omega_j \Sigma_j(\lambda)$ [which are related to R(λ) through Equations (3), (4), and (5)] in our laboratory; for the 0.5 - 3Å and 1-8 Å photometers they were found to be about 50% less than the calculated values previously used.

For a range of temperatures between 2×10^6 K and 32×10^6 K the value of the ratio S(T) is calculated. These ratios, which are given in Table 1, can be used to select the temperature that best fits the X-ray emission measurements during a solar flare. When the temperature is determined in this way, the graybody flux values can be corrected to the more accurate single temperature values by means of the formulas

$$F_A (0.5 - 3 \text{ Å}) = C (0.5 - 3 \text{ Å}, T) F_{GB} (0.5 - 3 \text{ Å})$$

$$F_A (1 - 8 \text{ Å}) = C (1 - 8 \text{ Å}, T) F_{GB} (1 - 8 \text{ Å})$$

(11)

where F_A is the absolute flux at temperature T, and C is the correction factor shown in Table 1. During a flare it is necessary to correct the 1-8 Å flux for background radiation from lower temperature regions outside the flare area. This correction is calculated on the basis of a temperature of 2×10^6 K from measurements taken before the flare.

Once the temperature has been determined from the graybody flux ratios, the emission measure can be obtained by using the last column in Table 1. A good approximation of the actual spectral distribution of the solar emission at less than 10 Å can then be obtained by using the equations of Culhane and Acton (1970) for free-free and free-bound continuum emission:

$$E(\lambda, T) = 4.1 \times 10^{-49} \lambda^{-1.7} T^{0.2} \exp\left[-c_2/\lambda T\right]$$

$$\times \left[1 - (7.1\lambda)^{-2.84 \times 10^8 T}\right]^{-1} \int n_e^2 \, dV \quad \text{erg cm}^{-2} \text{ sec}^{-1} \text{ Å}^{-1} .$$

(12)

This expression agrees to within 15% with the more detailed calculations of Culhane (1969) in the energy range from 1.5 to 15 keV and for temperatures between 4×10^6 K and 20×10^6 K.

The calculations by Tucker and Koren (1971), Mewe (1972), and Landini and Monsignori Fossi (1970) can be used to derive the line emission. Line radiation is significant in flares, during which it contributes about half the total emission below 10 Å for temperatures of 10×10^6 K. However, during quiet periods it can be neglected. The spectra thus derived are shown in Figure 2 for temperatures of 2, 8, and 32×10^6 K and demonstrate the relative importance of line and continuum emission.

By examining the values of C found in the table we find that the 1-8 Å graybody fluxes for temperatures near 2×10^6 K are smaller by a factor of three than the absolute flux values. At very high temperatures they are about three times greater than the absolute fluxes. The 0.5 - 3 Å graybody fluxes are five times too great at 2×10^6 K, are nearly equivalent at 3×10^6 K, and are too small by a factor of two at approximately 7×10^6 K.

TABLE 1

Temperature (10^6 K)	$R(T)$	C(0·5–3 Å, T)	C(1–8 Å, T)	$\dfrac{F_{GB}(0\cdot5\text{–}3\text{ Å})}{\int n_e^2\,dV}$
2·00	4·28(−6)*	0·220	3·29	1·15(−60)
2·38	1·97(−5)	0·453	3·00	2·47(−59)
2·83	6·57(−5)	0·798	2·57	3·06(−58)
3·36	1·80(−4)	1·19	2·16	2·59(−57)
4·00	4·33(−4)	1·56	1·82	1·65(−56)
4·76	9·52(−4)	1·82	1·55	8·34(−56)
5·66	1·95(−3)	1·97	1·33	3·52(−55)
6·73	3·76(−3)	2·01	1·15	1·27(−54)
8·00	6·86(−3)	1·96	1·00	4·01(−54)
9·51	1·18(−2)	1·84	0·874	1·13(−53)
11·3	1·93(−2)	1·70	0·761	2·85(−53)
13·5	2·99(−2)	1·55	0·660	6·53(−53)
16·0	4·39(−2)	1·41	0·572	1·37(−52)
19·0	6·10(−2)	1·27	0·494	2·64(−52)
22·6	8·06(−2)	1·16	0·429	4·72(−52)
26·9	1·02(−1)	1·07	0·376	7·82(−52)
32·0	1·24(−1)	0·995	0·335	1·20(−51)

* Numbers in parentheses indicate powers of 10, i.e. $4\cdot28(-6) = 4\cdot28 \times 10^{-6}$.

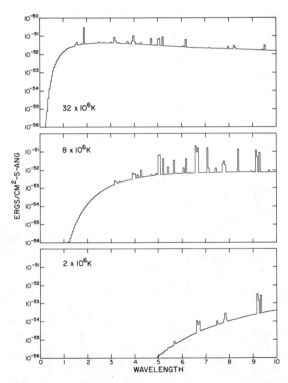

Fig. 2 Calculated solar X-ray spectra. The flux is given in terms of unit emission measure. The amplitude of the lines is representative of the energy contained in the line and not of the peak intensity.

Differential Emission Measure Calculation. With ionization chambers for a number of overlapping wavelength bands we are able to calculate the emission measure as a function of temperature rather than for a fixed temperature determined by the ratio of two photometer signals. A number of authors (Batstone *et al.*, 1960; Parkinson, 1973; Chambe, 1971; Walker, 1972; and Meekins, 1973) have described the nonflaring solar corona with multithermal models. They used various temperature distributions, all having the characteristic of decreasing emission measure with increasing temperatures above about 2×10^6 K. There is, however, some disagreement about the validity of these models, especially for solar flares where the limitation or requirement that $d/dT \int n_e^2 dV$ be less than zero may be unrealistic.

Dere, Horan, and Kreplin (1974b) describe solar X-ray emission with a multithermal analysis applied to the interpretation of data from the ionization chamber photometers aboard the Naval Research Laboratory SOLRAD 10 satellite. They specify the differential emission measure as a function of the temperature.

$$\frac{dB}{d\ln T} = \frac{d}{d\ln T} \int n_e^2 \, dV \, , \qquad (13)$$

where n_e is the coronal electron density, T is the electron temperature, and dV is the volume element in the corona.

The current developed in the ionization chamber photometer is given by

$$I = e\omega A \int \varepsilon(\lambda) \int \frac{d}{dB} E(\lambda, T) \frac{dB}{d\ln T} d\ln T \, d\lambda \, , \qquad (14)$$

where e is the electron charge, ω is the number of ion pairs produced per erg of absorbed energy, A is the detector window area, $\varepsilon(\lambda)$ is the detector efficiency, and $E(\lambda, T)$ is the solar flux per unit wavelength interval. Values of ω and $\varepsilon(\lambda)$ are determined by laboratory calibration. Here, as in the case of the single temperature calculation, the coronal emission spectrum includes both continuum and lines.

The outputs of six photometers aboard SOLRAD 10 were used in an iterative procedure to obtain $dB(T)/d\ln T$ by adjusting several free parameters to yield a minimum value for

$$\phi = \sum_{j \text{ detectors}} (I_j \text{ (measured)} - I_j \text{ (expected)})^2 / \Delta_j^2 \, , \qquad (15)$$

where Δ_j is the estimated experimental error in the measurement of the absolute efficiency of detector j. This technique has been applied to the analysis of both nonflare and flare X-ray measurements made by SOLRAD 10. Figure 3 shows the differential emission measure for two nonflare periods that represent extreme cases of solar activity for such periods.

The calculation of differential emission measure with the best available spectra

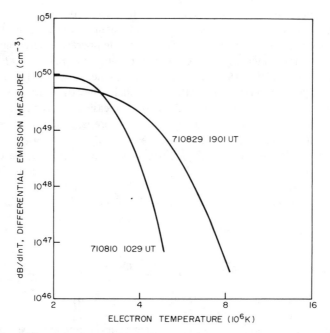

Fig. 3. Differential emission measure derived for two periods during which the sun was relatively quiet.

probably provides the most accurate description now available of the solar spectrum below 10 Å. The calculation is, however, quite time consuming and as a result has only been applied to a few events. When the flux values for the region below 10 Å are calculated from the differential emission measure, we find at most a 20% difference from those calculated on the basis of comparison of two photometer responses below 10 Å. Therefore, for the purposes of the ionospheric physicist the single temperature description is probably adequate.

The general behavior of the differential emission measure with time for the full disk during a small solar flare is presented in Figure 4. Curve 1 represents the early rising portion of the flare before peak temperature is attained. Curve 2 shows the behavior at the time when peak temperature is reached. Curve 3 is representative of the time of peak flux, and curve 4 is typical of the decay phase.

B. PROPORTIONAL COUNTER SPECTROMETER

The proportional counter is between the Bragg crystal spectrometer and the ionization chamber photometer with respect to energy or wavelength resolution. In the region above 1 Å, i.e., below 13 keV, its resolution is rather poor but improves as energy increases. At 2 Å a resolution of 16% FWHM can be achieved. Like the ionization chamber or Geiger counter photometer, the limits of wavelength

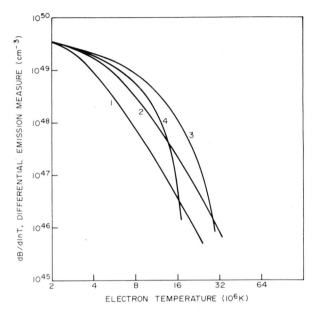

Fig. 4 Differential emission measure as a function of temperature for a small X-ray flare. Curves 1 and 2 represent the increasing phase of the flare before maximum and curves 3 and 4 are for two times during the decay phase.

sensitivity are set by the window material and thickness and by the nature of the gas filling and its pressure. The proportional counter has some distinct advantages over band-sensitive devices. Because the discharge produced by the absorption of an X-ray photon is usually limited to multiplication factors of 10^6 to 10^8 and does not spread throughout the volume of the counter, its response time is much faster than that of a Geiger counter. Typical pulse rise times are on the order of tens of nanoseconds as compared to tens of microseconds for the Geiger counter. As a result, the dynamic range is much greater than that attainable with a Geiger counter, although pileup of small amplitude pulses (Kane and Hudson, 1970) is a problem with steep spectral distributions and high counting rates, especially when the proportional counter is being used with a pulse height analyzer to measure the spectral distribution of incident X-ray radiation. Because the ionization produced by an absorbed photon is proportional to the photon energy, the device can be used to measure the energy of the photon if the gain remains constant. However, because the resolution is poor in the region above 1 Å, the conversion of counting rate to flux requires that an assumed spectral distribution be used, as is necessary with the Geiger counter or ionization chamber photometer. Culhane *et al.* (1969) used a power series in λ to describe the X-ray spectrum and an iterative procedure to fit this distribution, with moderate success, to the measurements of the proportional counter. They realized, however, that there were difficulties with this distribution and Culhane (1969) developed a distribution based on calculations of the free-free and free-bound emission mechanisms. Culhane and Acton (1970) then developed an empirical expression that has been found to be quite convenient for reduction of proportional counter measurements. This is the same formula that is described in Equation (12).

C. BRAGG CRYSTAL SPECTROMETER

Measurement of solar X-ray emission by Bragg crystal spectrometers has been the subject of review papers by Parkinson (1975), Walker (1972, 1975), Doschek (1972), Culhane and Acton (1974), and Neupert (1971). Ideally, a Bragg crystal spectrometer with absolute calibration would provide the solar spectral irradiance below 10 Å. In general, the wavelength calibrations have been more than adequate to identify the spectra of highly ionized ions present in the solar atmosphere, but absolute calibration for measurement of intensity has in some cases not been performed on the flight instruments. The majority of papers that discuss spectrometer results describe intensity only in terms of relative counting rates. In the usual arrangement, where the Bragg crystal spectrometer is used to obtain solar spectra, three different quantities enter the determination of overall instrument efficiency. These are the detector quantum efficiency, the transmission of any protective filters, and the crystal reflectivity in the first and higher orders. Kreplin and Meekins (1974) compared results of different authors who published measurements of the intensity of the O VIII Lyman-alpha line at 18.97 Å and found large differences that could not be explained by expected variations in solar activity. Such differences are probably a result of errors in the assumed or measured crystal reflectivity.

Bragg spectrometers have been designed primarily for the study of solar physics. Consequently, the more recent experiments have been fitted with collimators that restrict the instrument field of view to areas smaller than several arcmin. Unless the solar disk is scanned so that all areas are examined it is not possible to obtain a total flux measurement that would be useful to the ionospheric physicist.

Limitation of the field of view by the collimator also introduces another problem. Recent Skylab X-ray photographs (Vaiana et al., 1974) show the majority of X-ray emitting regions to be rather diffuse. As the spatial resolution of the collimator is increased, the number of photons collected per unit time becomes less, statistical fluctuations in the counting rate become large, and long periods are required to build up an X-ray image with the raster scanning technique. Thus, the collimated Bragg crystal spectrometer cannot be used to follow even moderately rapid variations in the total disk emission.

Conversely, from the point of view of the solar physicist, data from an uncollimated Bragg spectrometer is often difficult to analyze when multiple sources are present on the disk. Unless each source produces a resolved spectral line, the line profiles are very confused and measured intensities cannot be easily used to determine coronal temperatures from line ratios.

As mentioned above, the Bragg spectrometer does provide the highest wavelength resolution of any of the instruments used for the wavelength range below 10 Å. The study of solar physics demands wavelength resolution of 0.001 Å or less, which introduces the same kind of difficulty for the ionospheric physicist as did the high spatial resolution of the collimated spectrometer. That is, since the number of photons per wavelength interval is limited, wavelength scans must be slow enough to assure detection of a sufficient number of photons to give good counting statistics. However, the same slow scan that is required by the high wavelength resolution compromises the ability of the instrument to follow rapid changes in X-ray emission.

II. SOLAR X-RAY EMISSION BELOW 10 Å

A. PHOTOMETER MEASUREMENTS

Long-Term Variability. The NRL SOLRAD program has provided nearly continuous observation of the sun's emission in the X-ray region over the past solar cycle. The data provided by those spacecraft instruments have been published in *Solar Geophysical Data* over the past decade and have become the basis for the M, C, X classification of solar flares. This new classification is a better index of the geophysical significance of the flare than the older optical classification of 1, 2, and 3 (*Solar Geophysical Data*, 1976). Other data sources that should be mentioned are the Vela satellite measurements of Los Alamos Scientific Laboratory, which have been made more or less continuously since July 1965, and the Explorer 33 and 35 measurements of the Iowa group under Van Allen (Drake, Gibson, and Van Allen, 1969). More recently a photometry experiment was placed at synchronous altitude by the National Oceanic and Atmospheric Administration (NOAA) to provide measurements in the region below 10 Å in two bands (Grubb, 1975). Only with the NOAA and the most recent SOLRAD 11 A/B experiments has an effort been made to calibrate the actual flight photometers. The following description of the variability and absolute flux in the region below 10 Å is based primarily on the SOLRAD photometry experiments; these experiments extend over most of the past solar cycle.

The SOLRAD photometers were standardized as early as 1960 and all data have been processed for publication by using a standardized method for converting the measured photometer current to flux units. Therefore, it is possible to go back over the past solar cycle and obtain an accurate picture of the variability of the sun's emission if we are willing to accept some systematic error in the measurements below 10 Å. Kreplin (1970) has described the X-ray emission and its variability in the period from the last solar minimum, July 1964, to solar maximum in 1969.

The early SOLRAD satellites, 7A (1964-O1D), 7B (1965-16D), and 8 (1965-93A), were all spin stabilized on injection into earth orbit. Precession resulted from the interaction of the earth's magnetic field with the net dipole moment of the "broom" magnets used to protect the ionization chambers from the energetic electrons in the earth's radiation belts. Data loss occurred when the sun was out of the photometers' field of view. It was also necessary to make an aspect angle correction to the data when the sun's direction fell away from the satellite's spin plane by more than a few degrees. Additional data was lost because of the dependence on real-time data transmission and because the satellite passed through the earth's shadow.

SOLRADs 8, 9, and 10 were actively oriented to eliminate data loss from precession, and SOLRADs 9 and 10 carried memory systems to eliminate dependence on real-time data transmission. Identical detectors were flown aboard NASA's OGO-4 (Kreplin et al., 1969) and OSO-5 spacecraft. Thus, since August 1967 there has been a continuous record of the sun's X-ray emission below 10 Å, interrupted only by satellite passage into the earth's shadow and by charged particle interference from the earth's radiation belts. After the failure of the SOLRAD 9 orientation and memory systems in April 1974 the only real-time data available has been from SOLRAD 10. Since the launch of SOLRAD 11 A/B in March 1976, we have again provided continuous measurement of the sun's X-ray emission. The 119,000 km circular orbit reduces data loss from charged particle interference, and with two spacecraft 180° apart, only short periods of data loss occur despite dependence on real-time data transmission.

Using the standardized flux conversion, we show yearly plots of the daily average X-ray flux in Figures 5a, b, c, d, e, f, and g which cover the period from solar maximum (1969) to the present solar minimum (1975). During solar minimum the amount of 1-8 Å flux was less than the detection threshold of 1.5×10^{-5} erg cm^{-2}s^{-1} (2×10^{6} K graybody). The amount of flux rose rapidly in early 1966 and remained at about 10^{-3} erg cm^{-2}s^{-1} during the last part of 1967 and 1968. In 1969 periods were seen in which the background or nonflare average flux remained at about 5×10^{-3} erg cm^{-2}s^{-1} for periods of a week or two.

The variability of the 1-8 Å flux depends greatly on the presence or absence of active regions on the visible disk. Because active regions usually persist for longer than a single rotation, some evidence of the 28-day solar rotation period is seen. This periodicity is evident in the data for the last half of 1969.

The solar cycle variability is also evident in the 1-8 Å data. Solar maximum occurred in 1969 and 1970, but the fluctuations from day to day are large, and a more specific time for the maximum cannot be determined. In 1971 and 1972 the fluxes decreased, again showing large fluctuations up to a factor of ten. In 1973 the solar 1-8 Å emission dropped below the threshold of the SOLRAD 9 photometer, i.e., 1.3×10^{-4} erg cm^{-2}s^{-1}. After failure of the SOLRAD 9 orientation system in February 1974, SOLRAD 10 real-time data were used to form the daily averages. Emission dropped below the SOLRAD 10 threshold of 1×10^{-5} erg cm^{-2}s^{-1} for short periods late in 1974 and for longer times—a month or more— in 1975. A few observations from SOLRAD 11A in 1976 indicate flux of 4 to 7 $\times 10^{-5}$ erg cm^{-2}s^{-1} during relatively quiet periods. Thus, the flux below 10 Å shows a solar cycle variation of up to a factor of 500 as determined from the 1-8 Å photometer meaurements.

Short-Term Variability. The short-term variability of the sun's X-ray emission is shown in Figure 6. The background 1-8 Å flux is about 10^{-3} erg cm^{-2}s^{-1}. The impulsive events are solar X-ray flares that usually last from several minutes to several hours and, in the 1-8 Å band, cause flux enhancements up to several hundred times the background level. Below 3 Å, the enhancement may be several thousand times the background. The rise and decay of emission at shorter wavelengths is always sharper than that at longer wavelengths (Kreplin et al., 1969). The characteristics of radiation at the shortest wavelengths have been described by Kane (1974) for the region > 10 keV. His study, limited to quite small flares because of pulse pileup, indicated burst rise times of 2-5 seconds and somewhat longer decay times of 3-10 seconds. The maximum energy flux varied between 10^{-7} and 10^{-5} erg cm^{-2}s^{-1} in the 10 - 100 keV region.

Datlowe, Hudson, and Peterson (1974) studied 197 X-ray bursts in the range 5-15 keV with the use of a proportional counter aboard OSO-7. A typical event associated with an optical -N flare had a rise time of two minutes and a decay time, observed in the lower energy part of the spectrum, of about ten minutes. These bursts in the region > 10 keV occur during the period when the softer X-ray flare emission is rising most rapidly towards its maximum.

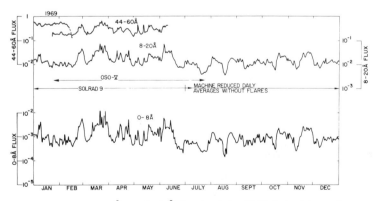

Fig. 5a Daily averages for 1-8 Å and 8-20 Å flux levels for 1969 derived from photometers aboard SOLRAD 9 and OSO-5. Flare X-ray flux enhancements have been removed from the averages as far as possible.

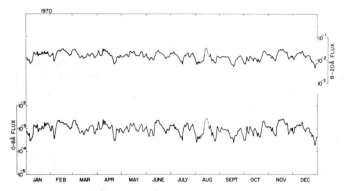

Fig. 5b   Daily averages for 1-8 Å and 8-20 Å flux levels for 1970 derived from SOLRAD 9 and OSO-5 photometers.

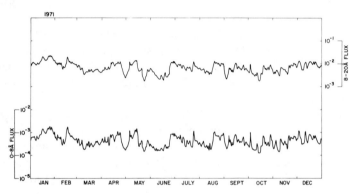

Fig. 5c   Daily averages for 1971 from SOLRAD 9 showing a significant decrease in the average levels as solar minimum is approached.

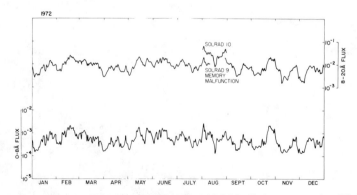

Fig. 5d   Daily averages for 1972 from SOLRAD 9. SOLRAD 10 data was used to fill a gap in the data due to a malfunction of the SOLRAD memory. The 1-8 Å data from the two satellites agree but there is a systematic error in the 8-20 Å flux levels.

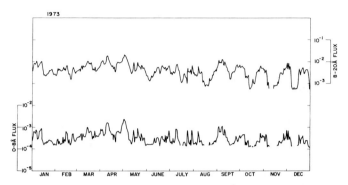

Fig. 5e Daily averages for 1973 from SOLRAD 9. The 1-8 Å flux is below the threshold for that detector in early June, in August, and in November. The SOLRAD 10 memory failed on 11 June 1973.

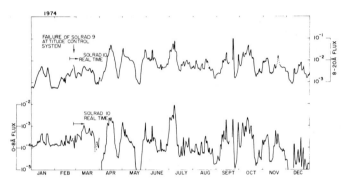

Fig. 5f Daily averages for 1974 of X-ray flux derived primarily from the SOLRAD 10 real-time data. Due to failure of the SOLRAD 9 attitude control system in February, no useful data was obtained from the storage system after April 1974. Threshold of SOLRAD 10 real-time system is 1×10^{-5} erg cm^{-2} s^{-1}. There are a number of periods when the 1-8 Å flux falls below this value.

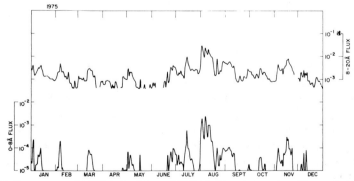

Fig. 5g Daily averages of 1975 derived from SOLRAD 10 real-time data. The period between February and June is very quiet.

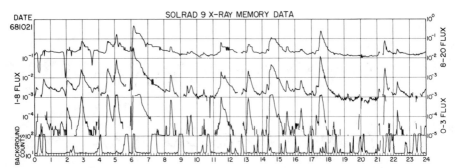

Fig. 6 Solar X-ray flux plot for 21 October 1968 showing the large variations that occur at times of X-ray flares. The uppermost curve is the 8-20 Å flux level, below it is 1-8 Å data, and next is the data from the 0.5-3 Å photometer. The bottom trace indicates periods of radiation belt particle interference.

B. PROPORTIONAL COUNTER AND BRAGG CRYSTAL SPECTROMETER MEASUREMENTS OF X-RAY CONTINUUM EMISSION

In this section, data obtained from proportional counter spectrometers has been assembled to illustrate the great variability and steepness of the solar spectrum below 10 Å. Included are some continuum measurements made with Bragg crystal spectrometers. All data are plotted in Figure 7.

Wolff (1974) measured the X-ray continuum with a Bragg crystal spectrometer with large area but moderate wavelength resolution. The integrated reflectivity of the graphite crystals was carefully measured and the authors claim an overall accuracy of ±15%. The measurement was made on 27 June 1972 at 1900 UT when two active regions were present on the disk. SOLRAD 9 1-8 Å X-ray flux was 3×10^{-4} erg cm^{-2} s^{-1} for the same time, which indicates only a moderate level of activity. Wolff's data, published in terms of counts per second, has been converted to erg cm^{-2} s^{-1} Å^{-1}. The data from the low-energy spectrometer are labeled W-L and those from the high-energy spectrometer are labeled W-H. Because of the great variability of the emission in this part of the spectrum we can only observe that the agreement between the proportional counter measurements and the Bragg spectrometer measurement is reasonable.

Walker, Rugge, and Weiss (1974a) derived continuum measurements from their EDDT crystal spectrometer, which was flown aboard the OV1-17 satellite. These are designated WA-1, WA-2, WA-3, and WA-4 in Figure 7. The measurements were made during an active period in which X-ray flares were observed. Again, the variability of the solar X-ray flux below 10 Å prevents a close comparison with other measurements. However, it is evident that these measurements agree rather well with those that Culhane et al. (1969) made with a proportional counter aboard OSO-4.

Bowles et al. (1967) used a photon spectrum described by

$$N(\lambda) \propto \exp\left[-hc/\lambda kT_e\right] \quad (16)$$

to analyze data from two proportional counters flown on 15 May 1966. The resulting spectrum is labeled B-5. The flux in the band 0-8 Å is 1.4×10^{-4} erg cm^{-2}s^{-1}, which is in close agreement with the SOLRAD 8 graybody value of 1.3×10^{-4} erg cm^{-2}s^{-1} in the same band. Data from other authors were also taken from the paper by Bowles et al. (1967) and included in Figure 7. Curves BC-1 and BC-2 are from measurements of Culhane et al. (1964) made on 27 April 1962 and 3 May 1962, respectively. Curve BM-3 comes from Manson (1964) and was derived from data taken in a rocket flight made on 25 September 1962. Spectrum BCH-4 was obtained on 12 June 1965 by Chodil et al. (1965), and BU-6 was made by Underwood (*private communication*) on 20 May 1966. All of these are nonflare measurements and demonstrate the great variability and steepness of the solar spectrum in this wavelength region even during quiet periods. At 3 Å there is a factor of 10^3 between the measurement of Chodil et al. (1965) on 12 June 1965 and that of Acton and Catura (1975) for 19 February 1968.

Culhane et al. (1969) obtained X-ray continuum measurements (labeled CU-1, CU-2 and CU-3) below 10 Å with a proportional counter aboard OSO-4. An expression of the form

$$I_\lambda = a_0 + a_1 \lambda + a_2 \lambda^2 + \ldots \quad (17)$$

was used to represent the spectral distribution in order to derive the flux values by an iterative procedure. Their results show a quiet period on 9 November 1967 as well as two other periods during an optical class 1B solar flare. The 0610 UT spectrum occurs during the increasing phase of the flare and that taken at 0613 UT is near the maximum. In the 0610 UT record the 1.9 Å iron lines and 3.2 Å calcium lines are evident although they are poorly resolved. The authors compared the quiet spectrum with calculated free-free and free-bound continuum spectra (Culhane, 1969) and found that a single temperature model is not adequate to describe their results.

Acton and Catura (1975) used large-area proportional counters (0.23 m^2) with pulse height analysis to study transient features of the quiet sun. Their measurements were converted to flux units through the use of the semiempirical expression of Culhane and Acton (1970) [Eq. (12)] for free-free bremsstrahlung and free-bound emission. Their three rocket flights sampled different amounts of activity but none was made during an X-ray flare. Their flux data has been translated into units of erg cm^{-2}s^{-1} Å^{-1} and labeled AC-1, AC-2, and AC-3. The extreme variability of the sun's emission at short wavelengths is evident. At 2 Å there is a difference of five orders of magnitude between the quiet spectrum and the flare spectrum that was measured by Culhane et al. (1969). Acton and Catura (1975) also found it necessary to include the contribution of the 1.9 Å emission line in the spectral function.

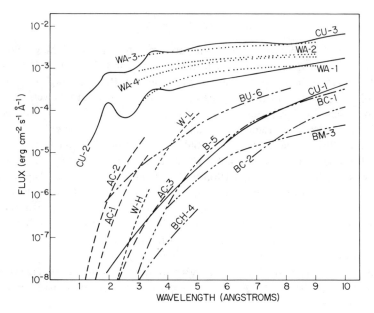

Fig. 7 Solar continuum spectra obtained with proportional counter and Bragg crystal spectrometers by various authors. The measurements are described in the text.

C. MEASUREMENT OF ABSOLUTE INTENSITIES OF SPECTRAL LINES BELOW 10 Å WITH BRAGG CRYSTAL SPECTROMETERS

Few measurements of absolute fluxes in spectral lines below 10 Å have been published, mostly because of the lack of absolute calibrations of the instruments. Even when these measurements have been made there is often a long period between calibration and spacecraft launch, and unfavorable environmental conditions can change calibrations by large factors.

In their experiment, flown on the OV1-17 satellite, Walker, Rugge, and Weiss (1974a) measured X-ray emission from the whole disk in the wavelength range below 8.5 Å. The reflectivity of the EDDT crystal used in their spectrometer was carefully measured, as was the efficiency of the entire instrument. A list of their more intense lines is given in Table 2. Their measurements were made at four different times on the same day (20 March 1969) and covered a variety of conditions of solar activity. The 0602 measurements were made at a relatively quiet time, while the 1529 measurement was made in the declining phase of a small X-ray flare. Measurements at 1642 and 1706 came after a somewhat larger flare. SOLRAD 9 data show similar 1-8 Å emissions for 1529 and 1706, thereby confirming the line intensity measurements.

Neupert, Swartz, and Kastner (1973) used an uncollimated crystal spectrometer in the OSO-5 satellite to measure line spectra between 6 and 25 Å. They used the

TABLE 2

Intensities of Emission Lines Below 10 Å

		20/3/69*				27/2/69+	18/5/69+	20/8/71**
		0602	1529	1642	1706	1407	0555	
3.950	Ar XVII	0.9°	2.8	6.4	2.5			
3.989	Ar XVI / Ar XVII	0.7	4.3	8.5	1.5			
4.303	S XV	0.7	3.2	5.8	1.7			
5.036	S XV	4.9	19.7	43.0	12.7			
5.066	S XIV / S XV	2.5	9.8	10.6	6.5			
5.088	S XVI	1.5	3.5	7.4	1.4			
5.105	S XIV / S XV	2.4	12.8	26.7	10.4			
5.219	Si XIII / Si XIV	0.4	3.9	16.6	2.7			
5.282	Si XIII	–	0.8	5.2	1.2			
5.401	Si XIII	0.9	4.6	12.8	4.9			
5.680	Si XIII	3.2	13.0	24.7	14.6			
5.810	Si XII	1.6	4.3	6.2	4.3			
6.179(6.18)++	Si XIV	3.1	25.5	47.7	15.1	520	3	
6.588(6.58)	Mg XII	0.9	1.2	7.9	2.1	–	4	
6.609		1.1	6.1	8.7	5.8			
6.649(6.65)	Si XIII / Al XII	49.5	157.1	226.7	168.5	670	26	
6.688	Si XII	18.9	47.7	65.9	59.7			
6.719	Si XII	13.2	26.8	32.3	21.1			
6.739(6.74)	Si XII / Si XIII / Mg XII	29.3	85.0	138.4	110.0	240	15	
6.765		1.3	3.3	6.0	8.0			
6.788	Si XII	0.9	3.3	6.6	3.6			
7.101(7.11)	Mg XII	2.0	20.3	27.6	17.1	180	5	
7.173(7.17)	Al XIII	0.8	5.1	7.6	4.4	150	3	
7.316	Mg XI	1.0	2.9	5.7	4.0			
7.474(7.47)	Mg XI	2.0	7.0	10.2	7.7	130	4	
7.751(7.76)	Al XII	5.4	19.6	21.3	19.4	80	5	
7.769	Al XII	0.4	4.4	6.2	3.4			
7.807	Al XI / Al XII	1.8	9.3	8.2	5.8			
7.830		2.0	6.0	5.8	4.4			
7.852(7.86)	Mg XI / Al XI	11.0	37.2	44.4	29.9	170	8	2.1
7.873	Al XI / Al XII	4.5	12.0	21.2	9.5			
7.893(8.00)		0.8	1.4	2.1	1.4	200	2	
8.421	Mg XII	17.4	102.8	190.8	113.0	1410	34	5.9
(8.83)	Fe XXIII?					120	1	
(8.98)	Fe XXII?					120	4	
(9.18)	Mg XI					}730	89	19
(9.22)	Mg XI							6.3
(9.32)	Mg XI					400	45	13
(9.49)						210	–	1.6
(9.73)						110	5	0.8

° $\times 10^{-5}$ erg cm^{-2} s^{-1}
++ Wavelengths from Neupert's line list
* Walker, Rugge, and Weiss (1974)
+ Neupert, Swartz, and Kastner (1973)
** Biegman et al (1974)

published reflectivities of Evans and Pounds (1968) for KAP fitted to their experimental measurements at three wavelengths. The line intensities are also listed in Table 2. The measurements for 27 February 1969 at 1407 were made during an optical class 2B flare that saturated the SOLRAD 9 1-8 Å photometer at about 6×10^{-1} erg cm^{-2} s^{-1}. The measurement of 18 May 1969 was also made during an X-ray flare (optical classification SN) which, however, had a 1-8 Å flux of about 2×10^{-2} erg cm^{-2} sec^{-1} according to the SOLRAD 9 1-8 Å photometer. This was larger than the 1-8 Å flux at 1642 on 20 March 1969; but Walker, Rugge, and Weiss's (1974a) measurements are about four times greater than those of Neupert, Swartz, and Kastner (1973) for the same lines. The authors did not publish estimates of the limits of error of their absolute flux values.

Biegman et al. (1974) made measurements of the emission late in a small flare (optical classification SN) on 20 August 1971 between 0322 and 0331 with a KAP spectrometer flown on a "Vertical-2" rocket. SOLRAD 9 indicated 1-8 Å flux in excess of 6×10^{-3} erg cm^{-2} s^{-1} at the same time. In comparison with the measurements of Walker, Rugge, and Weiss, their measurements seem to be rather low. They calibrated their detector in the laboratory with an accuracy of 30% and used Meekins' (1973) KAP reflectivity measurement tied to a laboratory measurement at 9.9 Å.

During solar flares, emission line spectra appear down to 1.9 Å. Very few measurements of line intensities are available in this region. Neupert et al. (1967) and Neupert and Swartz (1970) observed lines from the ions Ni XXVII, Fe XXVI and XXV, Ca XX and XIX as the most important lines below 3.5 Å during an optical class 2B flare. Although the lines appeared to be quite intense, these authors published no flux values. Meekins et al. (1968) also made observations down to 0.5 Å from OSO-4 and detected the 1.9 Å Fe line and the 3.2 Å calcium line during a 3B flare on 16 November 1967. They also did not publish the absolute intensities of their lines. Doschek et al. (1971) published observations of the 1.9 Å iron line and the Ca XX and XIX lines made from the Bragg crystal spectrometer aboard OSO-6. Again, intensities were only given in terms of counting rate.

Only three authors have published absolute intensities of the 1.9 Å line (Phillips, 1971; Phillips, Neupert, and Thomas, 1974; and Neupert, Thomas, and Chapman, 1974). Because the line is present in any measurable intensity only in solar flares, it can be expected to exhibit wide variations in intensity with time during flares. The peak values thus far published range from 0.7×10^{-4} to 6.19×10^{-4} erg cm^{-2} s^{-1}. This is a small fraction of the energy below 10 Å during a flare. Barletti and Tagliaferri (1973) calculated the possible effects that this line might have on the ion production rate in the lower ionosphere and found them to be very small.

III. SUMMARY

Measurements of solar X-ray radiation in the region below 10 Å show great variability. The amount of flux is strongly dependent on the presence of active regions on the sun's visible hemisphere. During solar maximum in 1969, 1-8 Å flux measured by ionization chamber photometers aboard SOLRAD satellites showed maximum nonflare emissions of about 5×10^{-3} erg cm^{-2} s^{-1} and minimum emissions of about 3×10^{-4} erg cm^{-2} s^{-1}. The 28-day rotation period can be seen occasionally but is usually masked by distribution in longitude of the active regions. The 11-y solar cycle variation is clearly described by the SOLRAD measurements. Average 1-8 Å emissions during maximum, 1969 to 1970, are about 10^{-3} erg cm^{-2} s^{-1}. Solar minimum values are at least a factor of 500 lower, i.e., less than 1×10^{-5} erg cm^{-2} s^{-1}.

These flux values are based on calculated photometer efficiencies and on the assumption of a graybody spectral distribution for conversion of photometer current to X-ray flux. Efficiency of the band-sensitive photometers can be measured in the laboratory to ±10%. The results indicate that flux values quoted above may be too small by about a factor of two. The measurement of photometer current can be made to ±2% in the laboratory, but with temperature variations and digitization for telemetry, current measurements in flight are usually made with an absolute accuracy of only ±5%.

The largest source of error in the photometer measurements comes from the spectral assumptions used to calculate the X-ray flux from the photometer current. Errors of a factor of three to five may appear in the calculated graybody fluxes when they are compared with fluxes calculated from the single temperature spectral distribution of Culhane and Acton (1970). When multithermal emission measure models are used, flux values differ by only about 20% from those based on the single temperature analysis.

Proportional counter and Bragg crystal spectrometers offer the possibility of obtaining spectral information as well as intensity. Resolution of proportional counters is at best about 16% at 2 Å; therefore a spectral distribution must be assumed. Temperatures of the emitting plasma can be derived by fitting the spectral distribution to the distribution of counts registered in the energy channels of a pulse height analyzer. Laboratory calibrations of efficiency can be made to ±10% or less and the absolute flux can be calculated to within a factor of two, dependent on the accuracy with which the assumed single temperature spectrum describes the actual solar output.

While the possibility of calibrating the Bragg crystal spectrometer to ±10% in the laboratory exists, this has not usually been done by most experimenters. As a result, very few absolute intensity measurements are available from these instruments. While line emission is found to be not very important at quiet times, it contributes significantly during flares. Line emission is highly variable and for this reason is one of the most useful diagnostic tools in solar physics.

ACKNOWLEDGMENTS

We would like to thank Dr. Hugh Hudson and Dr. A.B.C. Walker, Jr., for their very helpful criticisms of the early draft of this paper. We also acknowledge the support of the Naval Electronic Systems Command, the Air Systems Command, and NASA for their support of the SOLRAD program.

REFERENCES

Acton, L.W., and Catura, R.C., 1975, submitted to *Solar Phys.*
Barletti, R., and Tagliaferri, G.L., 1973, *J. Atmos. Terr. Phys. 35*, 821.
Batstone, R.M.; Evans, K.; Parkinson, J.H.; and Pounds, K.A., 1970, *Solar Phys. 13*, 389.
Biegman, I.L.; Grineva, Yu. I.; Korneev, V.V.; Krutov, V.V.; Mandel'stam, S.L.; Vainstein, L.A.; Vasilyev, B.N.; and Zhitnik, I.A., 1974, *P.N. Lebedev Physical Institute, Preprint No. 67*, Academy of Sciences of the U.S.S.R., Moscow.
Bowles, J.A.; Culhane, J.L.; Sanford, P.W.; Shaw, M.L.; Cooke, B.A.; and Pounds, K.A., 1967, *Planet. and Space Sci. 15*, 931.
Chambe, G., 1971, *Astr. and Ap. 12*, 210.
Chodil, G.; Jopson, R.C.; Mark, H.; Seward, F.D.; and Swift, C.D., 1965, *Phys. Rev. Letters 15*, 605.
Culhane, J.L., 1969, *M.N.R.A.S. 144*, 375.
_____ and Acton, L.W., 1970, *ibid. 151*, 141.
_____ and _____, 1974, *Annu. Rev. Astr. and Ap. 12*, 359.
_____; Willmore, A.P.; Pounds, K.A.; and Sanford, P.W., 1964, *Space Res. 4*, 741.
_____; Sanford, P.W.; Shaw, M.L.; Phillips, K.J.H.; Willmore, A.P.; Bowen, P.J.; Pounds, K.A.; and Smith, D.G., 1969, *M.N.R.A.S. 145*, 435.
Datlowe, D.W., Hudson, H.S., and Peterson, L.E., 1974, *Solar Phys. 35*, 193.
Dere, K.P.; Horan, D.M., and Kreplin, R.W., 1974a, *J. Atmos. Terr. Phys. 36*, 989.
_____, _____, and _____, 1974b, *Solar Phys. 36*, 459.
Doschek, G.A., 1972, *Space Sci. Rev. 13*, 765.
_____; Meekins, J.F.; Kreplin, R.W.; Chubb, T.A.; and Friedman, H., 1971, *Ap. J. 170*, 573.
Drake, J.F., Gibson, J., and Van Allen, J.A., 1969, *Solar Phys. 10*, 433.
Evans, K., and Pounds, K.A., 1968, *Ap. J. 152*, 319.
Grivena, Yu. I.; Korneev, V.V.; Krutov, V.V.; Mandel'stam, S.L.; Sylwester, I.A.; Vainstein, L.A.; Urnov, A.M.; and Zhitnik, I.A., 1975, *P.N. Lebedev Physical Institute Preprint No. 66*, Academy of Sciences of the U.S.S.R., Moscow.
Grubb, R.N., 1975, *NOAA Technical Memorandum ERL SEL-42*, U.S. Government Printing Office, Washington, D.C.

Horan, D.M., 1970, Ph.D. thesis, Catholic University of America, University Microfilms, 70-22133.
Kane, S.R., 1974, *Coronal Disturbances,* ed. G.Newkirk, D. Reidel, Dordrecht, 105.
_____ and Hudson, H.S., 1970, *Solar Phys. 14,* 414.
Kreplin, R.W., 1970, *Ann. Geophys. 26,* 567.
_____ and Meekins, J.F., 1974, *Space Res. 14,* 431.
_____; Horan, D.M.; Chubb, T.A.; and Friedman, H., 1969, *Solar Flares and Space Research,* ed. Z. Svestka and C. de Jager, North-Holland, Amsterdam, 121.
Landini, M., and Monsignori Fossi, B.C., 1970, *Astr. and Ap. 6,* 468.
Letfus, V., 1972, *Bull. Astr. Inst. Czechoslovakia 23,* 223.
Manson, J.E., 1964, *Environmental Research Paper No. 68 (AFCRL-64-932),* Air Force Cambridge Research Laboratories, Bedford, Mass.
McMaster, W.H.; Kerr DelGrande, N.; Mallett, J.H.; and Hubbell, J.H., 1969, *Compilation of X-Ray Cross Sections, UCRL-50174 (Sect. 2, Rev. 1) TID-4500, UC-34 Physics,* Lawrence Radiation Laboratory, Livermore, Calif.
Meekins, J.F., 1973, Ph.D. thesis, Catholic University of America, University Microfilms 73-21740.
_____; Kreplin, R.W.; Chubb, T.A.; and Friedman, H., 1968, *Science 162,* 891.
_____; Unzicker, A.E.; Dere, K.P.; and Kreplin, R.W., 1974, *NRL Report 7698,* Naval Research Laboratory, Washington, D.C.
Mewe, R., 1972, *Solar Phys. 22,* 459.
Neupert, W.M., 1971, *Physics of the Solar Corona,* ed. C.J. Macris, D. Reidel, Dordrecht, 237.
_____ and Swartz, M., 1970, *Ap. J. (Letters) 160,* L189.
_____, _____, and Kastner, S.O., 1973, *Solar Phys. 31,* 171.
_____, Thomas, R.J., and Chapman, R.D., 1974, *Solar Phys. 34,* 349.
_____; Gates, W.; Swartz, M.; and Young, R., 1967, *Ap. J. (Letters) 149,* L79.
Parkinson, J.H., 1973, *Solar Phys. 28,* 487.
_____, 1975, *Solar Gamma, X-, and EUV Radiation,* ed. S.R. Kane, D. Reidel, Dordrecht, 45.
Phillips, K.J.H., 1971, *Physics of the Solar Corona,* ed. C.J. Macris, D.Reidel, Dordrecht, 254.
_____, Neupert, W.M., and Thomas, R.J., 1974, *Solar Phys. 36,* 383.
Tucker, W.H., and Koren, M., 1971, *Ap. J. 168,* 283; *erratum 170,* 621.
U.S. Department of Commerce, 1976, *Solar-Geophysical Data 378* (Supplement), 6.
Unzicker, A.E., and Donnelly, R.F., 1974, *NOAA Technical Report ERL 310-SEL-31,* U.S. Government Printing Office, Washington, D.C.
Vaiana, G.S.; Davis, J.M.; Giacconi, R.; Krieger, A.S.; Silk, J.K.; Timothy, A.F.; and Zombeck, M., 1974, *Ap. J. (Letters) 185,* L47.
Walker, A.B.C., Jr., 1972, *Space Sci. Rev. 13,* 672.
_____, 1975, *Solar Gamma, X-, and EUV Radiation,* ed. S.R. Kane, D. Reidel, Dordrecht, 73.
_____, Rugge, H.R., and Weiss, K., 1974a, *Ap. J. 188,* 423.
_____, _____, and _____, 1974b, *ibid. 192,* 169.
Wende, C.D., 1969, *J. Geophys. Res. 74,* 4649.
_____, 1971, *Goddard Space Flight Center Report No. X-601-71-166,* Goddard

Space Flight Center, Greenbelt, Md.
Whyte, G.N., 1963, *Rad. Res. 18*, 265.
Wolff, R.S., 1974, *Solar Phys. 34*, 163.

THE AVAILABILITY AND DEVELOPMENT OF NBS RADIOMETRIC STANDARDS

R. P. Madden
National Bureau of Standards
U.S. Department of Commerce

From the desired accuracies established at this workshop it is apparent that the transfer radiometric standards available from NBS at the present time are inadequate in several respects. The accuracy of the *in-house* standards of visible, UV, and VUV irradiance above 120 nm needs to be improved and the spectral region should be extended from 18.6 nm to join the X-ray capability. Further, the technology for maintaining these improvements in accuracy from laboratory to field must be developed.

In the near future, NBS expects to establish the deuterium lamp between 105-200 nm and the argon miniarc between 114 and 350 nm as spectral irradiance standards. NBS also expects very soon to extend the calibration range of the VUV windowless diode as a spectral irradiance transfer detector standard down to 10 nm and ultimately down to 5 nm with a 10% uncertainty. Another NBS project has as its goal the establishment of an ionization-chamber standard detector which will cover the X-ray region and up to 5 nm in wavelength to overlap with the VUV windowless diode.

In the event increased funding becomes available, an expansion of effort would allow substantial reduction in the uncertainties stated in the following table. It is estimated that uncertainties in the transfer standards approaching 1% above 110 nm and 5% below are achievable within a four-year period.

NBS is also improving the state of the art in blackbody temperature-based radiometry. Here we are aiming at a factor of about 3 or 4 improvement (to 0.33% in the visible and 1% at 200 nm). With current staff and funding, this development will probably take five years. Additional support would allow these goals to be achieved in about three years.

The source and detector standards currently available from the U.S. National Bureau of Standards are listed below.

A. Spectral radiance transfer source standards

Source	Wavelength region (nm)	Uncertainty
argon miniarc	114-140	10.0%
argon miniarc	140-350	5.0%
deuterium lamp	165-350	10.0%
tungsten strip lamp	225-2400	
tungsten strip lamp	at 225	2.0%
tungsten strip lamp	at 650	0.7%
tungsten strip lamp	at 2400	0.6%

B. Spectral irradiance transfer source standards

Source	Wavelength region (nm)	Uncertainty
deuterium lamp	200-350	6.0%
mercury lamp	253.7	5.0%
tungsten halogen	250-1600	
tungsten halogen	at 250	3.0%
tungsten halogen	at 555	1.2%
tungsten halogen	at 1600	1.2%

C. Spectral irradiance nonportable transfer source standard

Source	Wavelength region (nm)	Uncertainty
SURF storage ring	5→IR	~5.0%

D. Spectral irradiance transfer detector standards

Detector	Wavelength region (nm)	Uncertainty
VUV windowless diode	18.6-125	5-10%
VUV window diode	115.0-253.7	5-10%
UV-B diode	200.0-320	10%
silicon photovoltaic	257.0	5%
silicon photovoltaic	351.0	5%
silicon photovoltaic	364.0	5%
silicon photovoltaic	420.0-700	2%
silicon photovoltaic	700.0-1150	5%

E. Spectral irradiance high-level primary detector standards

Detector	Wavelength region (nm)	Uncertainty
Elect.-calib.	257-2000	1%
pyroelectric	2000-14000	2%

NEP (1 sec) = 1 µW (air), 100 nW (vacuo)

PROBLEMS AND CONTROVERSIES IN THE MEASUREMENT OF THE SOLAR SPECTRAL IRRADIANCE

R. F. Donnelly
Space Environment Laboratory
National Oceanic and Atmospheric Administration
U.S. Department of Commerce

In our programs to understand the ionization and chemical balance of the earth's upper atmosphere and its temperature structure, we face the immediate problem of specifying the solar radiation input over large wavelength spans from the infrared to the X-ray region. This means the collection of many diverse measurements and their combination into a single empirical solar spectrum that is both internally consistent and complete, with a realistic specification of uncertainties. This collection and comparison process shows us certain difficulties in existing measurements as well as pitfalls in measurement procedures that may be applied in the future. The following summary of these problems is presented in the sincere hope that their recognition now will lead to their solution and avoidance in measurement programs during solar cycle 21.

A. CONFLICT BETWEEN ACHIEVED AND REQUESTED ACCURACIES

Some of the differences between achieved and desired accuracies are so great that refinements of current measurement techniques would probably not achieve sufficient accuracy. For example, it appears that detection of temporal variations of less than 1% in the solar optical and infrared radiation will require measurements from space. This involves new instrumental problems because of the harsh space environment. The requirements for some future measurements also imply improvements in standards. It is therefore extremely important that the precision and accuracy requirements for atmospheric and solar studies be scrutinized by atmospheric, ionospheric, and solar physicists.

B. SPECIALIZATION OF SPACE MEASUREMENTS

Instruments for measuring the solar flux with rockets and satellites are specialized according to differing needs of solar or terrestrial scientists. Since about 1970, most UV and EUV solar flux measurements for solar physics have tended toward high spatial and spectral resolution. Absolute flux accuracy seems unnecessary from the viewpoint of solar physics, because solar physicists are more interested in solar radiance measurements than in full-disk measurements. On the other hand, terrestrial scientists are mainly interested in full-disk measurements with high spectral resolution, high absolute accuracy, and, at X-ray wavelengths, high time resolution. As the specialization of measurements proceeds, there is an increasing need for improving communications among the various groups of scientists involved — solar physicists, ionospheric physicists, and atmospheric physicists. The results from each of the different types of measurements made by one of these groups can have important benefits for the other groups. For example, full-disk spectral radiation measurements with high absolute accuracy are still needed by the solar physics community.

At X-ray, EUV, and UV wavelengths, there are currently three types of satellite measurements: (1) solar research measurements with high spatial and spectral resolution intended for analyzing the physical structure of the solar atmosphere and its evolution; (2) atmospheric physics measurements of the full-disk absolute flux at most wavelengths of interest in the upper atmosphere, for example, the *Atmospheric Explorer (AE)* measurements made concurrently with numerous atmospheric experiments; and (3) solar monitoring measurements over broad wavelength bands with emphasis on real-time full-disk data on a continuous basis, high time resolution, and moderate absolute accuracy. Scientists with atmospheric experiments on the *AE* satellites are probably not as aware of recent solar flux measurements made in solar physics experiments as were their counterparts of a decade ago. One of the values of communications about the three types of measurements is that the increased knowledge of solar physics gained by the special solar physics instruments can help improve our interpretation of the full-disk flux. For example, in our workshop discussion on the size of the solar cycle variation at ultraviolet wavelengths, an argument was raised that an observed factor of two variation in the 1300-1700 Å range over a solar cycle must be too large because the observed variation with solar rotation for high levels of activity is only about ±15%. There are not enough active regions at solar maximum to achieve a factor of two variation. However, recent measurements with high spatial resolution have shown long term trends in the net flux from the solar network structure outside of solar active regions, which may contribute significantly to the UV solar cycle variation. Several years ago some satellite full-disk measurements were corrected for instrument degradation by assuming that the long-term drift in the measured Lyman-alpha flux was a measure of the degradation; such long-term changes would now have to be considered as possibly being truly solar. Because of disputes about instrumental errors, the size of the solar cycle variation of EUV and 1300-1700 Å flux is not clear. On the other hand, the evidence for about a factor of two variation of Lyman-alpha is now quite substantial. One would expect that other emissions from the upper chromosphere would have a comparable solar cycle variation, though further measurements will be necessary to determine the variations accurately as a function of wavelength. New solar physics measurements of phenomena such as X-ray bright points have not clearly specified their contribution below 10 Å, which is of interest in the D-region of the ionosphere.

C. FLUX VARIATIONS CAUSED BY SOLAR ROTATION AND ACTIVE REGION EVOLUTION

Analyses of X-ray, EUV, and UV temporal variations do not distinguish the evolutionary effects—the birth, growth, and decay of active regions—from the effects of the time-varying occultation of active regions caused by solar rotation. It is difficult to separate the two processes in full-disk measurements. Measurements with high spatial resolution are more helpful. It is important to separate these causes of time variation, because their terrestrial effects are different. To achieve good predictions of solar flux variation several days in advance, both of these processes must be better treated than in the past. For example, simple Fourier analyses of full-disk EUV or X-ray data result in relatively weak spectral power at the solar rotation period. This is caused by the phase shifts introduced in solar variation every few rotations when the growth of new active regions at different longitudes begins to dominate the variation.

D. DEGRADATION OF SATELLITE INSTRUMENTS

EUV measurements from satellites suffer marked degradation of efficiency. Preflight calibrations are, therefore, of little value a few weeks after launch. Furthermore, no one knows the extent of degradation that occurs between the calibration time and the initial observing time. Improvements have been made by flying rocket experiments during the satellite program to provide a recalibration of the satellite experiment. During the shuttle era, postflight calibrations should be possible. Furthermore, the difficult task of analyzing the problems causing instrument degradation must be pursued with vigor in order that future satellite measurements can achieve, for at least a year, the high absolute accuracies desired.

E. INSTRUMENT CONTAMINATION

The new accuracies being asked of spectral irradiance measurements made from space platforms (of order 5% below 120 nm, 1% above 120 nm, and better than 1% in the visible) probably will not be realized using prevailing instrument package and vehicle preparation techniques. The future for high accuracy space radiometry will involve taking over the technology for ultraclean component selection and preparation (which has been developed for ultrahigh vacuum applications such as surface research) and applying it to spectrometer, electronics, and vehicle preparation. In the EUV region, monolayer surface contamination significantly affects optical and photoemission properties of surfaces. In the visible, where the highest accuracies are being asked, experimenters must contend with the fact that contaminants that normally lie dormant at atmospheric pressure in the laboratory will volatize and spread under space atmospheric conditions.

This "cleanliness" factor is probably the explanation for the serious aging of a number of space experiments early in life. There is growing suspicion that contamination, together with the high solar UV flux levels on contaminated surfaces, such as mirrors, leads to rapid degradation of surface reflectance. Interestingly, the *SKYLAB* experiments—which received an unprogrammed protection from contamination for the first two weeks in space as a result of a mechanical malfunction (during which time volatile components were outgassed into space)— achieved very good stability. If the desired high accuracies are to be achieved, then an ultrahigh vacuum cleanliness protection program encompassing all aspects of the flight must be developed and used.

F. MISUSE OF THE 10.7 CM RADIO INDEX

The daily full-disk 10.7 cm flux measured near Ottawa at about 1700 UT is today the most commonly used and the best available indicator of the level of solar activity. It is highly correlated with many ionospheric and upper atmosphere parameters because it is usually also well correlated with the X-ray, EUV, and UV flux variations that cause the changes in the terrestrial indicators. However, a very high correlation coefficient is required between two quantities before one can be used to precisely predict the other. Current correlations between EUV and the 10.7 cm flux indicate the latter can be used to predict the EUV flux to an accuracy of only about ±50%. This low accuracy does not result from the accuracy of the measured 10.7 cm flux, which is quite sufficient. Rather, it occurs because the correlations are not extremely high. The 10.7 cm flux describes only a part of the solar atmosphere, a part that differs from active regions to the quiet sun. The EUV flux originates from a wide range of source-region temperatures and densities, including the chromosphere, chromosphere-corona transition region, the corona, and active-region loops with very high temperatures. It is physically unrealistic that all these physical regimes can be estimated by using one full-disk measurement at one wavelength and at only one time per day. But some scientists are trying to use the 10.7 cm flux to estimate the solar EUV flux to a much higher accuracy than is reasonable. This problem also arises when two sets of EUV flux measurements taken on different days but with similar 10.7 cm flux levels are compared. The differences between the two observations are often considered to be due to their combined measurement errors, when most of the difference could well be a real difference in solar EUV flux.

G. BROAD-BAND VERSUS HIGH WAVELENGTH RESOLUTION

There are three types of X-ray, EUV, and UV detectors according to their wavelength resolution—high wavelength resolution, medium resolution, and broad-band. At EUV wavelengths, these resolutions are about 0.1 Å, 10 Å, and 100-500 Å, respectively. At X-ray wavelengths, they are 0.001 Å, 0.1 Å, and 3 to 20 Å, respectively. Each of the three types of instruments for a given wavelength range has its advantages and disadvantages, as discussed below.

High-resolution spectrometer measurements are excellent for resolving minor emission lines and for identifying spurious light from higher orders that must be removed in absolute flux evaluations. They also provide better measures of scattered light. The data must be analyzed by deconvolving the instrument passband. These instruments are very suitable for rocket calibration flights, but they involve high telemetry rates and low scan-to-scan time resolution.

Medium resolution spectrometers are used for satellite measurements to separate strong line groups, major continua, etc., and are also used with high spatial resolution. In the discussions of apparent differences in flux measurements, experimenters using high spectral resolution question the accuracy of removing overlapping orders from the medium-resolution data and ask for more details on the data analysis. Clearly, it is now necessary for all experimenters measuring solar radiation to publish full details of the instruments, calibrations, and corrections to the data. When simultaneous measurements from both types of instruments are available, a direct comparison of flux data should be made to determine whether systematic errors are being made in corrections for overlapping orders and scattered light. These differences should be negligible when results for strong lines are compared. Recent measurements tend to agree for strong lines but disagree for minor lines and continua, which collectively involve a significant portion of the EUV flux. Broad-band experimenters question whether overcorrections were being made such that real minor continua and emission lines are, in effect, discounted as merely scattered light, resulting in reported spectrometer fluxes being too low.

Broad-band measurements require knowledge of the spectral shape of the incident radiation in order to interpret the absolute intensity of the broad-band flux. Since the shape of the solar spectrum is observed to vary in detail, high-resolution spectroscopists tend to be skeptical of the absolute accuracy of broad-band UV, EUV, or X-ray measurements. On the other hand, the sensitivity of the broad-band results to the variations in solar spectra may be quite low. It is important that the high- and medium-resolution experimenters *provide a set of spectra representing different levels of solar activity to be used to test the sensitivity of the broad-band results to spectral variations.* It is particularly important since the routine solar monitoring measurements tend to be broad-band because of the instruments' long life, higher time resolution, and lower telemetry rates. We should also note that since the terrestrial effects of solar radiation at a given altitude usually result from a few broad bands of radiation, many terrestrial programs can use the coarser measurements in monitoring and prediction schemes.

Broad-band filter measurements are free from scattered light problems and measure weak continua and emission lines just as well as strong lines. Pinholes in the filter window could pass strong longer wavelength radiation, causing erroneously high flux level measurements; however, some recent experiments have used an optical test to monitor for pinhole problems, and no evidence of significant problems was detected. Even though windows can be checked for pinholes that penetrate through the window, there may still be problems with non-uniformity of window thickness or partial degeneration of the filters. During the Shuttle era, postflight tests of the window transmission may aid in evaluating such problems. Questions have been raised about possible errors caused by the effects of strong radiation outside the main passband, where the broad-band detector sensitivity was low but not negligible. Such effects can be evaluated by using available high-resolution data and are not then a problem for some broad-band detectors.

Recently, problems in a broad-band X-ray measurement were encountered because of a weak indirect coupling of a current caused by solar UV inducement of photoelectrons with the measured detector current. The erroneous current only appeared when the instrument observed the sun and therefore behaved as if it were caused by the radiation intended to be measured. Similar errors could be occurring in some broad-band UV and EUV detectors where weak currents are measured.

In view of the many questions that passed between the different experimenters at the workshop, questions which have been around for many years, it was evident that the calibration and data interpretation should be documented in much greater detail. A major problem here is that the recent economic pressures for journal papers to be short have tended to remove these details from publication. On the other hand, some groups complained that their questions concerning whether another group had made a certain correction had been unanswered for years. It is clear that as the absolute accuracies are moved from a factor of two towards

10%, the details of calibration and data interpretation must be documented for scrutiny by other scientists using the data. The study of the intercomparison and compilation of solar flux data relevant to aeronomical research being made by Working Group IV of COSPAR is an important step in the solution of this problem.

H. FACTOR OF TWO UV AND EUV PROBLEMS

Discrepancies of about a factor of two have appeared among various UV and EUV measurements. The causes of these differences need to be resolved in order to achieve confirmed accuracies of the order of 1% at UV wavelengths and 10% at EUV wavelengths.

At UV wavelengths below 1700 Å, Heath reported at the workshop, the variation over a solar cycle was roughly a factor of two. This result is very important to atmospheric physics. His results seemed consistent with those for Lyman-alpha at 1215.6 Å reported by Vidal-Madjar and derived from results obtained by many different scientists. Some experimenters felt their own UV radiance measurements did not support such a large solar-cycle variation. In determining the solar irradiance from their radiance measurements, a correction for limb darkening was made; but the correction was not adjusted for different periods of the solar cycle. It is important to determine the variation of this function to establish whether it could cause the current discrepancy.

Some physicists modeling the ionosphere have concluded that to explain the observed electron density and temperature, the EUV fluxes for a moderate level of solar activity should be increased by a factor of two (e.g., Donnelly and Pope, NOAA Technical Report ERL 276-SEL 25). Recent EUV measurements for moderate levels of solar activity have been higher, and some of the earlier measurements have been revised upward. But all the possible causes of this factor of two discrepancy should be examined. The ionospheric results were determined on the basis of typical data for a moderate level of activity as judged by the 10.7 cm radio flux. The amount of data analyzed to determine "typical" may have been insufficient.

The possibility of errors in the ionospheric calculations or data was not discussed in detail. These calculations were made for days different from those of the EUV flux measurements used, but the days had comparable levels of solar activity as measured by the 10.7 cm radio flux. Much of the apparent discrepancy may result from the solar EUV flux differing significantly from the day of the EUV flux measurements to the day of the ionospheric measurements. The satellite EUV experiment suffered degradation with time, although a rocket calibration flight indicated little degradation near the time of the reported measurements. Unknown instrumental errors owing to changes in the instrument between the preflight calibration and the reported measurements are a problem in general. Agreement was achieved within the experimental errors for two cases in which rocket flight measurements of the solar EUV flux were used to compute the ionization as a function of altitude, and the results were then compared with concurrent ionospheric measurements. Truly valid comparisons of ionospheric data and EUV flux values can be made only when such concurrent measurements are available.

On the AE satellites, there are concurrent broad-band and spectrometer measurements for which there appears to be a factor of two difference in results. Actually, the experimental errors of ±30% in each instrument, when combined, can explain a factor of two difference because such "best estimate" errors are closer to 1 than 3 σ. In addition, there are problems in interpreting each type of data (see section G) and also problems caused by calibrating the AE instruments at different times because both instruments varied with age. The broad-band detector was calibrated before flight, while the spectrometer was calibrated with a high-resolution spectrometer on a rocket flown after both satellite instruments had degraded somewhat. Consequently, if the broad-band results are compared with the rocket flight, part of the difference may have resulted from aging of the broad-band instrument. On the other hand, if the results are compared at a time shortly after launch of the satellite, when the broad-band measurements are thought to have been accurate, problems arise from the aging of the spectrometer between then and the time of the rocket calibration flight and from unknown drifts in the broad-band detectors between their preflight calibration and their first solar measurements. It is therefore difficult to determine what are the real differences, if any, between the two sets of measurements from the same spacecraft.

Clearly, satellite instruments should include on-board calibration tests sufficient to determine the degradation as a function of time. Preflight calibrations combined with rocket or postflight calibrations may also be necessary to achieve ±10% accuracy at EUV wavelengths over periods of several months to a year. In summary, the factor of two discrepancies in EUV measurements may well be explained by calibration errors, degradation problems, misuse of the 10.7 cm flux as a precise scaler of EUV flux, problems in interpreting the data, and other sundry problems.

Comment

Each individual experimenter or group reporting new absolute measurements has an obligation, as do all scientists, to give sufficient information on the experimental techniques and data collection methods to convince even a skeptical reader that the necessary precautions to insure the integrity of the experiment have been taken. At the very least, we need a clear statement of the corrections that have been applied, indications of the size of these corrections, and the estimated systematic and random errors in the final results. Similarly, once such data have been presented in the professional literature, and in the absence of a modification or retraction by the original reporting author or authors, the data must be used as presented or else rejected as unreliable. Even minor adjustments, such as the reduction to 1 AU, should be cleared with the original author; and a statement of such clearance should be included by the user when he or she publishes modified data. Such careful accounting of the corrections to flux measurements is absolutely necessary if we are ever to track down sources of discrepancy between measurements made by different experimenters and subsequently published in compilations of solar flux data.

James E. Manson
U.S. Air Force Geophysics Laboratory

Comment

Since discrepancies in the measurements of absolute solar fluxes from the full disk are now obvious in the wavelength range below 3000 Å, Working Group IV (Upper atmosphere) of COSPAR decided to support a study to compare the various sets of solar EUV data. The following scientists are working on this intercomparison study for the indicated wavelength ranges: 3000-2000 Å, P. Simon; 2000-1000 Å, J.P. Delaboudiniere; 1000-100 Å, H.E. Hinteregger and G. Schmidtke; below 100 Å, R.F. Donnelly. We gave a preliminary report at the COSPAR meeting in Philadelphia (1976). The written summary of our study will contain an intercomparison of solar EUV data with error bars included. The time scale has to be taken into account for measurements below 1250 Å. Our report will also contain short descriptions of the experiments with technical details important in the absolute calibration, a list of the relevant publications on the solar flux data, a tabulation and description of current absolute solar EUV data available to users, and the addresses of experimenters who are willing to supply data on request.

Gerhard Schmidtke
Institut fur Physikalische Weltsramforschung

Editor's Note Added in Proof

The first report, "Intercomparison/Compilation of Relevant Solar Flux Data Related to Aeronomy," of Working Group IV was prepared by Dr. Schmidtke and his collaborators in May 1976. This group's summary comment on the variation of the solar spectral irradiance with the solar cycle is especially pertinent, and I quote the last two paragraphs from their report:

"With the exception of the very intense emission of hydrogen Lyman-alpha, *the expected variability of the absolute fluxes with solar cycle seems to be within the errors of the data. Typical (sometimes too optimistic) accuracy figures nowadays are 10-20% between 300-200 nm, 15-25% between 200-100 nm, 20-35% between 100-20 nm, and 50% below 20 nm. The corresponding accuracy figures for the period of the last solar maximum have been worse—they are not well known and could not be used to set upper limits for the variability of the spectral irradiance with solar cycle."*

"It is hoped that the application of in-flight calibration for absolute spectrophotometry will considerably improve the accuracy of the measurements. Radiation monitoring from satellites and measurements of the absolute solar fluxes with in-flight calibration aboard Spacelab should be accomplished to answer one of the key questions in aeronomy and related fields: What is the variability of the solar spectral irradiance with solar cycle?"

Oran R. White

MODELS OF THE SOLAR ATMOSPHERE

Eugene H. Avrett
*Center for Astrophysics
Harvard College Observatory and
Smithsonian Astrophysical Observatory*

The sun is extensively studied because of its influence on the earth and because it can be observed in sufficient detail to allow us to test physical theories in a wide variety of fields: nuclear, atomic, and molecular physics; high-temperature plasma physics; and stellar structure and evolution. Models of the solar atmosphere provide an essential link between the observational data and the physical description of the atmosphere. This chapter reviews the various theoretical and semiempirical models that are used to determine the structure of the photosphere, chromosphere, and transition region between the chromosphere and corona.

One of the immediate goals in solar research is to understand the mechanical heating that causes the outward temperature rise in the chromosphere and corona. It is thought that these outer regions are heated by the dissipation of wave energy generated in the convection zone beneath the photosphere, but the detailed processes are not well understood. A discussion of energy and momentum balance, wave generation, and mechanical heating is beyond the scope of this review. A comprehensive summary of recent research on these topics is given by Athay (1976).

The emphasis in the present chapter is on observational determinations of the temperature-density structure of the solar atmosphere.

There are two main classes of model solar atmospheres: theoretical models and semiempirical models, which differ according to the way the temperature distribution is obtained. In the first case the temperature distribution is calculated according to physical theory. In the second case it is obtained directly by trial and error until the computed spectrum agrees with observations. The temperature distribution obtained by the semiempirical method is the same as the one which the theoretical model calculation would predict if all important physical mechanisms were taken into account.

I. THEORETICAL MODELS

A theoretical model atmosphere is based on only three essential parameters:

the effective temperature of the star, characterizing the total flux of energy passing through the atmosphere; the surface gravity of the star; and the chemical composition of its atmosphere. Normally a model atmosphere must be simplified in a number of ways to make it tractable. In constructing a standard model, the modeller will generally: assume that the atmosphere is in either radiative equilibrium, without any mechanical transport of energy, or radiative-convective equilibrium, based on a convective mixing-length theory; assume that the atmosphere is in hydrostatic equilibrium; adopt an approximation method for treating line opacity; prescribe the velocity fields; choose a simplifying one-dimensional geometry— usually semi-infinite, plane-stratified layers, but sometimes spherically symmetric shells—thus ignoring any horizontal inhomogeneities; and, if the statistical equilibrium equations are not solved in detail, assume local thermodynamic equilibrium (LTE) in order to calculate atomic and molecular populations.

An example of a theoretical model is the one calculated for the sun by Kurucz (1974a). He chooses T_{eff} = 5770 K, log g = 4.44, the chemical abundances compiled by Withbroe (1971), a mixing-length to scale-height ratio of 2, doppler broadening velocity of 2 km/sec, plane-parallel geometry and LTE. The line opacity is in the form of distribution functions computed by Kurucz, Peytremann, and Avrett (1974) from a list of 1,760,000 atomic lines, but with no molecular lines included.

Kurucz's model represents the most extensive attempt so far to include line opacity of adequate detail in the theoretical calculations. Gustafsson et al. (1975) have computed similar models but included fewer lines. Details of their solar model are given by Bell et al. (1976).

Because of the difficulty of treating line opacity and because of the lack of atomic data, procedures have been adopted for treating line opacity in which the opacity is adjusted in such a way as to bring an otherwise theoretical model into approximate agreement with observation. Such procedures have been used by Carbon and Gingerich (1969), Böhm-Vitense (1970), Mutschlecner and Keller (1972), Bell (1973), and Peytremann (1974a,b).

The models described above are based on an assumption of LTE. Athay (1970) has calculated a line-blanketed solar model that includes the effects of departures from LTE for selected strong lines and for representative weak lines.

All of these theoretical models predict a reasonable temperature structure for the solar photosphere, but they fail to account for the outward temperature increase in the chromosphere. The reason for this deficiency is that the calculations do not include mechanical energy dissipation in the chromosphere, since a sufficiently quantitative theory for heating by wave dissipation is not presently available. One of the ultimate goals of solar model—atmosphere research is to incorporate the energy dissipated by wave motions in a line-blanketed, non-LTE theoretical model calculation. Ideally, the computed structure of the atmosphere should be such that the predicted spectrum agrees with the observed one. Some years of further work will be required to reach this goal.

II. SEMIEMPIRICAL MODELS

A more immediate goal is to establish agreement between predicted and observed thermal structures by a semiempirical method. While the temperature distribution in a semiempirical model is adjusted deliberately to yield a good match between these spectra, density and other parameters are calculated in the same way as for the theoretical models. We now describe briefly the semiempirical models that have been determined recently to fit various solar observations.

In 1967 an international conference was convened at the Hotel Bilderberg near Arnhem, Netherlands to establish an acceptable model of the quiet parts of the solar photosphere and low chromosphere. The papers from this conference appear in *Solar Physics,* Vol. 3, No. 1 (1968) and are reprinted as a book (deJager, 1968). Gingerich and deJager (1968) provide a model based on the continuum data available at that time.

This model, known as the Bilderberg Continuum Atmosphere, was subsequently revised by Gingerich *et al.* (1971) to take into account rocket observations by Parkinson and Reeves (1969) in the far ultraviolet around 0.16 μm and airborne observations by Eddy *et al.* (1969) in the far infrared near 300 μm. The revised model, called the Harvard-Smithsonian Reference Atmosphere (HSRA), also used the results of non-LTE chromospheric calculations by Noyes and Kalkofen (1970) and by Cuny (1971), and adopted the chromospheric boundary pressure suggested by Athay (1969).

Vernazza *et al.* (1973) proposed a modified version of the HSRA chromospheric temperature distribution based on further, more extensive non-LTE calculations, and extended the model into the chromosphere-corona transition region. Vernazza *et al.* (1976) also proposed modifications in the structure of the photosphere and temperature-minimum region based on a reexamination of the available central intensity and flux observations throughout the wavelength range 0.125-500 μm. This paper includes a comparison between various photospheric models. References to other semiempirical work are given there, and in the following sections.

III. THE LOWER PHOTOSPHERE

Wöhl (1975) has recently analyzed the center-to-limb variation of several infrared carbon lines and that of the infrared continuum near 1.75 μm. He found that the Gingerich *et al.* (1971) model predicts a greater center-to-limb variation than is observed, that the Kurucz (1974a) model predicts a variation that is smaller than observed, and that the model of Holweger and Müller (1974) is in closest agreement with the observations. This last model is a slightly modified version of the one constructed earlier by Holweger (1967) to fit both line and continuum data. Conclusions similar to those of Wöhl were reached by Vernazza *et al.* (1976), who give a temperature distribution for the lower photosphere based on the observations of the center-to-limb variation in the range 1-2 μm obtained by Pierce (1954), as recalibrated by Mitchell (1959) and David and Elste (1962).

Figure 1 shows temperatures as functions of $\tau_{0.5}$ (the continuum optical depth at 0.5 μm) in the range $1 \lesssim \tau_{0.5} \lesssim 10$, for the four models mentioned above. Also shown is the temperature distribution obtained by Allen (1976) using new center-to-limb observations in the infrared. This run of temperature can be expressed by the formula

$$T(K) = 6528 + 2000 \log \tau_{0.5}$$

in the range $\tau_{0.5} \gtrsim 1$. The 1-2 μm continuum is formed in the optical depth range $1 \lesssim \tau_{0.5} \lesssim 4$. Note that the models of Holweger and Müller, of Vernazza et al., and of Allen have a similar temperature gradient in this range. The earlier models of Elste (1968) and Lambert (1968) closely resemble these three models for $\tau_{0.5} \gtrsim 1$. However, the temperature gradients of the Gingerich et al. and Kurucz models are higher and lower, respectively, leading to greater and smaller center-to-limb variations.

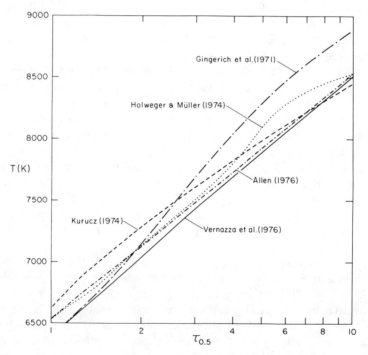

Fig. 1 Temperature in the lower photosphere as a function of the optical depth at 0.5 μm, according to several recent models.

The radiation near 1.6 μm originates in the deepest observable layers of the atmosphere. Relative measurements of the central intensity in the range 1-2.5 μm are given by Pierce (1954). The only absolute-intensity measurements in this range are those of Labs and Neckel (1967) for $\lambda < 1.25$ μm. These and other observations are reviewed by Labs (1975). In Figure 2 we plot the brightness temperatures corresponding to these intensities. Also we plot Pierce's relative measurements normalized in three ways, such that the brightness temperature at $\lambda = 1.3$ μm is 6200 K, 6300 K, and 6400 K, respectively. The solid curve in Figure 2 is the brightness temperature distribution computed from the model of Vernazza et al. (1976).

The 6200 K normalization at 1.3 μm is essentially the one chosen by Vernazza et al. (1976) to join Pierce's data with those of Labs and Neckel at $\lambda \approx 1.0$ μm. The resulting values, however, are not consistent with the Labs and Neckel data at 1.232 μm and 1.246 μm, and are not in good agreement with the predictions of

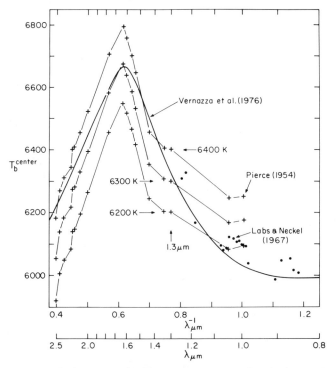

Fig. 2 Observed and calculated central brightness temperature values in the wavelength range from 0.8 to 2.5 μm. The relative measurements of Pierce are normalized in three alternative ways, to the values 6200K, 6300K, and 6400K at 1.3 μm. A normalization between 6300K and 6400K seems in best agreement with the absolute values of Labs and Neckel near $\lambda = 1.24$ μm and with calculated results based on center-to-limb observations.

models based on center-to-limb observations. It seems better to adopt one of the two higher curves, and to disregard Pierce's data at 0.99 μm and 1.05 μm. Models can be used in this way to establish the normalization of relative observations and to resolve inconsistencies between different data sets.

IV. THE UPPER PHOTOSPHERE AND TEMPERATURE–MINIMUM REGION

Figure 3 shows a comparison between several recently published models of the quiet solar atmosphere in the region $10^{-5} \leq \tau_{0.5} \leq 1$. The Kurucz (1974a) model is theoretically determined; the others in Figure 3 are semiempirical. All of the models included in this figure have already been described above except for the one by Ayres and Linsky (1976), which will be discussed at the end of the next section. Other models proposed by Athay (1970), Altrock and Cannon (1972), Lites (1973), and Mount and Linsky (1974) are discussed in the paper by Vernazza et al. (1976).

The models by Gingerich et al. (1971) and Vernazza et al. (1976) both were constructed to fit a wide variety of ultraviolet, visible, and infrared continuum

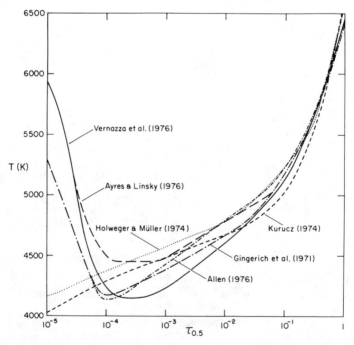

Fig. 3 Temperature as a function of the optical depth at 0.5 μm in the upper photosphere and temperature minimum region according to several recent models.

observations. The main difference between the two is that the more recent model is based on a larger sample of continuum data and on a less restrictive set of physical assumptions. The broad, 4150 K minimum temperature of the Vernazza *et al.* model is derived from continuum observations near 0.16 μm and on those in the 100–200 μm range. Figures 4 and 5 (reproduced from Vernazza *et al.*, 1976) show the basis for the adopted temperature distribution.

The solid curve in the upper panel of Figure 4 is the temperature distribution with T_{min} = 4150 K and the dashed curve is the distribution for a model with a minimum temperature 300 K higher. The brightness temperature distributions in the range 2.5–500 μm corresponding to these two models are shown in the lower part of the same figure, together with the available observations. Figure 5 shows the computed and observed central intensities in the range 0.125–0.2 μm. The computed energy distribution for $\lambda > 0.168$ μm in Figure 5 is highly uncertain because the continuum cross sections are not well known and because the line data are incomplete. In the range $\lambda < 0.168$ μm, however, the computed and observed intensities both refer to the continuous spectrum (due principally to Si I) and may be compared directly.

These comparisons in the far infrared and far ultraviolet indicate that the model with a 4150 K temperature minimum agrees better with available continuum observations than does a model with a minimum temperature of 4450 K. On the basis of these comparisons with continuum observations, the minimum temperature seems to lie in the range 4050–4250 K. However, other evidence suggests that the value is near 4450 K.

V. EVIDENCE FOR A HIGHER MINIMUM TEMPERATURE

The Ca II and Mg II resonance-line profiles can be used to infer the temperature structure of the upper photosphere and low chromosphere (Linsky and Avrett, 1970; Lemaire and Skumanich, 1973). These are deep, high-opacity lines with broad wings. The intensity distribution at various positions in these lines arises from a range of depths extending over much of the photosphere and low chromosphere. The intensity decreases towards line center as the effective depth of formation moves higher in the photosphere, because the line opacity increases towards line center. As shown in Figure 6, there are emission peaks located just outside the line core. This emission is due to the chromospheric rise in temperature. The central absorption is caused by a decrease in the excitation of the line higher in the chromosphere. The intensity minimum in the line wing (called K_1) just beyond the emission peak (K_2) is formed in the temperature minimum region.

Given an adequate theory of line formation, such observations should enable us to construct an atmospheric model. In particular, we should be able to determine

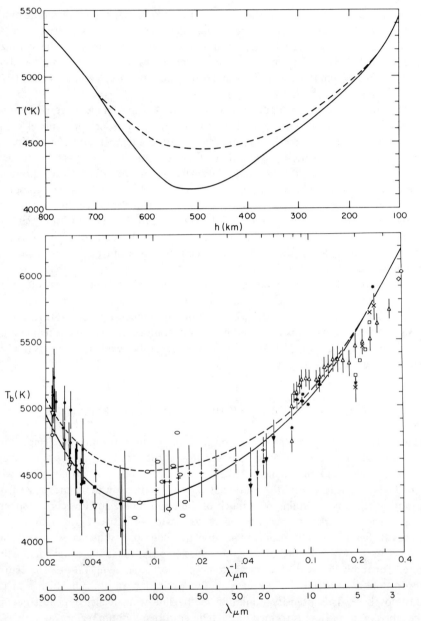

Fig. 4 Top: The temperature as a function of height determined by Vernazza, Avrett, and Loeser (1976), and that of a model with a minimum temperature 300K higher. Bottom: Brightness temperature calculated from the two models in the wavelength range from 2.5 to 500 μm, compared with observations.

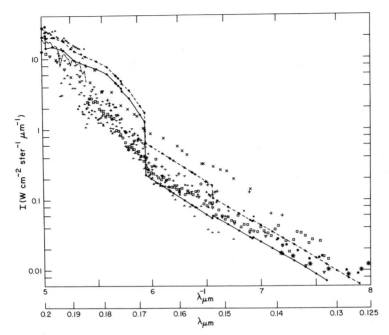

Fig. 5 Central intensities in the range from 0.125 to 0.2 μm, corresponding to the same temperature distributions as in Figure 4, compared with observations (from Vernazza, Avrett, and Loeser, 1976).

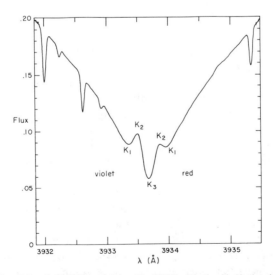

Fig. 6 Flux profile of the Ca II K line obtained in 1976 with the McMath telescope at Kitt Peak National Observatory by William Livingston and Oran White (private communication).

the minimum temperature from the observed intensity minimum.

In the theory of line formation in stellar atmospheres it is normally assumed that the line absorption and emission coefficients have the same frequency dependence and that atoms emit radiation with no memory of how they were excited. Thus the line source function, which is the ratio of emission and absorption coefficients, is independent of frequency within the narrow bandwidth of the line, and the scattering of line radiation is said to be completely noncoherent. This assumption of frequency independence, also called complete redistribution, has been regarded as a reasonable approximation for use in source-function calculations in most astrophysical problems (see Thomas, 1957; Jefferies and White, 1960; Avrett and Hummer, 1965; Jefferies, 1968; Hummer and Rybicki, 1971; Athay, 1972). The assumption of complete redistribution is clearly valid in the doppler core of a line owing to random doppler shifts. It was thought to be valid also in the line wings for cases of practical interest because of high collisional rates which cause redistribution. It is now recognized, however, that there are important cases in which the density is so low that a significant amount of coherent scattering takes place in the line wings, i.e., photons are absorbed and re-emitted without a change of frequency.

In attempting to match computed and observed intensities in the wings of the Lyman alpha line, Vernazza et al. (1973) noted that the observed intensities could be explained only by assuming that the scattering is mostly coherent. They estimated 93% coherent and 7% noncoherent scattering. Soon afterwards, Milkey and Mihalas (1973a,b) studied the Lyman alpha partial redistribution problem in detail and verified these estimates theoretically.

The theory of partial redistribution has been developed further and applied to the Mg II and Ca II resonance lines in a series of papers by Milkey and Mihalas (1974), Milkey et al. (1975a,b,c), Shine et al. (1975b,c) and Mihalas et al. (1976). See also Heasley and Kneer (1976) and an earlier paper by Hummer (1969). These theoretical developments have been supplemented by recent experimental work (Carlsten and Szöke, 1976; Driver and Snider 1976).

Ayres and Linsky (1976) have compared the Mg II resonance line profiles by Kohl and Parkinson (1976) and the Ca II profiles obtained by Brault and Testerman (1972) with computed spectra based on a partial redistribution formalism. They conclude, in agreement with Milkey and Mihalas (1974) and with Shine et al. (1975 b,c), that the minimum temperature derived from the Ca II lines is 4450 ± 130 K and that the Mg II analysis implies a minimum temperature about 50 K higher. These results disagree with the minimum temperature of 4150 ± 100 K estimated by Vernazza et al. (1976) from the continuum data in Figures 4 and 5. The temperature distribution proposed by Ayres and Linsky is shown in Figure 3.

The uncertainty in the estimate T_{min} = 4450 ± 130 K is exclusive of possible deficiencies in the partial redistribution theory. A smaller amount of coherence lowers the estimated temperature, but a certain minimum amount is necessary to explain the observed limb darkening in the H and K line wings. See Shine et al. (1975b).

Athay (1970) has derived T_{min} = 4330 ± 150 K from a radiative-equilibrium model which includes the effects of non-LTE line blanketing. Kurucz (1974a) found that T = 4300 K at $\tau_{0.5} \approx 10^{-4}$ where the minimum temperature is located, from his line-blanketed, radiative-equilibrium LTE calculation. One would expect that the minimum temperature obtained by these theoretical models would be less than the value found empirically since there is likely to be mechanical heating in this region of the atmosphere, an effect which is not taken into account in these theoretical model calculations. Thus, the available radiative-equilibrium calculations tend to support the higher minimum temperature obtained from the resonance-line analysis, rather than the lower one based on continuum observations. The possibility remains that these models do not include sufficient line opacity. More line blanketing would produce a lower minimum temperature.

Ayres and Linsky (1976) suggest that possible systematic errors in absolute calibration result in so much uncertainty that the model with the higher minimum temperature in Figure 4 is consistent with continuum observations. This possibility needs careful study, as do the partial redistribution theory and the predictions of line-blanketed radiative equilibrium models. Ayres and Linsky point out that the role of atmospheric inhomogeneities also must be assessed as we try to understand the two contradictory determinations of the solar minimum temperature.

We conclude this discussion of the photosphere by noting that models can also be based on line observations. Jefferies (1968) has reviewed various methods involving central line intensities. A model derived from the observed center-to-limb behavior of the Mg I 4571 Å line profile is described by Altrock (1974). Ramsey and Johnson (1976) show how a model can be obtained from analyses of a series of lines in a multiplet under the assumption of source function equality (see Waddell, 1962, 1963; Avrett, 1966; Avrett and Kalkofen, 1968).

VI. THE CHROMOSPHERE

The temperature distribution in the chromosphere can be determined from continuum observations in the ranges $\lambda < 0.16$ μm and $\lambda > 200$ μm, from the profiles of lines formed in the chromosphere, and from eclipse measurements. Of these observations, the ultraviolet continuum data contain the greatest amount of information about the thermal structure of the chromosphere. The most important chromospheric features in this range are the hydrogen Lyman continuum, the C I and Si I continua, and the far wings of the hydrogen Lyman alpha line; see Noyes and Kalkofen (1970), Cuny (1971), Noyes (1971), Gingerich et al. (1971), Vernazza and Noyes (1972), Vernazza et al. (1973). The continuum observations in the range $\lambda > 200$ μm provide important constraints indicating whether a model is acceptable or not, but the errors are large enough to permit a fairly wide choice of models. A comparison between the observations and model predictions in this range is given by Vernazza et al. (1973).

There is such a large opacity change with wavelength across the profile of a line that different portions of a line can be formed in very different regions of the atmosphere. The wings of the Ca II and Mg II resonance lines are formed in the photosphere while the central portions have their depths of formation in the chromosphere. The resonance lines of C I and O I have their outer and central portions formed in the lower and middle chromosphere (6000 K and 8000 K), respectively. The outer portions of the C II resonance lines originate in the middle chromosphere (8000 K) while the centers are formed in the transition region between the chromosphere and corona (20,000 K).

Comparisons between the calculated and observed profiles of such lines should provide much information about the structure of the atmosphere. However, a calculation of these non-LTE, optically thick lines requires the numerical solution of a very complex system of equations, and the calculations depend on atomic data which are often poorly known. Also, the non-LTE effects tend to decouple the emergent radiation field from the thermal structure of the atmosphere where the radiation is emitted, a property which complicates the interpretation of such line data. Thus, chromospheric lines can be used only to test various models to determine whether they lie within an acceptable range of validity.

Similar qualifications apply to eclipse observations. Off-limb measurements give unique information about the variation of atmospheric emission with geometrical height, particularly in the low chromosphere. Outside this region, other types of measurements are needed in order to determine a model. Tanaka and Hiei (1972) have used chromospheric continuum data, observed at the total eclipses of 12 November 1966 and 7 March 1970, to construct a model of the low chromosphere.

Athay and Canfield (1970) used the values of $\langle n_e n_p T^{-3/2} \rangle$ determined from eclipse observations by Henze (1969) to establish several alternative models of the chromosphere. They also computed profiles and total intensities for the chromospheric resonance lines of O I at λ 1302 and λ 1305 to find out which model gives O I results in best agreement with observations. Chipman (1971) carried out the same kind of analysis for the resonance lines of Mg II, O I, C II, and C III, and deduced several properties of the chromosphere and of the chromosphere-corona transition region (which we discuss in Section VIII).

Rutten (1976) has used the profiles of five Ba II lines observed at five positions on the solar disk, as well as eclipse observations of the λ 4554 resonance line to test several models of the photosphere and low chromosphere. He finds that partial frequency redistribution must be taken into account in the analysis of these lines.

All of the models described so far are representations of the quiet solar atmosphere. Shine and Linsky (1974) propose models of chromospheric plage regions, while Machado and Linsky (1975) discuss flare models of the chromosphere and photosphere.

VII. SYNTHESIS OF PHOTOSPHERIC AND CHROMOSPHERIC SPECTRA

Here we illustrate the way in which a detailed synthetic spectrum can be obtained for comparison with observed photospheric and chromospheric spectra. The synthesis program written by Kurucz (1974b) now includes the line data compiled by Kurucz and Peytremann (1975). This program has been modified to use the number densities and other atmospheric parameters determined by the non-LTE, semiempirical model calculations of Vernazza *et al.* (1973, 1976).

Sample preliminary results obtained by Kurucz (*private communication*) are shown in Figure 7. The two dark curves represent two sets of photoelectric rocket observations obtained by Kohl, Parkinson, and Reeves (to be published). Their observations extend from 0.13 to 0.32 μm, but we show only the narrow section between 0.2145 μm and 0.2165 μm. The lighter curve is the theoretically determined spectrum, which has not been instrumentally broadened. A small portion of this figure is shown in Figure 8 with line identifications.

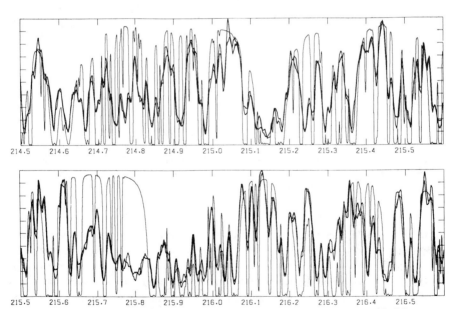

Fig. 7 Observed and computed spectra in the wavelength range from 214.5 to 216.5 nm (2145 to 2165 Å) normalized separately on a linear vertical scale.

IV SOLAR SPECTRUM

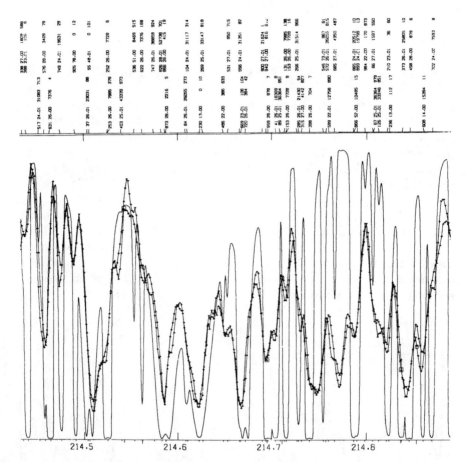

Fig. 8 A small section of Figure 7 showing line identifications. The identifications at the top consist of: the last three digits of the wavelength, the element identification and charge (atomic number before decimal point, charge after; thus 14.00 = Si I and 28.01 = Ni II), then the lower energy level in cm^{-1}, and finally the fraction of the full scale in per mil for the center of each line if it was in isolation. Strong lines thus have small central intensities and weak lines have values near the continuum, which appears close to the top border.

In this sample, the theoretical spectrum does not always match the observed one. Further work is needed to improve the line data, i.e., to add missing atomic and molecular lines and to obtain better wavelengths, opacities, and line widths.

Detailed comparisons between observed and computed lines in different parts of the spectrum are used to establish the chemical composition of the solar atmosphere. Recent studies of solar chemical composition are reviewed by Ross and Aller (1976) and by Withbroe (1976).

VIII. THE CHROMOSPHERE–CORONA TRANSITION REGION

Various spectroscopic data indicate that the corona ($T \gtrsim 10^6$ K) is separated from the upper chromosphere ($T \lesssim 8000$ K) by a transition region only a few hundred kilometers thick. Chipman (1971) found evidence, based on the center-to-limb behavior of the C II resonance lines, for a temperature plateau in the lower part of the transition region. Vernazza and Noyes (1972) determined that such a plateau increases the calculated intensities at short wavelengths in the Lyman continuum and leads to better agreement with observations. Vernazza et al. (1973) proposed a model having a 20,000 K plateau in the lower transition region in order that sufficient optical thickness is produced in the Lyman lines to account for the observed central reversals.

The structure of the upper part of the transition region ($T \gtrsim 10^5$ K) is determined mainly from the intensities of optically thin emission lines in the extreme ultraviolet. Empirical models based on low-resolution data have been proposed in recent years by Dupree (1972), Withbroe and Gurman (1973), and Jordan (1975). The observations and theoretical interpretation of the transition region have been discussed in great detail in a number of reviews. See Pottasch (1964, 1970), Athay (1971), Jordan and Wilson (1971), Frisch (1972), Moore and Fung (1972), Withbroe (1975), and Athay (1976). These discussions are based mainly on ultraviolet line data; properties of the transition region determined from radio data are reviewed by Dubov (1971), by Noci (1971), and by Kanno and Tanaka (1975). Skylab observations of emission lines in the extreme ultraviolet have provided new opportunities to study the inhomogeneous structure of the transition region and the energy balance in this part of the atmosphere. See Mariska and Withbroe (1975), Reeves et al. (1976) and Gabriel (1976). A recent model of the transition region and corona based on energy balance is given by McWhirter et al. (1975). See also Kopp (1972). The temperature and density structure of the corona and inner solar wind is discussed by Kopp and Orrall (1976).

Several recent papers have dealt with the unusual properties of the helium lines. Jordan (1975) pointed out that the theoretical intensities of the He I and He II resonance lines are anomalously low when they are treated as optically thin lines. Milkey et al. (1973) previously carried out a detailed calculation for He I, treating the optically thick resonance lines, and found that similar results are obtained unless a high-temperature plateau is introduced into the transition-region model. Zirin (1975) suggested that photoionization of helium by coronal lines could enhance the helium line strengths, while Shine et al. (1975a) described a diffusion mechanism that would produce a similar effect. Avrett et al. (1976) showed that helium ionization is determined by coronal line radiation but that the He II resonance line is formed in the transition region by collisional excitation and resonance scattering. It has been observed that, unlike typical chromospheric or transition region lines, the helium lines are less intense in coronal holes than in quiet regions of the solar disk (Munroe and Withbroe 1972); but this property is poorly understood. These lines and others formed in the lower part of the transition region need further study.

IX. INHOMOGENEITIES

Almost every proposed temperature-density model of the solar atmosphere pertains to an idealized motionless atmosphere stratified in plane layers with no horizontal inhomogeneities, despite the fact that the sun exhibits pronounced small-scale and large-scale structure. There are several reasons for this preoccupation with one-dimensional models: (1) Important information about temperature-density stratification remains to be obtained from the disk-center spectrum.(Many features of the spectrum are poorly understood and need to be studied initially in a relatively simple way.)(2) It seems likely that various small-scale regions (such as supergranulation cells and the chromospheric network) and various large-scale regions (such as quiet and active regions and coronal holes) can be usefully studied individually with separate one-dimensional models. (3) It is possible that the temperature-density stratification obtained by a one-dimensional analysis can be meaningfully assigned to layers that have local height variations changing with horizontal position and time in response to motions in the atmosphere and to the constraints of the local magnetic field. (4) One-dimensional models of the sun are needed in order to compare solar models with one-dimensional stellar models.

There is clear evidence from center-to-limb observations that the upper chromosphere and the transition region have an irregular geometrical height distribution. Kundu and Liu (1975) have recently shown that there is limb darkening at a wavelength of 1.2 mm in contrast to the brightening predicted by any spherically symmetric model with an outward increase in temperature. Lantos and Kundu (1972) derived a model in which chromospheric spicules are included that accounts for the observed center-to-limb behavior between 1.2 mm and 9.5 mm. Vernazza and Noyes (1972) found that spicules help account for the center-to-limb observations of the Lyman continuum. Mariska and Withbroe (1975) studied the limb-brightening curves for O VI and Mg X in the extreme ultraviolet and found that spicules should be included in the model of the transition region and corona in order to account for the penetration of transition-region material into the corona up to 20,000 km above the limb. As noted earlier, Gabriel (1976) gives a two-dimensional model of the upper chromosphere and corona based on magnetic flux concentrations that occur at the boundaries of super-granulation convection cells.

Despite the importance of spicules and other inhomogeneities at the limb, the atmospheric stratification determined from the spectrum at disk center still seems to be well defined. Athay (1976) estimates that if one percent of the solar surface were covered with spicule-like objects, such objects would be responsible for nearly all of the chromospheric emission seen at the limb, but that their presence would not greatly affect the disk-center spectrum. Even so, one-component models must at least account for the contrast between intensities observed in the chromospheric network and the intensities within supergranulation cells. Reeves (1976) finds that the contrast between the network and the centers of cells is the greatest for transition-region lines with $\log T \approx 5.2$; there the network contributes approximately 75% of the intensity of quiet solar regions.

The various layers of the low chromosphere and photosphere are much smoother in their geometrical height distribution than are the transition-region layers. This lower portion of the atmosphere thus can be approximately represented by plane or spherical shells. However, it is still necessary to account for the observed contrast between structures of the size of supergranules (c.f. Worden, 1975). A review of recent inhomogeneous models of the photosphere is given by Turon (1975). A two-component model of the low chromosphere is proposed by Nakayama (1976).

X. VELOCITY DISTRIBUTIONS

The solar atmosphere exhibits a variety of motions, both large and small in scale, having both organized and random patterns. Recent reviews by Beckers and Canfield (1975) and by Canfield and Beckers (1975) divide the subject according to whether the motions are spatially resolved or unresolved. They point out that this division is somewhat arbitrary since what is a resolved velocity field for one observer is an unresolved one for another.

From a spectroscopic point of view, most types of atmospheric motions can be described in terms of turbulent-velocity distributions. Spectroscopic turbulence refers to the effect of motions on observed spectral lines and is introduced to account for the observed doppler widths, which exceed the widths due to thermal doppler broadening. This concept of turbulence then includes such organized motions as convection and acoustic waves in addition to various random, small-scale motions.

In model atmosphere calculations, turbulence is usually separated into microvelocity and macrovelocity distributions. The microvelocity distribution is used to specify local doppler widths in the calculation of the radiation field, and pertains to eddies of optical thickness much less than unity. The macrovelocity distribution pertains to the average broadening due to instrumentally unresolved doppler shifts caused by motions of macroscopic elements. Observations at high spatial resolution can be used to distinguish between such small-scale and large-scale motions. The distinction is important because only the microvelocity component should be used in determining the doppler linewidth of the local radiation field. In some model calculations the microvelocity distribution also is used to determine a turbulent pressure which is added to the gas pressure in the equation of hydrostatic equilibrium (see Athay and Canfield 1970; Vernazza et al. 1973, 1976).

The review articles by Beckers and Canfield and by Canfield and Beckers cited earlier give extensive references to the various papers dealing with motions in the solar atmosphere. Also, see Evans and Testerman (1975).

XI. CONCLUSIONS

We have given a summary of the types of model atmospheres that are used for interpreting solar observations and for understanding the basic physical processes that give rise to the chromosphere and corona.

Further theoretical work needs to be done to analyze the observations now available. The development of theoretical models has only barely kept up with the succession of new observations of the sun. Further theoretical developments should lead to a greatly improved understanding of the structure and behavior of the solar atmosphere.

ACKNOWLEDGMENTS

I thank T. Ayres, R. Kurucz, R. Loeser, J. Mariska, and J. Vernazza for their comments on the manuscript. This work was supported in part by a grant from the Max C. Fleischmann Foundation.

REFERENCES

Altrock, R.C., 1974, *Solar Phys. 34*, 37.

——————— and Cannon, C.J., 1972, *ibid. 26*, 21.

Athay, R.G., 1969, *ibid. 9*, 51.

———————, 1970, *Ap. J. 161*, 713.

———————, 1971, *Physics of the Solar Corona*, ed. C.J. Macris, D. Reidel, Dordrecht, 36-65.

———————, 1972, *Radiation Transport in Spectral Lines*, D. Reidel, Dordrecht.

———————, 1976, *The Solar Chromosphere and Corona*, D. Reidel, Dordrecht.

——————— and Canfield, R.C., 1970, *Spectrum Formation in Stars with Steady State Extended Atmospheres*, ed. H.G. Groth and P. Wellmann, NBS Special Pub. 332, 65.

Avrett, E.H., 1966, *Ap. J. 144*, 59.

——————— and Hummer, D.G., 1965, *M.N.R.A.S. 130*, 295.

_____ and Kalkofen, W., 1968, *J. Quant. Spectrosc. and Rad. Transf. 8*, 219.

_____, Vernazza, J.E., and Linsky, J.L., 1976, *Ap. J. (Letters)*, in press.

Ayres, T.R., and Linsky, J.L., 1976, *Ap. J. 205*, 874.

Beckers, J.M., and Canfield, R.C., 1975, *Colloquium on Physics of Motions in Stellar Atmospheres, Nice, France*, AFCRL-TR-75-0592, September.

Bell, R.A., 1973, *M.N.R.A.S. 164*, 197.

_____; Eriksson, K.; Gustafsson, B.; and Nordlund, Å., 1976, *Astr. and Ap. Suppl. 23*, 37.

Böhm-Vitense, E., 1970, *Astr. and Ap. 8*, 283.

Brault, J., and Testerman, L., 1972, *Preliminary Edition of the Kitt Peak Solar Atlas*, Kitt Peak National Observatory, Tucson, Ariz.

Canfield, R.C., and Beckers, J.M., 1975, *Colloquium on Physics of Motions in Stellar Atmospheres, Nice, France, AFCRL-TR-75-0592*, September.

Carbon, D.F., and Gingerich, O.J., 1969, *Theory and Observation of Normal Stellar Atmospheres*, ed. O. Gingerich, M.I.T. Press, Cambridge, Mass., 377-400.

Carlsten, J.L., and Szöke, A., 1976, *Phys. Rev. Letters*, in press.

Chipman, E., 1971, *Smithsonian Ap. Obs. Spec. Rept. No. 338*.

Cuny, Y., 1971, *Solar Phys. 16*, 293.

David, K.H., and Elste, G., 1962, *Z. Ap. 54*, 12.

de Jager, C., 1968, *The Structure of the Quiet Photosphere and the Low Chromosphere*, Springer-Verlag, New York.

Driver, R.D., and Snider, J.L., 1976, *Phys. Rev.*, in press.

Dubov, E.E., 1971, *Solar Phys. 18*, 43.

Dupree, A.K., 1972, *Ap. J. 178*, 527.

Eddy, J.A., Léna, P.J., and MacQueen, R.M., 1969, *Solar Phys. 10*, 330.

Elste, G., 1968, *ibid. 3*, 106.

Evans, J.C., and Testerman, L., 1975, *ibid. 45*, 41.

Frisch, H., 1972, *Space Sci. Rev. 13*, 455.

Gabriel, A.H., 1976, *Phil. Trans. Roy. Soc. London A 281*, 339.

Gingerich, O.J., and de Jager, C., 1968, *Solar Phys. 3*, 5.

_____; Noyes, R.W.; Kalkofen, W.; and Cuny, Y., 1971, *ibid. 18*, 347.

Gustafsson, B.; Bell, R.A.; Eriksson, K.; and Nordlund, Å., 1975, *Astr. and Ap. 42*, 407.

Heasley, J.N., and Kneer, F., 1976, *Solar Phys.*, in press.

Henze, W., 1969, *ibid. 9*, 65.

Holweger, H., 1967, *Z. Ap. 65*, 365.

_____ and Müller, E.A., 1974, *Solar Phys. 39*, 19.

Hummer, D.G., 1969, *M.N.R.A.S. 145*, 95.

_____ and Rybicki, G., 1971, *Ann. Rev. Astr. and Ap. 9*, 237.

Jefferies, J.T., 1968, *Spectral Line Formation*, Blaisdell Publishing Co., Waltham, Mass.

_____ and White, O.R., 1960, *Ap. J. 132*, 767.

Jordan, C., 1975, *M.N.R.A.S. 170*, 429.

_____ and Wilson, R., 1971, *Physics of the Solar Corona*, ed. C.J. Macris, D.Reidel, Dordrecht, 219-236.

Kanno, M., and Tanaka, R., 1975, *Solar Phys. 43*, 63.

Kohl, J.L., and Parkinson, W.H., 1976, *Ap. J. 205*, 599.

Kopp, R.A., 1972, *Solar Phys. 27*, 373.

_____ and Orrall, F.Q., 1976, *Astr. and Ap.*, in press.

Kundu, M.R., and Liu, S.-Y., 1975, *Solar Phys. 44*, 361.

Kurucz, R.L., 1974a, *ibid. 34*, 17.

_____, 1974b, *Ap. J. (Letters) 188*, L21.

_____ and Peytremann, E., 1975, *Smithsonian Ap. Obs. Spec. Rept. No. 362*.

_____, _____, and Avrett, E.H., 1974, *Blanketed Model Atmospheres for Early-Type Stars*, Smithsonian Institution, Washington, D.C.

Labs, D., 1975, *Problems in Stellar Atmospheres and Envelopes*, ed. B. Baschek, W.H. Kegel, and G. Traving, Springer-Verlag, New York.

_____ and Neckel, H., 1967, *Z. Ap. 65*, 133.

Lambert, D.L., 1968, *M.N.R.A.S. 138*, 143.

Lantos, P., and Kundu, M.R., 1972, *Astr. and Ap. 21*, 119.

Lemaire, P., and Skumanich, A., 1973, *Astr. and Ap. 22*, 61.

Linsky, J.L., and Avrett, E.H., 1970, *Pub. Astr. Soc. Pacific 82*, 169.

Lites, B.W., 1973, *Solar Phys. 32*, 283.

Machado, M.E., and Linsky, J.L., 1975, *ibid. 42*, 395.

Mariska, J.T., and Withbroe, G.L., 1975, *ibid. 44*, 55.

McWhirter, R.W.P., Thonemann, P.C., and Wilson, R., 1975, *Astr. and Ap. 40*, 63.

Mihalas, D.; Shine, R.A.; Kunasz, P.B.; and Hummer, D.G., 1976, *Ap. J. 205*, 492.

Milkey, R.W., and Mihalas, D., 1973a, *Ap. J. 185*, 709.

_____ and _____, 1973b, *Solar Phys. 32*, 361.

_____ and _____, 1974, *Ap. J. 192*, 769.

_____, Ayres, T.R., and Shine, R.A., 1975a, *ibid. 197*, 143.

―――――, Heasley, J.N., and Beebe, H.A., 1973, *ibid. 186*, 1043.

―――――, Shine, R.A., and Mihalas, D., 1975b, *ibid. 199*, 718.

―――――, ―――――, and ―――――, 1975c, *ibid. 202*, 250.

Mitchell, W.E., Jr., 1959, *ibid. 129*, 93.

Moore, R.L., and Fung, P.W., 1972, *Solar Phys. 23*, 78.

Mount, G.H., and Linsky, J.L., 1974, *ibid. 41*, 17.

Munro, R.H., and Withbroe, G.L., 1972, *Ap. J. 176*, 511.

Mutschlecner, J.P., and Keller, C.F., 1972, *Solar Phys. 22*, 70.

Nakayama, K., 1976, *Pub. Astr. Soc. Japan 28*, 141.

Noci, G., 1971, *Physics of the Solar Corona*, ed. C.J. Macris, D.Reidel, Dordrecht, 13-28 and 308-316.

Noyes, R.W., 1971, *Physics of the Solar Corona*, ed. C.J. Macris, D. Reidel, Dordrecht, 192-218.

――――― and Kalkofen, W., 1970, *Solar Phys. 15*, 120.

Parkinson, W.H., and Reeves, E.M., 1969, *ibid. 10*, 342.

Peytremann, E., 1974a, *Astr. and Ap. 33*, 203.

―――――, 1974b, *Astr. and Ap. Suppl. 18*, 81.

Pierce, A.K., 1954, *Ap. J. 119*, 312.

Pottasch, S.R., 1964, *Space Sci. Rev. 3*, 816.

―――――, 1970, *Ultraviolet Stellar Spectra and Ground-Based Observations, IAU Symposium No. 36*, ed. L. Houziaux and E.H. Butler, 241-249.

Ramsey, L.W., and Johnson, H.R., 1976, *Solar Phys.*, in press.

Reeves, E.M., 1976, *ibid.*, in press.

―――――, Vernazza, J.E., and Withbroe, G.L., 1976, *Phil. Trans. Roy. Soc. London A 281*, 319.

Ross, J.E., and Aller, L.H., 1976, *Science 191*, 1223.

Rutten, R., 1976, Ph.D. thesis, University of Utrecht.

Shine, R.A., and Linsky, J.L., 1974, *Solar Phys. 39*, 49.

―――――, Gerola, H., and Linsky, J.L., 1975a, *Ap. J. (Letters) 202*, L101.

―――――, Milkey, R.W., and Mihalas, D., 1975b, *Ap. J. 199*, 724.

―――――, ―――――, and ―――――, 1975c, *ibid. 201*, 222.

Tanaka, K., and Hiei, E., 1972, *Pub. Astr. Soc. Japan 24*, 323.

Thomas, R.N., 1957, *Ap. J. 125*, 260.

Turon, P. 1975, *Solar Phys. 41*, 271.

Vernazza, J.E., and Noyes, R.W., 1972, *Solar Phys. 22*, 358.

———————, Avrett, E.H., and Loeser, R., 1973, *Ap. J. 184*, 605.

———————, ———————, and ———————, 1976, *Ap. J. Suppl. 30*, 1.

Waddell, J., 1962, *Ap. J. 136*, 231.

———————, 1963, *ibid. 138*, 1147.

Withbroe, G.L., 1971, *The Menzel Symposium on Solar Physics, Atomic Spectra, and Gaseous Nebulae*, ed. K.B. Gebbie, NBS Spec. Pub. 353, 127.

———————, 1975, *Solar Phys. 45*, 301.

———————, 1976, *ibid.*, in press.

——————— and Gurman, J.B., 1973, *Ap. J. 183*, 279.

Wöhl, H., 1975, *Solar Phys. 43*, 285.

Worden, S.P., 1975, *ibid. 45*, 521.

Zirin, H., 1975, *Ap. J. (Letters) 199*, L63.

V. THE SOLAR PLASMA AND ENERGETIC PARTICLES

The preceding two chapters have been concerned with the contemporary picture of the amount of radiant energy from the sun and its possible variation with time. Even though electromagnetic radiation energetically dominates the solar input into the earth's atmosphere, the plasma and high-energy particles emitted from the sun play crucial roles in determining the physical conditions within the earth's magnetosphere. Thus this chapter is devoted to the properties of the solar wind and the high-energy particle flux as measured in the vicinity of the earth. The contributions by Feldman *et al.* and Lanzerotti give a current empirical picture of the plasma and particle flux from the sun. We again step outside traditional solar physics as Heymann and Damon discuss the possibility of variations in both the solar wind and the energetic particle flux deduced from lunar and meteoritic samples as well as from the ^{14}C record in tree rings. The coupling of the empirical picture covering the last 30 years with inferences about the solar wind flow and the degree of solar activity over the last 5 billion years of the sun's life yields another set of clues to the physics of the solar corona and solar activity.

The picture developed by the four authors in this chapter is simultaneously simple and complex. When averaged over periods of months, the mean solar wind properties—the average density, the average flow speed, the average magnetic field strength, and the average energy density—show very little variation during the solar cycle; but when examined in detail, solar wind measurements give a picture of streams spraying out from local areas on the sun along spiral tracks that finally define the interaction with the earth's magnetosphere. This is not simply an outward flow distorted by the solar rotation; it is a magnetohydrodynamic flow system possessing observable leading-edge shock fronts at $\sim$3 AU, occasional disturbances propagating outward from the sun, and associations with large-scale magnetic structure on the solar surface. Similarly, when measured over a coarse time scale of $\sim 10^5$ y, the average proton flux from the sun appears to have been constant to within a factor of 2 for the last several million years; but, as Lanzerotti carefully points out, the contributions to this high-energy particle flux appear to come from a relatively small number of solar flares during each solar cycle. Even though averages over long time scales indicate that the sun has been relatively stable in its solar wind flow and degree of solar activity over the last 4 billion years, within any given solar cycle and perhaps from one solar cycle to the next, these plasma and particle emissions undergo fluctuation.

Since we do not yet fully grasp the solar/terrestrial interaction in terms of the combined radiation and particle input into the earth's magnetosphere, we are unable to relate the established solar variations to changes in the lower atmosphere of the earth that are of importance to man. Nevertheless, physical interaction with the earth's magnetic field and the upper atmosphere is clearly apparent, and we continue to seek the meaning of its influence at the earth's surface. One important distinction between the radiation and plasma fields emitted by the sun is that the solar wind plasma carries magnetic fields originally generated by the solar dynamo. Consequently, the solar wind is capable of carrying magnetic polarity information about the direction of the solar magnetic fields from the sun to the earth. If there is then a real connection between the 22-y solar magnetic cycle and any tropospheric effects, the solar wind and its effect on the interplanetary magnetic field appear to be a key to our understanding of it. Since the solar wind influences both the solar and galactic cosmic ray fluxes at the earth through a modulation process and through magnetic field-line reconnection at the magnetopause, the effects of the solar wind and the high-energy particle flux in the terrestrial atmosphere must be described together.

We must also consider the relative magnetic geometry of the sun and the earth in our attempt to understand the interaction between the solar wind and the earth's magnetosphere. The existence of large regions of one magnetic polarity at the solar poles, coupled with the observations showing that these unipolar fields are carried into interplanetary space above and below the solar equatorial plane from the north and south solar poles, respectively, implies that the position of the earth relative to the effective neutral plane of the solar magnetic field is important. As we seek a physical picture of the overall solar/terrestrial interaction, a general understanding of the consequences of the basic sun-earth magnetic geometry and its variation may lead to simplifications in our concepts of interplanetary magnetic sector structure, modulation of cosmic rays, the precipitation and injection of particles into the earth's atmosphere, and the interpretation of geomagnetic activity.

In contrast to the solar radiation field, the solar wind, particles, and fields suffer substantial changes in their properties during their propagation from the sun to the earth. As Lin, Lanzerotti, Holzer, and Hundhausen emphasize, we do not yet have the unfolding or deconvolution procedure required to invert measurements at 1 AU to retrieve the physical state of the solar wind and energetic particle flux in the solar atmosphere proper. Conclusions about detailed properties of the solar atmosphere (as inferred from solar wind, and field measurements at the earth) must be regarded with caution until the sun-to-earth propagation effects and the effects of the sun-earth geometry are properly dealt with in data analyses.

Oran R. White

PLASMA AND MAGNETIC FIELDS FROM THE SUN

William C. Feldman
John R. Asbridge
Samuel J. Bame
John T. Gosling
University of California
Los Alamos Scientific Laboratory

The purpose of this chapter is to summarize present knowledge of the interplanetary plasma and magnetic field at 1 AU. Since different aspects of this information are useful to research workers active in various scientific disciplines, an attempt was made to organize the data according to potential applications. Towards this end, three categories of usage are considered. The first deals with parameters of the solar wind that are useful as input conditions for other physical systems such as the magnetosphere and atmosphere of the earth and other interplanetary bodies. The next category treats the solar wind as an extension of the solar corona and concentrates on information that is useful for a fuller understanding of solar and interplanetary physics. The final category treats the solar wind as a readily accessible example of a collisionless, high-beta plasma, useful as a laboratory for learning plasma physics.

The instrumentation used to observe the solar wind at 1 AU, as well as the form of the measured data, are briefly described in section I. In section II, average values and ranges of plasma parameters that characterize the interplanetary medium at 1 AU are summarized in tabular form. This is followed in section III by a short discussion of observed microstructure. Latitude variations of the flow at 1 AU are summarized in section IV, descriptions of solar wind stream structures and interplanetary shock-wave disturbances observed at 1 AU are given in sections V and VI, and possible long-term solar wind variations observed in association with the 11-y solar activity cycle are presented in section VII. Section VIII is a summary of this paper.

It is useful at this point to define some of the common symbols used in the following text and data tabulations. The symbols N, $\overline{V}$, T, and $\overline{Q}$ denote the density, velocity, temperature, and heat flux of a particle component of the plasma and $\overline{B}$ denotes the magnetic field vector. Subscripts p, α, and ε on these and other symbols specify the proton, alpha-particle, and electron components, respectively. Boltzmann's constant is denoted by k, particle mass by m, the charge on the electron by e, and the speed of light by c.

I. INSTRUMENTATION

Direct and long-term measurements of the properties of the interplanetary plasma began with the Mariner 2 flight to Venus in 1962 (Snyder et al., 1963). In that mission and since, most plasma measurements have been made using two basic types of instruments: the modulated-potential Faraday cup and the curved-plate electrostatic analyzer (e.g., Hundhausen, 1968; Vasyliunas, 1971). Both types of instruments measure, with varying degrees of precision, the energy per unit charge of incident particles as well as the direction of their arrival. A limited amount of solar wind data has been measured using a combined Wien filter and electrostatic analyzer (Ogilvie, McIlwraith, and Wilkerson, 1968), which is capable, in principle, of separately measuring the particle energy and charge.

The basic plasma data set from which most properties of the solar wind are derived thus consists of electron and ion particle (or charge) flux measured over a two- or three-dimensional matrix whose axes are defined by particle energy per unit charge and one or two angles of incidence. Two- or three-dimensional particle velocity distributions, $f(\underline{V})$, can be derived from these matrices with varying degrees of precision or ambiguity, depending on the type of instrumentation, the detailed hardware design, and the data accumulation cycle (see discussions in Hundhausen, 1968, and Vasyliunas, 1971). Most interplanetary plasma data that have been published and interpreted consist of the various lower velocity moments of f and, in some cases, the detailed shapes of f (which include information concerning the higher velocity moments). A basic set of velocity moments which has been tabulated and interpreted extensively in the literature consists of the density, N; the bulk velocity, $\underline{V}$; the pressure tensor or, effectively, the temperature matrix, (T); and the heat flux, $\underline{Q}$.

At least two attempts have been made, using a linear regression analysis, to cross-check the measurements from two or more plasma analyzers in space (when possible) in order to establish parameter error limits (Moreno and Signorini, 1973; Neugebauer, 1976*b*). This method of calibration can yield only rough values of the measurement errors for several reasons. Most importantly, detector efficiency may be a nonlinear function of particle energy, and the various parameters determined from data returned by some of the plasma analyzers are nonlinearly interrelated in a way that depends on the values of some of the parameters. This is so because reduction of some of the existing plasma data relies on assumed models of particle velocity distributions.

Because many of the derived plasma quantities desired for the present summary have not been presented previously in a unified fashion and because of the above intercalibration uncertainties, we have decided to use the 3-h-averaged Los Alamos plasma data set combined from IMPs 6, 7, and 8 (e.g., Feldman et al., 1976*a*) as the basic data source. This procedure has the advantages that all three Los Alamos instruments have been intercalibrated in space, all three data sets are reduced using identical computer codes, and the data span approximately 3.5 y, longer than any other homogeneous source. These data have been supplemented where necessary by already published results, which are cited appropriately. Estimated errors in determining the basic fluid parameters from individual particle spectra are: ±30% for the density, N; ±2% for the speed, V; ±1.5° for the bulk flow direction, ϕ_V; ±15% for the proton and electron temperature, T_P and T_ε, respectively; ±0.2 for the proton and electron temperature anisotropy, $(T_\parallel/T_\perp)_P$ and $(T_\parallel/T_\perp)_\varepsilon$, respectively; and ±30% for the proton and electron heat flux, Q_P and Q_ε, respectively, with a noise level of 2×10^{-5} ergs cm^{-2} s^{-1} for Q_P and 10^{-3} ergs cm^{-2} s^{-1} for Q_ε. Although the determination uncertainties of averages of T_P, T_ε, $(T_\parallel/T_\perp)_P$, and $(T_\parallel/T_\perp)_\varepsilon$ are somewhat less than stated above, the remaining error limits result from systematic effects and so are not reduced by averaging. Similar errors for the alpha-particle parameters are difficult to estimate because they depend (at least for the Los Alamos IMP instruments) on the accuracy of the procedure used to subtract the proton component. During times when the protons make a negligible contribution to the alpha parameters, the estimated uncertainty is approximately ±15% for the helium abundance, N_α/N_P; approximately ±2% of V_P for the helium/hydrogen velocity difference, $V_{\alpha p}$; and approximately ±20% for the alpha-particle temperature, T_α.

Interplanetary magnetic fields have also been measured using two types of instruments: the flux-gate and vector helium magnetometers (Ness, 1970; King, 1975). The parameters of interest are the three components of the DC magnetic field, $\overline{B}$, along with their characteristic time variations. The time variability of $\overline{B}$ is often presented in the form of power density spectra. Since the magnetic field is carried away from the sun by the solar wind at speeds large compared to hydromagnetic wave speeds, power density spectra of the basic field components provide a measure of the energy density in field variations over interplanetary distance scales, d, which are related to the solar wind speed and frequency of observed variation, ω, through the relation $d \cong V/\omega$.

The accuracy of solar wind magnetic field data has been estimated from cross-checks of field values determined simultaneously by different instruments in interplanetary space (Hedgecock, 1975a; King, 1976). The field and angle values from most simultaneously operating magnetometers are in agreement to within about 0.2 γ and 2°, respectively.

II. THE INTERPLANETARY PLASMA CONFIGURATION AT 1 AU

A. STATISTICAL DESCRIPTION

Statistical properties of the bulk flow characteristics of the solar wind at 1 AU are summarized in Table 1. Included in the tabulation are the mean, most probable, and median values of the various physical quantities directly related to the solar-

TABLE 1

BULK FLOW PARAMETERS OF THE AVERAGE SOLAR WIND

PARAMETER	MEAN	σ	MOST PROBABLE	MEDIAN	5-95% RANGE LIMIT		DATA SOURCE
N (cm^{-3})	8.7	6.6	5.0	6.9	3.0	to 20.0	IMP
V (km s^{-1})	468	116	375	442	320	to 710	IMP
ϕ_v (degrees)	-0.64^*	2.6	-0.75^*	0.67	-5.0	to $+3.5$	IMP
θ_v (degrees)	-0.45^*	---	$-0.6^\dagger$	---	$(-4$	to $+3)^\dagger$	a
(NV) (cm^{-2} s^{-1})	3.8×10^8	2.4×10^8	2.6×10^8	3.1×10^8	1.5×10^8	to 7.8×10^8	IMP
$\left[(NV)m_p V\right]$ (g cm^{-1} s^{-2})	2.9×10^{-8}	1.8×10^{-8}	2×10^{-8}	2.5×10^{-8}	1×10^{-8}	to 5.8×10^{-8}	IMP
$(NV)(\frac{1}{2}m_p v^2)$ erg cm^{-2}s^{-1})	0.70	0.52	0.33	0.55	0.1	to 1.45	IMP
M_A	10.7 (6.5)	4.8 (3.3)	---	10.1 (5.9)	4.4	to 20.0	b (c)
M_S	7.7	1.5	7.5	7.7	5.6	to 10.0	IMP

*These values are not different from zero by more than instrumental systematic uncertainties. The polar angles, ϕ_v and θ_v are defined in a right handed coordinate system with z axis pointing to the north eclipse and x axis pointing radially away from the sun.

$\dagger$These values are determined by eye from the published histogram.

Data Source Codes:

(a) Mihalov and Wolfe, 1971
(b) Ness et al., 1971
(c) Formisano et al., 1974
IMP The combined Los Alamos IMP 6, 7 and 8 plasma data between March 1971 and July 1974.

wind density and velocity observed at 1 AU. Also tabulated is the parameter range within which each physical quantity is measured 90% of the time (i.e., the 5 to 95% range limits). References containing the data sources are listed at the right. An identical format is used for presenting most of the data tabulated in this section. Similar statistical properties of the internal fluid state of the solar wind are summarized in Table 2, with the various internal energy densities listed separately in Table 3.

Some cautionary remarks are necessary for proper use of some of the parameter values listed in tables 1 through 3. First, all entries for the most probable parameter values have been determined by eye from respective histograms and so are accurate only to about ±0.1 of the widths of the distributions (approximately σ, the rms parameter variations, also listed in the tables). It is emphasized that all quantities requiring a determination of the absolute magnitude of the density are uncertain by at least ±30%. Although the IMP 6 and 7 densities both agree roughly with those measured using the Jet Propulsion Laboratory OGO 5 and Apollo 15 analyzers (see Neugebauer, 1976b, but note that the IMP 6 densities used there are too low by 20%), they are about 60% higher than those determined during a similar period in 1971 using the Explorer 35 and HEOS 2 instruments and published in Diodato et al. (1974). It is possible that this uncertainty contributes, at least in part, to the different magnitudes of the average Alfven speed, V_A, and hence the Alfven mach number, M_A, reported in the literature (Ness et al., 1971; Formisano et al., 1974) and listed in Table 1. Another caution is that the alpha-particle entries in the tables

TABLE 2

INTERNAL FLUID STATE OF THE AVERAGE SOLAR WIND

PARAMETER	MEAN	σ	MODE	MEDIAN	5-95% RANGE LIMIT	DATA SOURCE
B (γ)	6.2	2.9	5.1	5.6	2.2 to 9.9	a,b,c
T_p (°K)	1.2×10^5	0.9×10^5	0.5×10^5	0.95×10^5	0.1×10^5 to 3.0×10^5	IMP
T_ε (°K)	1.4×10^5	0.4×10^5	1.2×10^5	1.33×10^5	0.9×10^5 to 2.0×10^5	IMP
T_α (°K)	5.8×10^5	5.0×10^5	1.2×10^5	4.5×10^5	0.6×10^5 to 15.5×10^5	IMP
(N_α/N_p)	0.047	0.019	0.048	0.047	0.017 to 0.078	IMP
$<\delta v^2>^{1/2}$ (km s^{-1})	20.5	12.1	13.5	17.0	6.4 to 42.9	IMP
T_ε/T_p	1.9	1.6	0.7	1.5	0.37 to 5.0	IMP
T_α/T_p	4.9	1.8	4.8	4.7	2.3 to 7.5	IMP

Data Source Codes:

(a) Ness et al., 1971
(b) Hedgecock, 1975b
(c) King, 1976
IMP The combined Los Alamos IMP 6, 7 and 8 plasma data betwween March 1971 and July 1974.

TABLE 3

ENERGY DENSITY IN THE AVERAGE SOLAR WIND (in units of 10^{-10} ergs cm^{-3})

PARAMETER	MEAN	σ	MODE	MEDIAN	5-95% RANGE LIMIT	DATA SOURCE
$N_p(\tfrac{1}{2} m_p v^2)$	144	9	100	125	50 to 310	IMP
$B^2/8\pi$	1.7	1.2	---	1.2	---	a
$1.5 N_p kT_p$	1.9	2.3	0.7	1.3	0.4 to 5.5	IMP
$1.5 N_p kT_\varepsilon$	2.5	2.2	1.1	2.0	0.7 to 7.3	IMP
$1.5 N_\alpha kT_\alpha$	0.4	0.5	0.06	0.28	0.04 to 1.2	IMP
$1.5 N_p m_p <\delta v^2>$	1.1	3.3	0.3	0.37	0.05 to 3.3	IMP

Data Source Codes:

(a) Formisano et al., 1974
IMP The combined Los Alamos IMP 6, 7 and 8 plasma data between March 1971 and July 1974.

are not truly representative of the full range of possible observations, since they were derived from a biased data set. Because the data were measured using an instrument that discriminates only particle energy per unit charge, strict selection criteria were imposed on the alpha-parameter determinations before they were accepted into the present data set. For example, alpha data were rejected if either the spectra of measured particle energy per unit charge did not exhibit two resolved peaks (corresponding to hydrogen and helium, respectively) or the transverse (relative to the magnetic field direction) component of the hydrogen/helium velocity difference was larger than 5 km s^{-1}, the average uncertainty in Vαp (Asbridge et al., 1976). Thus measurements of the helium component of the solar wind were preferentially rejected during hot, high-speed flow conditions. However, uncertainties in the helium parameters (especially the helium temperature, T_α) may also be large during cold flow conditions. Since for low T_P conditions T_α/T_P tends to be low also (Feldman, Asbridge, and Bame, 1974; Hirshberg, Asbridge, and Robbins, 1974), determinations of T_α will be affected more by finite instrumental widths than will determinations of T_p. This effect will cause a systematic overestimate of the measured temperature ratio, T_α/T_p, at low T_p. In addition, all helium parameters were determined using a simple two-dimensional bi-Maxwellian model after subtraction of the proton component. Whenever the subtraction and/or the helium model fit was poor, the resulting parameters may be unreliable. In particular, values for T_α will tend to be high. Lastly, the entries for $<\delta V^2>^{1/2}$ were derived from $<\delta V_e^2>^{1/2}$ (the rms value, for each 3-h interval, of deviations of the ecliptic component of the bulk flow velocity from the 3-h average) using the relation $<\delta V^2>^{1/2} = 1.5 <\delta V_e^2>^{1/2}$.

Although values for N, V, and ϕ_V have been listed in Table 1, the reader is cautioned for several reasons against using them to calculate the angular momentum flux of the solar wind and hence to evaluate the torque applied back on the sun. Not only are the systematic uncertainties in ϕ_V greater than the measured deviation from 0° (here positive ϕ_V denotes a flow corotating with the sun) but all the spacecraft measurements have been made in a single plane (the ecliptic), thus yielding results that may not be representative of the entire spherical shell at 1 AU. For example, it is possible that during extended time intervals, the solar wind measured in the ecliptic originates from that polar region of the sun which happens to extend closest to the equator or to be topologically connected to the ecliptic plane at 1 AU by virtue of the coronal magnetic geometry (Wagner, 1976). Interactions of the measured solar wind with ambient gas from the opposite pole in both latitude and longitude may systematically bias the average direction of the flow measured in the ecliptic plane. Since all measurements to date of the angular momentum flux (Strong et al., 1967; Egidi et al., 1969; Hundhausen et al., 1970; Brandt and Heise, 1970; Lazarus and Goldstein, 1971; Wolfe, 1972) suffer systematic or random uncertainties that are similar in magnitude to those inherent in the combined IMP data set, we conclude with Hundhausen (1972a) that no determination to date has been precise enough to establish definitely an average nonradial flow direction that differs from 0°.

B. THE STRUCTURELESS SOLAR WIND

To date, three attempts have been made to identify the structure-free state of the solar wind. Initially this state was associated with low speed ($V \cong 320$ km s^{-1}) and low proton temperature because those flow conditions appeared to be most quiet and steady (Coon, 1966; Neugebauer and Snyder, 1966; Strong et al., 1966; Hundhausen 1968). A later analysis generalized the previous identification to include all flow conditions for which the density, flow speed, and proton temperature were observed to be constant over a time interval corresponding to the time required for a volume of plasma travelling at an average speed (400 km s^{-1}) to traverse a density scale height (0.5 AU), or about 50 hours (Neugebauer, 1976b). A third analysis has suggested that the structure-free solar wind may be observed at high speed (Gosling et al., 1976; Feldman et al., 1976a) because this state appears to originate from large, uniform, open-field regions in the solar corona (Hundhausen, 1972b; Krieger, Timothy, and Roelof, 1973, 1974; Nolte et al., 1976; Sheeley, Harvey, and Feldman, 1976).

The plasma characteristics of two of these flow conditions are compared with those of the average solar wind in tables 4 and 5. Values for a basic set of plasma quantities that define the "average" state of the solar wind at 1 AU are assembled in the first column of Table 4. Listed in the second column are values for the same set of quantities averaged over those time intervals between March 1971 and July 1974 when the solar wind speed was below 350 km s^{-1} but not increasing with time. This last requirement was imposed to minimize the effects of interplanetary dynamical processes on the observed parameters. In all, 61 time intervals were found. A similar listing for the high-speed state of the solar wind is given in the right-hand column of Table 4. Here data were averaged over those times during the

TABLE 4

PLASMA CHARACTERISTICS OF VARIOUS TYPES OF SOLAR WIND FLOWS

PARAMETER	AVERAGE			LOW SPEED			HIGH SPEED		
	MEAN	σ	% VAR.	MEAN	σ	% VAR.	MEAN	σ	% VAR.
N (cm^{-3})	8.7	6.6	76	11.9	4.5	38	3.9	0.6	15
V (km s^{-1})	468	116	25*	327	15	5*	702	32	5*
NV (cm^{-2}s^{-1})	3.8x10^8	2.4x10^8	63	3.9 x10^8	1.5 x10^8	38	2.7 x10^8	0.4x10^8	15
ϕ_V (degrees)	-0.6	2.6	430	+1.6	1.5	94	-1.3	0.4	31
T_p (°K)	1.2x10^5	0.9x10^5	75	0.34x10^5	0.15x10^5	44	2.3 x10^5	0.3x10^5	13
T_ϵ (°K)	1.4x10^5	0.4x10^5	29	1.3 x10^5	0.3 x10^5	20	1.0 x10^5	0.1x10^5	8
T_α (°K)	5.8x10^5	5.0x10^5	86	1.1 x10^5	0.8 x10^5	68	14.2x10^5	3.0x10^5	21
T_ϵ/T_p	1.9	1.6	84	4.4	1.9	43	0.45	0.07	16
T_α/T_p	4.9	1.8	37	3.2	0.9	28	6.2	1.3	21
$<\delta V^2>^{1/2}$ (km s^{-1})	20.5	12.1	59	9.6	2.9	31	34.9	6.2	18
N_α/N_p	0.047	0.019	40	0.038	0.018	47	0.048	0.005	10
Average			98			45			17

*Not included in the average percentage variation

TABLE 5

RADIAL ENERGY FLUX OF VARIOUS TYPES OF SOLAR WIND FLOWS (in ergs cm^{-2} s^{-1})

PARAMETER	AVERAGE MEAN	σ	LOW SPEED MEAN	σ	HIGH SPEED MEAN	σ
$NV(GM_s m_p/R_s)$	1.21	0.8	1.24	0.5	0.86	0.1
$NV(\frac{1}{2}m_p V^2)$	0.70	0.5	0.35	0.1	1.13	0.2
$N_\alpha V_\alpha (GM_s m_\alpha/R_s)$	0.056	0.05	0.045	0.02	0.042	0.009
$N_\alpha V_\alpha (\frac{1}{2}m_p V_\alpha^2)$	0.033	0.03	0.013	0.008	0.054	0.01
$2.5NVkT_p$	0.016	0.02	0.0043	0.002	0.023	0.05
$2.5NVkT_e$	0.018	0.02	0.016	0.07	0.010	0.002
$2.5N_\alpha V_\alpha kT_\alpha$	0.0035	0.005	0.0005	0.0004	0.0066	0.003
$V(B^2/8\pi)^*$	0.008	----	0.0056	-----	0.012	-----
$1.5NV(m_p\langle\delta V^2\rangle)$	0.0057	0.02	0.0010	0.0006	0.0096	0.006
$\overline{Q}_\epsilon \cdot \hat{r}$	0.0043	0.003	0.0027	0.001	0.0032	0.0006
$\overline{Q}_p \cdot \hat{r}$	1.3×10^{-4}	2×10^{-4}	2.9×10^{-5}	2×10^{-5}	2.3×10^{-4}	9×10^{-5}
Total at 1 AU	2.05		1.68		2.15	
Total at Sun**	9.5×10^4		7.8×10^4		9.9×10^4	

* Calculated by assuming $\langle VB^2/8\pi \rangle = \langle V \rangle B^2/k\pi$ where $\langle B^2/8\pi \rangle$ is assumed to be the same for all three flow conditions and given by Formisano et al., 1974.

** Assuming radial flow

Definition of Constants: G is the gravitational constant, M_s is the mass of the sun, and R_s is the solar radius.

same period when the solar wind speed was above 650 km s^{-1} but beyond the speed maximum of individual high-speed streams (again to minimize the effects of dynamical processes). A total of 19 acceptable high-speed time intervals were found (Feldman et al., 1976a). Fractional variations about each average parameter value are listed separately to the immediate right.

A few explanatory comments concerning the parameter values in Table 4 are in order to aid in their proper use. It is noted that the density listed for low-speed conditions is substantially higher than that tabulated previously for similar solar wind conditions (Hundhausen, 1970). It is possible that some of the difference between the Vela 3 and IMP low-speed density values may be caused by systematic instrumental uncertainties such as counter overflows in the case of the Vela 3 analyzer or uncertainties in the dead-time correction in the case of the IMP instruments. However, it is also possible that the difference may be related to their different periods of observation with respect to the solar activity cycle. For example, inspection of the 61 low-speed intervals accepted from the IMP data reveals that this high average results from many anomalously high noncompressional density enhancements associated with very low proton temperatures. The noncompressional nature of many of these enhancements is confirmed by the fact that variations in N and Tp often appear to be out of phase. Such noncompressional density enhancements were not common in the Vela 3 data.

For comparison purposes, the average density listed in the high-speed column in Table 4 was calculated from the densities measured at 1 AU without correction for the plasma rarefaction so prominently observed within high-speed streams. This accounts for the fact that the average high-speed proton particle flux listed below (2.7×10^8 cm^{-2} s^{-1}) is somewhat less than the value (3.3×10^8 cm^{-2} s^{-1}) listed previously for the same data set (Feldman et al., 1976a), which did include such a correction.

Several interesting facts emerge from a comparison of corresponding parameter values in Table 4 (see also Table 1 of Neugebauer, 1976b). The proton flux at 1 AU is observed to be smaller at the higher speeds. Although it is likely that there is a systematic instrumental effect in the bulk flow direction for the average data set in Table 4, the difference in directions between the low- and high-speed sets is significant. There appears to be more corotation at the low speeds than at the high speeds. The puzzling aspect of this fact is that in assembling the low- and high-speed data sets summarized in Table 4 an attempt was made to eliminate fast stream/slow stream interaction zones (which would tend to produce the same effect). As noted many times (Burlaga and Ogilvie, 1970; Hundhausen et al., 1970; Pizzo et al., 1973; Burlaga and Ogilvie, 1973; Formisano et al., 1974; Neugebauer, 1976b), the proton temperature increases with the bulk speed. Not only is the electron temperature, T_e, lower than the proton temperature during high-speed flows (Feldman et al., 1975, 1976a) but it is lower than the average electron temperature measured both generally and during low-speed flow conditions. Inspection of the magnitude of T_e for each of the 61 low-speed events reveals that, for these conditions, values of T_e are very nonuniform and span nearly the full range of electron temperature values observed in general. Observed values of T_α, T_p, and $\langle \delta V^2 \rangle^{1/2}$ are higher at high speeds. Although they are not shown here, it is true that these increases are such that $T_p \propto T_\alpha \propto \langle \delta V^2 \rangle^{1/2}$. The ratio T_α/T_p is also higher at higher speeds as noted previously (Feldman, Asbridge, and Bame, 1974; Feynman, 1975) and its value at high speeds is significantly greater than 4, the value at which proton and alpha-particle thermal speeds are equal. The helium abundance, N_α/N_p, is lower on the average during low speeds as noted previously (Robbins, et al., 1970; Hirshberg, Asbridge, and Robbins, 1972, 1974b; Ogilvie, 1972; Moreno and Palmiotto, 1973). However, inspection of individual high-speed stream events reveals that N_α/N_p is unusually steady and close to 5%, independent of the value of the maximum speed. In contrast, although the average value of N_α/N_p is lower at the lowest speeds, its value for individual low-speed events is neither steady during a particular event nor uniform from event to event. Whereas the full range of 3-h average values of N_α/N_p during the 19 high-speed events used to construct Table 4 spans less than a factor of 2, its range during the 61 low-speed events spans close to a factor of 100. This fact is reflected in the large (factor of 5) ratio in percentage rms variation of N_α/N_p for the two different data sets tabulated. As seen in Table 4, the average fractional variation of all parameters (excluding speed) is substantially larger during low-speed flows than during the structure-free, high-speed flows. With the exception of the bulk velocity, this fact is true for each individual parameter as well. Thus, viewed as a statistical ensemble of events, the only thing steady and uniform about low-speed conditions is the bulk velocity. In all other respects, structure-free, high-speed flows form a much more uniform emsemble of interplanetary conditions than do the interaction-free, low-speed flows. It is therefore suggested that theoretical models of the structure-free solar wind should concentrate on predicting the 1 AU parameters of high-speed interplanetary flows rather than those of the low-speed solar wind.

The partition of the total solar-wind energy flux for the three flow conditions characterized in Table 4 is summarized in Table 5. Note that although the partition between the flux of work done against solar gravity and the kinetic convective energy flux at 1 AU is inverted for low- and high-speed flow conditions, the total energy fluxes are similar to each other and to that observed on the average. If the solar wind expands from only ~20% of the solar corona (Bohlin, 1976) then the specific energy loss from the corona to the solar wind must be close to 5×10^5 ergs cm^{-2} s^{-1} (±~30%) and, from Table 4, the particle flux must be close to 1×10^{14} protons cm^{-2} s^{-1} (±30%).

C. THE SOLAR WIND AS A COLLISIONLESS PLASMA

As seen in Tables 3 and 5, the solar wind is dominated energetically by its bulk convection. However, the characteristics of the internal state of the interplanetary medium are of interest to those studying the physics of collisionless plasmas. Before proceeding, it is first necessary to indicate length and time scales typical of the average solar wind. Therefore a selected subset of the various plasma scales is summarized in Table 6 and discussed in order from top to bottom in the following paragraphs.

The Debye length, $\lambda_D = 6.9 \times 10^{-2} (T_\varepsilon/N_P)^{1/2}$ m, is the minimum distance scale over which a plasma can exhibit collective behavior. In other words, the physics of phenomena with characteristic distance scales that are less than about 10 m at 1 AU is described best in terms of single or few particle interactions. The next three lengths are the thermal electron, proton, and alpha-particle gyroradii. It is readily seen that all of these characteristic plasma lengths are very small compared to the correlation length of transverse magnetic field variations observed at 1 AU, $\ell_B \cong 10^7$ km, and the local plasma density scale height, $\ell_N = (d\ln N/dr)^{-1} = 0.5$ AU $= 7.5 \times 10^7$ km. Similarly, all of the local frequencies characteristic of the interplanetary plasma at 1 AU—the electron and proton plasma frequencies

$$\omega_{p\varepsilon} = \left(\frac{4\pi Ne^2}{m_\varepsilon} \right)^{1/2} \text{ and } \omega_{pp} = \left(\frac{m_e}{m_p} \right)^{1/2} \omega_{p\varepsilon}, \qquad (1)$$

as well as the electron and proton gyrofrequencies

$$\Omega_\varepsilon = \frac{eB}{m_\varepsilon c} \text{ and } \Omega_p = \left(\frac{m_\varepsilon}{m_p}\right)\Omega_\varepsilon, \tag{2}$$

are very much higher than the inverse of the expansion scale time,

$$\tau_x^{-1} = V/\ell_N. \tag{3}$$

A quantity that indicates the degree of collective behavior of a plasma is the number of particles in a Debye sphere, $N\lambda_D^3$. If $N\lambda_D^3$ is much greater than 1, collective processes will dominate. In the solar wind at 1 AU, $N\lambda_D^3 \cong 7 \times 10^9$, which is indeed much greater than unity, so that collective behavior is expected.

Another parameter useful for indicating the degree of departure expected from energy equipartition and/or pressure isotropy is the ratio of the coulomb energy transfer and/or collision time, τ_c, to the relevant expansion times, τ_x. Median values of representative ratios for the three most abundant solar-wind constituents are as follows (Spitzer, 1956): the electron-electron energy equipartition time relative to the time required for a thermal electron to traverse ℓ_N is $(\tau_c/\tau_x)_{\varepsilon\varepsilon} = 5.31 \times$

TABLE 6

LENGTH AND TIME SCALES TYPICAL OF THE AVERAGE SOLAR WIND

PARAMETER	MEAN	σ	MEDIAN	DATA SOURCE
λ_D (meters)	9.9	3.0	9.8	IMP
R_ε^* (km)	2.5	---	---	a, IMP
R_p (km)	78	36	76	a
R_α^* (km)	177	---	---	a, IMP
ℓ_B (km)	1.0×10^7	0.4×10^7	0.94×10^7	b
ℓ_N (km)	7.5×10^7	---	---	
$\omega_{p\varepsilon}$ (s^{-1})	1.6×10^5	0.5×10^5	1.5×10^5	IMP
ω_{pp} (s^{-1})	3.7×10^3	1.2×10^3	3.5×10^3	IMP
Ω_ε (s^{-1})	1.1×10^3	0.5×10^3	0.95×10^3	a
Ω_p (s^{-1})	0.57	0.25	0.52	a
τ_x^{-1} (s^{-1})	6.2×10^{-6}	1.5×10^{-6}	5.9×10^{-6}	IMP
$N\lambda_D^3$	6.7×10^9	3.0×10^9	6.2×10^9	IMP
$(\tau_c/\tau_x)_{\varepsilon\varepsilon}$	1.7	1.4	1.4	IMP
$(\tau_c/\tau_x)_{pp}$	34	76	11	IMP
$(\tau_c/\tau_x)_{\alpha p}$	53	89	18	IMP

*R_ε and R_α are evaluated by multiplying the value for R_p by the average IMP value of $(m_\varepsilon/m_p)^{\frac{1}{2}} <T_\varepsilon/T_p>$ and $<T_\alpha/T_p>^{\frac{1}{2}}$, respectively.

Data source codes: (a) Formisano et al., 1974
(b) Hedgecock, 1975b
IMP The combined Los Alamos IMP 6, 7 and 8 plasma data between March 1971 and July 1974.

10^{-10} $(T_\varepsilon^2/N) \cong 1.4$. This value is actually an upper limit since a thermal electron must travel a longer distance than ℓ_N along $\hat{B}$ in order to traverse a density scale height because the interplanetary magnetic field is not radial. The proton-proton self-collision time relative to the time required for a proton to convect through ℓ_N is $(\tau_c/\tau_x)_{pp} = 5.85 \times 10^{-9}$ $(VT_p^{1.5}/N) \cong 11$. The helium/hydrogen energy equipartition time relative to the time required for an ion to convect through ℓ_N is $(\tau_c/\tau_x)_{\alpha p} = 3.13 \times 10^{-9}$ $(T_\alpha/4 + T_p)1.5$ $(V/N) \cong 18$. The fact that the median value of $(\tau_c/\tau_x)_{\varepsilon\varepsilon}$ is close to unity at 1 AU suggests that velocity distributions of interplanetary thermal electrons should be nearly isotropic. On the other hand, $(\tau_c/\tau_x)_{pp}$ [which in turn is smaller than $(\tau_c/\tau_x)_{\varepsilon p}$] and $(\tau_c/\tau_x)_{\alpha p}$ [which in turn is smaller than $(\tau_c/\tau_x)_{\alpha\alpha}$] are sufficiently large that interplanetary hydrogen and helium velocity distributions should be anisotropic and thermal energy equipartition is not expected.

Selected characteristics of the detailed internal state of the average solar wind observed at 1 AU are summarized in Table 7. Defining symbols from top to bottom, the angles ϕ_B and θ_B specify the orientation of the interplanetary magnetic field, $\bar{B}$, in the same coordinate system used above for the bulk flow direction. Therefore $\phi_B = \theta_B = 0°$ corresponds to $\bar{B}$ in the ecliptic plane pointing radially away from the sun. Since for this tabulation, all that is required is the orientation of $\bar{B}$, ϕ_B was averaged only over those intervals when $\hat{B}$ was pointing in the general direction away from the sun. The remaining plasma quantities were

TABLE 7
DETAILED INTERNAL STATE OF THE AVERAGE SOLAR WIND: GENERAL

PARAMETER	MEAN	σ	MODE	MEDIAN	5-95% RANGE LIMIT		DATA SOURCE
ψ_B (degrees)*	43	40	---	-45	---		a,b,c,d
θ_B (degrees)*	0.3	25	---	0.0	---		a,b,c,d
$(T_\parallel/T_\perp)_p$	1.5	0.7	1.3	1.4	0.7	to 2.8	e
$(T_\parallel/T_\perp)_\varepsilon$	1.18	0.16	1.09	1.15	1.0	to 1.44	e
$(T_\parallel/T_\perp)_\alpha$	1.3	0.9	1.0	1.2	0.3	to 3.4	e
T_ε/T_p	2.0	1.3	1.2	1.6	0.6	to 4.6	e
T_α/T_p	3.8	2.8	3.5	3.6	1.6	to 6.6	e
$(N_\alpha T_\alpha/N_p T_p)$	0.23	0.14	0.25	0.23	0.06	to 0.49	e
$V_{\alpha p}$	7.0	12.5	2.5	5.0	-9.4	to 30.0	e
q_ε	0.26	0.17	0.20	0.23	0.03	to 0.56	e
q_p	0.35	0.27	0.27	0.3	0.04	to 0.75	e
β_p	0.7	0.9	0.4	0.5	0.11	to 1.70	e
β_ε	1.3	1.9	0.8	0.8	0.3	to 3.1	e
$\beta_p + \beta_\varepsilon$	2.1	2.4	1.3	1.7	0.5	to 4.6	e
V_A (km s^{-1})	50	24	50	46	30	to 100	e
C_s (km s^{-1})	63	15	59	61	41	to 91	e

*The coordinate system for magnetic field longitude, ϕ_B, and latitude, θ_B, is right handed with the x axis pointing radially away from the sun and the z axis pointing to the north ecliptic.

Data Source Codes: (a) Formisano et al., 1974 (c) Ness et al., 1971
(b) Hedgecock, 1975b (d) King, 1976
(e) The combined IMP 6 plasma (Los Alamos) and magnetic field (Goddard Space Flight Center) solar wind data between March 1971 and February 1972.

evaluated using the first year of combined IMP 6 plasma (Los Alamos) and magnetic field (Goddard Space Flight Center) solar-wind data (e.g., Feldman et al., 1976c). In this way, plasma vector and tensor quantities could be (and have been) deprojected from their measured two-dimensional values according to the simultaneously measured three-dimensional orientation of $\hat{B}$.

Before proceeding to a definition of the symbols, a note of caution is offered with respect to the values of the helium parameters in Table 7. In this tabulation, a more stringent set of acceptance criteria was imposed on the helium data than was imposed in the previous tabulations. In particular, data were accepted only if the parameters determined by a fully nonlinear analysis of the hydrogen and helium velocity distributions together were in agreement with corresponding parameters determined using an iterated linear analysis of the helium and hydrogen distributions separately (and used in the previous tabulations), within roughly the uncertainties specified in section II (see Asbridge et al., 1976). In this way, a greater accuracy on the accepted helium parameters is traded for a more restricted data set. Specifically, helium data in high-speed regions were preferentially rejected relative to those accepted for the previous tabulations. This explains in part why the average value for T_α/T_p in Table 7 is 3.8 whereas in Table 2 it is 4.9.

The temperature ratio parameters in Table 7 provide measures of the degree of pitch-angle isotropy and energy equipartition. Here the symbols for parallel and perpendicular refer to directions relative to the local magnetic field vector. In evaluating the relative helium/hydrogen velocity difference, $V_{\alpha p}$, a sign was affixed according to the direction of $V_{\alpha p}$ relative to the proton heat-flux vector, $\hat{Q}_p$. The dimensionless heat-flux vector, defined by $q = Q/[1.5NkT(kT_\parallel/m)^{1/2}]$ provides a measure of the third-moment distortion from symmetry along $\hat{B}$. The ratio of particle thermal pressure to the magnetic field pressure is given by $\beta = 8\pi NkT/B^2$. Tabulated values of the Alfvén speed $V_A = B/(4\pi N_p m_p)^{1/2}$ were evaluated using only the measured proton mass density instead of the total plasma mass density. The average value of V_A is therefore systematically high by about 10% ($<N_\alpha/N_p> \cong 5\%$). Lastly, in determining the sound speed $C_S = [k(T_e + 3T_p)/m_p]^{1/2}$, it was assumed that the electrons respond isothermally and the protons adiabatically in one dimension to a longitudinal sound wave.

Inspection of Table 7 reveals several facts of interest. Although on the average $\hat{B}$ is observed to point in the expected spiral direction ($\phi_B \sim -45°$, $\theta_B \cong 0°$), variations from the average are large. Furthermore, ϕ_B is observed to vary with solar wind bulk speed as predicted by Parker (1958) during structure-free conditions (Neugebauer, 1976b). As expected, the proton and alpha-particle thermal distributions are not generally isotropic. In addition, a limited analysis of contour levels of measured proton velocity distributions shows that often in high-speed streams, although the overall thermal anisotropy is small, the anisotropy of the peak of the proton distribution is such that $T_\perp \sim 2.4 T_\parallel$ on the average (Bame et al., 1975b). Solar-wind electrons are generally isotropic although, as shown in

Table 8, departures from equilibrium are observed for electrons with energies above about 50 eV. Energy is not equipartitioned among the various particle species. The helium component is, on the average, hotter than the electrons which, in turn, are generally hotter than the protons. The solar wind helium component generally moves faster than the protons in a direction aligned with $\hat{Q}_P$, which generally points away from the sun (Asbridge et al., 1976). Although on the average $V_{\alpha p}$ is less than V_A, it is largest in high-speed regions (Hirshberg, Asbridge, and Robbins, 1974; Neugebauer, 1976a; Asbridge et al., 1976). A preliminary analysis (Asbridge et al., 1976) of one particular high-speed stream in 1971 showed that variations in the three-dimensional helium/hydrogen velocity difference, $|U_{\alpha p}|$, were correlated with the Alfvén speed, V_A; although the scatter was large, the correlation could be represented by the relation $|U_{\alpha p}| \cong V_A - 26$. The magnitudes of both q_e and q_p are comparable and large. This fact is but another manifestation of the collisionless nature of the interplanetary plasma at 1 AU. In particular, q_p has been observed at times to exceed unity and to be highly correlated with the proton thermal anisotropy (Feldman et al., 1973a, b). These observations, coupled with extensive surveys of measured shapes of proton distributions, suggest an intimate dependence of the local proton thermal state on the phenomenon of interpenetrating proton streams (Feldman et al., 1973c, 1974). The origin of such streams is, at present, not known. Detailed properties of measured electron velocity distributions are summarized in Table 8 and are discussed in the next few paragraphs. The total plasma β appears to be enough greater than unity, on the average, that it seems unlikely that internal energy is partitioned equally between the plasma and magnetic field. A comparison of the average magnitudes of the Alfvén and sound speeds shows that they are comparable in the solar wind at 1 AU.

Present descriptions of solar-wind electrons are not as complete as those of ions,

TABLE 8
DETAILED INTERNAL STATE OF THE AVERAGE SOLAR WIND: ELECTRONS

PARAMETER	BIMAXWELLIAN MODEL		NUMERICAL MODEL	
	MEAN	σ	MEAN	σ
N_H/N	0.065	0.027	0.038	0.017
ΔV_H (km s^{-1})	689	369	1215	579
$(N_H \Delta V_H / N_c)$ (km s^{-1})	49	33	49	30
T_c (°K)	1.25×10^5	0.29×10^5	---	---
T_H (°K)	6.9×10^5	1.1×10^5	8.7×10^5	1.4×10^5
$(T_\parallel / T_\perp)_c$	1.08	0.08	---	---
$(T_\parallel / T_\perp)_H$	1.22	0.18	1.29	0.28
(T_H / T_c)	5.7	1.3	7.2	1.4
$(N_H T_H)/(N_c T_c)$	0.36	0.13	0.27	0.11
E_B (eV)	57	15	---	---

in part because the electrons are subsonic at 1 AU whereas the ions are supersonic. Only recently have the problems associated with electron measurements and their analysis begun to be understood. In fact there is still some disagreement over the interpretation of solar-wind electron observations, primarily because of possible effects of spacecraft charging upon the measurements and their interpretation. It is our opinion, however, that the description provided below adequately and correctly quantifies present information concerning the electron component of the solar wind. This description does provide agreement between electron and proton densities and bulk velocities (as required for zero space charge and zero current outflow from the sun). It also satisfies the expectation that relative velocities between the various parts of measured electron velocity distributions are well aligned with the simultaneously measured magnetic field. Two of many possible quantitative descriptions of measured solar-wind electron velocity distributions are summarized in the next two paragraphs.

Although electron velocity distributions are expected and measured to be more isotropic than the ion distributions, they, too, exhibit characteristics that betray their collisionless state at higher energies. It has been found that solar-wind electron distributions, f, can be adequately characterized by a superposition of two distributions, $f = f_c + f_H$ (Montgomery et al., 1968; Feldman et al., 1975). Here f_c, or the cold component, is to a good approximation a bimaxwellian function that provides an excellent fit to interplanetary electrons with energy $E \lesssim 50$ eV. The hot component, f_H, is what remains at higher energies when f_c is subtracted from f. Hot and cold components as defined above generally move relative to one another along the total distribution symmetry axis (which is aligned with $\hat{B}$) in such a way that no net current flows in the frame of reference moving with the protons, i.e., $N_c \Delta \bar{V}_c + N_H \Delta \bar{V}_H = 0$ within experimental errors. Here N_c and N_H are the densities of f_c and f_H, respectively, and $\Delta \bar{V}_c$ ($\Delta \bar{V}_H$) is the difference in velocity between the cold (hot) component and the total electron bulk velocity. This bulk velocity is observed to be equal to the solar wind proton bulk velocity to within about 50 km s^{-1}. A result of the above separation of electrons according to energy into hot and cold components is that the total electron heat flux is carried primarily by the convection of the hot electrons relative to the plasma bulk velocity along $\hat{B}$. The fact that variations in this convection velocity are observed at times to follow variations in the local Alfvén speed has been used as evidence that the interplanetary heat flux is locally regulated at least some of the time at 1 AU (Feldman et al., 1976b, c). Other parameters that describe the full velocity distribution are the temperatures of the hot and cold components (T_H and T_c, respectively) and the energy above which f begins to rise above f_c, E_B. The parameter E_B is therefore close to the energy at which $f_c = f_H$.

In order to quantify the above description, two models of interplanetary electron distributions were used to characterize those measured using the Los Alamos IMP plasma analyzers (Feldman et al., 1975). The first assumes that both

f_C and f_H are essentially bimaxwellian functions and the best-fit parameters are determined using the method of least squares. The second model cuts off the bimaxwellian interpolation of the hot-component distribution at energies lower than E_B, and the corresponding parameters are determined by numerical integration. Average parameters characterizing each of these models are summarized in Table 8 (see also Feldman *et al.*, 1975 for more details).

To conclude this section, the reader is cautioned for several reasons against using the set of average parameter values summarized in Tables 6, 7, and 8 for a generalized plasma stability analysis. Most important, because of resonant wave-particle interactions, the instability thresholds of many kinetic processes depend critically on the detailed shapes of simultaneously observed electron, proton, and alpha-particle velocity distributions. However the different quantitative parameters necessary for such an analysis are not presently known in sufficient detail for the ions and moreover are at times highly intercorrelated in such a way that not all the parameters are equal to their average values simultaneously. Nevertheless, the range of observed values may provide a basis for future experimental and theoretical studies.

D. HEAVY ION CONTENT OF THE SOLAR WIND

Although the solar wind consists primarily of protons, electrons, and alpha particles, it carries with it a trace of heavier elements. The several most abundant of these elements are sometimes measurable at 1 AU and provide information concerning both the physical state of the intermediate corona from which the solar wind originates and the dynamics of the coronal expansion.

A number of experimental methods have been used to study solar wind ion abundances. The most common of these is the measurement of energy per charge spectra, E/q, of the ions. Beginning with Mariner 2 (Neugebauer and Snyder, 1966), E/q measurements have shown two prominent groups of ions, interpreted to be hydrogen and helium ions with a nearly common bulk velocity. The helium abundance is quite variable. Vela and IMP determinations have shown that the abundance ranges as low as 0.2% and as high as 25%. The higher values are often associated with solar flare activity, but large variations occur at other times as well. High-resolution measurements of E/q spectra made with Vela analyzers during special times of low ion temperatures have shown additional multiple groups of ions above the helium peak. These groups are interpreted to be various ion species of oxygen, silicon, and iron, and significant variations in their abundances have been found in different solar-wind samples.

A possible ambiguity in the identification of the two most prominent ion groups was eliminated by using a combined crossed-field velocity selector and E/q analyzer. It was thus shown that the most prominent solar-wind ion group was composed of protons and the second, having a mass-to-charge ratio twice that of the first group, is most likely composed of other ionic species, probably alpha particles, as previously assumed (Ogilvie and Wilkerson, 1969).

Measurements of abundance ratios of the light noble gases and their isotopes have been made by Geiss and his coworkers (see Geiss, 1973, for a review). The ions were trapped in aluminum foils which were exposed on the moon during the Apollo missions and returned to earth. When the foils were heated the gases were released and analyzed with a mass spectrometer to give ratios of the isotopes of helium, neon, and argon. These results also show some variations in the abundances. Further measurements have been made of abundances of solar-wind ions implanted in lunar materials brought from the moon by Apollo. However, interpretations of these measurements are greatly complicated by variations in the loss rates of the various elements from the lunar materials.

A determination of the iron abundance during the Apollo 17 mission has been made by Zinner and his coworkers (1974). They have analyzed pits etched into mica samples which were exposed on the lunar surface. Only solar-wind iron and heavier ions register with this technique. The experiment determined the average flux of iron ions during the exposure. The iron flux combined with the average of the solar-wind proton flux, measured separately during the same period, gives the abundance. It was also determined that during the exposure the flux of ions heavier than iron was not anomalously high.

Average abundances from the Vela and Apollo measurements are shown in Table 9 and compared with abundance values for the most common elements in the solar system, photosphere, and corona. Considering the accuracies of the various determinations (roughly a factor of 2) the agreement is surprisingly good. However, as previously mentioned, there are variations of two orders of magnitude in the solar-wind helium abundance. Smaller variations over a range of three and more orders of magnitude occur in the abundances of oxygen, silicon, and iron. These results show that the solar atmosphere has an inhomogeneous composition, at least in the outer regions, which perhaps extends all the way into the transition region.

Because of its special importance, we add one last note concerning the solar wind helium abundance. The value of (N_α/N_p) listed in Table 9 is an average over many years of solar-wind observations. As mentioned above, the salient feature of this body of observations is the large measured variability of (N_α/N_p), which is especially marked at the lower bulk speeds. However, it has been shown above that values of the helium abundance are remarkably uniform during high-speed flow conditions. It is therefore tempting to suggest that a best estimate for (N_α/N_p) in the solar wind is 0.048 ± 0.005, the value observed in the structure-free high-speed solar wind.

Quantitative determinations of the thermal state of the intermediate corona can be made from the measured relative densities of the heavy-element ionization states observed in the solar wind at 1 AU (Hundhausen et al., 1968a, b). In this way, estimates of the temperature and temperature gradient in the corona in the

TABLE 9

SOLAR WIND AVERAGE ABUNDANCES AND COMPARISONS WITH ABUNDANCES OF THE MOST COMMON ELEMENTS IN THE SOLAR SYSTEM, PHOTOSPHERE, AND CORONA

ELEMENT (ISOTOPE)	SOLAR SYSTEM (a)	PHOTOSPHERE (b)	CORONA (b)	SOLAR WIND E/q SPECTRA (c)	SOLAR WIND MASS SPECTROMETER (d)	SOLAR WIND ETCH PITS (e)
H	1	1	1	1	1	---
He^3	9.0×10^{-6}(f)	---	---	---	1.7×10^{-5}	---
He^4	6.9×10^{-2}	---	---	4.0×10^{-2}(g)	4.0×10^{-2}(g)	---
C	3.7×10^{-4}	3.7×10^{-4}	5.6×10^{-4}	---	---	---
N	1.2×10^{-4}	1.2×10^{-4}	1.1×10^{-4}	---	---	---
O	6.8×10^{-4}	6.8×10^{-4}	5.6×10^{-4}	5.2×10^{-4}	---	---
Ne^{20}	9.8×10^{-5}	---	---	---	7.0×10^{-5}	---
Ne^{21}	3.0×10^{-7}	---	---	---	1.7×10^{-7}	---
Ne^{22}	1.2×10^{-5}	---	---	---	5.1×10^{-6}	---
Ne	1.1×10^{-4}	2.8×10^{-5}	3.5×10^{-5}	---	7.5×10^{-5}	---
Na	1.9×10^{-6}	1.7×10^{-6}	2.8×10^{-6}	---	---	---
Mg	3.3×10^{-5}	3.5×10^{-5}	4.5×10^{-5}	---	---	---
Al	2.7×10^{-6}	2.5×10^{-6}	2.8×10^{-6}	---	---	---
Si	3.1×10^{-5}	3.6×10^{-5}	4.5×10^{-5}	7.5×10^{-5}	---	---
S	1.6×10^{-5}	1.6×10^{-5}	1.4×10^{-5}	---	---	---
Ar	3.7×10^{-6}	4.5×10^{-6}	5.6×10^{-6}	---	3.0×10^{-6}(h)	---
Ca	2.3×10^{-6}	2.1×10^{-6}	2.5×10^{-6}	---	---	---
Fe	2.6×10^{-5}	2.5×10^{-5}	4.7×10^{-5}	5.3×10^{-5}	---	4.1×10^{-5}
Ni	1.5×10^{-6}	1.9×10^{-6}	3.4×10^{-6}	---	---	---

(a) Cameron, 1973

(b) Adopted values from Withbroe, 1971

(c) Bame et al., 1975a

(d) Values derived from ratios given by J. Geiss, 1973

(e) Zinner et al., 1974

(f) Terrestrial value of 0.00013 for He^3/He^4 used

(g) The He^4 abundance average of the 17 examples from which the E/q abundances were derived was .003. However, in a survey of experimental determinations Ogilvie and Hirshberg (1974) have found the average to vary between .035 and .045 over solar cycle 20, so a mean value of .040 has been adopted here, and used to convert the ratio data given by J. Geiss to abundance values.

(h) This value was derived from measurements of the abundance of Ar^{36} by assuming an isotopic abundance of 0.842 (Cameron, 1973).

range of heliocentric distances between 1.5 and 3.9 solar radii were made using heavy-ion data measured with the Vela 5 and 6 plasma analyzers at times of quiet solar wind (low speed, low temperature; see Bame et al., 1974). The following results were obtained. The average iron freezing-in temperatures ranged near 1.5×10^6 K. The measured ratio of observed oxygen and nitrogen ionization state densities indicated freezing-in temperatures of approximately 2.1×10^6 K. If the electron densities tabulated by Newkirk (1967) are applicable to the region in the corona where the low-speed solar wind originates, and if a spherically symmetric expansion is assumed, then the temperature gradient derived from the above temperature values is consistent with the value predicted for a heat-conduction-dominated, spherically symmetric corona, $T \propto r^{-2/7}$ (Chapman, 1957). However, since the derived gradient value depends on the validity of the above two assumptions, not much confidence should be placed in this result until a self-consistent model of the coronal expansion is found.

III. SOLAR WIND MICROSTRUCTURE

The interplanetary plasma appears basically noisy over a broad spectrum of wavelengths. By way of description, the term "microstructure" was first coined to denote those plasma and magnetic field variations observed, in a frame of reference stationary with respect to the sun, on time scales shorter than one hour (Burlaga and Ness, 1968; Burlaga, 1969). However, a comprehensive study of solar-wind variability measured with Mariner 5 instrumentation (Goldstein and Siscoe, 1972) showed that a natural break in the physics of solar-wind variability occurs on a time scale of about one day. Therefore it seems natural to consider separately those phenomena that have time scales longer than a day (the high-speed streams and large shock-wave disturbances of the solar wind described in sections V and VI) from those that have characteristic time scales shorter than a day (denoted here, as in Goldstein and Siscoe, by the term "microstructure").

To date, three categories of microstructure with time scales between tens of minutes and one day have been isolated. The first consists of constant total-pressure fluctuations convected away from the sun by the solar wind (Burlaga *et al.*, 1969; Belcher and Davis, 1971; Goldstein and Siscoe, 1972). Particularly large-amplitude examples of these waves (characterized by variations in density and temperature that are out of phase) are often observed between high-speed streams at 1 AU. The second category consists of large-amplitude, transverse velocity and magnetic field variations identified as Alfvén waves (Coleman, 1967; Unti and Neugebauer, 1968; Belcher *et al.*, 1969; Belcher and Davis, 1971; Goldstein and Siscoe, 1972) propagating away from the sun within the frame of reference of the solar wind. Wave amplitudes are observed to be largest ($\delta V \lesssim V_A$, $\delta B \lesssim B$) within high-speed regions at 1 AU. Lastly, because of the collisionless nature of the interplanetary plasma, longitudinal velocity variations (with respect to $\hat{B}$) in the solar wind probably develop into the interpenetrating streams of ions which, at 1 AU, appear to be convecting relative to each other along the magnetic field direction (Feldman *et al.*, 1973c, 1974; Asbridge *et al.*, 1974, 1976). They are most clearly resolved in velocity on the trailing edge of high-speed streams but appear to be present as unresolved proton velocity components throughout high-speed flow conditions (Feldman *et al.*, 1976b). The lower flux-density component usually has the higher speed, so the resultant third velocity moment, $\overline{Q}_p$, points away from the sun (Feldman *et al.*, 1974; Asbridge *et al.*, 1976).

On time scales shorter than 1 to 10 min, the observed microstructure consists predominantly of directional discontinuities (Ness *et al.*, 1966; Burlaga, 1969, 1972). This type of discontinuity is perceived as a sudden change in the orientation of the field and may or may not be accompanied by changes in the field strength or plasma properties. Events in which $\overline{B}$ is seen to rotate by more than 30° in less than 30 s occur, on the average, about once per hour (Burlaga, 1969, 1972). It is found that during low-speed flow conditions most of these structures are convected transverse pressure balances (tangential discontinuities), but during high-speed conditions roughly equal numbers of tangential discontinuities and sharply crested, propagating Alfvén disturbances are observed (Burlaga, 1972; Belcher, 1975; Belcher and Solodyna, 1975; Siscoe, 1974).

IV. LATITUDE VARIATIONS IN THE SOLAR WIND

There is reason to believe that the plasma outflow from the sun is not uniform at all latitudes. The most convincing evidence for a regular latitude variation in flow comes from the technique of interplanetary scintillation of radio sources. Although early studies by this technique provided conflicting results (e.g., Hewish, 1972), more recent observations (Coles, 1973; Coles and Maagoe, 1972; Coles *et al.*, 1974) indicate that the solar wind speed tends to be higher and more variable at high solar latitudes. The observations suggest that the average gradient away from the solar equatorial plane is about 2 km s^{-1} deg^{-1}. This gradient is sufficiently small that within the ecliptic plane (which is inclined approximately 7.5° to the solar equator) variations associated with changes in solar latitude would likely be difficult to detect. Nevertheless, long-term averages of the speed measured in the ecliptic plane during much of 1965-1967 were consistent with a speed gradient that increased away from the solar equatorial plane (e.g., Hundhausen *et al.*, 1971; Rhodes and Smith, 1975). However, the gradient derived from these observations is about 10 km s^{-1} deg^{-1}, considerably larger than that inferred from the scintillation measurements. Thus, either the gradient near the solar equator is larger than elsewhere, or the 1965-1967 observations are anomalous. Examination of data taken during other epochs should help settle the question. Our preliminary analysis of data obtained in the ecliptic plane after 1967 indeed suggests that the 1965-1967 period may be anomalous.

The dominant polarity of the interplanetary magnetic field (i.e., its direction toward or away from the sun) appears to depend on heliographic latitude. At latitudes above the equator the polarity is predominantly (~68% of the time at 7° north) that of the north polar region. At southern latitudes it is primarily that of the south polar region (Rosenberg and Coleman, 1969; Wilcox and Scherrer, 1972; Hedgecock, 1975c; Rosenberg, 1975).

The situation with regard to latitude effects associated with the polar (north-south) component of the interplanetary magnetic field is less clear. On the one hand, several analyses (e.g., Rosenberg, 1970; Coleman and Rosenberg, 1971; Rosenberg et al., 1973) have indicated that the polar component of the field is generally nonzero and on the average tends to be oriented in a manner consistent with a global flow pattern that diverges from the equatorial regions. Contrary to this result, the analysis of approximately 5.5 years of data led Hedgecock (1975b) to conclude that there was no evidence for such an effect.

On the other hand, a considerable body of evidence supports the idea that high-speed streams of solar wind originate in coronal holes (Krieger et al., 1973, 1974; Nolte et al., 1976; Sheeley et al., 1976). Since the polar regions are nearly always covered by coronal holes, it may be expected that on the average, the global flow pattern should converge towards, rather than diverge from, the equatorial regions (Hundhausen, McIntosh, and Wagner; open discussion at the NASA Skylab Workshop on Coronal Holes held in Boulder, Colorado, February 1976). Such a convergent flow pattern is consistent with the conclusion of Wagner (1976) that during 1972 and 1973, the observed solar-wind sector structure had its origin in alternate polar cap regions of the sun.

V. LARGE-SCALE VARIATIONS IN THE FLOW

Variations in solar wind bulk flow properties on time scales longer than a day are coherent (Goldstein and Siscoe, 1972) and organized into series of high-speed streams. Each solar wind stream is unique, yet almost all high-speed streams share common characteristics near 1 AU. Important aspects of stream profiles observed in the ecliptic at 1 AU (e.g., Gosling et al., 1972) include the rapid rise and slow decay of the speed and ion temperatures; the concentration of mass, momentum, and energy near the leading edge of the streams; the presence of a high-pressure ridge near the leading edge; and an east-west change in flow direction at the pressure ridge (Siscoe et al., 1969; Siscoe, 1972). These characteristic profiles result from the steepening of the streams in transit from the sun (Carovillano and Siscoe, 1969; Burlaga et al., 1971; Siscoe, 1972; Siscoe and Finley, 1972; Gosling et al., 1972; Matsuda and Sakurai, 1972; and Hundhausen, 1972b) and thus only indirectly reflect flow conditions closer to the sun.

Variations in electron properties across high-speed streams are not as well documented as are the proton properties. However, a preliminary study indicates that both the electron cold thermal core temperature and the hot-component temperature generally vary together, reaching a maximum between the density and velocity maxima and a minimum near the center of streams (Feldman et al., 1975, 1976a). The thermal anisotropy is generally largest within the high-speed regions, as is the ratio of hot-to-cold component densities and the difference in velocity between the hot component (as well as the core) and the total electron population.

The helium component of the solar wind has been found to vary in a characteristic fashion across solar wind streams (e.g., Hirshberg, Asbridge, and Robbins, 1974; Asbridge et al., 1976). The average helium/hydrogen abundance ratio increases on the average and becomes much less variable in going from low to high speeds, the relative velocity between helium and hydrogen increases almost linearly with bulk flow speed according to the empirical relation, $|V_{\alpha p}| = .08 V_p - 23$ (in km s^{-1}), and the helium/hydrogen temperature ratio reaches a minimum value of about 2 at the center of the particle-density peak. Some of these helium-stream effects can be attributed to the fact that coulomb collisions are important in the dense regions at the leading edges of streams but relatively unimportant elsewhere (Feldman, Asbridge, and Bame, 1974; Neugebauer, 1976a).

Variations in the interplanetary magnetic field strength, B, are also organized by the high-speed stream structure. Numerous studies have established that B maximizes almost simultaneously with the density at the leading edges of streams (e.g., Wilcox and Ness, 1965; Neugebauer and Snyder, 1967; Ness et al., 1971; Belcher and Davis, 1971; Burlaga et al., 1971). As with the bulk flow properties, this characteristic of the field is caused by the steepening of high-speed solar wind streams in interplanetary space.

The interplanetary magnetic field tends to point predominantly toward or away from the sun (along the basic Archimedian spiral) for several days at a time. This aspect of the interplanetary field is known as magnetic sector structure (Wilcox and Ness, 1965). There are from two to four sectors per solar rotation (Svalgaard and Wilcox, 1975; Hedgecock, 1975b). Magnetic sectors are extremely persistent features of the interplanetary solar wind at 1 AU, and often recur at the same solar longitude for many solar rotations (Svalgaard and Wilcox, 1975). There is a close association between magnetic sector structure and solar wind streams. Boundaries between magnetic sectors almost always occur in the low-speed gas between streams; that is, high-speed streams are almost always unipolar (Wilcox and Ness, 1965; Coleman et al., 1966; Gosling et al., 1976). However, successive streams often have identical, rather than opposite, polarities. Further, high-speed streams are not nearly as stable as the magnetic sectors within which they are imbedded. Many streams do not endure for as long as one solar rotation; those that do endure often vary significantly in shape, amplitude, and Carrington longitude from one rotation to the next (Gosling and Bame, 1972; Gosling et al., 1972; Gosling et al., 1976). The recurrence period of stable stream structures is close to

27.13 days, comparable to most estimates of the rotation period of equatorial regions on the sun.

VI. INTERPLANETARY SHOCK WAVE DISTURBANCES

Shock fronts represent a special class of discontinuity that propagate through the solar wind plasma. At 1 AU they occur about once a month on the average. Association of shocks with solar activity is difficult and subject to uncertainties. However, more than half of all shocks observed near 1 AU can be reasonably associated with large solar flares accompanied by Type II and Type IV radio bursts (Hundhausen, 1972b; Chao and Lepping, 1974). Most (but not all) interplanetary shocks are of intermediate strength. The dynamical properties of the typical interplanetary shock have been tabulated by Hundhausen (1972c) and are presented in Table 10. Occasionally (once every two years or so) much stronger shocks involving speed jumps of 200 or 300 km s^{-1} are observed (e.g., Dryer et al., 1975; Gosling et al., 1975). More complete tabulations of observed shocks can be found elsewhere (Hundhausen, 1970; Chao and Lepping, 1974).

Because most interplanetary shocks are not strong, the motion of a shock through interplanetary space is largely the result of the outward flow of the solar wind plasma rather than the propagation of the shock relative to the plasma. Thus shock geometries become controlled and distorted by solar wind stream structure (e.g., Heinemann and Siscoe, 1974; Hirshberg, Nakagawa, and Wellck, 1974). The result is that, although most interplanetary shocks observed in the earth probably originate within ±45° of the central meridian of the sun, shock normals exhibit a wide range of directions (e.g., Taylor, 1969; Bavassano et al., 1973, Chao and Lepping,

TABLE 10

DYNAMICAL PROPERTIES OF THE TYPICAL INTERPLANETARY SHOCK OBSERVED NEAR 1 AU*

Flow Speed of Preshock Plasma	390 km s^{-1}
Flow Speed of Postshock Plasma	470 km s^{-1}
Shock Propagation Speed Relative to Stationary Observer	500 km s^{-1}
Shock Propagation Speed Relative to Ambient Solar Wind	110 km s^{-1}
Mach Number (Sonic or Alfvén)	2 to 3
Transit Time from the Sun	55 hr.

*Table 1 of Hundhausen, 1972c

1974) and cannot be used reliably to ascertain the origin of any particular disturbance.

The flow pattern behind interplanetary shocks varies considerably from event to event. At least two characteristic patterns have been identified (Hundhausen *et al.*, 1970)—one in which the speed and density continue to rise for many hours after the passage of the shock and another in which those parameters decrease steadily following the abrupt increases at the shock itself. It has been suggested that these flow patterns correspond to "driven" and "blast" waves, respectively. Another aspect of some, but certainly not all, shock-wave disturbances is the brief appearance of plasma unusually rich in helium (~10-15% by number relative to hydrogen) some 5-12 h after the shock passage (Gosling *et al.*, 1967; Bame *et al.*, 1968; Ogilvie, Burlaga, and Wilkerson, 1968; Lazarus and Binsack, 1969; Hirshberg *et al.*, 1970; Hirshberg, 1972; Hirshberg, Asbridge, and Robbins, 1971; Hirshberg, Bame, and Robbins, 1972). It has been suggested that this helium-rich plasma identifies gas ejected from the lower corona or chromosphere coincident with large solar flares. Following, or simultaneous with, helium enrichments the kinetic temperature of the solar wind plasma reaches anomalously low values (~6 $\times$ 10^4 K for electrons, ~2 $\times$ 10^4 K for protons). These low kinetic temperatures suggest the presence of magnetic loops in interplanetary space either closed upon themselves or rooted in the sun (Gosling *et al.*, 1973; Montgomery *et al.*, 1974). However, this field geometry has not yet been confirmed (or denied) by direct measurements of the interplanetary field.

VII. LONG-TERM SOLAR WIND VARIATIONS

The visible sun has been observed to undergo characteristic variations in recent times with a fundamental 11-y half-period. A full description of the different manifestations of this variability is given elsewhere in this volume (Eddy, in Chapter II). Basically, though, the most conspicuous change in the appearance of the solar surface concerns the number of visible sunspots, chromospheric plages, and solar flares, which can be together called *solar activity;* the greater the relative number of spots, plages, and flares, the greater the activity.

There are several established variations in solar wind structure observed in association with the 11-y solar activity cycle. Variations in the number of sudden commencements of geomagnetic storms (and hence the number of interplanetary shock disturbances, if the results of Chao and Lepping are generally valid) follow variations in the smoothed sunspot number (Mayaud, 1975). Solar-wind stream structure varies characteristically in association with the 11-y activity cycle (Bame *et al.*, 1976; Gosling *et al.*, 1976). Large-amplitude, high-speed solar wind streams and streams with maximum speeds in excess of 700 km s^{-1} are far more common in years of declining and minimum solar activity than near solar maximum. Both the low-frequency power density in transverse magnetic field fluctuations and the correlation length of the interplanetary magnetic field vary in phase with the solar activity cycle (Hedgecock, 1975c). It should be noted that the latter two long-term interplanetary variations have been established for solar cycle 20 only.

It has been suggested that the proton density of the solar wind (Egidi et al., 1969; Diodato et al., 1974) and helium abundance (Robbins et al., 1970; Ogilvie and Hirshberg, 1974) also vary characteristically with phase during the solar cycle. Whereas the proton density appears to be lowest at solar maximum, the helium abundance appears to be highest. However, because the magnitudes of the two effects are not larger than possible systematic measurement uncertainties, and because both studies were based on insufficiently long time intervals, we believe that although characteristic long-term proton density and helium abundance variations are possible, they are, at present, only marginally established.

Conspicuous for the absence of significant long-term variability are the magnitude of the interplanetary magnetic field, B (Hedgecock, 1975c; King, 1976), and the solar-wind sector structure (Svalgaard and Wilcox, 1975). However, it should be noted that although the average magnitude of B remains constant in time, as solar activity increases the distribution of B changes in such a way that a decrease in the most probable magnitude of B is just compensated for by an increase in the number of observed high field values. Similarly, although the phases of the interplanetary magnetic sectors appear continuous between 1926 and 1973, small characteristic variations in the sector recurrence period (between 27 and 28.5 days) are observed in association with the phase of the solar activity cycle. No similar variation in the recurrence period of high-speed streams has been observed during the latest activity cycle (solar cycle 20).

VIII. SUMMARY

In the first part of this chapter, various characteristics of the interplanetary plasma and magnetic field observed directly at 1 AU were summarized in tabular form. An extensive and unified set of solar wind data was organized with the specific aim of being useful for research workers active in the fields of magnetospheric, atmospheric, solar, interplanetary and plasma physics. The presentation included a statistical description of *in situ* solar wind observations between March 1971 and July 1974, summarized in tables 1, 2, and 3. Although the tabulated parameters may not be truly representative of the state of the interplanetary medium at 1 AU averaged over a long time period (compared to the 11-y solar activity cycle), the tables provide the most complete and homogeneous description presently available. An attempt was made to define and characterize the structure-free state of the solar wind in tables 4 and 5. This state should be most directly applicable to theoretical models of the steady-state coronal expansion. Since the parameters of high-speed flows form a much more uniform ensemble of interplanetary conditions than do those of low-speed flows, we have tentatively identified the structure-free interplanetary state with the high-speed solar wind. This identification is consistent with the suggestion that high-speed flows originate in large regions of the solar corona where strong, closed magnetic field structures are absent. Our best estimate, at present, of the specific energy loss to the solar wind from that portion of the corona which expands into interplanetary space is 5×10^5 ergs cm^{-2} s^{-1} (±~30%). The accompanying particle flux is 1×10^{14} protons cm^{-2} s^{-1} (±30%). A statistical description of the internal state of the average solar wind of importance to plasma physics was given in summary in tables 6, 7, and 8. Many aspects of this state betray the collisionless nature of the interplanetary plasma. Average heavy-ion abundances in the solar wind were compared in Table 9 with their respective solar system, photospheric, and coronal values. No differences in average abundances have been observed that are larger than experimental uncertainties (roughly a factor of 2). Of special importance was the helium abundance. Since values of the helium abundance during high-speed flow conditions are remarkably uniform, it was tempting to suggest that a best estimate for (N_α/N_p) in the solar wind is 0.048 ± 0.005.

In the last half of this review we discussed various types of observed solar wind variability. This variability, on all time scales, is perhaps the most salient feature of the interplanetary plasma. Particularly important for the purposes of this volume are those variations observed in association with the 11-y (half-period) solar activity cycle. Three clear associations with the solar activity cycle have been made. Variations in the number of sudden commencements of geomagnetic storms follow variations in the smoothed sunspot number (Mayaud, 1975), suggesting more interplanetary shock waves during solar maximum. Large-amplitude, high-speed streams are more pronounced during years of declining and minimum solar activity than during years near solar maximum. Finally, the amplitude of low-frequency transverse magnetic field fluctuations varies in phase with the solar sunspot cycle.

ACKNOWLEDGMENTS

We wish to thank A. J. Hundhausen and B. Abraham-Shrauner for many useful discussions and suggestions and N. F. Ness and D. H. Fairfield for the use of their IMP 6 magnetic field data. This work was performed under the auspices of the U.S. Energy Research and Development Administration and supported in part by NASA.

REFERENCES

Asbridge, J.R., Bame, S.J., and Feldman, W.C., 1974, *Solar Phys. 37*, 451.

_____; _____; _____; and Montgomery, M.D., 1976, *J. Geophys. Res. 81*, 2719.

Bame, S.J.; Asbridge, J.R.; Feldman, W.C.; and Gosling, J.T., 1976, *Ap. J. 207*, 977.

_____; _____; _____; and Kearney, P.D., 1974, *Solar Phys. 35*, 137.

_____; _____; _____; and _____, 1975a, *ibid. 43*, 463.

_____; _____; Hundhausen, A.J.; and Strong, I.B., 1968, *J. Geophys. Res. 73*, 5761.

_____; _____; Feldman, W.C.; Gary, S.P.; and Montgomery, M.D., 1975b, *Geophys. Res. Letters 2*, 373.

Bavassano, R., Mariani, F., and Ness, N.F., 1973, *J. Geophys. Res. 78*, 4535.

Belcher, J.W., 1975, *ibid. 80*, 4713.

_____ and Davis, L., Jr., 1971, *ibid. 76*, 3534.

_____ and Solodyna, C.V., 1975, *ibid. 80*, 181.

_____, Davis, L., Jr., and Smith, E.J., 1969, *ibid. 74*, 2302.

Bohlin, J.D., 1976, *AGU International Symposium on Solar-Terrestrial Physics, Boulder, Colorado*, June 7-18, in press.

Brandt, J.C., and Heise, J., 1970, *Ap. J. 159*, 1057.

Burlaga, L.F., 1969, *Solar Phys. 7*, 54.

_____, 1972, *Solar Wind*, ed. P.J. Coleman, Jr., C.P. Sonett, and J.M. Wilcox, NASA SP-308, 309.

_____ and Ness, N.F., 1968, *Canadian J. Phys. 46*, 5962.

_____ and Ogilvie, K.W., 1970, *Ap. J. 159*, 659.

_____ and _____, 1973, *J. Geophys. Res. 78*, 2028.

_____, _____, and Fairfield, D.H., 1969, *Ap. J. (Letters) 155*, L171.

_____; _____; _____; Montgomery, M.D.; and Bame, S.J., 1971, *Ap. J. 164*, 137.

Cameron, A.G.W., 1973, *Space Sci. Rev. 15*, 121.

Carovillano, R.L., and Siscoe, G.L., 1969, *Solar Phys. 8*, 401.

Chao, J.K., and Lepping, R.P., 1974, *J. Geophys. Res. 79*, 1799.

Chapman, S., 1957, *Smithsonian Contrib. Ap. 2*, 1.

Coleman, P.J., Jr., 1967, *Planet. Space Sci. 15*, 953.

V SOLAR PLASMA AND ENERGETIC PARTICLES

———————— and Rosenberg, R.L., 1971, *J. Geophys. Res. 76*, 2917.

————————; Davis, L.; Smith, E.J.; and Jones, D.E., 1966, *ibid. 71*, 2831.

Coon, J.H., 1966, *Radiation Trapped in the Earth's Magnetic Field*, ed. B.M. McCormac, D. Reidel, Dordrecht, 231.

Coles, W.A., 1973, *Proceedings of Solar Terrestrial Relations Conference*, University of Calgary, Calgary, 653.

———————— and Maagoe, S., 1972, *J. Geophys. Res. 77*, 5622.

————————, Rickett, B.J., and Rumsey, V.H., 1974, *Solar Wind Three*, ed. C.T. Russell, Institute of Geophysical and Planetary Physics, University of California, Los Angeles, 351.

Diodato, L., Moreno, G., and Signorini, C., 1974, *J. Geophys. Res. 79*, 5095.

Dryer, M.; Eviatar, A.; Fröhlich, A.; Jacobs, A.; Joseph, J.; and Weber, E., 1975, *J. Geophys. Res. 80*, 2001.

Egidi, A., Pizzella, G., and Signorini, C., 1969, *ibid. 74*, 2807.

Feldman, W.C., Asbridge, J.R., and Bame, S.J., 1974, *ibid. 79*, 2319.

————————; ————————; ————————; and Gosling, J.T., 1976a, *ibid.*, in press.

————————; ————————; ————————; and Lewis, H.R., 1973a, *Phys. Rev. Letters 30*, 271.

————————; ————————; ————————; and Montgomery, M.D., 1973b, *J. Geophys. Res. 78*, 6451.

————————; ————————; ————————; and ————————, 1973c, *ibid. 78*, 2017.

————————; ————————; ————————; and ————————, 1974, *Rev. Geophys. and Space Phys. 12*, 715.

————————; ————————; ————————; Gary, S.P.; and Montgomery, M.D., 1976b, *J. Geophys. Res. 81*, 2377.

————————; ————————; ————————; Montgomery, M.D.; and Gary, S.P., 1975, *ibid. 80*, 4181.

————————; ————————; ————————; Gary, S.P.; Montgomery, M.D.; and Zink, S.M., 1976c, *ibid.*, in press.

Feynman, J., 1975, *Solar Phys. 43*, 249.

Formisano, V., Moreno, G., and Amata, E., 1974, *J. Geophys. Res. 79*, 5109.

Geiss, J., 1973, *Proceedings of the 13th International Cosmic Ray Conference*, Vol. 5, University of Denver, Denver.

Goldstein, B., and Siscoe, G.L., 1972, *Solar Wind*, ed. P.J. Coleman, Jr., C.P. Sonett, and J.M. Wilcox, NASA SP-308, 506.

Gosling, J.T., and Bame, S.J., 1972, *J. Geophys. Res. 77*, 12.

————————, Pizzo, V., and Bame, S.J., 1973, *ibid. 78*, 2001.

————————; Asbridge, J.R.; Bame, S.J.; and Feldman, W.C., 1976, *ibid.*, in press.

_____; Hundhausen, A.J.; Pizzo, V.; and Asbridge, J.R., 1972, *ibid.* 77, 5442.
_____; Asbridge, J.R.; Bame, S.J.; Hundhausen, A.J.; and Strong, I.B., 1967, *ibid.* 72, 1813.
_____; Hildner, E.; MacQueen, R.M.; Munro, R.H.; Poland, A.I.; and Ross, C.L., 1975, *Solar Phys. 40*, 439.
Hedgecock, P.C., 1975a, *Space Sci. Instr. 1*, 83.
_____, 1975b, *Solar Phys. 42*, 497.
_____, 1975c, *ibid. 44*, 205.
Heinemann, M.A., and Siscoe, G.L., 1974, *J. Geophys. Res. 79*, 1349.
Hewish, A., 1972, *Solar Wind*, ed. P.J. Coleman, Jr., C.P. Sonett, and J.M. Wilcox, NASA SP-308, 477.
Hirshberg, J., 1972, *ibid.*, 582.
_____, Asbridge, J.R., and Robbins, D.E., 1971, *Solar Phys. 18*, 313.
_____, _____, and _____, 1972, *J. Geophys. Res. 77*, 3583.
_____, _____, and _____, 1974, *ibid. 79*, 934.
_____, Bame, S.J., and Robbins, D.E., 1972, *Solar Phys. 23*, 467.
_____, Nakagawa, Y., and Wellck, R.E., 1974, *J. Geophys. Res. 79*, 3726.
_____; Alksne, A.; Colburn, D.S.; Bame, S.J.; and Hundhausen, A.J., 1970, *ibid.* 75, 1.
Hundhausen, A.J., 1968, *Space Sci. Rev. 8*, 690.
_____, 1970, *Rev. Geophys. and Space Phys. 8*, 729.
_____, 1972a, *Solar Wind*, ed. P.J. Coleman, Jr., C.P. Sonett, and J.M. Wilcox, NASA SP-308, 261.
_____, 1972b, *Coronal Expansion and Solar Wind*, Springer-Verlag, New York.
_____, 1972c, *Solar Wind*, ed. P.J. Coleman, Jr., C.P. Sonett, and J.M. Wilcox, NASA SP-308, 393.
_____, Bame, S.J., and Montgomery, M.D., 1971, *J. Geophys. Res. 76*, 5145.
_____, Gilbert, H.E., and Bame, S.J., 1968a, *Ap. J. (Letters) 152*, L3.
_____, _____, and _____, 1968b, *J. Geophys. Res. 73*, 5485.
_____; Bame, S.J.; Asbridge, J.R.; and Sydoriak, S.J., 1970, *ibid. 75*, 4643.
King, J.H., 1975, *Interplanetary Magnetic Field Data Book*, NASA-TM-X-72544 (NSSDC 75-04).
_____, 1976, *J. Geophys. Res. 81*, 653.
Krieger, A.S., Timothy, A.F., and Roelof, E.C., 1973, *Solar Phys. 29*, 505.
_____; _____; Vaiana, G.S.; Lazarus, A.J.; and Sullivan, J.D., 1974, *Solar Wind Three*, ed. C.T. Russell, Institute of Geophysical and Planetary Physics, University of California, Los Angeles, 132.
Lazarus, A.J., and Binsack, J.H., 1969, *Ann. IQSY 3*, 378.
_____ and Goldstein, B.E., 1971, *Ap. J. 168*, 571.
Matsuda, T., and Sakurai, T., 1972, *Cosmic Electrodyn. 3*, 97.
Mayaud, P.N., 1975, *J. Geophys. Res. 80*, 111.
Mihalov, J.D., and Wolfe, J.H., 1971, *Cosmic Electrodyn. 2*, 32b.
Montgomery, M.D., Bame, S.J., and Hundhausen, A.J., 1968, *ibid. 73*, 4999.

————————; Asbridge, J.R.; Bame, S.J.; and Feldman, W.C., 1974, *ibid. 79*, 3103.
Moreno, G., and Palmiotto, F., 1973, *Solar Phys. 30*, 207.
————————, and Signorini, C., 1973, *ELDO-CEDES/ESRO-CERS Scient. and Tech. Rev. 5*, 401.
Ness, N.F., 1970, *Space Sci. Rev. 11*, 459.
————————, Hundhausen, A.J., and Bame, S.J., 1971, *J. Geophys. Res. 76*, 6643.
————————, Scearce, C.S., and Cantarano, S., 1966, *ibid. 71*, 3305.
Neugebauer, M., 1976a, *ibid. 81*, 78.
————————, 1976b, *ibid. 81*, 4664.
———————— and Snyder, C.W., 1966, *ibid. 71*, 4469.
———————— and ————————, 1967, *J. Geophys. Res. 72*, 1823.
Newkirk, G., Jr., 1967, *Annu. Rev. Astr. and Ap. 5*, 213.
Nolte, J.T.; Krieger, A.S.; Timothy, A.F.; Gold, R.E.; Roelof, E.C.; Vaiana, G.; Lazarus, A.J.; Sullivan, J.D.; and McIntosh, P.S., 1976, *Solar Phys.*, in press.
Ogilvie, K.W., 1972, *J. Geophys. Res. 77*, 3957.
———————— and Hirshberg, J., 1974, *ibid. 79*, 4595.
———————— and Wilkerson, T.D., 1969, *Solar Phys. 8*, 435.
————————, Burlaga, L.F., and Wilkerson, T.D., 1968, *J. Geophys. Res. 73*, 6809.
————————, McIlwraith, N., and Wilkerson, T.D., 1968, *Rev. Sci. Instr. 39*, 441.
Parker, E.N., 1958, *Ap. J. 128*, 664.
Pizzo, V.; Gosling, J.T.; Hundhausen A.J.; and Bame, S.J., 1973, *J. Geophys. Res. 78*, 6469.
Rhodes, E.J., Jr., and Smith, E.J., 1975, *ibid. 80*, 917.
Robbins, D.E., Hundhausen, A.J., and Bame, S.J., 1970, *ibid. 75*, 1178.
Rosenberg, R.L., 1970, *J. Geophys. Res. 75*, 5310.
————————, 1975, *ibid. 80*, 1339.
———————— and Coleman, P.J., Jr., 1969, *ibid. 74*, 5611.
————————, ————————, and Ness, N.F., 1973, *ibid. 78*, 51.
Sheeley, N.R., Jr., Harvey, J.W., and Feldman, W.C., 1976, *Solar Phys.*, in press.
Siscoe, G.L., 1972, *J. Geophys. Res. 77*, 27.
————————, 1974, *Solar Wind Three*, ed. C.T. Russell, Institute of Geophysical and Planetary Physics, University of California, Los Angeles, 151.
———————— and Finley, L.T., 1972, *J. Geophys. Res. 77*, 35.
————————, Goldstein, B., and Lazarus, A.J., 1969, *ibid. 74*, 1759.
Snyder, C.W., Neugebauer, M., and Rao, V.R., 1963, *ibid. 68*, 6361.
Spitzer, L., Jr., 1956, *The Physics of Fully Ionized Gases*, Interscience, New York.
Strong, I.B.; Asbridge, J.R.; Bame, S.J.; and Hundhausen, A.J., 1967, *Zodiacal Light and Interplanetary Medium*, ed. J. Weinberg, NASA SP-150, 365.
————————; ————————; ————————; Heckman, H.H.; and Hundhausen, A.J., 1966, *Phys. Rev. Letters 16*, 631.
Svalgaard, L., and Wilcox, J.M., 1975, *Solar Phys. 41*, 461.
Taylor, H.E., 1969, *ibid. 6*, 320.
Unti, T.W.J., and Neugebauer, M., 1968, *Phys. Fluids 11*, 563.
Vasyliunas, V.M., 1971, *Methods Exp. Phys. 9*, 49.

Wagner, W., 1976, *Ap. J. 206*, 583.
Wilcox, J.M., and Ness, N.F., 1965, *J. Geophys. Res. 70*, 5793.
—————— and Scherrer, P.H., 1972, *ibid. 77*, 5385.
Withbroe, G.L., 1971, *The Menzel Symposium on Solar Physics, Atomic Spectra and Gaseous Nebulae*, ed. K.B. Gebbie, NBS Spec. Pub. 353, 127.
Wolfe, J.H., 1972, *Solar Wind*, ed. P.J. Coleman, Jr., C.P. Sonett, and J.M. Wilcox, NASA SP-308, 170.
Zinner, E.; Walker, R.M.; Borg, J.; and Maurette, M., 1974, *Proceedings of the Fifth Lunar Science Conference, Vol. 3*, ed. W.A. Gose, Pergamon, Elmsford, N.Y., 2975.

Comment

Magnetic field variations observed at the surface of the earth are produced, in part, by ionospheric and magnetospheric currents associated with the interaction between the solar wind and the magnetosphere. It is thus tempting to suggest that something can be learned about the solar wind through the study of geomagnetic indices, which are frequently used to represent geomagnetic variations (see below). Any information about the solar wind gained from such a study would be especially useful because direct observations of the solar wind have been available only for about fifteen years, whereas geomagnetic data have been obtained for more than a century. However, to acquire such information it is generally necessary to have a description of the important physical processes occurring in the solar wind/ magnetosphere/ionosphere system that represents them far better than they are currently understood.

A geomagnetic observatory measures the vector magnetic field and its variations at the observatory location. Geomagnetic indices, which presumably reflect the effects of ionospheric and magnetospheric currents, are constructed by taking the observed components of the geomagnetic field, making corrections for local conditions at the observing site (and in some cases for regular variations of the field), and averaging the corrected values over time and over space (i.e., over several observatories). Different corrections and different averaging techniques lead to a multiplicity of geomagnetic indices which are not *simply* related to any of the physical processes that are important in the solar wind/ magnetosphere/ionosphere system.

To illustrate some of the problems involved in drawing inferences about the solar wind from geomagnetic indices, let us consider the especially simple case of the Dst index. Apparently, Dst provides a good measure of the strength of the ring current, which is a magnetospheric current carried by energetic particles drifting around the earth at an altitude of a few earth radii. The ring current strength is decreased in time by atmospheric precipitation and charge exchange of energetic particles and is increased in time by the injection of energetic particles from the outer magnetosphere. The injection process is directly related to the mass and energy transfer from the solar wind to the magnetosphere that is associated with field-line reconnection at the magnetopause. In turn, reconnection is apparently associated with the bulk-flow energy density of the solar wind and with the direction of the interplanetary magnetic field. An increase in the solar wind bulk-flow energy density or in the magnitude of the southward component of the interplanetary magnetic field will, on a time scale determined by the state of the magnetosphere, tend to increase the ring current strength. Thus, though it may be possible to predict (at least qualitatively) changes in Dst from a knowledge of solar wind parameters (e.g., Russell, 1974), it is virtually impossible to reverse the process and deduce solar wind properties from a knowledge of Dst.

The physical processes associated with geomagnetic indices other than Dst are generally much more complex, so the problem of deducing solar wind parameters from these indices would appear to be a hopeless task at the present time. Of course, certain very general properties of the solar wind might in some cases be inferred from the indices. For example, the appearance of an 11-y cycle in certain indices may imply that at least some solar wind parameters have a solar-cycle dependence. Yet even in this case, it is entirely possible that the solar cycle dependence of a particular geomagnetic index can be explained without invoking any variation of solar wind parameters.

Readers interested in geomagnetic indices and the physical processes with which they are related are referred to the basic text by Chapman and Bartels (1940), the compendium edited by Matsushita and Campbell (1967), and the recent review paper by Russell (1974).

Thomas E. Holzer
High Altitude Observatory

References

Chapman, S., and Bartels, J., 1940, *Geomagnetism*, Clarendon Press, Oxford.

Matsushita, S., and Campbell, W.H., 1967, *Physics of Geomagnetic Phenomena*, Academic Press, New York.

Russell, C.T., 1974, in *Correlated Interplanetary and Magnetospheric Observations*, ed. D.E. Page, D. Reidel, Dordrecht.

MEASURES OF ENERGETIC PARTICLES FROM THE SUN

L.J. Lanzerotti
Bell Laboratories
Murray Hill, New Jersey

The purpose of this review is to discuss the character and nature of particle fluxes from solar flare events that have been measured at one astronomical unit (1 AU). That the sun occasionally emits elementary particles (protons) of relativistic energies was only discovered about three solar cycles ago when Lange and Forbush (1942) and Forbush (1946) reported the appearance of anomalously high count rates in ionization chambers at ground level. It is interesting to note that the 1942 solar events from which energetic particles were first discovered were also the events in which solar radio bursts were first observed by British army groups studying the problems of radar jamming during the defense of Britain (Hey, 1973).

Although solar flare particles have only been observed for about three solar cycles, it became evident quite early after Forbush's reports that the intensity or number density (or both) of energetic flare particles varies greatly from flare to flare. Thus, it is not possible to define a rigorous "measure" of energetic particles from the sun that is independent of flare configuration or of the time period within the solar cycle. That is, although much work is being done, we do not yet know enough about the mechanisms of production of the energetic particles to predict expected particle levels from a particular flare size. Consequently, this chapter presents the particle flux data at 1 AU from various flares that have occurred during the past three solar cycles. It has only been during the last (twentieth) solar cycle that continuous measurements of solar particles at very low energies ($\lesssim 10$ MeV/nucleon) have been possible outside the earth's magnetosphere on spacecraft. The data from cycle 19 are mainly inferential, obtained from measurements of ionization in the polar regions produced by solar particles incident on the earth at high latitudes.

In addition to the solar cosmic ray events, whose frequency depends somewhat on the solar cycle, the galactic cosmic rays incident on the earth are also related to the solar cycle. It appears that some mechanism controlled by the sun produces a modulation of these cosmic rays, more or less in phase with the solar cycle. Thus, we devote a small portion of this chapter to discussions of the dependence of the galactic cosmic ray flux on the solar cycle.

1. SOLAR PARTICLE EVENTS

The duration of a solar particle event depends on the energy of the particles observed. A solar particle event also has a duration and a time-intensity profile that are somewhat dependent on the location of the particle-producing flare on the sun and on the propagation conditions in the interplanetary medium. Furthermore, the number of events that might be classified as solar particle events is strongly correlated to the detected particle energies and therefore is a function of the available measurement technology.

The dependence of the time-intensity profile for a solar particle event on particle energy as well as on propagation conditions is schematically illustrated in Figure 1 for an idealized event with a flare-produced interplanetary shock wave accompanying the particles (part A) and for an event without an accompanying interplanetary shock wave (part B). For illustrative purposes this figure was drawn for a solar event occurring on the western hemisphere of the sun; the event accelerated particles to relativistic energies. As is discussed below, only 23 relativistic solar

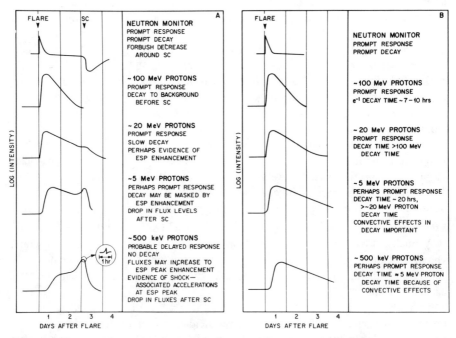

Fig. 1 (A): Schematic illustration of proton fluxes that might be measured at 1 AU after a western hemisphere flare when a flare-produced shock wave is observed at the earth. (B): Schematic illustration of proton fluxes that might be measured at 1 AU after a western hemisphere flare if no flare-produced shock wave were observed at the earth.

particle events have been observed since Forbush's discovery in 1942. Although the particle time-intensity profiles in this figure are schematic and not representative with respect to intensities, nevertheless the profiles do provide an indication of the variability in solar particle events. The onset times of the various particle energies exhibit a velocity dispersion appropriate for the energy being detected.

The time-intensity profiles show that relativistic particle enhancements may last for only a few tens of minutes to a few hours after the onset of the flare. However, not all relativistic particle events have a prompt response after the flare maximum (usually taken as the Hα brightness maximum); some flare events that are of longer duration take a longer time to build up to a maximum intensity (Pomerantz and Duggal, 1974a). The greatest variability in the time duration and intensity of solar particles occurs at the lowest energies. For discrete particle events the particle fluxes of lowest energy ($\lesssim$ 1 MeV) tend to increase in intensity until arrival of the interplanetary shock wave. (This shock wave is usually the cause of geomagnetic storm disturbances in the earth's magnetosphere.)

It is a rather straightforward matter to determine the number of discrete relativistic solar particle events; with the detection technology and spacecraft capabilities of the last solar cycle, it is now easy to determine the number of events in which particles with energies of at least 10 MeV were created. But the determination of the number of discrete low-energy ($<$ a few MeV) solar particle events is quite ambiguous; the sun has been found to be a more or less steady emitter of particles at these energies (Fan *et al.*, 1968).

An example of the variability in the time-intensity profiles of solar flare particles with respect to particle energy and time is shown in Figure 2, using data for the month of March 1970 (Lanzerotti, 1972). These data, obtained in interplanetary space by the proton monitoring experiment on *Explorer 41*, show that as the energy of the detected particle decreases, the number of interplanetary proton enhancements increases and the complexity of the time-intensity profiles greatly increases. For example, if only the high-energy channel ($>$ 60 MeV) were available for analysis, only one large event (March 27) and a small event (March 24) would have been apparent. Although the decay times are somewhat longer, both of these events had a temporal profile similar to that shown in Figure 1. As the detected particle energy decreases, more solar events are observed. The solar fluxes in the 1 to 10 MeV channel of Figure 2 are observed to remain above their background level throughout the 31 d period. At the lowest energy, the great amount of structure observed in the time-intensity profiles precludes the correlation of each particle enhancement with a discrete flare event.

Three other basic classifications of solar particle enhancements, in addition to the more classical flare-associated events, have been discussed extensively in the literature. These are: (a) Particle enhancements associated with solar active centers display no significant velocity dispersion at onset and appear to be co-rotating with the centers. Such enhancements are often observed to occur over many successive

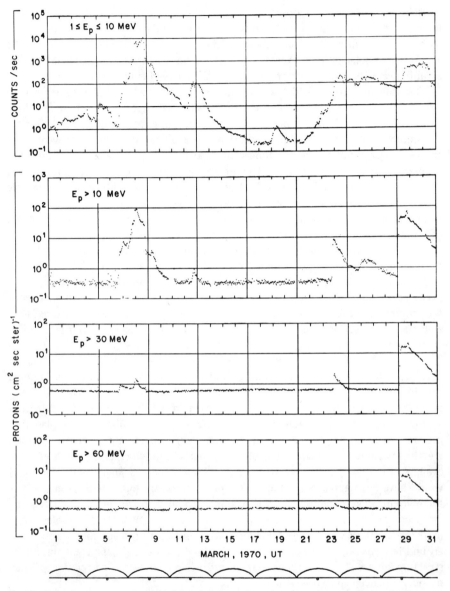

Fig. 2 Solar protons measured in interplanetary space by the solar proton monitoring instrument on Explorer 41 during March 1970.

solar rotations (Fan et al., 1968; McDonald and Desai, 1971). (b) Recurrent particle increases occasionally occur in the next solar rotation after a discrete flare event. They appear to originate in the same active region that produced the flare (Bryant et al., 1965). (c) Energetic storm particles are enhancements of low-energy protons that appear for several hours around the time of arrival of the interplanetary shock wave (Axford and Reid, 1963; Bryant et al., 1965; Rao, McCracken, and Bukata, 1967). Proton enhancements lasting for several minutes, apparently resulting from acceleration directly at the shock front, have been reported (Singer, 1970; Lanzerotti, 1969; Armstrong, Krimigis, and Behannon, 1970; Ogilvie and Arens, 1971; Sarris and Van Allen, 1974).

The large solar particle events from solar active centers or events that look more or less diffusive (Fig. 1B) and that derive from discrete flares are the events most discussed in the literature. However, as shown in Figure 2, interplanetary space is often filled with low-energy solar protons. The relatively small number of large events that have dominated the literature are put into perspective relative to the large number of low-energy events in a recent compilation of solar particle events between 1955 and 1969 (Svestka and Simon, 1975). This catalog is complete insofar as possible and contains a list of some 732 particle events of solar origin recorded at the earth and a record of attempts to relate these events to possible sources on the sun.

II. MEASURES OF SOLAR PARTICLES

Since the low-energy particles are not always easily related to specific solar events and have only been measured in interplanetary space over the past solar cycle, and since no extensive compilation or integrated study of their intensities has yet been reported, this section is devoted to the characteristics of discrete particle events for protons of energy $\gtrsim$ 10 MeV. The lower energy solar protons are not unimportant, but relatively less is known of them than of particles at somewhat higher energies. Later we will discuss solar particle composition and the particle spectra of various events.

A. GROUND-LEVEL EVENTS

It is noted in the introduction that energetic solar particles were first detected by ionization chambers located on the earth. Until the satellite era, ground-based measurements with ionization chambers and, somewhat later, neutron monitors were the primary means for studying solar particles. However, a solar proton with momentum of the order of 1 GV rigidity* ($\sim$500 MeV) is required to produce the

---

* *Magnetic rigidity* = $B\rho$ = $pc/300Ze$ *in eV, where B is the magnetic field and* ρ_p, *and Z are the particle gyro radius, momentum, and charge number, respectively.*

upper atmosphere interactions that result in count increases in polar region ground-level neutron monitors; a rigidity of several tens of GV is required for protons to be detected near the equator.

A plot of the percent increase in the count rate of ground-level instruments for each of the ground-level solar particle events is shown in Figure 3 as a function of year of occurrence (Pomerantz and Duggal, 1974a). Also plotted in the figure are the sunspot numbers for the past three solar cycles, beginning in 1940. The number of detected ground-level events increased quite drastically after about 1954 when the use of the neutron monitors and the super neutron monitors (about 1962) became common. These increases are all normalized to what a polar station would observe. Since the background proton intensity at a polar station is of the order of 5×10^2 protons cm^{-2} s^{-1}, the intensity of protons above about 500 MeV at the time of maximum intensity of each event can be estimated. However, it must be remembered that the percent increase is for the maximum, which may last only a few minutes, while the total event may take a few hours.

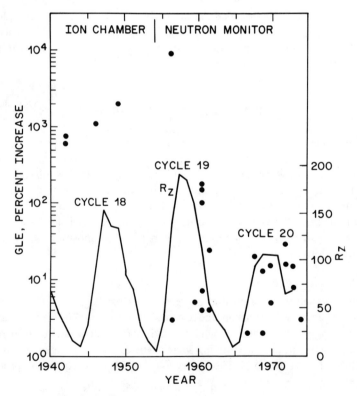

Fig. 3 Magnitudes of ground-level solar cosmic ray events since their discovery in 1942. Also plotted are yearly sunspot numbers. Adapted from Pomerantz and Duggal, 1974a.

B. INDIRECT MEASURES

Indirect measures of the solar proton fluxes have been obtained over the last two solar cycles, beginning in about 1952 using riometer measures of the level of cosmic noise absorption in the ionosphere (Bailey, 1962). When solar protons incident on the ionosphere/upper atmosphere produce enhanced absorption of cosmic noise as a result of the increased ionization, the event is called a polar cap absorption (PCA) event. A plot of the number of PCA events per year for the last two solar cycles for equivalent 30 MHz absorption >2.5 dB is shown in the top half of Figure 4 (adapted from Pomerantz and Duggal, 1974a). Also shown is the yearly sunspot number since 1954. In the bottom half of the figure the amount of cosmic noise absorption produced in each event has been summed for each year and plotted as the yearly total of the absorption (Σ dB) as a function of the solar cycle. Sunspot numbers are also shown here.

An active research area five or six years ago was the search for simple relationships between the absorption A in dB and the flux J cm^{-2} s^{-1} of solar protons incident on the ionosphere (Potemra et al., 1967, 1969; Potemra, 1972). Potemra

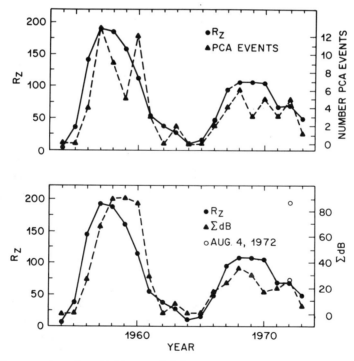

Fig. 4 Top, annual number of PCA events >2.5 dB and annual sunspot numbers during two solar cycles. Bottom, integrated yearly PCA magnitude (Σ dB) and annual sunspot number during two solar cycles. Adapted from Pomerantz and Duggal, 1974a.

(1972) determined that an estimate of proton fluxes for energies >7 MeV is least sensitive to the power law exponent of the spectrum of the incident protons. He derived the empirical relationship

$$J(E > 7 \text{ Mev}) = (0.083)^{-2} A^2 \quad . \tag{1}$$

This relationship can be used to estimate the yearly flux of solar protons >7 MeV from the data in Figure 4.

The PCA data of Figure 4 indicate that the yearly incidence of solar protons as measured by the amount of cosmic noise absorption roughly follows the sunspot cycle. The major exception to this during these two solar cycles was the large solar event of August 1972 during the declining phase of solar cycle 20. The absorption provided by this event alone was as large as the total yearly absorption during any of the years of cycle 19 and more than twice as large as the absorption during any one of the years of cycle 20.

A caveat must be entered on the interpretation of the PCA data and its relationship to the actual fluxes of solar protons that might exist in interplanetary space outside the earth's magnetosphere. The PCA data records solar protons incident on the top of the earth's atmosphere at about 80 or 90 km in altitude. These protons had to enter the magnetosphere and be precipitated down magnetic field lines. There still remains some controversy and uncertainty as to the mechanism (or mechanisms) by which such energetic solar protons enter the magnetosphere (Morfill and Scholer, 1973; Michel and Dessler, 1975); thus the protons measured over the polar caps may not accurately reflect actual flux levels outside the magnetosphere.

C. DIRECT MEASURES

As discussed above, continuous measures of low-energy solar protons outside the magnetosphere only occurred during the last solar cycle. For cycle 19, at times when direct measures do not exist, various investigators used PCA data (Bailey, 1962) and polar auroral ("glow") data (Webber, 1966) to draw conclusions on the fluxes of incident protons. An evaluation and compilation of these data was presented by Yucker (1972) for the period from February 1956 through 1962. These solar event data, together with actual satellite measurements of proton fluxes greater than 30 MeV taken outside the magnetosphere during cycle 20, were compiled by King (1974) and are plotted in the top of Figure 5. The data for the February 1965 event are taken from a compilation of Webber (1966). The data in Figure 5 include all of the events, periods of time of about a week, for which the time-integrated flux of protons above 30 MeV exceeded 10^6 particles cm^{-2}. The data during solar cycle 20 result primarily from measurements made by instruments flown by groups at the Johns Hopkins University Applied Physics Laboratory, Bell Laboratories, NASA/Goddard Spaceflight Center, and University of Chicago on the *Explorer 34* and *Explorer 41* spacecraft. The period prior to May 1967 is not as

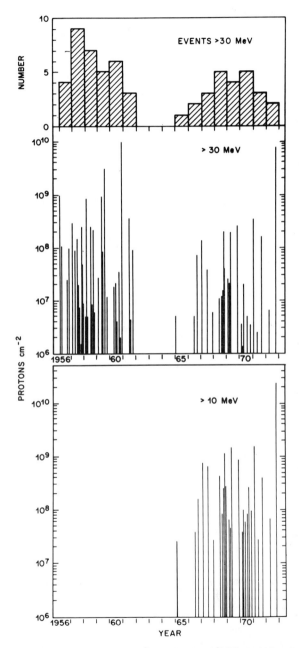

Fig. 5 Event-integrated solar proton fluxes >10 MeV and >30 MeV for the major solar events of the last two solar cycles. Adapted from King, 1974.

well covered by satellite measurements as the time after this date. The total number of events per year with fluxes $>10^6$ protons cm^{-2} are plotted above the 30 MeV event-integrated proton fluxes.

The bottom half of Figure 5 contains the event-integrated fluxes of protons $>10^6$ cm^{-2} for particles of energy >10 MeV for the same two cycles. For protons >10 MeV the event of 4 August 1972 had a total integrated intensity about 20 times that of any of the other events in solar cycle 20. The intensity of the >30 MeV fluxes in this August event approximately matched the intensity of the series of particle events that occurred in November 1960 during the declining phase of solar cycle 19.

Thus, the two largest particle events, the series in November 1960 and in August 1972, occurred during the declining phases of cycles 19 and 20. In 1970, commenting on past solar cycles, Dodson and Hedeman (1970) predicted that a major center of solar activity might occur in cycle 20, perhaps "in early 1973." That such a center occurred in late 1972 is essentially a successful prediction on the part of these authors, particularly in view of our current level of understanding of the solar cycle and its development.

D. PARTICLE SPECTRA

In general, for solar particle events with energies between about 0.5 MeV/nucleon to about 50-100 MeV/nucleon the event-integrated particles can be represented to a good approximation by a power law in energy. That is,

$$J(E) = J_0(E/A)^{-\gamma}, \qquad (2)$$

where A is the number of nucleons. The event-integrated proton flux and α particle flux spectra for seven of the largest events in the 1967-1969 time interval (during the maximum activity of cycle 20) are shown in Figure 6 (Lanzerotti and Maclennan, 1973). Similar spectra for eight events in 1970-1972 (the declining phase of cycle 20) were recently published by Stroscio et al. (1976). For all the events in Figure 6 γ was larger for the α particle fluxes than for the proton fluxes. That is, the α particles all tended to have steeper, or softer, energy spectra than did the protons. The exponent ranged from $\gamma_p = 3.0$, $\gamma_\alpha = 3.3$ for days 188 to 200 in 1968 to $\gamma_p = 1.3$, $\gamma_\alpha = 1.5$ for days 306 to 310 in 1969.

The α to proton ratios are plotted in Figure 7 as a function of energy for the seven events. These ratios show clearly that, for the individual events as well as for the average of the seven events, the spectra for the α particles are softer than are the spectra for the protons. Power-law fits to the average spectra for the seven events gave $\gamma_\alpha = 1.9$, $\gamma_p = 1.7$.

The particle fluxes from the seven events (Fig. 6) can be used to obtain the average α and proton fluxes in the energy range ~ 0.5 MeV/nucleon to ~ 20 MeV/nucleon measured in interplanetary space in the interval from day 148 (28 May) in 1967 to day 310 (6 November) in 1969. More than half of the alphas and

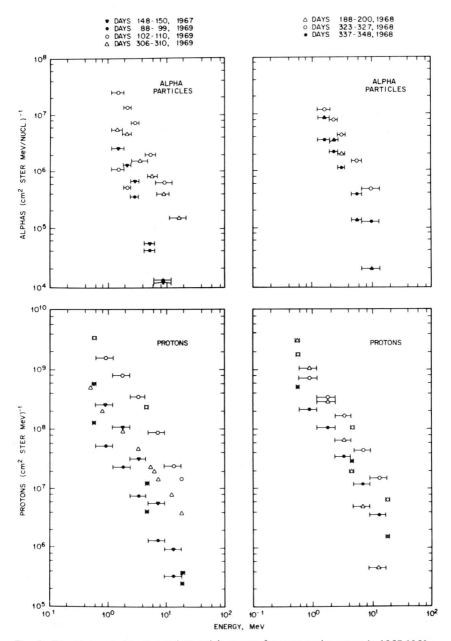

Fig. 6 Event-integrated proton and α particle spectra for seven major events in 1967-1969. Adapted from Lanzerotti and Maclennan, 1973.

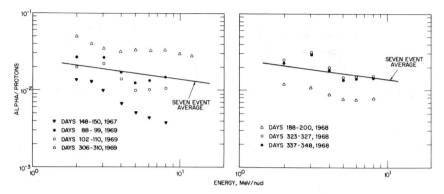

Fig. 7 Alpha particle to proton flux ratios for the event-integrated spectra in Figure 6.

protons in these averaged spectra was produced in the three largest events (days 102 in 1969, 323 in 1968, and 188 in 1968). The inclusion of the additional fluxes of flare-produced protons and alphas from the numerous smaller flares that also occurred in the time interval of the spectra of Figure 6 could increase the total fluxes by perhaps a factor of two (see also Stroscio et al., 1976). However, as can be seen from the >10 MeV data of Figure 5, the absolute fluxes from 1967 through 1972 (excluding the August 1972 event) could perhaps be a factor of about four larger than the total flux for the seven events.

E. SOLAR PARTICLE COMPOSITION

In addition to the protons and α particles, solar particle events are also composed of substantial fluxes of electrons. Extensive reviews of the characteristics of solar electron events, both relativistic (Simnett, 1974) and nonrelativistic (Lin, 1974), have been published. Many of the nonrelativistic electron events are observed at earth very promptly after a solar flare; these events, called "scatter-free" (Lin, 1970), often cannot be associated with low-energy, interplanetary solar protons. Although these events have been called "pure" electron events, the purity should not be construed to mean that protons may not be accelerated in the flares. The technological problem of the detection of protons $\lesssim 0.5$ MeV and the interplanetary propagation conditions probably both conspire to produce the "pure" electron events as observed at earth. In general, the large proton events are accompanied by electrons (Simnett, 1974).

The number of nonrelativistic (>45 KeV) electron events (flux > 10 cm^{-2} sr^{-1} s^{-1}) for the period 1964 to 1972 are plotted in Figure 8 together with the smoothed sunspot numbers (after Lin, 1974). There is a general solar cycle dependence even though substantial variability exists. An indication of the flux levels from electron (>45 KeV) events is given in Figure 9 where the maximum fluxes for each event are plotted against the solar longitude of the originating flare

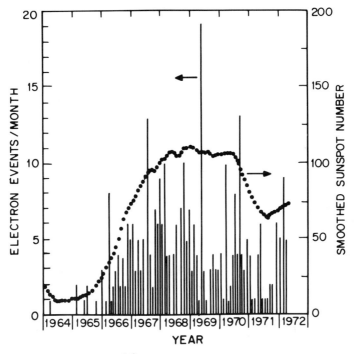

Fig. 8 Number of nonrelativistic (E >45 keV) electron events per month with flux >10 $cm^{-2} sr^{-1} s^{-1}$ and smoothed sunspot number for 1964-1972. Adapted from Lin, 1974.

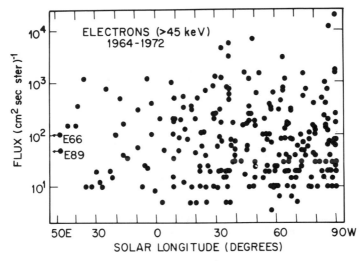

Fig. 9 Maximum flux intensity for nonrelativistic (E >45 keV) solar electron events plotted against the solar longitude of the originating flare. Adapted from Lin, 1974.

(after Lin, 1974). There is a wide variability in the flux, with no strong dependence on solar longitude.

The electron (0.3-0.9 MeV) to proton (~11-100 MeV) flux ratios for events in 1967-1971 are plotted against the solar longitude of the parent flare in Figure 10 (from Simnett, 1974). The wide range of variability in the ratio probably arises from variations in the numbers of the particles accelerated in the various flares as well as from variations in the rigidity dependence of the solar and interplanetary propagation conditions.

In terms of particle ionization in the upper atmosphere, solar electrons generally are unimportant contributors to PCA events except for a few cases at the very beginning of solar events (Lanzerotti and Maclennan, 1973). One of these was the event of 2 November 1969, in which an initial large spike of ionization (measured by riometers) that lasted for a few hours was shown to be produced by the arrival of large fluxes of solar electrons.

In addition to solar electrons, elements with $Z > 2$ and isotopes of helium and hydrogen have been measured during solar events beginning with the use of rocket-flown emulsion packages in the late 1950s (Biswas and Fichtel, 1965). In the last few years the measurement of solar particle composition has greatly expanded.

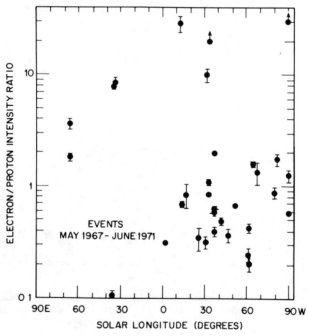

Fig. 10 Ratio of electron (0.3 - 0.9 MeV) flux intensity maximum to proton (11 - 100 MeV) flux intensity maximum for major events in 1967-1971. Adapted from Simnett, 1974.

It is now clear that, unlike the solar particle composition at higher energies, the intensities of nuclei with energies of a few MeV/nucleon are highly variable from one flare to another (Armstrong and Krimigis, 1971; Hovestadt, 1973). The heavier elements such as iron are often quite "enriched" with respect to the helium abundance in comparison to that observed in the solar photosphere (Price et al., 1972; Lanzerotti, Maclennan, and Graedel, 1972; Mogro-Campero and Simpson, 1972; Price, 1973; Fan et al., 1975; Gloeckler et al., 1975; Hurford et al., 1975). Furthermore, the isotopes of hydrogen, helium, and heavier atoms have been observed in flares (Tilles, DeFelice, and Fireman, 1963; Hsieh and Simpson, 1970; Garrard, Stone, and Vogt, 1973; Anglin, Dietrich, and Simpson, 1973; Anglin, 1975) and some flare events have been observed in which the ratio $He^3/He^4 > 1$ (Balasubrahmanyan and Serlemitsos, 1974; Hurford et al., 1975). The next few years should bring rapid advance in the information available on the atomic and isotopic composition of solar cosmic rays and their relationship to the solar wind composition and solar photospheric composition. Although knowledge of the composition of solar cosmic rays is undoubtedly important in furthering our understanding of solar acceleration processes and questions of solar composition, as well as questions on the composition of the earth's radiation belts, the energy content of $Z > 2$ solar cosmic rays at 1 AU is insignificant in comparison to the energy carried by the fluxes of solar hydrogen and helium nuclei.

F. PROPAGATION

As noted in the discussion of Figure 1, the time-intensity profiles of solar cosmic rays are dependent on the propagation conditions in interplanetary space. The profiles also depend on the location of the originating flare on the sun with respect to the earth. Although it is not my purpose to dwell on interplanetary propagation conditions, it should be noted that, statistically, the distribution of the originating flares for the 23 ground-level events is centered around approximately 60° west helio longitude and approximately 15° north helio latitude (Pomerantz and Duggal, 1974a). This distribution means that several relativistic events observed on earth originated from flares beyond the western limb of the sun. It also means that relativistic solar protons can be observed most readily at the earth when they originate from flares that occur near the interplanetary field lines that map from the sun to the earth. However, two of the more intense ground-level solar particle events (17 July 1959 and 4 August 1972) could have been produced by adiabatic Fermi acceleration in interplanetary space (Pomerantz and Duggal, 1974b; Levy, Duggal, and Pomerantz, 1976).

The situation is somewhat different for low-energy solar protons and for the solar protons that produce the PCA events. For these particles the centroid of the originating flare is near 0° helio longitude (Fritzova-Svestkova and Svestka, 1971). This solar longitudinal distribution of the flares that produce PCA events is similar to the longitudinal distribution of solar active centers that produce large geomagnetic storms on the earth (Akasofu and Chapman, 1972). The relation between these two facts is that the large enhancements of the low-energy solar protons that contribute to PCA events accompany the interplanetary, flare-produced shock waves that strike the magnetosphere and produce the geomagnetic storms.

III. GALACTIC COSMIC RAYS

The galactic cosmic rays that form the background nuclear radiation that is incident on the earth have been studied since their discovery by Hess (1912). Continuous measures of this radiation exist for only the last two solar cycles, although some ionization chamber measurements exist from the late 1930s. Plotted as a function of time in Figure 11 is the cosmic ray ionization rate in ion pairs $cm^{-2} s^{-1}$ at 20 g cm^{-2} altitude corrected to polar latitudes (Neher, 1971; Anderson, 1973; Pomerantz and Duggal, 1974a). The ionization rate at this altitude varies by about a factor of two or more during the solar cycle as well as from one solar cycle to another. This atmospheric cutoff corresponds to protons of energy about 500 MeV; the data thus demonstrates that the solar cycle modulation at this energy is about 100%. However, for the particles that reach the ground at middle-latitudes (cosmic ray muons of energy ~ 10 GeV), the maximum peak-to-peak variation over the last four solar cycles is only about four percent (Beach and Forbush, 1969).

The integral incident energy spectra of galactic cosmic rays as a function of particle energy/nucleon are shown in Figure 12 (Waddington, 1972). The data for protons and α particles are plotted for both solar maximum and minimum. The principal reason that the solar cycle variation is so small when the integral energies are plotted as in Figure 12 is because the galactic cosmic ray spectra have broad

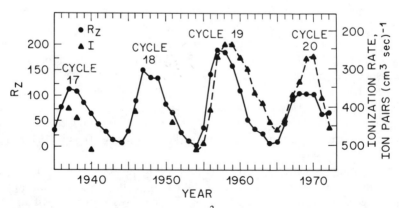

Fig. 11 Ionization rate at an altitude of 20 g cm^{-2} for the last four solar cycles. All of the data have been normalized to geomagnetic location of Thule, Greenland. Adapted from Pomerantz and Duggal, 1974a.

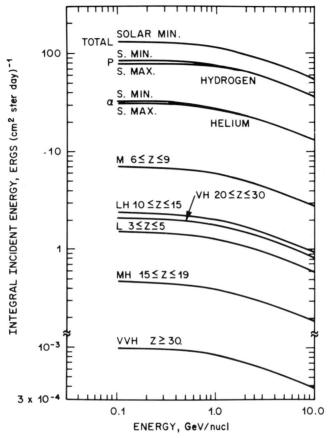

Fig. 12 Integral energy intensities for galactic cosmic ray nuclei. Adapted from Waddington, 1972.

peaks at proton and α particle energies of about 200-300 MeV/nucleon. While the solar cycle modulation for particles of energies $\lesssim$ 200 MeV/nucleon approaches a factor of five or more, the relative contribution of these energy particles to the total incident flux is small. Although it is believed that processes occurring on the sun are the cause of the modulation of galactic cosmic rays as they enter the solar system from intersteller space (Rao, 1972) it is also quite true that the most reliable measurements of galactic cosmic rays are limited basically to the last two solar cycles. Indeed, Pomerantz and Duggal (1974a) have shown that a correlation coefficient of 0.92 is obtained between the yearly cosmic ray intensity and the Dow Jones industrial average for solar cycle 19. This gives one some cause to consider that perhaps we do not yet understand all of the relevant physical laws and/or parameters that affect galactic cosmic ray propagation in the solar system.

IV. SUMMARY

It has been my intention to compile relevant data on both measured and inferred solar and galactic cosmic rays for recent solar cycles at 1 AU outside the earth's atmosphere and magnetosphere and to point out possible problems in interpretion and characterization of solar cosmic ray events. To investigate specific individual time periods it is necessary to refer to original satellite measurements and the original literature in order to obtain (where they exist) the actual particle intensities and spectra as well as the species measured. Several references for these include, in addition to those cited already, the data catalog of solar cosmic rays compiled by Van Hollebeke, Wang, and McDonald (1974), the data published regularly in *Solar-Geophysical Data* from the solar proton monitoring experiments on the various Explorer (IMP) spacecraft, and the *Catalog of Solar Particle Events, 1955-1969* (Svestka and Simon, 1975).

It is my belief that the low-energy ($\lesssim 5$ MeV) solar protons that accompany the interplanetary shock waves that produce the geomagnetic storms may be important in the energy input mechanism or mechanisms in the magnetosphere. One of the unknown factors at the present time is the relationship of the very low-energy (few tens of keV) solar cosmic rays to the solar wind intensities at the time of interplanetary shock waves. How does the energy carried by these particles compare with energy input into the magnetosphere at other times by the solar wind alone? This is an area of investigation that needs further work in order to determine the total energy input into the magnetosphere/ionosphere/atmosphere system on the arrival of an interplanetary shock wave at the earth.

As noted in Section II*E*, detailed measurements of solar particle composition over a wide energy range are only beginning to be available. The measurements of these particles in the future will be useful for studies of particle propagation and for relating the composition ratios to photospheric abundances in different solar active regions.

In summary, the sun emits energetic particles over a wide range of species and energies into interplanetary space. The emission of these particles appears to be related to the solar cycle as represented by the temporal distribution of solar flares, sunspots, and active regions. The intensities of particles measured at the earth that might occur from any solar disturbance configuration are quite unpredictable, however. Extrapolations of solar proton fluxes at 1 AU to those expected in a future solar cycle have been made, for example, by King (1974), and predictions of solar activity have been made by, for example, Dodson and Hedeman (1970). However, to obtain a measure of solar particle fluxes similar to the measures of solar active regions and sunspots that have existed for decades, extensive future monitoring of the interplanetary particle population will be necessary.

REFERENCES

Akasofu, S.I., and Chapman, S., 1972, *Solar Terrestrial Physics*, Oxford University Press, London.

Anderson, H.R., 1973, *J. Geophys. Res. 78*, 3958.

Anglin, J.P., 1975, *Ap. J. 198*, 733.

―――――, Dietrich, W.F., and Simpson, J.A., 1973, *Ap. J. (Letters) 186*, L41.

Armstrong, T.P., and Krimigis, S.M., 1971, *J. Geophys. Res. 76*, 4230.

―――――, ―――――, and Behannon, K.W., 1970, *ibid. 75*, 5980.

Axford, W.I., and Reid, G.C., 1963, *ibid. 68*, 1743.

Bailey, D.K., 1962, *J. Phys. Soc. Japan 17*, Suppl. AI, 106.

Balasubrahmanyan, V.K., and Serlemitsos, A.T., 1974, *Nature 252*, 460.

Beach, L., and Forbush, S.E., 1969, *Carnegie Inst. Wash. Pub. 175*, 22.

Biswas, S., and Fichtel, C.E., 1965, *Space Sci. Rev. 4*, 709.

Bryant, D.A.; Cline, T.L.; Desai, U.D.; and McDonald, F.B., 1965, *Ap. J. 141*, 478.

Dodson, H.W., and Hedeman, E.R., 1970, *Solar Terrestrial Physics/1970*, Part I, ed. E.R. Dyer, D. Reidel, Dordrecht, 151.

Fan, C.Y., Gloeckler, G., and Hovestadt, D., 1975, *Solar Gamma-, X-, and EUV Radiation*, IAU Symposium No. 68, ed. S.R. Kane, D. Reidel, Dordrecht, 411.

―――――; Pick, M.; Pyle, R.; Simpson, J.A.; and Smith, D.R., 1968, *J. Geophys. Res. 73*, 1555.

Forbush, S.E., 1946, *Phys. Rev. 70*, 771.

Fritzova-Svestkova, L., and Svestka, Z., 1971, *Solar Phys. 17*, 212.

Garrard, T.L., Stone, E.C., and Vogt, R.E., 1973, *High Energy Phenomena on the Sun*, ed. R. Ramaty and R.G. Stone, NASA publication SP-342, 341.

Gloeckler, G.; Hovestadt, D.; Vollmer, O.; and Fan, C.Y., 1975, *Ap. J (Letters) 200*, L45.

Hess, V.F., 1912, *Phys. Zs. 13*, 1084.

Hey, J.S., 1973, *The Evolution of Radio Astronomy*, Watson Academic, New York, 14.

Hovestadt, D., 1973, *Proceedings 13th International Cosmic Ray Conference*, Vol. 5, 3685.

Hsieh, K.C., and Simpson, J.A., 1970, *Ap. J. (Letters) 162*, L191.

Hurford, G.J.; Mewaldt, R.A.; Stone, E.C.; and Vogt, R.E., 1975, *ibid. 201*, L95.

King, J.H., 1974, *J. Spacecraft and Rockets 11*, 401.

Lange, I., and Forbush, S.E., 1942, *Terr. Mag. 47*, 185.

Lanzerotti, L.J., 1969, *J. Geophys. Res. 74*, 2451.

──────────, 1972, *Proceedings National Symposium Natural and Manmade Radiation in Space*, ed. E.A. Warman, NASA publication TM-X-2440, 193.

──────────, and Maclennan, C.G., 1973, *J. Geophys. Res. 78*, 3935.

──────────, ──────────, and Graedel, T.E., 1972, *Ap. J. (Letters) 173*, L39.

Levy, E.H., Duggal, S.P., and Pomerantz, M.A., *J. Geophys. Res. 81*, 51.

Lin, R.P., 1970, *ibid. 75*, 2583.

──────────, 1974, *Space Sci. Rev. 16*, 189.

McDonald, F.B., and Desai, U.D., 1971, *J. Geophys. Res. 76*, 808.

Michel, F.C., and Dessler, A.J., 1975, *ibid. 80*, 2309.

Mogro-Campero, A., and Simpson, J.A., 1972, *Ap. J. (Letters) 177*, L37.

Morfill, G., and Scholer, M., 1973, *Space Sci. Rev. 15*, 267.

Neher, H.V., 1971, *J. Geophys. Res. 76*, 1637.

Ogilvie, K.W., and Arens, J.F., 1971, *ibid. 76*, 13.

Pomerantz, M.A., and Duggal, S.P., 1974a, *Rev. Geophys. Space Phys. 12*, 343.

────────── and ──────────, 1974b, *J. Geophys. Res. 79*, 913.

Potemra, T.A., 1972, *Radio Sci. 7*, 571.

──────────, Zmuda, A.J., Shaw, B.W., and Haave, C.R., 1967, *J. Geophys. Res. 72*, 6077.

──────────, ──────────, Haave, C.R., and Shaw, B.W., 1969, *ibid. 74*, 6444.

Price, P.B., 1973, *High Energy Phenomena on the Sun*, ed. R. Ramaty and R.G. Stone, NASA publication SP-342, 377.

──────────; Hutcheon, I.; Cowsik, R.; and Barber, D.J., 1971, *Phys. Rev. Letters 26*, 916.

Rao, U.R., 1972, *Space Sci. Rev. 12*, 719.

──────────, McCracken, K.G., and Bukata, R.P., 1967, *J. Geophys. Res. 72*, 4325.

Sarris, E.T., and Van Allen, J.A., 1974, *ibid. 79*, 4157.

Simnett, G.M., 1974, *Space Sci. Rev. 16*, 257.

Singer, S., 1970, *Intercorrelated Satellite Observations Related to Solar Events*, ed. V. Manno and D.E. Page, D. Reidel, Dordrecht, 571.

Stroscio, M.A.; Katz, L.; Yates, G.K.; Sellers, B.; and Hanser, F.A., 1976, *J. Geophys. Res. 81*, 283.

Svestka, Z., and Simon, P., 1975, *Catalog of Solar Particle Events, 1955-1969*, D. Reidel, Dordrecht.

Tilles, D., DeFelice, J., and Fireman, E.L., 1963, *Icarus 2*, 258.

Van Hollebeke, M.A., Wang, J.R., and McDonald, F.B., 1974, *A Catalogue of Solar Cosmic Ray Events IMPS 4 and 5*, NASA publication X-661-74-27.

Waddington, C. J., 1972, *Proceedings National Symposium Natural and Manmade Radiation in Space*, ed. E.A. Warman, NASA publication TM-X-2440, 209.

Webber, W.R., 1966, Report D2-84274-1 (Seattle: The Boeing Co.).

Yucker, W.B., 1972, *Proceedings National Symposium Natural and Manmade Radiation in Space*, Ed. E. A. Warman, NASA publication TM-X-2440, 345.

PARTICLE OUTPUT OF THE SUN IN THE PAST

D. Heymann
Department of Geology
Department of Space Physics and Astronomy
Rice University
Houston, Texas

Atomic particles emitted by the sun interact with objects in their path in two ways. On entering the surfaces of solids, the particles are always scattered as well as slowed down by ionization. Alternatively, the particles can engage in nuclear reactions with atoms of the target material. Which of the two is the dominant process depends principally on the kinetic energy of the incoming particles. At relatively low energies (KeV/amu range), no detectable nuclear reactions occur; all particles not scattered or reflected back into space come to rest beneath the surface, generally within approximately 1000Å in depth. A sharp lower energy boundary above which nuclear reactions become detectable cannot be specified. For the purpose of general orientation, 1 MeV/amu is a reasonable value. Even so, most ions of 1 MeV/amu do not participate in nuclear reactions but come to rest at a fairly well-defined depth called range. It is not until energies on the order of GeV/amu are reached that nuclear interaction lengths become significantly shorter than the particle's ranges and that the overwhelming majority of the particles become involved in nuclear reactions.

The records from solar particles now extant in objects such as meteorites and lunar samples are detectable in three forms: 1) ions from the sun trapped in these objects, 2) stable and radioactive products of nuclear reactions, and 3) radiation damage effects. These records are initially "skin-deep." Solar wind ions come to rest within approximately 1000Å. Effects from solar flare ions are detectable to roughly 1 g/cm^2. Accordingly, the initial records are in permanent danger of being obliterated by a variety of known processes. For instance, micrometeorite impacts and solar wind ions themselves, by way of ion sputtering, whittle surfaces down, thereby erasing records deposited at earlier times. Violent, crater-forming impacts on the surfaces of objects such as the moon or meteorite parent objects may cause heating by shock, and the heating, in turn, may anneal nuclear tracks (see below), liberate volatile products from nuclear reactions, or liberate volatile elements trapped from the solar wind.

Fortunately, nature has provided niches in which ancient records have been preserved, if not perfectly, at least detectably. Foremost among these is a rock type called breccia. Imagine the effect of a crater-forming impact on the dust-covered surface of the moon. The surfaces of the dust particles are known to be impregnated with elements derived from the solar wind and solar flares. Among these are the very volatile inert gases. Shock waves traveling through the dust can result in the formation of tangible, fairly coherent rocks called shock breccias. Luckily, the heating caused by the shock has been mild enough in many cases such that the inert gases have not been quantitatively expelled during or after the shock. As a result, the ancient record, which was contained on the vulnerable surfaces of small grains prior to the shock, was transferred to the interior of much larger rocks, where it is better protected against ion sputtering or micrometeorite erosion. Similar breccias are known to exist among meteorites. Another mode of preservation is made possible by the dust layer on the surface of the moon itself. We know from studies of the drill cores taken by *Apollo 15, 16,* and *17* that much of the soil below the lunar surface was deposited millions to hundreds of millions of years ago, was covered, and has remained undisturbed ever since (cf. Russ, Burnett, and Wasserburg, 1972). Therefore, samples from depth often contain records from the distant past.

The chemical and mineralogical composition of the surfaces on which the solar particles impinge are important parameters. A detailed discussion of this topic is beyond the scope of this paper. All but a few of the ancient solar records have been found in stony meteorites or in lunar samples. Both are essentially silicate materials in which the major elements are oxygen, silicon, aluminum, magnesium, iron, calcium, sodium, potassium, and titanium. Individual minerals contain these elements in different proportions. Obviously, the nature and concentrations of the products of nuclear reactions from solar cosmic rays depend greatly on the chemical composition of the target material (cf. Walton, 1974). For instance, flare protons can produce copious quantities of radioactive ^{22}Na in magnesium-rich minerals, but will produce little if any ^{22}Na in pure ilmenite ($FeTiO_3$). Nuclear track registration depends on the material in which tracks are produced (Fleischer, Price, and Walker, 1976). And, while the stopping properties of minerals for solar wind ions are grossly similar, there are vast differences in their ability to retain the trapped particles, particularly those of volatile elements, for periods of millions or billions of years. Studies of mineral retentivities have only begun in the last few years. These have already revealed a few general trends. Closely packed lattices such as those of metals and oxides have much better retentivity than the more open lattices of silicate minerals. Among the latter, high iron and titanium content usually enhance retentivity.

The topic of this paper has been presented and discussed at a number of conferences and workshops in recent years. I refer the reader to the proceedings by Sonnett, Coleman, and Wilcox, (1972), Russell (1974), and Criswell and Freeman (1974).

I. CHRONOLOGY

The theme of this paper is the record in meteorites and lunar samples of particle emission from the sun in the past, including the recent as well as the very distant past. This raises the question of how we know when the record was acquired by these objects.

The number of firmly established dates is small, and all of these fall in the very recent past. Nuclear effects of solar cosmic rays have been studied both in meteorites (Davis, Stoenner, and Schaeffer, 1963; Fireman, DeFelice, and Tilles, 1963; Fireman and Goebel, 1970) and in lunar samples (cf. Eldridge, O'Kelley, and Northcutt, 1973) that were exposed to particles from known solar flares and were also studied by way of satellite instrumentation (e.g., the August 1972 flare). The satellite observations of particle fluxes and energy spectra of the known flares provide the basic information needed for theoretical calculations (cf. Reedy and Arnold, 1972; Walton, 1974) of concentration profiles of nuclides produced by solar cosmic rays in meteorites and lunar samples. The good agreement between theoretical predictions and observations for very recent times is an essential ingredient for our conclusions about the more distant past (cf. Finkel et al., 1971).

A variety of chronometers is being used in our attempts to ascertain the past exposure dates of meteorites and lunar samples. The most straightforward among these are radioactive species with a broad spectrum of mean lifetimes. In Table 1 are listed radioactive nuclides that have been widely used in solar cosmic ray studies. Their half lives range from 5.7 days (^{52}Mn) to ~ 2-3 million years (^{10}Be, ^{53}Mn).

TABLE I
Half-lives of radionuclides measured in lunar samples and in meteorites

Nuclide	$t_{1/2}$		Nuclide	$t_{1/2}$	
^{52}Mn	5.7	d	^{3}H	12.26	y
^{51}Cr	27.8	d	^{44}Ti	47	y
^{37}Ar	35.1	d	^{63}Ni	92	y
^{59}Fe	45	d	^{39}Ar	325	y
^{7}Be	53	d	^{14}C	5730	y
^{58}Co	71	d	^{59}Ni	8×10^{4}	y
^{56}Co	77.3	d	^{60}Fe	$\sim 10^{5}$	y
^{46}Sc	83.8	d	^{36}Cl	3×10^{5}	y
^{57}Co	272	d	^{26}Al	7.4×10^{5}	y
^{54}Mn	312	d	^{53}Mn	2×10^{6}	y
^{49}V	330	d	^{10}Be	2.7×10^{6}	y
^{55}Fe	2.4	y			
^{22}Na	2.60	y			
^{60}Co	5.24	y			

Solar flares are transient phenomena, thus the specific activities of radioactive nuclides in meteorites and lunar samples fluctuate considerably on the time scales of larger flares. However, if we consider very long time periods, for example, 10^6 years (roughly 10^5 solar cycles, if the length of the cycles has not changed since 10^6 years ago), we are dealing with a mode of irradiation that can be compared to the bombardment of a target with protons in a cyclotron whose proton beam is not perfectly stable but fluctuates around some mean value. The mean value itself may show long-term drifts, sudden systematic shifts, or random shifts, depending on the conditions in the source that provides the protons for acceleration. The specific activities of radioactive nuclides in the target reflect such drifts and sudden shifts; however, their mean lifetimes are important buffers in this respect. A simple example illustrates the point. Suppose that the mean proton flux from solar cosmic rays was substantially greater 10^6 years ago than it is now. Meteorites and lunar samples studied now contain virtually no ^{37}Ar atoms (mean lifetime, ~ 50 days) that were produced 10^6 years ago, but they do contain ^{27}Al ($\bar{t} \sim 4 \times 10^6$ years), ^{10}Be ($\bar{t} \sim 4 \times 10^6$ years), and ^{53}Mn ($\bar{t} \sim 3 \times 10^6$ years) atoms that were produced during the era of enhanced mean proton flux. The resolution power of this chronometer depends mainly on the spacing of the mean lifetimes. Inspection of Table 1 shows that 14 nuclides have $0 < t_{1/2} < 10$ years; three have $10 < t_{1/2} < 100$ years, one has $100 < t_{1/2} < 1000$ years, and the remaining seven have $1000 < t_{1/2} < 10^7$ years. Obviously, the resolution power is increasingly poorer for the more distant past, and information from more than a few million years ago cannot be gleaned from the nuclides listed in Table 1. For time scales of $\sim 10^9$ years a few nuclides, such as ^{40}K, are available; however, ^{40}K produced by solar cosmic rays has not yet been detected.

The third chronometer relies on nuclear reactions induced by galactic cosmic rays. Their nuclear interaction length in matter exceeds that of solar cosmic rays by roughly 2 to 3 orders of magnitude. Dates of exposure of matter to galactic protons have been determined both for meteorites and for lunar samples; these range from $<10^6$ to about 2×10^9 years ago. A time stratigraphy of lunar cores is based on secondary neutrons produced by galactic cosmic rays in the lunar surface (cf. Russ et al., 1972). For meteorites, the galactic proton chronometer is interpreted in a fairly simple manner: an exposure to galactic cosmic rays 10^7 years ago is taken to mean that the meteorite was broken off from a much larger parent body 10^7 years ago. Thus solar flare records in this meteorite were acquired at some time during the last 10^7 years. For lunar samples the interpretation is often much more complex, as such samples are known to have complex histories of excavation and burial on the lunar surface. Records from solar wind and solar flares are most easily decipherable in lunar rocks with simple irradiation histories, i.e., rocks that were excavated only once and were deposited on the lunar surface where they have remained undisturbed ever since. Only a handful of such lunar rocks are now known. Even so, there exist on the lunar surface time markers in the form of dated ejecta blankets from large craters such as South Ray Crater (~10^6 years) or North Ray Crater (~ 50 ×10^6 years). We have firm reasons to assume that most materials (rocks, soils) in these ejecta blankets contain solar wind or solar flare records which postdate the times of formation of these craters.

For times more than ~2×10^9 years ago we must rely on much more indirect chronometers, such as the K-Ar, Rb-Sr, etc. formation ages of the materials in which the solar records are contained. A typical argument is as follows: a stony meteorite is known to contain noble gases, including argon, derived from the solar wind. Its K-Ar age of 4.6×10^9 years is interpreted as a gas retention age, that is, 4.6×10^9 years ago the object had cooled to a temperature where near-quantitative retention of radiogenic ^{40}Ar (from ^{40}K decay) began. Therefore it is very unlikely that the gases in this object derived from the solar wind could have been acquired prior to 4.6×10^9 years ago. This leaves a vast unaccounted time during which the solar wind gases might have been acquired, i.e., from 4.6×10^9 years ago to the onset of the much more recent cosmic ray bombardment, which rarely exceeds ~10^8 years for stony meteorites. Especially when the meteorite or the lunar sample is a breccia, we need to know the time at which the brecciation occurred, not the time of formation of the parent materials from which the breccia is composed. We do not know much about the dates of brecciation of meteorites or of lunar samples except in a few isolated cases. It is widely assumed that brecciated meteorites and lunar samples rich in trapped gas formed at the time indicated by their K-Ar (or more recently ^{39}Ar-^{40}Ar) ages or very soon thereafter. Evidence in support of this assumption is the presence of nuclear tracks from fission of extinct ^{244}Pu (cf. Fleischer *et al.*, 1976) in several of these breccias.

In summary, the chronology of solar particle records is relatively firmly established for the last ~10^9 years, but is increasingly ill-defined for the more distant past. Lunar breccias may contain solar wind records from ~3-4×10^9 years ago, and brecciated meteorites may extend this record to ~ 4.6×10^9 years ago.

II. THE SOLAR WIND

It is well known that ions with kinetic energies of a few hundred eV/amu or more become trapped when they impinge on solid surfaces. The trapping efficiency depends on the target and the energy of the ion (cf. Meister, 1969). In general, trapping efficiencies are low at ~100 eV/amu but increase rapidly with increasing kinetic energy, and reach essentially 100% between 1 and 2 KeV/amu. Studies of grains exposed to the solar wind have shown that these are often jacketed by amorphous coatings caused by intense solar wind radiation damage (cf. Borg *et al.*, 1971). The thickness of such amorphous coatings is proportional to the kinetic energy of the ions; hence the coatings afford a means of studying solar wind speeds.

The peculiar geochemical characteristics of the solar system are the reason that only a few elements have, thus far, been studied for ions implanted by the solar wind. These are hydrogen, carbon, nitrogen, and all the inert gases. Elements that are probably still fairly abundant in the solar wind, such as oxygen, magnesium, silicon, and iron, are also usually very abundant in the catcher minerals. Elements that occur only in trace quantities in minerals (e.g., Au, Pt) are also rare in the solar wind. It is only recently that, by a clever choice of minerals and the use of the ion microprobe, ^{54}Fe, ^{52}Cr, ^{48}Ti, ^{31}P, ^{24}Mg, and ^{12}C derived from the solar wind have begun to be studied (Zinner et al., 1976).

The volatile elements, once implanted, are not retained quantitatively by the minerals but tend to be lost by diffusion processes. Since diffusion is a mass-dependent phenomenon, these losses lead to mass fractionation of both the lost and the retained particles implanted by the solar wind. In general, the fractionation causes a greater depletion of lighter than of heavier atoms. Fractionations among elements (e.g., helium versus neon) can be substantial. Fractionations among stable isotopes of a particular element are, in general, less severe. The upshot, however, is that the solar wind composition record in lunar samples and meteorites is seldom, if ever, identical to that in the solar wind itself. Studies of lunar samples have shown that the records in the mineral ilmenite ($FeTiO_3$) are, in this respect, the most reliable (Eberhardt et al., 1970).

The evidence suggests that the solar wind or a similar entity was present at 1 AU some $3\text{-}4 \times 10^9$ years ago and perhaps as much as 4.6×10^9 years ago. A major achievement from the Apollo program is the establishment of the fact that krypton and xenon occur in the solar wind and the determination of the isotopic compositions of these two elements in the sun. There are suggestions for secular increases of $^3He/^4He$, $^{13}C/^{12}C$, and $^{15}N/^{14}N$ ratios in the solar wind during at least the last 4×10^9 years. In addition, there are hints that the mean solar wind velocity may have varied substantially in the past.

Suess, Wanke, and Wlotzka (1964) were the first to suggest that the "gas-rich" meteorites, which are breccias, contain inert gas atoms trapped from the solar wind. Their hypothesis has been amply confirmed by the experiments of Eberhardt, Geiss, and Grogler (1965a,b), who showed that the concentrations of trapped inert gases in the Khor Temiki meteorite, as predicted, are inversely proportional to grain size. (Stony meteorites are made up of individual grains ranging in size from less than 1 μm to about a few mm.) A similar dependency of trapped gas content on grain size has been found by several investigators of lunar soils (cf. Eberhardt et al., 1970). The current fluxes of ^{3}He, ^{4}He, ^{20}Ne, ^{21}Ne, ^{22}Ne, and ^{36}Ar at 1 AU have been determined by the Apollo solar wind composition experiments (Geiss et al., 1970, 1971, 1972a,b). A summary is given in Table 2.

Two fundamental problems arise when meteorites and lunar samples are used to investigate the ancient solar wind. The solar wind record is preserved in the

TABLE 2
Solar wind composition from Apollo solar wind composition experiments

Mission	Avge. solar wind ^{4}He flux, 10^6 cm^{-2} sec^{-1}	^{4}He/^{3}He	^{4}He/^{20}Ne	^{20}Ne/^{22}Ne
Apollo 11	6.2 ± 1.2	1860 ± 140	430 ± 90	13.5 ± 1.0
Apollo 12	8.1 ± 1.0	2450 ± 100	620 ± 70	13.1 ± 0.6
Apollo 14	4.2 ± 0.8	2230 ± 140	550 ± 70	13.65 ± 0.50
Apollo 15	17.7 ± 2.5	2310 ± 120	550 ± 50	13.65 ± 0.30
Apollo 16	12.0 ± 1.8	2260 ± 100	570 ± 50	13.80 ± 0.40

Mission	^{20}Ne/^{36}Ar
Apollo 14	37^{+10}_{-5}
Apollo 15	20^{+8}_{-5}
Apollo 16	29 ± 6

Weighted Averages

^{4}He/^{3}He	^{4}He/^{20}Ne	^{20}Ne/^{22}Ne	^{20}Ne/^{36}Ar
2350 ± 120	570 ± 70	13.7 ± 0.3	28 ± 9

Source: Geiss et al., 1972 b.

first micron or so beneath the exterior surfaces of solid objects (for evidence see Eberhardt et al., 1970). Because of this, the record is in permanent danger of being altered or wholly erased by destruction of particle surfaces (erosion by ion sputtering and micrometeorite bombardment) or by the selective escape of trapped ions from these surfaces (diffusive loss of inert gases, loss of carbon as CH_4, etc.). Such processes usually result in changes of relative element abundances as well as of relative isotopic abundances. Consequently, every reconstruction of a primary record is fraught with uncertainties about mass fractionations that may have occurred since the deposition of the record itself.

The second problem stems from the fact that we do not now have well-developed, reliable techniques for determining trapping times. For example, most grains in lunar soils contain inert gases derived from the solar wind. The amounts of trapped ^{4}He correspond to exposure to solar wind fluxes of present-day intensity for several hundreds of years (Eberhardt et al., 1970). An individual grain could have trapped all of its ^{4}He during the last few hundred years (i.e., since 1600 AD), but it could also have trapped its gas billions of years ago, have become and remained buried in the lunar soil cover, and have been thrown onto the surface by a meteorite impact only a few days before it was picked up by the astronauts.

Despite these uncertainties, we do have a rough timetable for the solar wind or, more cautiously, an entity resembling the solar wind. Bogard and Nyquist (1972) concluded from a study of the *Apollo 15* deep drill core that the solar wind at 1 AU shows little variation for the last 400×10^6 years (original time estimate from Russ et al., 1972; a more recent paper by Curtis and Wasserburg [1975] suggests 500×10^6 years). Heymann and Yaniv (1970), Yaniv and Heymann (1972), and Heymann (1975) argue that many lunar soils contain grains whose surfaces were exposed to ^{40}Ar implantation from the lunar atmosphere as much as $3-4 \times 10^9$ years ago (Heymann, 1975). These authors assumed that the acceleration mechanism proposed by Manka and Michel (1970) is valid. This mechanism requires the presence of an interplanetary magnetic field and a more-or-less permanently flowing solar plasma at 1 AU. The objection of Baur et al. (1972) that the predicted low implantation energy of ~0.5-1.5 KeV for ^{40}Ar ions is not observed (the experiments by Baur et al. suggest 30 KeV or more) has been answered by Yaniv and Heymann (1972), who pointed out that the solar wind mechanism could have accelerated ions to 30 KeV or more if the near-lunar electric field gradient in the past was roughly an order of magnitude greater than it is now. Such a condition could have occurred either because the strength of the interplanetary magnetic field at 1 AU was once, on the average, ten times greater or because the average solar wind propagation velocity was greater by roughly a factor of three. Thus the ^{40}Ar record from lunar samples can be interpreted to mean that the average solar wind velocity was considerably greater in the distant past ($3-4 \times 10^9$ years ago, according to Yaniv and Heymann [1972]) than it is now.

Some lunar samples contain trapped gases of the solar wind variety and appear to have become isolated systems (breccias) about 4×10^9 years ago. Among these are certain *Apollo 14* breccias (LSAPT, 1971). Unfortunately, the time of formation of *Apollo 11* breccias has not been determined; these rocks contain abundant evidence (Hintenberger and Weber, 1973) that the large quantities of trapped gases in them originated in the solar wind.

The K-^{40}Ar and U, Th-^{4}He gas retention ages of many stony meteorites are about 4.6×10^9 years (Zähringer, 1964). The dark, gas-rich portions of the gas-rich meteorites cannot be reliably dated by these techniques because the radiogenic ^{40}Ar and ^{4}He in these are masked by their respective trapped components. The light, gas-poor portions, however, show ages between 4 and 4.6×10^9 years. The ^{40}K-^{40}Ar ages of the important carbonaceous chondrites are usually short, $1-2 \times 10^9$ years (Mazor, Heymann, and Anders, 1970), but these are only apparent ages since these meteorites have suffered appreciable loss of radiogenic ^{40}Ar. The presence of excess ^{129}Xe from the decay of extinct ^{129}I in stony meteorites is not conclusive evidence for an early trapping of solar wind. Hence the question of whether a solar wind was present at the time of formation of the meteorites some 4.6×10^9 years ago or shortly thereafter cannot be definitely answered. The formation ages of the meteorites are merely a strong indication that this was the case.

Whether or not the average isotopic and elemental composition of the solar wind has changed over long time periods is more difficult to answer in view of the fractionation processes mentioned above. Bogard and Nyquist (1972) concluded that the composition of the solar wind inert gases seems to have undergone little if any variation during the last 400×10^6 years, and the trapped inert gases in the *Apollo 11* and *14* breccias indicate little if any variation during the last 3.5-4.0×10^9 years.

Trapped inert gases in carbonaceous and gas-rich meteorites, particularly trapped neon, appear to be mixtures of several components. A thorough discussion of this topic is beyond the scope of this review; the reader is referred to Black (1972a, b), Jeffery and Anders (1970), and Eberhardt (1974). One of these components, Neon-B, with $^{20}Ne/^{22}Ne = 12.52 \pm 0.18$ and $^{21}Ne/^{22}Ne = 0.0335 \pm 0.0015$, has been assigned to the normal solar wind, i.e., solar wind of present-day composition (Black, 1972a). Neon-D, with $^{20}Ne/^{22}Ne = 14.5 \pm 1.0$ and $^{21}Ne/^{22}Ne$ undetermined, has been associated by Black (1972a) with the primitive solar wind. According to Black (1972a), Helium-B ($^3He/^4He = 3.9 \times 10^{-4}$) and Argon-B ($^{36}Ar/^{38}Ar = 6 \pm 1$), are associated with Neon-B.

Eberhardt et al. (1972) compared data on the trapped $^3He/^4He$ ratio in different reservoirs of trapped solar wind particles and concluded that these data strongly suggest but do not firmly establish a secular increase, by about 40%, of the $^3He/^4He$ ratio in the solar wind.

Epstein and Taylor (1972) and Kaplan, Smith, and Ruth (1970) were the first to conclude that carbon in lunar soils contains an addition from the solar wind, from carbonaceous meteorites or comets, or from some unknown lunar source. Epstein and Taylor (1972) discussed three models, each of which requires solar wind carbon of variable isotopic composition. Even if this interpretation is correct, it is not yet clear how the $^{12}C/^{13}C$ ratio has changed through time. However, Kerridge, Kaplan, and Lesley (1974) argue that the variations of $^{12}C/^{13}C$ ratios in lunar samples are caused by lunar surface processes such as hydrogen stripping, agglutination, and impact comminution. Epstein and Taylor (1974) have found that carbon in grain surfaces of lunar soils is extremely enriched (10-20%) in ^{13}C. They conclude that these extreme enrichments are most likely a result of lunar surface processes, but concede that the ^{13}C content of solar wind carbon may have increased through time. Becker and Clayton (1975) conclude that the value of $^{15}N/^{14}N$ of nitrogen in the lunar soil cover has increased by 15% over a period of somewhere between 4.5×10^8 and about 4×10^9 years. The increase may be a result of a change in this ratio of solar wind nitrogen or it may have resulted from a decreasing contribution of an isotopically light, indigenous (i.e., genuinely lunar) lunar nitrogen recycled through the lunar atmosphere (see ^{40}Ar discussion above). Kerridge, Kaplan, and Petrowski (1975) have, independently, made the same observation about a secular increase of the $^{15}N/^{14}N$ ratio.

III. SOLAR FLARES, PART I: RADIOACTIVE PRODUCTS

Records from solar flares in meteorites have been reported occasionally (Fireman and Goebel, 1970; Davis et al., 1963; Fireman et al., 1963); however, these have usually been restricted to measurements of relatively short-lived activities such as ^{3}H or ^{37}Ar, which are enhanced in some meteorites that collide with the earth soon after the occurrence of a relatively intense flare. After the return of the *Apollo 11* samples, several investigators deduced the presence of radioactivities produced by solar flares in lunar rocks and soils (Begemann et al., 1970; D'Amico, DeFelice, and Fireman, 1970; Herzog and Herman, 1970; O'Kelley et al., 1970; Perkins et al., 1970; Shedlovsky et al., 1970; Stoenner, Lyman, and Davis, 1970; Wrigley and Quaide, 1970). The concentrations of ^{3}H, ^{22}Na, ^{26}Al, and ^{39}Ar were found to be enhanced in soils relative to rocks, partly because of chemical differences, but also because of input by solar flares. Concentration (depth) gradients in rocks were measured by O'Kelley et al. (1970), Perkins et al. (1970), and Shedlovsky et al. (1970). Shedlovsky et al. (1970) presented a theoretical treatment, later amplified by Reedy and Arnold (1972), and reported that the predicted and observed activities of the short-lived species ^{56}Co, ^{57}Co, ^{54}Mn, ^{55}Fe, and ^{22}Na in *Apollo 11* rock 10017 are in good agreement. The interpretation of long-lived isotopes such as ^{53}Mn, ^{36}Cl, and ^{26}Al is more complicated because various models for the long-term irradiation history of lunar rocks are possible (Shedlovsky et al., 1970). However, a few tentative quantitative conclusions have been drawn.

At the second Lunar Science Conference in 1971, Finkel et al. (1971) concluded from the depth profiles of ^{56}Co, ^{54}Mn, ^{22}Na, ^{55}Fe, ^{26}Al, and ^{53}Mn in *Apollo 12* rock 12002 that the solar flare activity at the moon, when averaged over various periods of the past and up to a few million years ago, appears to have been similar in intensity and spectral shape to that observed at present. The best fit apparent mean flux J is 100 protons cm^{-2} s^{-1}, and the shape parameter R_o is 80 MV (megavolts); however, these numbers are uncertain. Rancitelli et al. (1971) concluded that ^{26}Al to ^{22}Na ratios in *Apollo 12* samples are consistent with constant solar activity during the past million years. D'Amico et al. (1971), from a comparison of ^{3}H and fossil track data on *Apollo 12* rock 12002, concluded that the solar flare intensity over the past 5×10^6 years appears to have been somewhat smaller than the average during the past 30 years, i.e., 4×10^9 protons cm^{-2} y^{-1} from ^{3}H measurements.

In contradistinction, Fields et al. (1971) reported a very high ^{236}U/^{238}U ratio in the *Apollo 12* soil 12070. They argued that the ^{236}U was produced by solar flare proton-induced reactions on ^{238}U. With a proton spectrum whose rigidity is 100 MV, the flux required to produce the equilibrium ^{236}U/^{238}U ratio in 12070 is 5000 protons cm^{-2} s^{-1} (E > 10 MeV), two orders of magnitude greater than the flux inferred from ^{26}Al. This requires an unusual flux history with very high fluxes from 100 = 2 × 10^6 years ago and then, around a few million years ago, a rapid decline by almost two orders of magnitude.

Rancitelli et al. (1972) have reiterated the theme of constancy of solar flares during the past 10^6 years from measurements of *Apollo 14* and *15* samples. The average proton flux was deduced as 60 protons cm^{-2} s^{-1} with a shape factor γ (from $\frac{dj}{dE} = kE^{-\gamma}$) = 3.1. Begemann et al. (1972) measured ^{14}C in a rock as a function of depth and concluded that a flux of 300-500 protons cm^{-2} s^{-1} with R_o = 100 MV was required to account for the ^{14}C. They pointed out that some, if not all, of the apparent excess ^{14}C might be produced in the solar surface by way of ^{14}N (n,p) ^{14}C, with the ^{14}C carried to the moon in flares. D'Amico et al. (1971) had already suggested a similar mechanism, i.e., ^{3}He (n,p) ^{3}H for excess ^{3}H. The required fluxes are ^{14}C = 1 × 10^{-3} atoms cm^{-2} s^{-1} and ^{3}H = 5 × 10^{-3} atoms cm^{-2} s^{-1}.

Wahlen et al. (1972) revised their model calculations from the case of infinite plane geometry to a more realistic hemispherical geometry. From studies of two *Apollo 14* rocks they deduced

Nuclide	Shape factor R_o (MV)	J (E > 10MeV) protons cm^{-2} s^{-1} (4 π)
^{22}Na	85	110
^{56}Fe	100	100
^{26}Al	100	80
^{53}Mn	100	90

The differences are not significant, hence the results strengthen the earlier observations of constant solar cosmic rays. Bhandari, Bhattacharaya, and Padia (1976), from studies of ^{26}Al in four lunar rocks, concluded that the average solar cosmic ray intensity has been constant within ±20% during the period from ~4× 10^6 to ~0.6 × 10^6 years ago. Fireman et al. (1972) concluded from ^{3}H and ^{39}Ar measurements that the flare intensity averaged over the past 30 years has been the same as that averaged over the past 100 years. Fireman et al. (1973) quantified this statement with the deduction that the average proton flux, E > 50 MeV, was approximately 8 × 10^8 protons cm^{-2} y^{-1}. Walton (1974) reviewed the data in the light of new cross section data in Ca and Ti and concluded that the ^{39}Ar data imply that, on the average, one flare per year with an intensity similar to that of the flare on 4 August 1972 occurred during the last 1000 years, or several flares of lesser intensity, adding up to the same number of protons, occurred.

Accounts of short-lived radioactivities (^{3}H, ^{22}Na, ^{48}V, ^{51}Cr, ^{7}Be, ^{56}Co, ^{46}Sc, etc.) have been made for samples from all six Apollo missions. The results, in terms of numbers of protons and their spectral distribution, have in general conformed with observations made by way of satellites.

In summary, theoretical models for the production of radionuclides in lunar rocks and soil covers by solar cosmic rays have been developed. The agreement between calculated and observed production rates of several short-lived activities is good. Application to long-lived activities shows that the average solar flare intensity and spectral distribution does not seem to have changed substantially (e.g., by no more than a factor of two) during the past 2×10^6 years. Apparent increases during the past 15,000-20,000 years and the past 100 years are indicated, respectively, by ^{14}C and ^{34}Ar. The puzzling $^{236}U/^{238}U$ result has not been resolved.

IV. SOLAR FLARES, PART II: STABLE PRODUCTS

The detection of stable nuclides produced by solar flares is more difficult than the detection of radioactive species because the solar flare component is more easily masked by indigenous atoms and by the galactic proton-produced component. Measurements of ^3He, ^4He, ^{20}Ne, ^{21}Ne, ^{22}Ne, ^{36}Ar, and ^{38}Ar, cross sections in magnesium, aluminium, silicon, and calcium, made by Walton (1974), who used protons in the energy range 10-45 MeV, have provided a quantitative basis for inert gas studies. Walton concluded that the $^3He/^4He$ ratio produced by solar flares in most silicate materials is too similar to the ratio produced by galactic cosmic rays and therefore that helium is not suitable for solar flare studies. He concluded furthermore that the $^{21}Ne/^{22}Ne$ ratio produced by solar flares in magnesium-rich materials (a few percent or more of magnesium) is too similar to the ratio produced by galactic cosmic rays. However, aluminium-rich materials such as plagioclase feldspar are potentially useful because of the small (<0.1) $^{21}Ne/^{22}Ne$ ratio produced by solar flares in aluminium. Walton also concluded that $^{36}Ar/^{38}Ar$ ratios are of little use but that $^{21}Ne/^{38}Ar$ ratios may be useful, particularly in minerals poor in magnesium and rich in aluminium and calcium, because of the copious solar flare production of ^{38}Ar in calcium.

Significant exposure to solar flare protons should be indicated by unusually low $^{21}Ne/^{38}Ar$ ratios. Unfortunately, diffusive loss of inert gases also decreases this ratio for the gas remaining behind in minerals. Consequently, the $^{21}Ne/^{38}Ar$ record will always remain somewhat ambiguous.

Walton (1974) reviewed evidence that helium, neon, and argon are produced by solar flares in meteorites and lunar samples. He concluded that the evidence is nearly always ambiguous. This is particularly true for meteorites and lunar samples with low $^{21}Ne/^{38}Ar$ ratios, which may be a result of either copious solar flare production of ^{38}Ar or preferential ^{21}Ne diffusion loss.

Walton et al. (1975) reviewed the neon systematics of gas-rich meteorites as reported by Black (1972a). From heating experiments, Black (1972a) concluded that these meteorites contain a neon component, called Neon-C, with $^{20}Ne/^{22}Ne$

= 10.6 ± 0.3 and ^{21}Ne/^{22}Ne = 0.042 ± 0.003, which he attributed to neon directly implanted by solar flares. This conclusion implies that solar flare neon is significantly different, isotopically, from solar wind neon. Walton et al. (1975) argued that Neon-C may only be an apparent component, i.e., a mixture of solar wind neon and neon produced by solar flares in the aluminium-rich minerals of the meteorites. They concluded further that the gas-rich meteorites must, in that case, have been subjected to very large proton fluxes (E > 10 MeV) of at least 630 protons cm^{-2} s^{-1}, and perhaps up to 10^4 times greater, depending on the trapping efficiency of solar wind neon in the meteoritic minerals. In the context of current thinking, the possible exposure to these high fluxes occurred as much as ~4.6 $\times 10^9$ years ago. Walton et al. (1975) pointed out that their interpretation of the neon systematics in gas-rich meteorites allows the isotopic compositions of solar wind and solar flare neon to be identical.

Rao et al. (1971) reported strong evidence for solar flare-produced ^{132}Xe in lunar near-surface material. Their data seem to agree with a flux of 100 protons cm^{-2} s^{-1} (E > 10MeV). However, the pertinent cross sections in barium are not known; therefore, the conclusion cannot be firm.

V. SOLAR FLARES, PART III: TRACKS

Latent nuclear tracks are produced in solids when high energy ions are slowed down and stopped in them. The damaged regions of the solid can be revealed by a number of techniques, such as electron micrography or etching. The issues are too complex to be discussed here; the interested reader is referred to a recent book by Fleischer et al. (1976).

The tracks provide a variety of information about the radiation that produced them. Track densities are related to radiation dose or fluence, track gradients (i.e., track density as a function of depth) carry information on energy spectra, and track length is, among other factors, related to the mass of the ion that produced the track.

Information about the ancient environment of the sun from track studies has great potential, particularly when such studies are combined with investigations of trapped solar wind particles and radioactive products. Such studies are in progress in several laboratories around the world.

Pellas, Poupeau, and Lorin (1968), Pellas et al. (1969), and Lal and Rajan (1969) were the first to raise the issue of solar flare tracks when they discovered very large track densities ($\simeq 10^9$ tracks cm^{-2}) in exterior surfaces of crystals and chondrules* in gas-rich meteorites. They concluded that the irradiating particles

---

* Chondrules are near-spherical objects, usually <1 mm in diameter, that are present in certain stony meteorites.

(i.e., nuclei of the iron group) were probably of solar origin. The deduced energy spectrum of the iron nuclei ($1/E$ to $1/E^2$) implies that the irradiation took place at a time when the thickness of the inner solar system was less than ~1 mg cm^{-2}, otherwise the ions could not have penetrated to the galactic plane. The principal argument in favor of an early irradiation is the observation that the crystals show tracks on all sides, which seems to imply that they were irradiated while they were still dispersed in space, i.e., prior to their incorporation into the more massive meteorites. The good correlation between tracks and large concentrations of inert gases implanted by the solar wind seems to be strong support for this hypothesis.

The accumulated experience from the Apollo program, from on-site geologic observations of the astronauts as well as from studies of lunar samples themselves, shows that the features observed by Pellas *et al.* can be produced in the soil cover of an object with no atmosphere prior to the formation of tangible rocks by impact processes. This is not to say that the tracks in gas-rich meteorites as observed by Pellas *et al.* are not of great antiquity, but to point out that the exact location of the grains at the time of exposure is unclear.

Recently Poupeau *et al.* (1974) investigated the solar wind and solar flare record in aubrites, a type of meteorite whose component grains must have been irradiated in the comminuted soil cover of their parent body. The track densities of individual crystals correspond to about 10^3-10^4 years of exposure to solar flares at a radial distance of ~3 AU from the sun with present-day solar flare intensity. This is about 100 times shorter than for typical lunar soils. One possible interpretation is that the aubritic materials were exposed to solar flares during a period of intense bombardment of their parent objects with solid debris, perhaps at the time of the so-called lunar cataclysm 4×10^9 years ago (Tera, Papanastassiou, and Wasserburg, 1973). High impact rates might limit the residence time of grains in soil covers, particularly on relatively small parent objects, by ejection of the soil into interplanetary space and by copious formation of soil breccias, rocks that shield the grains in their interiors against further exposure to solar flares.

The study of tracks in lunar samples has significantly broadened our understanding of solar flare and solar wind output. Price *et al.* (1973) concluded from a comparison of solar flare Fe tracks in the returned Surveyor camera glass, in several lunar samples, and in the Fayetteville meteorite that the distribution of energy in solar flares has not changed since 4×10^9 years ago. The implicit assumption is that the Fayetteville meteorite was assembled at least that long ago. Lee (1976), however, has recently stated that the modulation of solar flare particles between 1 and 3 AU cannot be ignored. His calculations, which are based on standard diffusion-convection-adiabatic deceleration theory with a diffusion coefficient independent of the distance from the sun, show that the spectrum at 3 - AU should be slightly less steep than that at 1 AU.

Poupeau *et al.* (1973) concluded from a comparison of track densities in the older lunar highland soils (*Luna 20*) and track densities in younger mare soils that the sun was more active at the beginning of the solar system. Crozaz et al. (1974), from studies of two mare soils, find no evidence for an early active sun. However, they point out that their results do not rule out such an epoch in the early solar history.

The absolute flux of heavy solar flare particles over long periods in the past is still poorly established (Walker, 1975).

VI. PARTICLES FROM THE SUN DURING VERY EARLY TIMES; NUCLEOSYNTHESIS IN THE EARLY SOLAR SYSTEM

Geochemical and isotopic studies of matter in the solar system have revealed variations in the isotopic composition of many elements. The great majority of such variations have been accounted for by known, nonnuclear chemical, physical, or biological fractionation processes, by the decay of natural radioactivities, or by the production of stable isotopes by bombardment with cosmic rays. The remaining variations, often called isotopic anomalies, have generated a variety of interpretations. These anomalies occur in the elements oxygen, neon, magnesium, mercury, and xenon.

It is widely accepted that the basic pattern of elemental and isotopic abundances in the solar system was established by contributions from big-bang nucleosynthesis (Reeves, 1974). The isotopic anomalies mentioned above are usually considered to be modulations superimposed on the basic pattern by events either just prior to the formation of the solar system or during the earliest phases of its evolution.

One group of investigators contends that the anomalies have their origins in special nucleosynthetic events originating outside the solar system (Reynolds, 1960; Black, 1972a; Clayton, Grossman, and Mayeda, 1973; Lee and Papanastassiou, 1974; Clayton, 1975). Because the sun is not involved in these theories, I will not discuss them here. Another theory is that the anomalies may be, at least in part, due to solar particle irradiation of matter in the early solar system.

Perhaps the most substantial study of "nucleosynthesis during the early history of the solar system" is the paper of this title by Fowler, Greenstein, and Hoyle (1962). These investigators considered the synthesis of ^{2}D, ^{6}Li, ^{7}Li, ^{9}Be, ^{10}B, and ^{11}B in icy planetesimals from 1 to 50 m in diameter, which were bombarded with charged particles (average energy 500 MeV/nucleon) accelerated in magnetic flares at the surface of the condensing sun. Reeves (1974) traced the further development of this idea. Bernas et al. (1965) showed that the role of neutrons was strongly overestimated in the Fowler, Greenstein, and Hoyle model. Gradsztajn (1965) and Bernas et al. (1967) suggested that the synthesis occurred during bombardment of the gaseous protosolar nebula. Ryter et al. (1970) also raised serious objections, and the Fowler, Greenstein, and Hoyle model and its variants were then abandoned in favor of nucleosynthesis of the light elements by big-bang, etc.

The neon anomalies are both complex and huge. We know that the isotopic composition of neon in the sun (^{20}Ne/^{22}Ne ≈ 12-13) is radically different from that in the earth's atmosphere (^{20}Ne/^{22}Ne = 9.8). Carbonaceous meteorites (stony meteorites that appear to be very primitive) are thought to contain Neon-A with ^{20}Ne/^{22}Ne ~ 8.2 (Pepin, 1967) and Neon-E (Black, 1972a) with ^{20}Ne/^{22}Ne $\lesssim$ 1.3 (Eberhardt, 1975). In these components, ^{21}Ne/^{22}Ne varies somewhat but is always less than 0.04. Several authors have suggested that Neon-E might be pure ^{22}Ne (Jeffery and Anders, 1970; Black, 1972a; Eberhardt, 1974). Neon with ^{20}Ne/^{22}Ne << 12 is known to be produced by galactic protons as well as by protons of the energy range from 10 to 45 MeV (Walton, 1974) in major target elements such as magnesium, aluminium, silicon, and iron. The main problem is that the bombardment of these elements by protons usually results in the production of ^{21}Ne/^{22}Ne >> 0.04. Jeffery and Anders (1970), who were among the first to consider the effects of proton bombardment of solid grains in the early solar system, concluded that the proton spectrum must have had a very steep slope or a cutoff near 7 MeV (no protons of higher energy) because of the onset of ^{24}Mg (p, α) ^{21}Na → ^{21}Ne. They deduced an integrated proton flux of 10^{15}-10^{16} cm^{-2}. A problem shared by all theories that call on the irradiation of the solar nebula is, in the words of Jeffery and Anders, "The range of 7 MeV protons is only 0.08 g/cm^2, far less than the predicted surface densities in the inner parts of the nebula 10^3-10^4 g/cm^2 (Cameron, 1963)." That is, if the sun were the source of the protons, only a very small fraction of the material in the solar nebula could have been irradiated, if one accepts the customary theories of evolution of the solar nebula. The alternative model of Jeffery and Anders, ^{22}Na production in hot silicate grains that lose any ^{21}Ne produced by proton bombardment, relaxes the constraint on the energy spectrum of the protons (in fact, almost any spectrum is acceptable) but encounters very serious difficulties of its own because of the short mean lifetime of ^{22}Na. Herzog (1972), who has reconsidered the problem, affirms that the irradiation hypothesis survives only under very restrictive conditions. More recently, Adouze et al. (1975) reported the exciting discovery of amorphous, heavily irradiated grains of submicron size in a material rich in Neon-E from the Orgeuil meteorite. Since the grains are rich in magnesium, the authors invoke the mechanisms originally proposed by Herzog (1972), whereby ^{22}Ne is produced by ^{25}Mg (p, α) ^{22}Na → ^{22}Ne. To avoid overproduction of ^{21}Ne by ^{24}Mg (p, α) ^{21}Na → ^{21}Ne and of ^{20}Ne by ^{23}Na (p, α) ^{20}Ne, the authors require a steeply falling proton energy spectrum with truncation at the lower end (below about 3 MeV) because ^{23}Na (p, α) ^{20}Ne has a threshold energy of about 1 MeV. The authors argue that the irradiation could have taken place in the solar cavity and would thus provide direct information on proton fluxes and energy spectra from the early sun. However, other astrophysical sites, such as the expanding envelope of a supernova remnant, have also been considered by Adouze et al.

Heymann and Dziczkaniec (1975) pointed out that several problems connected with the irradiation of solid grains can be alleviated by considering the proton bombardment of a cooling gas phase in which condensation takes place. These authors first considered the magnesium anomalies reported by Gray and Compston (1974) and Lee and Papanastassiou (1974) and observed that the positive and negative ^{26}Mg anomalies observed in the Allende meteorite could be explained by the ^{26}Mg (p,n) ^{26}Al reaction in a gas phase, with proton fluxes ranging from 5.7×10^{17} to 4.3×10^{19} protons cm^{-2} y^{-1} at 1 MeV, depending on the steepness of the assumed exponential energy spectra. The more recent observations on magnesium anomalies by Lee, Papanastassiou, and Wasserburg (1975) allow reduction of these fluxes by at least a factor of 50, and unpublished calculations in our laboratory that take into account the effects of relative aluminum and magnesium condensation rates indicate that fluxes $\sim 10^{14}$ protons cm^{-2} y^{-1} at 1 MeV may be possible. Heymann and Dziczkaniec (1975) pointed out that the ^{22}Ne(p,n) ^{22}Na scheme with sodium condensation predicts quite closely the amounts of Neon-E observed in the Orgueil meteorite by Eberhardt (1974), thus linking the neon and magnesium anomalies to a common origin. Heymann and Dziczkaniec (1975) advocate the sun as the source of the protons.

It would be premature to present any conclusive statement about particle flux from the sun. Suffice it to say that no serious attempts to account for the oxygen and xenon anomalies by particle irradiation from the sun have been reported. Even if the latter should turn out to be a result of processes outside the solar system, one would hope that at least the neon anomalies and perhaps the magnesium anomalies are related to the particle output of the very young sun. I do not know of any other record that would permit such detailed inferences about particle fluxes and energy spectra as that afforded by these anomalies.

BIBLIOGRAPHY OF PAPERS ON SOLAR FLARE
RADIONUCLIDES IN LUNAR SCIENCE CONFERENCE
PROCEEDINGS I—VI (1970-1975)

I. *Proceedings of the Apollo 11 Lunar Science Conference*, Vol. 2 (Suppl. 1, *Geochim. Cosmochim. Acta)*, ed. A.A. Levinson, Pergamon Press, Elmsford, New York, 1970.

Begemann, F.; Vilcsek, E.; Rieder, R.; Born, W.; and Wänke, H., pp. 995-1006.

D'Amico, J., DeFelice, J., and Fireman, E.L., pp. 1029-1036.

Fields, P.R.; Diamond, H.; Metta, D.N.; Stevens, C.M.; Rokop, D.J.; and Moreland, P.E., pp. 1097-1102.

Herzog, G.F., and Herman, G.F., pp. 1239-1246.

O'Kelley, G.D.; Eldridge, J.S.; Schonfeld, E.; and Bell, P.R., pp. 1407-1424.

Perkins, R.W.; Rancitelli, L.A.; Cooper, J.A.; Kaye, J.H.; and Wogman, N.A., pp. 1455-1470.
Shedlovsky, J.P.; Honda, M.; Reedy, R.C.; Evans, J.C., Jr.; Lal, D.; Lindstrom, R.M.; Delany, A.C.; Arnold, J.R.; Loosli, H.H.; Fruchter, J.S.; and Finkel, J.R., pp. 1503-1532.
Stoenner, R.W., Lyman, W.J., and Davis, R., Jr., pp. 1583-1594.
Wrigley, R.C. and Quaide, W.L., pp. 1751-1756.

II. *Proceedings of the Second Lunar Science Conference*, Vol. 2 (Suppl. 2 *Geochim. Cosmochim. Acta)*, ed. A.A. Levinson, M.I.T. Press, Cambridge, Mass., 1971.

Armstrong, T.W., and Alsmiller, R.G., pp. 1729-1746.
O'Kelley, G.D.; Eldridge, J.S.; Schonfeld, E.; and Bell, P.R., pp. 1747-1756.
Rancitelli, L.A.; Perkins, R.W.; Felix, W.D.; and Wogman, N.A., pp. 1757-1772.
Finkel, R.C.; Arnold, J.R.; Imamura, M.; Reedy, R.C.; Fruchter, J.C.; Loosli, H.H.; Evans, J.C.; Delany, A.C.; and Shedlovsky, J.P., pp. 1773-1790.
Wrigley, R.C., pp. 1791-1796.
Bochsler, P.; Eberhardt, P.; Geiss, J.; Loosli, H.; Oeschger, H.; and Wahlen, M., pp. 1803-1812.
Stoenner, R.W., Lyman, W., and Davis, R., Jr., pp. 1813-1824.
D'Amico, J.; DeFelice, J.; Fireman, E.L.; Jones, C.; and Spannagel, G., pp. 1825-1842.
Fields, P.R.; Diamond, H.; Metta, D.N.; Stevens, C.M.; and Rokop, D.J., pp. 1571-1576.
Marti, K., and Lugmair, G.W., pp. 1590-1606.

III. *Proceedings of the Third Lunar Science Conference*, Vol. 2 (Suppl. 3, *Geochim. Cosmochim. Acta)*, ed. D. Criswell et al., M.I.T. Press, Cambridge, Mass., 1972.

Fields, P.R.; Diamond, H.; Metta, D.N.; Rokop, D.J., and Stevens, C.M., pp. 1637-1644.
Wahlen, M.; Honda, M.; Imamura, M.; Fruchter, J.S.; Finkel, D.C.; Kohl, C.P.; Arnold, J.R.; and Reedy, J.S., pp. 1719-1732.
Yokoyama, Y., Auger, R.; Bibron, R.; Chesselet, R.; Guichard, F.; Leger, C. Mabuchi, H.; Reyss, J.L.; and Sato, J., pp. 1733-1746.
Fireman, E.L.; D'Amico, J.; DeFelice, J.; and Spannagel, G., pp. 1747-1762.
Eldridge, J.S.; O'Kelley, G.D.; Northcutt, K.J.; and Schonfeld, E., pp. 1651-1670.
Keith, J.E., Clark, R.S., and Richardson, K.A., pp. 1671-1680.
Rancitelli, L.A.; Perkins, R.W.; Felix, W.D.; and Wogman, N.A., pp. 1681-1692.
Begemann, F.; Born, W.; Palme, H.; Vilcsek, E.; and Wänke, H., pp. 1693-1702.

IV. *Proceedings of the Fourth Lunar Science Conference*, Vol. 2. (Suppl. 4, *Geochim. Cosmochim. Acta)*, ed. W.A. Gose, Pergamon Press, Elmsford, N.Y., 1973.

Clark, R.S., and Keith, J.E., pp. 2105-2114.
Eldridge, J.S., O'Kelley, G.D., and Northcutt, K.J., pp. 2115-2122.
Fields, P.R.; Diamond, H.; Metta, D.N.; and Rokop, K.J., pp. 2123-2130.
Fireman, E.L., D'Amico, J., and DeFelice, J., pp. 2131-2144.

Yokoyama, Y.; Sato, J.; Reyss, J.L.; and Guichard, F., pp. 2209-2228.

V. *Proceedings of the Fifth Lunar Science Conference*, Vol. 2. (Suppl. 5, *Geochim. Cosmochim. Acta*), ed. W.A. Gose, Pergamon Press, Elmsford, N.Y., 1974.

Imamura, M.; Nishiizumi, I.; Honda, M.; Finkel, R.C.; Arnold, J.; and Kohl, C.P., pp. 2093-2104.
Keith, J.E., Clark, R.S., and Bennett, L.J., pp. 2121-2138.
O'Kelley, G.D., Eldridge, J.S., and Northcutt, K.J., pp. 2139-2148.
Rancitelli, L.A.; Perkins, R.W.; Felix, W.D.; and Wogman, N.A., pp. 2185-2204.
Walton, J.R.; Heymann, D.; Jordan, J.L.; and Yaniv, A., pp. 2045-2060.
Yokoyama, Y., Reyss, J.L., and Guichard, F., pp. 2231-2248.

VI. *Proceedings of the Sixth Lunar Science Conference*, Vol. 2 (Suppl. 6, *Geochim. Cosmochim. Acta*), Pergamon Press, Elmsford, N.Y., 1975.

Yokoyama, Y., Reyss, J.L., and Guichard, F., pp. 1823-1844.
Keith, J.E., Clark, R.S., and Bennett, L.J., pp. 1879-1890.
Rancitelli, L.A.; Fruchter, J.S.; Felix, W.D.; Perkins, R.W.; and Wogman, N.A., pp. 1891-1900.
Bhandari, N.,Bhattacharaya,S.K., and Padia, J.T., pp. 1913-1926.
Leich, D.A.; Niemeyer, S.; Rajan, R.S.; and Srinivasan, B., pp. 2085-2095.

REFERENCES

Adouze, J.; Bibring, J.P.; Dran, J.C.; Maurette, M.; and Walker, R.M., 1975, in press, *Ap. J.*
Baur, H.; Frick, U.; Funk, H.; Schultz, L.; and Signer, P., 1972, *Proceedings Third Lunar Science Conference*, ed. D. Criswell *et al.*, M.I.T. Press, Cambridge, Mass., 1947.
Becker, R.H., and Clayton, R.N., 1975, *Proceedings Sixth Lunar Science Conference*, Vol. 2, Pergamon Press, Elmsford, N.Y., 2131.
Begemann, F.; Vilcsek, E.; Rieder, R.; Born, W.; and Wänke, H., 1970, *Proceedings Apollo 11 Lunar Science Conference* (Suppl. 1, *Geochim. Cosmochim. Acta*), ed. A.A. Levinson, Pergamon Press, Elmsford, N.Y., 995.
—————; Born, W.; Palme, H.; Vilcsek, E.; and Wanke, H., 1972, *Proceedings Third Lunar Science Conference* (Suppl. 3, *Geochim. Cosmochim. Acta*), ed. D. Criswell *et al.*, M.I.T. Press, Cambridge, Mass., 1693.
Bernas, R.; Epherre, M.; Gradsztajn, E.; Klapisch, R.; and Yiou, F., 1965, *Phys. Letters 15*, 147.
—————; Gradsztajn, E.; Reeves, H.; and Schatzmann, E., 1967, *Ann. Phys. New York 44*, 426.
Bhandari, N., Bhattacharaya, K.S., and Padia, J.T., 1976, *Lunar Science VII*, The Lunar Science Institute, Houston, Tex., 49.

Black, D.C., 1972a, *Geochim. Cosmochim. Acta 36*, 347.

——————, 1972b, *ibid. 36*, 377.

Bogard, D.D., and Nyquist, L.E., 1972, *The Apollo 15 Lunar Samples*, The Lunar Science Institute, Houston, Tex., 342.

Borg, J.; Maurette, M.; Durrieu, L.; and Jouret, C., 1971, *Proceedings Second Lunar Science Conference* (Suppl. 2, *Geochim. Cosmochim. Acta*), ed. A.A. Levinson, M.I.T. Press, Cambridge, Mass., 2027.

Cameron, A.G.W., 1963, *Icarus 1*, 339.

Clayton, D.D., 1975, *Nature 275*, 36.

Clayton, R.N., Grossman, L., and Mayeda, T.K., 1973, *Science 182*, 485.

Criswell, D.R., and Freeman, J.W., ed., 1974, *Proceedings, Conference on Interactions of the Interplanetary Plasma with the Modern and Ancient Moon*, The Lunar Science Institute, Houston, Tex.

Crozaz, G.; Taylor, G.J.; Walker, R.M.; and Seitz, M.G., 1974, *Proceedings Fifth Lunar Science Conference* (Suppl. 4, *Geochim. Cosmochim. Acta*), ed. W.A. Gose, Pergamon Press, Elmsford, N.Y., 2591.

Curtis, D.B., and Wasserburg, G.J., 1975, *The Moon 13*, 185.

D'Amico, J., DeFelice, J., and Fireman, E.L., 1970, *Proceedings Apollo 11 Lunar Science Conference* (Suppl. 1, *Geochim, Cosmochim. Acta*), ed. A.A. Levinson, Pergamon Press, Elmsford, N.Y., 1029.

——————; ——————; ——————; Jones, C.; and Spannagel, G., 1971, *Proceedings Second Lunar Science Conference* (Suppl. 2, *Geochim. Cosmochim. Acta*), ed. A.A. Levinson, M.I.T. Press, Cambridge, Mass. 1875.

Davis, R., Jr., Stoenner, R.W., and Schaeffer, O.A., 1963, *Radioactive Dating*, International Atomic Energy Agency, Vienna, 355.

Eberhardt, P., 1974, *Earth Planet. Sci. Letters 24*, 182.

——————, 1975, *Meteoritics*, 401.

——————, Geiss, J., and Grögler, N., 1965a, *Mineral. Petrog. Mitt. 10*, 535.

——————, ——————, and ——————, 1965b, *J. Geophys. Res. 70*, 4375.

——————; ——————; Graf, H.; Grögler, N.; Krähenbühl, U.; Schwaller, H.; Schwarzmüller, J.; and Stettler, A., 1970, *Proceedings Apollo 11 Lunar Science Conference*, ed. A.A. Levinson, Pergamon Press, Elmsford, N.Y., 1037.

——————; ——————; ——————; ——————; Mendia, M.D.; Morgeli, M., Schwaller, H.; and Stettler, A., 1972, *Proceedings Third Lunar Science Conference*, ed. D. Criswell *et al.*, M.I.T. Press, Cambridge, Mass., 1821.

Eldridge, J.S., O'Kelley, G.D., and Northcutt, K.J., 1973, *Proceedings Fourth Lunar Science Conference* (Suppl. 4, *Geochim. Cosmochim. Acta*), ed. W.A. Gose, Pergamon Press, Elmsford, N.Y. 2115.

Epstein, S., and Taylor, H.P., Jr., 1972, *Proceedings Third Lunar Science Conference*, ed. D. Criswell *et al.*, M.I.T. Press, Cambridge, Mass., 1429.

—————— and ——————, 1974, *Proceedings Fifth Lunar Science Conference*, ed. W.A. Gose, Pergamon Press, Elmsford, N.Y. 1839.

Fields, P.R.; Diamond, H.; Metta, D.N.; Stevens, C.M.; and Rokop, D.J., 1971, *Proceedings Second Lunar Science Conference* (Suppl. 2, *Geochim. Cosmochim. Acta*), ed. A.A. Levinson, M.I.T. Press, Cambridge, Mass., 1571.

Finkel, R.C.; Arnold, J.R.; Imamura, M.; Reedy, R.C.; Fruchter, J.C.; Loosli, H.H.; Evans, J.C.; Delany, A.C.; and Shedlovsky, J.P., 1971, *ibid.*, 1773.

Fireman, E.L., and Goebel, R., 1970, *J. Geophys. Res.* **75**, 215.

_____, D'Amico, J., and DeFelice, J., 1973, *Proceedings Fourth Lunar Science Conference* (Suppl. 4, *Geochim. Cosmochim. Acta*), ed. W.A. Gose, Pergamon Press, Elmsford, N.Y., 2131.

_____; D'Amico, J.; DeFelice, J.; and Spannagel, G., 1972, *Proceedings Third Lunar Science Conference* (Suppl. 3, *Geochim. Cosmochim. Acta*), ed. D. Criswell *et al.*, M.I.T. Press, Cambridge, Mass. 1747.

_____, DeFelice, J., and Tilles, D., 1963, *Radioactive Dating*, International Atomic Energy Agency, Vienna, 323.

Fleischer, R.L., Price, P.B., and Walker, R.M., 1976, *Nuclear Tracks in Solids: Principles and Applications*, University of California Press.

Fowler, W.A., Greenstein, J.L., and Hoyle, F., 1962, *Geophys. J.R.A.S.* **6**, 148.

Geiss, J.; Bühler, F.; Cerutti, H.; and Eberhardt, P., 1972a, *Apollo 15 Preliminary Science Report NASA SP-289*.

_____; _____; _____; _____; and Filleux, Ch., 1972b, *Apollo 16 Preliminary Science Report NASA SP-315*.

_____; _____; _____; _____; and Meister, J., 1971, *Apollo 14 Preliminary Science Report NASA SP-272*.

_____, Eberhardt, P.; Bühler, F.; Meister, J.; and Signer, P., 1970, *J. Geophys. Res.* **74**, 5972.

Gradsztajn, E., 1965, *Ann. Phys. Paris* **10**, 791.

Gray, C.M., and Compston, W., 1974, *Nature* **251**, 495.

Herzog, G.F., 1972, *J. Geophys. Res.* **77**, 6219.

_____ and Herman, G.F., 1970, *Proceedings Apollo 11 Lunar Science Conference*, ed. A.A. Levinson, Pergamon Press, Elmsford, N.Y., 1239.

Heymann, D., 1975, *Earth Planet. Sci. Letters* **27**, 445.

_____ and Yaniv, A., 1970, *Proceedings Apollo 11 Lunar Science Conference*, ed. A.A. Levinson, Pergamon Press, Elmsford, N.Y., 1261.

_____ and Dziczkaniec, M., 1975, *Science* **191**, 79.

Hintenberger, H., and Weber, H., 1973, *Proceedings Fourth Lunar Science Conference* (Suppl. 4, *Geochim. Cosmochim. Acta*), ed. W.A. Gose, Pergamon Press, Elmsford, N.Y., 2003.

Jeffery, P.M., and Anders, E., 1970, *Geochim. Cosmochim Acta* **34**, 1175.

Kaplan, I.R., Smith, J.W., and Ruth, E., 1970, *Proceedings Apollo 11 Lunar Science Conference*, ed. A.A. Levinson, Pergamon Press, Elmsford, N.Y., 1317.

Kerridge, J.F., Kaplan, I.R., and Lesley, F.D., 1974, *Proceedings Fifth Lunar Science Conference*, W.A. Gose, Pergamon Press, Elmsford, N.Y., 1855.

_____, _____, and Petrowski, C., 1975, *Proceedings Sixth Lunar Science Conference*, ed., Pergamon Press, Elmsford, N.Y., 2151.

LSAPT, 1971, *Apollo 14 Preliminary Science Report NASA SP-272*.

Lal, D., and Rajan, R.S., 1969, *Nature* **223**, 269.

Lee, M.A., 1976, in press.

Lee, T., and Papanastassiou, D., 1974, *Geophys. Res. Letters* **1**, 225.

————————, ————————, and Wasserburg, G.J., 1976, *Geophys. Res. Letters 3*, 109.
Manka, R.H., and Michel, F.C., 1970, *Science 169*, 278.
Mazor, E., Heymann, D., and Anders, E., 1970, *Geochim. Cosmochim. Acta 34*, 781.
Meister, J., 1969, Ph.D. thesis, University of Bern, Switzerland.
O'Kelley, G.D.; Eldridge, J.S.; Schonfeld, E.; and Bell, P.R., 1970, *Proceedings Apollo 11 Lunar Science Conference* (Suppl. 1, *Geochim. Cosmochim Acta*), ed. A.A. Levinson, Pergamon Press, Elmsford, N.Y., 1407.
Pellas, P., Poupeau, G., and Lorin, J.C., 1968, *Bull. Soc. Cedilla Française Mineral Cristallog. 91*, 1.
————————; ————————; ————————; Reeves, H.; and Adouze, J., 1969, *Nature 223*, 272.
Pepin, R.O., 1967, *Earth Planet Sci. Letters 2*, 13.
Perkins, R.W.; Rancitelli, L.A.; Cooper, J.A.; Kaye, J.H.; and Wogman, N.A., 1970; *Proceedings Apollo 11 Lunar Science Conference* (Suppl. 1, *Geochim. Cosmochim. Acta*), ed. A.A. Levinson, Pergamon Press, Elmsford, N.Y., 1455.
Poupeau, G.; Chetrit, G.D.; Berdot, J.L.; and Pellas, P., 1973, *Geochim. Cosmochim. Acta 37*, 2005.
————————; Kirsten, T.; Steinbrunn, F.; and Storzer, D., 1974, *Max-Planck Institut für Kernphysik, Heidelberg*, pub. MPIH-1974, Vol. 14.
Price, P.B.; Chang, J.H.; Hutcheon, K.D.; MacDougall, D.; Rajan, R.S.; Shirk, E.K.; and Sullivan, J.D., 1973, *Proceedings Fourth Lunar Science Conference* (Suppl. 4, *Geochim. Cosmochim. Acta*), ed. W.A. Gose, Pergamon Press, Elmsford, N.Y., 2347.
Rancitelli, L.A.; Perkins, R.W.; Felix, W.D.; and Wogman, N.A., 1971, *Proceedings Second Lunar Science Conference* (Suppl. 3, *Geochim. Cosmochim. Acta*), ed. A.A. Levinson, M.I.T. Press, Cambridge, Mass., 1757.
————————; ————————; ————————; and ————————, 1972, *Proceedings Third Lunar Science Conference* (Suppl. 3, *Geochim. Cosmochim. Acta*), ed. D. Criswell et al., M.I.T. Press, Cambridge, Mass. 1681.
Rao, M.N.; Gopalan, K.; Venkatavaradan, V.S.; and Wilkening, L., 1971, *Nature 233*, 114.
Reedy, R.C., and Arnold, J.R., 1972, *J. Geophys. Res. 77*, 537.
Reeves, H., 1974, *Annu. Rev. Astr. and Ap. 12*, 437.
Reynolds, J.H., 1960, *Phys. Rev. Letters 4*, 8.
Russ, G.P., III, Burnett, D.S., and Wasserburg, G.J., 1972, *Earth Planet. Sci. Letters 15*, 172.
Russell, C.T., ed., 1974, *Solar Wind Three*, Institute of Geophysics and Planetary Physics, University of California, Los Angeles.
Ryter, C.; Reeves, H.; Gradsztajn, E.; and Adouze, J., 1970, *Astr. and Ap. 8*, 389.
Shedlovsky, J.P.; Honda, M.; Reedy, R.C.; Evans, J.C., Jr.; Lal, D.; Lindstrom, R.M.; Delany, A.C.; Arnold, J.R.; Loosli, H.H.; Fruchter, J.S.; and Finkel, R.C., 1970, *Proceedings Apollo 11 Lunar Science Conference*, ed. A.A. Levinson, Pergamon Press, Elmsford, N.Y., 1503.
Sonnett, C.P., Coleman, P.J., Jr., and Wilcox, J.M., ed., 1972, *Proceedings Solar Wind Conference, Asilomar 1971*, NASA SP-308.
Stoenner, R.W., Lyman, W.J., and Davis, R., Jr., 1970, *Proceedings Apollo 11 Lunar Science Conference*, ed. A.A. Levinson, Pergamon Press, Elmsford, N.Y., 1583.
Suess, H.E., Wänke, H., and Wlotzka, F., 1964, *Geochim. Cosmochim. Acta 28*, 595.
Tera, F., Papanastassiou, D.A., and Wasserburg, G.J., 1973, *Lunar Science IV*, Lunar Science Institute, Houston, Tex., 723.

Wahlen, M.; Honda, M.; Imamura, M.; Fruchter, J.S.; Finkel, D.C.; Kohl, C.P.; Arnold, J.R.; and Reedy, J.S., 1972, *Proceedings Third Lunar Science Conference* Suppl. 3, *Geochim. Cosmochim. Acta*), ed. D. Criswell *et al.*, M.I.T. Press, Cambridge, Mass., 1719.

Walker, R.M., 1975, *Ann. Rev. Earth and Planet Sci. 3*, 99.

Walton, J.R., 1974, Ph.D. thesis, Rice University, Houston, Tex.

——————; Heymann, D.; Jordan, J.L.; and Yaniv, A., 1974, *Proceedings Fifth Lunar Science Conference* (Suppl. 5, *Geochim. Cosmochim. Acta)*, ed. W.A. Gose, Pergamon Press, Elmsford, N.Y., 2045.

Wrigley, R.C., and Quaide, W.L., 1970, *Proceedings Apollo 11 Lunar Science Conference* (Suppl. 1, *Geochim. Cosmochim. Acta)*, ed. A.A. Levinson, Pergamon Press, Elmsford, N.Y., 1751.

Yaniv, A., and Heymann, D., 1972, *Proceedings Third Lunar Science Conference*, ed. D. Criswell *et al.*, M.I.T. Press, Cambridge, Mass., 1967.

Zahringer, J., 1964, *Annu. Rev. Astr. and Ap.*, 121.

Zinner, E.; Walker, R.M.; Chaumont, J.; and Dran, J.C., 1976, *Lunar Science VII*, Lunar Science Institute, Houston, Tex., 965.

SOLAR INDUCED VARIATIONS OF ENERGETIC PARTICLES AT ONE AU

Paul E. Damon
*Laboratory of Isotope Geochemistry,
Department of Geosciences,
University of Arizona
Tucson, Arizona*

Solar wind particles, solar flare particles, and galactic cosmic rays compose the flux of energetic particles in interplanetary space. Solar flare particles can be distinguished by their relationship to solar activity and by their differential energy spectrum, which follows a power law in kinetic energy of the form

$$dJ/dE = k'E^{-\gamma} \quad . \tag{1}$$

with γ typically between two and four (Walker, 1975). They constitute the dominant flux of energetic particles in the energy range from 100 keV/amu to about 100 MeV/amu. The solar flare particle flux can be shown to vary with solar activity as measured by the Wolf sunspot number. However, radioisotope data from lunar samples, with the exception of radiocarbon (^{14}C), indicate that the average flux and energy spectrum integrated over periods of time varying from decades to millions of years has not changed by more than a factor of two. Radiocarbon measurements on lunar samples indicate a much larger variation in the flux of solar flare particles (Boeckl, 1972; Begemann *et al.*, 1972). However, these measurements do not agree with the results of thermoluminescence measurements on lunar rocks (Hoyt, Walker, and Zimmerman, 1973). It will also be shown that strict limits can be set on the variability of solar activity by measurement of the ^{14}C activity of dendrochronologically dated tree rings.

I. SUNSPOTS, FLARES, AND SOLAR WIND

Flares are often seen to develop from preexisting plages. The plage swells and brightens and erupts into a mound, cone, or loop. The important point, in the present context, is that the plages, which give rise to flares, are nearly always found close to sunspots. Often, part of the flare occurs close to the area of maximum field strength associated with sunspots, and it is this configuration that has been associated with the ejection of particles from flares (see E. Tandberg-Hanssen, 1967, for a review of flare phenomena). Not surprisingly, therefore, there exists, according to Tandberg-Hanssen, a strong correlation between the frequency of flare incidence and Wolf sunspot number R defined as

$$R = 10 g + f \quad , \qquad (2)$$

where f is the total number of sunspots regardless of size, and g is the number of sunspot groups. There also exists a close correlation between sunspot number and the monthly mean of the total area A of all sunspots on the visible hemisphere at any given time (Waldmeier, 1955). The relationship between A and R is

$$A = 16.7 R \quad . \qquad (3)$$

It follows, then, that flares and flare particle emission should vary with long- and short-term solar activity variations such as the 11-y solar cycle. Indeed, phenomena such as the Forbush minima and polar cap absorption, associated with non-relativistic particle ejection from flares, show a close correlation with the 11-y cycle; and, also, ground-level enhancement events associated with relativistic flare particles tend not to occur during the years of minimum sunspot number because of the paucity of large flares (see review by Pomerantz and Duggal, 1974).

Solar activity also measurably changes the intensity of galactic cosmic rays incident on the earth's atmosphere, as first reported by Forbush (1954). It was formerly considered that the non-quiet sun increment of the solar wind must vary in intensity and velocity with solar flare activity and, when the geomagnetosphere and heliomagnetosphere varied in response to the changing solar wind, the cosmic ray flux reaching the earth's atmosphere would also change. Consequently, solar activity would modulate the galactic cosmic ray intensity in antiphase with solar activity and an 11-y cycle of galactic cosmic rays would be observed. Thus, according to this model, sunspot activity causes flares that emit high-energy particles and at the same time change the intensity of the solar wind that supplies the "driving force" for modulating the flux of galactic cosmic rays impinging on the earth's atmosphere. However, spacecraft measurements of solar wind over the past decade (Feldman et al., 1976, this volume) do not show significant variation of solar wind parameters in accordance with the above model. Although the model is likely to be more complex, the empirical relationship between galactic cosmic ray modulation and solar activity is well established (see also Lanzerotti, 1976, this volume).

The solar-activity-modulated galactic cosmic rays produce neutrons which, in turn, produce ^{14}C by an n,p reaction on atmospheric ^{14}N. The ^{14}C equilibrates with atmospheric CO_2, participates in the carbon-oxygen cycle, and is stored in organic matter, e.g., the annual growth rings of trees. Thus tree rings contain a record of past atmospheric ^{14}C concentrations produced by neutrons resulting from the solar-activity-modulated galactic cosmic ray flux. This record can be retrieved by measuring the ^{14}C activity and stable carbon isotope ratios of dendrochronologically dated tree rings. This information will be used in § III of this paper to evaluate variations in solar activity during the last 7,500 y.

II. MEASUREMENTS OF PROTONS FROM SOLAR FLARES

Average values for the solar proton flux, measured from balloons and satellites, are given in Table 1. The average omnidirectional solar proton flux of energy greater than 10 MeV for solar cycle 19 and the first eight years of solar cycle 20 is $J_{10} = 120$ p cm^{-2} s^{-1} with a spectral shape parameter (R_0) of about 90 MV as derived from the exponential rigidity distribution,

$$\frac{dJ}{dR} = k\, e^{-R/R_0} \quad , \tag{4}$$

where R is the rigidity of the particle in units of momentum per unit charge or megavolts (MV) and k is an integration constant. It is surprising that the proton fluxes (J_{10}) for solar cycles 19 and 20 are so similar when one considers that 70% of the total solar protons with energy greater than 10 MeV during the period

TABLE 1. Average solar proton fluxes for Apollo missions 11–17 from measurements of short-lived radioisotopes in lunar rocks and soils and measurements from high altitude balloons and satellites.

		$J\ (E_p > 10\ \text{MeV})$			
		^{56}Co ($\bar{T}$=110d)	^{54}Mn ($\bar{T}$=1.2 y)	^{55}Fe ($\bar{T}$=3.75 y)	^{22}Na ($\bar{T}$=3.75 y)
Mission and date	Ref.	(protons cm^{-2} s^{-1}, 4π)			
Apollo 11, July 1969	(4)				97(100)
Apollo 12, Nov. 1969	(1)	130(80)	135(85)	100(100)	140(90)
	(4)				75(100)
Apollo 14, Feb. 1971	(2)	200(100)		100(85)	100(85)
	(4)				94(100)
Apollo 15, July 1971	(3)				120
	(4)				64(100)
Apollo 16, Apr. 1972	(4)				63(100)
Apollo 17, Dec. 1972	(4)				275(100)
Average solar proton flux for Apollo 11-17 from ^{22}Na:					114(95)
Average solar proton flux for A.D. 1963-1971, measured during balloon and satellite experiments (Hoyt et al., 1973, from compilations by King, 1972):					130(80)
Average solar proton flux for solar cycle 19, A.D. 1953-1963, measured during balloon and satellite experiments (Hoyt et al., 1973, from compilations by Webber et al., 1963):					110(100)

The number in parentheses is R_0 in MV units.

Ref. (1): Finkel et al. (1971)   Ref. (2): Wahlen et al. (1972)
Ref. (3): Rancitelli et al. (1972)   Ref. (4): Yokoyama et al. (1973)

1966 to 1972 were emitted in a solar flare that occurred early in August of 1972 (Walker, 1975). Characteristically, several large flares usually contribute the bulk of the high-energy particles during a solar cycle.

The absence of a strong magnetic field and atmosphere makes the moon an ideal place to monitor particles from the sun. A glass filter, exposed for nearly three years from the Surveyor III spacecraft before its return on the Apollo 12 mission, showed that the energy spectrum of solar flare particles followed the relationship $E^{-\gamma}$ (equation 1) with $\gamma = 3$ (Walker, 1975). Time-dependent phenomena in lunar rocks, such as the production of radioisotopes in nuclear collisions and the storage of trapped electrons from ionization, have been used to determine the integrated average flux during the mean life of the radioisotope or trapped electrons. Average integrated solar proton fluxes from short-lived isotopes are given in Table 1. Except for two anomalous values, the average J_{10} determined from the radio-isotopes is 101 ± 25 (σ) p cm^{-2} s^{-1}. This flux is integrated over a period of time comparable to solar cycle 19 and the first eight years of solar cycle 20 (1953–1971). This may be compared with the average J_{10} of 120 p cm^{-2} s^{-1} as measured from high-altitude balloons and satellites. If only the fluxes determined from ^{22}Na are included, the integrated flux, 114 p cm^{-2} s^{-1}, is in better agreement. The two anomalous values are correlated with unusually intense flare events. The value of 275 p cm^{-2} s^{-1} from samples collected during the Apollo 17 mission can be correlated with the anomalously large flare of August 1972. Also, the high value for ^{56}Co from a rock collected during the Apollo 14 mission may be compared with the flux measured by Bostrom, Williams, and Arens (1971) for the 24 January 1971 flare [$J_{10} = 160$ p cm^{-2} s^{-1}, $R_0 = 70$ MV]. An R_0 of 75 MV also fits the data of Wahlen et al. (1972), well within the limits of error. The corresponding flux of 175 p cm^{-2} s^{-1} is in excellent agreement with the satellite solar proton monitor experiment. Cobalt 56 in rock 14321 (Apollo mission 14) provided an effective monitor of the 24 January flare.

A tabulation of the data for longer lived isotopes, whose mean lives span the last five million years, is given in Table 2. The average J_{10} (excluding ^{14}C) determined from these isotopes is 100 p cm^{-2} s^{-1}, $R_0 = 100$ MV. Surprisingly, in view of the short-term variability of the solar flare J_{10}, this value is very close to the value of 120 p cm^{-2} s^{-1}, $R_0 = 90$ MV for 1953–1971 (solar cycles 19 and 20).

The fluxes calculated from ^{14}C are a factor of three to five higher. This anomalous value is not consistent with the J_{10} derived from other long-lived isotopes. Begemann et al. (1972) suspect that the excess ^{14}C was produced not in the rock but in the sun, as a result of an (n,p) reaction on solar ^{14}N. The excess ^{14}C is then erupted with solar protons during flares and implanted in lunar rocks. Recently, Fireman, D'Amico, and De Fleice (1976) presented evidence for derivation of the excess ^{14}C from the solar wind rather than from solar flares. Heating experiments demonstrate that the excess ^{14}C is readily degassed and hence superficially trapped like other solar wind particles. The results of Hoyt, Walker,

TABLE 2. Average solar proton fluxes during the past five million years from measurements of radioisotopes and thermoluminescence in lunar rocks.

Reference	Isotope	Mean-life ($\bar{T}$) Years	Spectral Parameter	4π flux E_p >10 MeV (prot. cm^{-2} s^{-1})
Wahlen et al. (1972)	^{26}Al	1.07×10^6	$R_o = 100$	80
	^{53}Mn	5.34×10^6	$R_o = 100$	90
Rancitelli et al. (1972)	^{26}Al	1.07×10^6	$\gamma = 3.1$	120
Yokoyama et al. (1973)	^{26}Al	1.07×10^6	$R_o = 150^*$	70^*
Boeckl (1972)	^{14}C	8.27×10^3	$R_o = 100$	200
Begemann et al. (1972)	^{14}C	8.27×10^3	$R_o = 80$ or $R_o = 100$	500
Hoyt et al. (1973)	Thermo-luminescence	3×10^3	$\gamma = 2.3$	40-80

Average solar proton flux from radioisotopes excluding ^{14}C ($R_o = 100$) = 100 p/(cm^{-2} s)$^{-1}$.
(*preferred value)

and Zimmerman (1973) from thermoluminescence measurements on lunar rock 14310 tend to support the J_{10} estimates from long-lived isotopes other than ^{14}C. The mean life of the trapped electrons is 3×10^3 y and the integrated flux (J_{10}) derived during a period of time comparable to the mean life, as determined by thermoluminescence experiments, is 40 to 80 p cm^{-2} s^{-1}.

Table 3 includes data for the α/p ratio for the solar wind and solar flares as measured from satellites. These measurements are in substantial agreement with spectroscopic measurements of solar prominences and the solar chromosphere and with measurements of the α/p flux ratio in solar cosmic rays from rockets and high-altitude balloons (Lanzerotti, Reedy, and Arnold, 1973). Lanzerotti, Reedy, and Arnold used Explorer 34 and Explorer 41 measurements to calculate the expected production rates of ^{57}Co ($\bar{T}$ = 390 d, where $\bar{T}$ = mean life), ^{58}Co ($\bar{T}$ = 2.82 y) and ^{59}Ni ($\bar{T}$ = 11.5 × 10^4 y) in lunar rocks. The reactions which produce these nuclides are ^{56}Fe (α, n) ^{59}Ni (threshold energy, 1.4 MeV/nucleon), ^{56}Fe (α,p2n) ^{57}Co (5.6 MeV/nucleon). The predicted production rates, based on the satellite measurements, were well within a factor of three of the measurements of ^{57}Co and ^{59}Ni on rock 12002 that were collected during the Apollo 12 landing. Lanzerotti, Reedy, and Arnold (1973, p. 1234) concluded "that the solar surface material that is accelerated by flare processes has had approximately the same composition ratios between hydrogen and helium for approximately 10^5 years."

However, there is evidence for short-term variability of the composition of solar flare particles. The reader may refer to Price (1973) for a review of this subject. Briefly, the composition of solar particles varies from flare to flare and changes with energy. There is a strong excess of heavy elements at energies below a few million electron volts per nucleon. This excess decreases with energy. For

TABLE 3. Measurements from satellites of the average relative abundance of helium to hydrogen (α/p) in solar wind and solar flares.

Source	Period	α/p	Reference	Comments
Mariner 2	Aug. 29 to Dec. 30, 1962	0.046	Hundhausen (1970)	Solar wind, energy-per-charge analysis only, assumptions about temperatures; 10% of data included in analysis; sample favors low solar wind speeds
Vela 3	July 1965 to July 1967	0.037	Hundhausen (1970)	Solar wind, energy-per-charge analysis only; 60% of data included in analysis
Explorer 34	May 30, 1967 to Jan. 1, 1968	0.051	Hundhausen (1970)	Solar wind, energy-per-charge and mass analysis; 5% of data included in analysis; sample favors high solar wind density
Explorer 34 & Explorer 41	May 28, 1967 to Jan. 1, 1970	0.02 to 0.04	Lanzerotti et al. (1973)	Solar flares (7 large and numerous small); E-1.9 energy dependence from 1 to 12 MeV/nucleon; α/p ratio is for particles of equal energy per nucleon.

example, the Fe to He ratio in the flares of 18 April 1972 and 4 August 1972 was enhanced by a factor of 10 to 20 at energies below a few million electron volts per nucleon, but approached the solar photosphere ratio at greater energies.

III. ATMOSPHERIC ^{14}C AND SOLAR ACTIVITY

The most fundamental assumption of ^{14}C dating is that the ratio of ^{14}C to ^{12}C for atmospheric CO_2 remains constant with time. Willard Libby, in his classic book *Radiocarbon Dating*, thoroughly analyzed this assumption and concluded that it had not changed significantly in historic times (Libby, 1955). This conclusion was based on the measurement of biogenic organic materials of historically known age ranging from A.D. 1072 to 2950 B.C., i.e., encompassing ~5,000 y of historic time. It soon became apparent that combustion of fossil fuels (Suess, 1955) and atomic bomb tests (Rafter and Ferguson, 1957) were changing the ^{14}C content of the atmosphere, but, except for these artificial perturbations, constancy was assumed prior to the twentieth-century technological explosion. However, shortly after the demonstration of artificial perturbations, de Vries (1958) was able to show that there had been measurable natural variations of the atmospheric ^{14}C concentration during the Little Ice Age (sixteenth through nineteenth centuries). This was accomplished by determining the ^{14}C concentration and stable carbon isotopic composition of tree rings and charred wheat of known age. These natural fluctuations in the atmospheric ^{14}C concentration are known as the de Vries effect.

Since the initial demonstration of the de Vries effect, a number of laboratories have systematically determined the ^{14}C concentration of tree rings in order to evaluate the fluctuation of the concentration of ^{14}C in the atmosphere. This work has been greatly aided by the establishment of a 7,484-y chronology for bristlecone pine (Ferguson, 1970). To determine the fluctuation of atmospheric radiocarbon, measurements are made of the ^{14}C concentration and stable isotopic concentration of dendrochronologically dated tree rings. The results are expressed as Δ values which can be thought of as the per mil difference between the ratio of ^{14}C to ^{12}C in the atmosphere at any time, t, relative to the atmospheric ratio of ^{14}C to ^{12}C for the year 1890, which is chosen as a standard prior to technological perturbation. I have discussed the basic assumptions and equations, methods of analyses, and evaluation of errors elsewhere (Damon, 1970).

Figure 1 shows the results from five laboratories of Δ measurements for dated tree rings supplied primarily by the University of Arizona Laboratory for Tree-Ring Research (Damon, Long, and Grey, 1970; Damon, Long and Wallick, 1972; Lerman, Mook, and Vogel, 1970; Ralph and Michael, 1970; Stuiver, 1969; Suess, 1965, 1967; and data of Suess in Houtermans, 1971). All samples that included

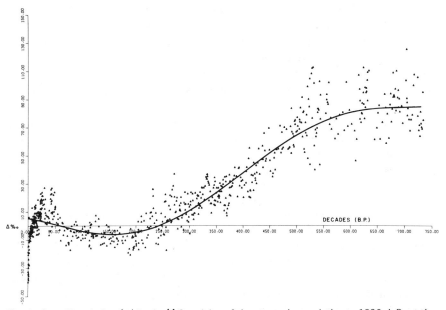

Fig. 1 Per mil variation (Δ) in the ^{14}C activity of the atmosphere relative to 1890 A.D. as the reference time vs. dendrochronologic age in decades B.P. (before 1950 A.D.). The Δ values are determined by measuring the ^{14}C activity and stable carbon isotope composition of dendrochronologically dated tree rings. The trend curve is a third-order orthogonal polynomial. The relatively large change represented by the trend curve results from a quasi-sinusoidal change in the intensity of the earth's dipole magnetic field. Carbon 14 production is also modulated by solar activity as can be seen for the Little Ice Age (100-500 years B.P.) during which the Maunder and Spörer minima occurred. Twentieth-century Δ values have been affected by combustion of fossil fuels.

more than 25 annual rings were rejected. Most of the remaining samples contained ten rings or less. The overall trend of the data shown by the solid line in Figure 1 is thought to be caused by a quasi-sinusoidal variation of the earth's magnetic field intensity with a period around 8,000 to 10,000 y. The reader may refer to the proceedings of the Twelfth Nobel Symposium for a thorough discussion, by various workers, of the geomagnetic field effect (Olsson, 1970). The Suess effect due to combustion of fossil fuels is evident after 1890, and the fluctuations observed by de Vries during the sixteenth to nineteenth centuries are also quite evident.

It can be seen that the de Vries effect consists of fluctuations around the trend curve that continue with greater or lesser intensity during the entire 7,484-y span of time. By using an electrical analog model proposed by de Vries (1959) and also by simply summing each solar cycle independently and comparing the resulting histogram, inverted, with the variation curve, Stuiver (1961, 1965) was the first to show a convincing relationship between the de Vries effect fluctuations and solar activity. The nature of the modulating mechanism of the galactic cosmic ray flux has been discussed in § I. Following Stuiver, Grey and others (Grey, Damon, and Long, 1966; Grey, 1969; Grey and Damon, 1970) used a computer model to demonstrate the relationship between Δ and solar activity. The model assumed that the ^{14}C content of the atmosphere during the jth year, n_{aj}, would be equal to the amount produced during that year, Q_j, plus the residuum from the previous year, $n_{a(j-1)}$, where Q_j was given by Lingenfelter's equation (Lingenfelter, 1963) relating Q_j to the sunspot number, R. The residuum was assumed to be lost from the atmospheric reservoir at a rate τ, the mean life of a disturbance in the ^{14}C content of the atmospheric reservoir. The equation, which was evaluated, year by year, by an iterative integration using an IBM 7092 computer, was as follows:

$$n_{aj} = n_{a(j-1)}e^{-1/\tau} + Q_j \quad . \tag{5}$$

Figure 2 shows a comparison between the theoretical curve derived from the above model and a smoothed curve through the measured Δ data where the sunspot number is derived from measurements by the Zurich Observatory (Waldmeier, 1961). The less accurate data of Schove (1955), derived from historical accounts of aurorae, which vary with solar activity, were used to test the agreement between measured and calculated values for the thirteenth through the nineteenth centuries (Figure 3). It seemed to us that the agreement could hardly be a matter of chance. The model predicts that the atmosphere will act as a low-pass filter attenuating 11-y solar-cycle-induced changes in Δ by two orders of magnitude (see Figure 2 for the theoretical magnitude of the 11-y solar cycle fluctuations). Damon, Long, and Wallick (1973) confirmed this great attenuation (see Figure 4). Further confirmation has been provided by Stuiver (1974, 1975).

Despite the good agreement between theory and measurement provided by the simple model discussed above, Ekdahl and Keeling (1973) point out that more

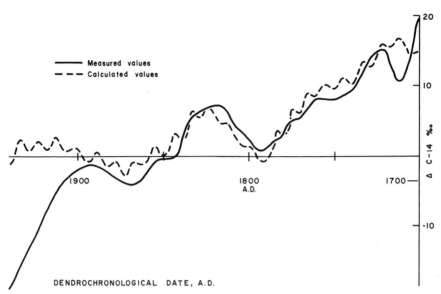

Fig. 2 A comparison of per mil ^{14}C variations (Δ) as determined by measurements of tree rings and as calculated from a simple solar modulation model based on Lingenfelter's (1963) relationship between sunspot number and ^{14}C production (after Grey and Damon, 1970). The theoretical curve shows the predicted 11-year ^{14}C cycle ($3°/oo$ peak to trough) which is near the limit of detection for precise gas counting measurements (see Fig. 4). It also shows the decline in the ^{14}C content of the atmosphere after the Maunder solar activity minimum and a distinct peak between 1800 and 1850 A.D. The measured data follow the theoretical curve closely until 1890 A.D. after which there is a divergence resulting from the combustion of fossil fuels.

complicated ^{14}C reservoir models, e.g., their five- and six-reservoir models, predict a significantly greater attenuation of changes in ^{14}C production with a period greater than 20 y, although the simple model does predict about the same attenuation for the 11-y solar cycle ^{14}C production variations. The disparity in predicted attenuations could be a fault of the reservoir models or an underestimation by Lingenfelter of the effect of solar cycle activity on ^{14}C production which was compensated in the iterative integration by choosing a value of τ that produced excellent agreement. Values of τ between 2 and 100 y were programmed. The model was not very sensitive for values greater than 40 years, but a τ of 100 years was used in the calculations shown in Figures 2 and 3. Two- and three-box reservoir models give similar atmospheric residence times, i.e., circa 50 y (Damon, 1970; Houtermans, Suess, and Oeschger, 1973).

Using the data set shown in Figure 1, the average value of Δ was computed (Table 4) for the Maunder and Spörer minima (Eddy, 1976) and preceding, intermediate, and succeeding intervals of time. It can be seen that values of Δ increase by $10°/oo$ during both the Spörer and Maunder minima. There is a 180-y interval

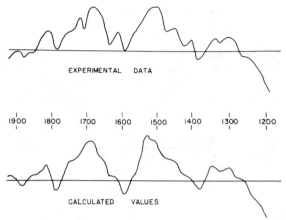

Fig. 3 Comparison of experimental and calculated values of Δ vs. time (years A.D.) from the simple solar modulation model based on Lingenfelter's (1963) relationship between sunspot number and ^{14}C production for the Little Ice Age using sunspot data from the Zurich Observatory (Waldmeier, 1961) and Schove's (1955) sunspot estimates based on historical accounts of aurorae (after Grey and Damon, 1970). The Δ values for the Spörer (1450-1550 A.D.) minimum and Maunder (1645-1715 A.D.) minimum are about 15 ‰ (see Table 4).

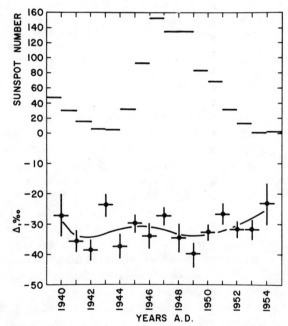

Fig. 4 Per mil ^{14}C measurements (Δ values) for annual rings from the Radio Ridge tree compared with sunspot numbers (Damon, Long, and Wallick, 1973). Note the small variation in ^{14}C (Δ) indicated by the orthogonal polynomial trend curve.

TABLE 4. Average relative atmospheric ^{14}C concentrations (Δ) during the Maunder and Spörer minima.

Time (A.D.)	Epoch	Δ	$\bar{\sigma}$	σ**
1716-1791	Following 75 years	6.1	±1.3	(±7.7)
1645-1715	Maunder minimum	16.8	±1.3	(±5.6)
1551-1644	Intermediate Period	6.3	±1.7	(±6.6)
1450-1550	Spörer minimum*	14.3	±1.6	(±8.4)
1374-1449	Preceding 75 years	2.9	±1.9	(±4.2)

*The Spörer minimum, according to Eddy (1976), extends from A.D. 1460-1550. Since its extent is not precisely known, I have taken a full century.

**The standard deviation (σ) claimed by the various authors who have provided the measurements is ± 5°/oo. It is interesting that somewhat more than half of the variance is due to measurement errors. The standard deviation of the mean ($\bar{\sigma}$) leaves no doubt as to the validity of the high Δ values during the Maunder and Spörer minima.

between the midpoints of the two minima. Damon, Long, and Grey (1966) showed that the best sinusoidal fit to the Δ data for the Little Ice Age had a period of 200 y and an amplitude of 10°/oo. Table 5 shows an analysis of the Δ data using the Blackman-Tukey (1958) Fourier analysis. For the time from 0 to 2,000 y, a 182-y periodicity is observed with a signal-to-noise ratio of 3.0. According to Schove (1955), average sunspot numbers vary from about 20 for the Maunder minimum to about 90 for the intermediate period. For a 180- to 200-y periodicity, the six-box model of Ekdahl and Keeling (1973) predicts an attenuation of 18. Using the most recent estimates by Lingenfelter and Ramaty (1970) of the dependence of ^{14}C production on sunspot number would have yielded a variation of about 100°/oo in ^{14}C production for the fluctuation from R = 20 to R = 90. An attenuation of 18 yields a Δ amplitude of 5.5°/oo within a factor of two of the observed value. A variation of R from 0 to 150 is required for perfect agreement. According

TABLE 5. Significant ^{14}C periodicities (Blackman and Tukey, 1958) for 2,000-y intervals.

Interval (years, B.P.)	Period	Signal to noise ratio
0-2,000	400	2.5
	182	3.0
	69	3.6
	56	1.8
2,000-4,000	500	3.0
	286	6.1
4,000-6,000	1000	1.2
	286	1.7
	100	1.9

to Eddy's (1976) analysis, there were virtually no sunspots during the Maunder Minimum. Is it possible that the sunspot number for the intermediate period was greater than previous estimates would suggest? In any case, agreement within a factor of two is adequate, although analysis of the eighteenth- and nineteenth-century sunspot and Δ data suggest to me that the more complicated reservoir models are predicting too great an attenuation for frequencies between 100 and 200 y.

Eddy (1976) points out that the 11-y cycle and longer cycles may not be continuous and, in fact, solar activity may be more erratic than many workers would like to believe. The Fourier power-spectrum analysis of Table 5 confirms his suggestion. The 180-y periodicity is quite evident from 0 to 2,000 years B.P. (B.P. = before present), but does not appear for the interval from 2,000 to 6,000 B.P. Irregular fluctuations do persist, as can be seen from Figures 5, 6, and 7. Next, we shall take up the question of whether or not these irregular fluctuations

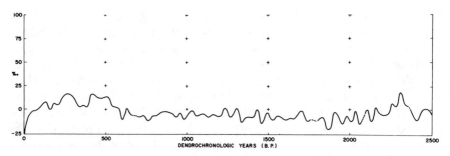

Fig. 5 Smooth curve drawn through the 25-y-interval averages of ^{14}C data (Δ) vs. dendrochronologic age (0-2,500 B.P.). Note that the solar activity modulation of ^{14}C production is relatively subdued during the earlier part of the twentieth century when the geomagnetic field intensity is most intense (dashed lines indicate extrapolation for intervals without data).

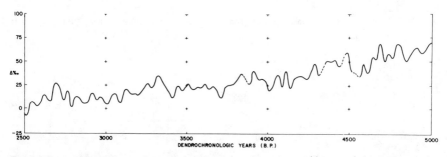

Fig. 6 Smooth curve drawn through the 25-y-interval averages of ^{14}C data (Δ) vs. dendrochronologic age (2,500-5,000 B.P.). Note that the solar activity modulation of ^{14}C is more intense as the Δ values rise near the geomagnetic field intensity minimum. (Dashed lines indicate extrapolation for intervals without data.)

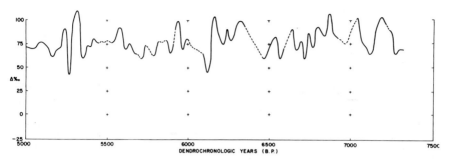

Fig. 7 Smooth curve drawn through the 25-y-interval averages of ^{14}C data (Δ) vs. dendrochronologic age (5,000-7,500 B.P.). Note that the solar activity of ^{14}C is markedly more intense during this period of low geomagnetic field intensity. (Dashed lines indicate extrapolation for intervals without data.)

vary greatly in amplitude. In order to accomplish this, we will use the scatter of measurements around the trend curve in Figure 1 as a measure of the intensity of these fluctuations.

As previously mentioned, the trend of the Δ values in Figure 1, shown by the smooth curve, is thought to be the result of a quasi-sinusoidal variation in the earth's dipole field. Bucha (1970) gives the following equation as the best fit to the magnetic field data:

$$H(t) = \left[2.8 \sin \frac{2\pi}{8900} (t + 405) + 7.7\right] \times 10^{25} \text{ gauss cm}^3 . \qquad (6)$$

The sinusoidal approximation to the geomagnetic field data is shown in Figure 8. It is intuitively obvious and has been quantitatively demonstrated by Lingenfelter and Ramaty (1970), according to a suggestion by Suess (*private communication*), that since the cutoff rigidities at all latitudes increase as H increases, the average global ^{14}C production becomes less sensitive to solar modulation variations at high field intensities and more sensitive at low field intensities. It can be seen in Figure 5 that the solar modulation effect on ^{14}C production is subdued at the beginning of the Christian era when the magnetic field intensity is high and the solar modulation effect increases as the magnetic field decreases (Figure 6), reaching distinctly higher values when the magnetic field is lowest during the seventh millennium B.P.

In two papers involving the dendrochronologic calibration of the radiocarbon time scale (Damon, Long, and Wallick, 1972; Damon et al., 1974), using the data in Figure 1, we determined the trend curve by curvilinear regression using Tchebyshev orthogonal polynomials. The ^{14}C dates were grouped in 25-y intervals as for the Δ values in Figures 5, 6, and 7. In order to estimate the uncertainty of the calibration, the standard deviations were also computed for 250 radiocarbon-year intervals. In Figure 8, we have converted these standard deviations in radiocarbon

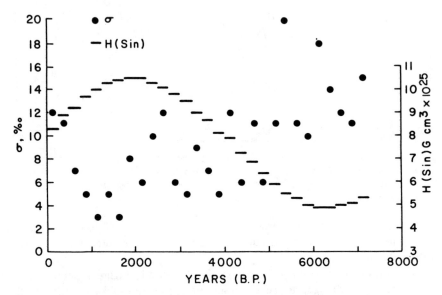

Fig. 8 Sinusoidal approximation to the geomagnetic field intensity and standard deviation of ^{14}C fluctuations about the Tchebyschev orthogonal polynomial trend curve versus time in years B.P. Note that the deviations are smaller when the magnetic field is most intense and vice versa.

years to standard deviations in the Δ values and plotted them along with the sinusoidal approximation to the geomagnetic field intensity. It can be seen, as one would expect, that the variations tend to be small when the geomagnetic field intensity is high and vice versa. Next, in Figure 9, these standard deviations about the trend curve are plotted against the geomagnetic field intensity. It can be seen that the values of σ group around the curve derived from Figure 3 of Lingenfelter and Ramaty (1970), which shows the predicted changes in ^{14}C production rate between solar minimum and solar maximum. In Figure 10, the residual standard deviation around the production rate variation curve is plotted. The residuum is now scattered randomly with no temporal correlation and residual σ is $\pm 3.2^o/oo$. The measurement precision is $\pm 5^o/oo$ and there are, on the average, two to three measured Δ values per 25-y interval. Thus, the expected experimental σ per 25-y interval is $\pm 3.2^o/oo$ and the residuum is virtually entirely the result of experimental error. Consequently, it is not necessary to invoke other factors, such as short-term variations of the geomagnetic field, changes in the galactic cosmic ray flux, or climate-induced perturbations of the ^{14}C reservoirs, to explain the data. By comparison with the ^{14}C fluctuations during the Little Ice Age, I believe it is safe to say that, when 250-y intervals are compared, changes in solar activity, as measured by the variability of the neutron flux which produces ^{14}C, have been much less than 20% during the last 7,500 y. Furthermore, fluctuations as extreme as those during the Little Ice Age, which includes the Maunder and Spörer minima, are the rule rather than the exception. It remains to be seen whether regular fluctuations of solar activity on a longer time scale occur. The coupling of low solar activity with abnormally low temperatures during the Maunder and Spörer minima suggests a causal relationship. If this relationship is not fortuitous, a prolonged episode similar to the Maunder minimum could produce much more severe conditions in this era than those prevailing during the Little Ice Age.

IV. CONCLUSIONS

A number of conclusions seem to be warranted by the data that have been evaluated in this paper.

First, one or several solar flares may account for the bulk of solar particle emission with energies greater than 10 MeV during a solar cycle. Consequently, it is surprising that the proton fluxes (J_{10}) were so similar during solar cycles 19 and 20.

Second, integrated proton fluxes (J_{10}) as determined by measurements of short-lived radioactive isotopes in lunar samples are in excellent agreement with

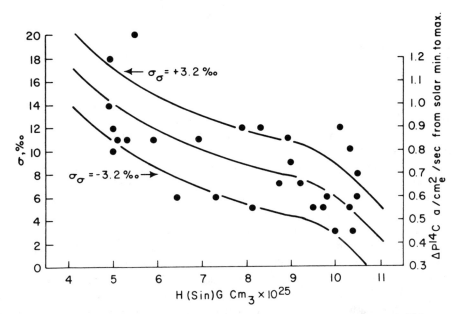

Fig. 9 Deviation about the Tchebyschev orthogonal polynomial curve and changes in ^{14}C production rate between solar minimum and solar maximum as a function of magnetic field intensity. Note the correlation between deviations about the trend curve and ^{14}C production variability.

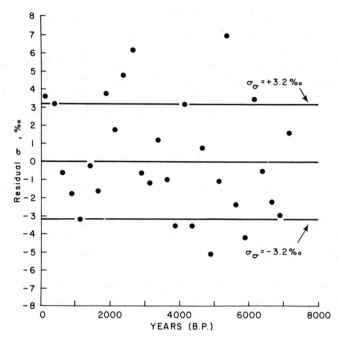

Fig. 10 Residual deviation after removal of variance due to the dampening effect of the geomagnetic field on heliomagnetic modulation. The residuum is compatible with the experimental error.

satellite solar proton monitor measurements.

Third, with the exception of ^{14}C measurements in lunar samples, measurements of other long-lived isotopes in lunar samples yield integrated proton fluxes (J_{10}) that are in surprisingly good agreement with the flux for solar cycles 19 and 20. The data suggest that proton fluxes integrated over the span of millennia or even millions of years have not varied by more than a factor of two in five million years.

Fourth, the excess ^{14}C content of lunar samples is probably not the result of ejection from the sun by flares as suggested by Begemann et al. (1972). Fireman et al. (1976) produced evidence that it is a component of the solar wind.

Fifth, analysis by Lanzerotti et al. (1973) suggests that "the solar surface material that is accelerated by flare processes has had approximately the same composition ratios between hydrogen and helium for approximately 10^5 years." However, there is evidence of short-term variability in the composition of solar flare particles. The composition varies with the energy of the particle.

Sixth, radiocarbon measurements on tree rings confirm historical evidence for great variability of solar activity within a time scale of a quarter of a millennium, e.g., during the Maunder and Spörer minima. This variability appears to be irregular, but averages out such that solar activity fluctuations on this time scale have not greatly changed during the past 7,500 years and probably have remained within limits of ±10%.

Finally, it is perhaps fortuitous that the solar cycles 19 and 20 seem to be so typical of long-term solar flux emission. The sun appears to be a variable star flickering within certain limits, but these limits still leave room for fluctuations of considerable extent and import to human events. It is well known that the Maunder and Spörer periods of low solar activity correspond with periods of low temperature on earth. If this relationship is causal and not fortuitous, a prolonged episode similar to the Maunder minimum could produce much more severe conditions than prevailed during the Little Ice Age. For this reason, it is highly desirable to extend measurements of ^{14}C fluctuations back still farther in time to the end of the Wisconsin glaciations when climatic conditions were more severe than those which have prevailed during the Holocene epoch.

Kepler was a worshipper of the sun, considering it to be a fit sensorium for God. Kepler has not been alone. Solar variability, even though circumscribed by definite limits during the past five million years, constitutes, in the words of John Eddy (1976), "but one more defeat in our long and losing battle of wanting to keep the sun perfect, and if not perfect, constant, and if inconstant, regular."

ACKNOWLEDGMENTS

I benefitted from correspondence and conversations with Drs. James R. Arnold of the University of California at San Diego; John A. Eddy of the High Altitude Observatory, Boulder, Colorado; Louis J. Lanzerotti of Bell Laboratories, Murray Hill, New Jersey; Robert M. Walker of Washington University, St. Louis; and Laurel L. Wilkening and Juan Carlos Lerman of the University of Arizona, Tucson. Critical reviews of the first draft by Chang-Yun Fan of the University of Arizona and Robert M. Walker of Washington University were particularly helpful. Ms. Sandra D. Harralson typed, carefully proofed, and corrected the first and second drafts. I am also particularly grateful to Dr. Edward I. Wallick of the Research Council of Alberta for the use of Blackman-Tukey Fourier spectral analysis data from our unpublished manuscript. This work was supported by National Science Foundation grant DES74-13362 and the State of Arizona. University of Arizona, Department of Geosciences Contribution No. 714.

REFERENCES

Begemann, F.; Born, W.; Plame, H.; Vilcsek, E.; and Wanke, H., 1972, in *Proceedings Third Lunar Science Conference*, Vol. 2 (Suppl. 3, *Geochim. Cosmochim. Acta*), ed. D. Criswell et al., M.I.T. Press, Cambridge, Mass., 1693.

Blackman, R.B., and Tukey, J.W., 1958, *The Measurement of Power Spectra*, Dover Press, New York.

Boeckl, R.S., 1972, *Earth Planet. Sci. Lett. 16*, 269.

Bostrom, C.O., Williams, D.J., and Arens, J.R., 1971, *Solar Geophys. Data No. 328*, part II.

Bucha, V., 1970, *Radiocarbon Variations and Absolute Chronology* (Proceedings 12th Nobel Symposium), ed. I.U. Olsson, John Wiley & Sons, New York, 501.

Damon, P.E., 1970, *ibid.*, p. 571.

———, Long, A., and Grey, D.C., 1966, *J. Geophys. Res. 71*, 1055.

———, Long, A., and Grey, D.C., 1970, *Radiocarbon Variations and Absolute Chronology* (Proceedings 12th Nobel Symposium), ed. I.U. Olsson, John Wiley & Sons, New York, 615.

———, Long, A., and Wallick, E.I., 1972, *Proceedings Eighth International Conference on Radiocarbon Dating*, Wellington, New Zealand, 44.

———, Long A., and Wallick, E.I., 1973, *Earth Planet. Sci. Lett. 20*, 300.

———; Ferguson, C.W.; Long, A.; and Wallick, E.I., 1974, *Am. Antiq. 39*, 350.

de Vries, Hl., 1958, *Proc. K. Ned. Akad. Wet.*, B61.

———, 1959, *Researches in Geochemistry*, ed. P.H. Abelson, John Wiley & Sons, New York, 169.

Eddy, J.A., 1976, *Science 112*, 1189.

Ekdahl, C.A., and Keeling, C.D., 1973, *Carbon and the Biosphere* (Proceedings 24th Brookhaven Symposium in Biology), ed. G.M. Woodwell and E.V. Pecan, Technical Information Services, U.S. Atomic Energy Commission, CONF-720510, 51.

Feldman, W.C.; Asbridge, J.R.; Bame, S.J.; and Gosling, J.T., 1976, this volume.

Ferguson, C.W., 1970, *Radiocarbon Variations and Absolute Chronology* (Proceedings 12th Nobel Symposium), ed. I.U. Olsson, John Wiley & Sons, New York, 237.

Finkel, R.C.; Arnold, J.R.; Imamura, M.; Reedy, R.C.; Fruchter, J.S.; Loosli, H.H.; Evans, J.C.; Delany, A.C.; and Shedlovsky, J.P., 1971, *Proceedings Second Lunar Science Conference*, Vol. 2, ed. D. Criswell et al., M.I.T. Press, Cambridge, Mass., 1773.

Fireman, E.L., D'Amico, J., and De Fleice, 1976, *Proceedings Seventh Lunar Science Conference, Lunar Science Institute*, Houston, Tex., March 1976, 257.

Forbush, S.E., 1954, *J. Geophys. Res. 59*, 525.

Grey, D.C., 1969, *J. Geophys. Res. 74*, 6333.

——— and Damon, P.E., 1970, *Scientific Methods in Medieval Archaeology*, ed. R. Berger, University of California Press, Berkeley, 167.

_____, Damon, P.E., and Long, A., 1966, paper presented at Sixth Western National Meeting, American Geophysical Union, Los Angeles, 7-9 Sept.

Houtermans, J.C., 1971, Ph.D. thesis, University of Bern, Switzerland (see appendix for data of H.E. Suess).

_____, Suess, H.E., and Oeschger, H., 1973, *J. Geophys. Res. 78*, 1897.

Hoyt, H.P., Jr., Walker, R.M., and Zimmerman, D.W., 1973, *Proceedings Fourth Lunar Science Conference*, Vol. 3 (Suppl. 4, *Geochim. Cosmochim. Acta*), ed. W.A. Gose, Pergamon Press, Elmsford, N.Y., 2489.

Hundhausen, A.J., 1970, *Rev. Geophys. and Space Phys. 8*, 729.

King, J.H., 1972, *National Space Science Data Center NSSCD 72-14*.

Lanzerotti, L.P., 1976, this volume.

_____, Reedy, R.C., and Arnold, J.R., 1973, *Science 179*, 1232.

Lerman, J.C., Mook, W.G., and Vogel, J.C., 1970, *Radiocarbon Variations and Absolute Chronology* (Proceedings 12th Nobel Symposium), ed. I.U. Olsson, John Wiley & Sons, New York, 275.

Libby, W.F., 1955, *Radiocarbon Dating*, 2d ed., University of Chicago Press, Chicago.

Lingenfelter, R.E., 1963, *Rev. Geophys. 1*, 35.

_____ and Ramaty, R., 1970, *Radiocarbon Variations and Absolute Chronology* (Proceedings 12th Nobel Symposium), ed. I.U. Olsson, John Wiley & Sons, New York, 513.

Olsson, I.U. (ed.), 1970 *Radiocarbon Variations and Absolute Chronology* (Proceedings 12th Nobel Symposium), John Wiley & Sons, New York.

Pomerantz, M.A., and Duggal, S.P., 1974, *Rev. Geophys. and Space Phys. 12*, 343.

Price, P.B., 1973, *Space Sci. Rev. 15*, 69.

Rafter, T.A., and Fergusson, G.J., 1957, *Science 126*, 557.

Ralph, E.K., and Michael, H.N., 1970, *Radiocarbon Variations and Absolute Chronology* (Proceedings 12th Nobel Symposium), ed. I.U. Olsson, John Wiley & Sons, New York, 619.

Rancitelli, L.A.; Perkins, R.W.; Felix, W.D.; and Wogman, N.A., 1972, *Proceedings Third Lunar Science Conference*, Vol. 2 (Suppl. 3, *Geochim. Cosmochim. Acta*), M.I.T. Press, Cambridge, Mass., 1681.

Schove, D.J., 1955, *J. Geophys. Res. 60*, 127.

Stuiver, M., 1961, *ibid. 66*, 273.

_____, 1965, *Science 149*, 533.

_____, 1969, *Radiocarbon 11*, 545.

_____, 1974, paper presented at Symposium on Quaternary Dynamics, Geological Society of America National meeting, Miami Beach, Fla., 18-20 Nov.

_____, 1975, paper presented at First Miami Conference on Isotope Climatology and Paleoclimatology, Key Biscayne, Fla., 16-22 Nov.

Suess, H.E., 1955, *Science 122*, 415.

_____ , 1965, *J. Geophys. Res. 70*, 5937.

_____ , 1967, *Radioactive Dating and Methods of Low-Level Counting*, International Atmospheric Energy Agency, Vienna, 143.

Tandberg-Hanssen, E., 1967, *Solar Activity*, Chap. 6, Blaisdell Publishing, Waltham, Mass., 239.

Wahlen, M.; Honda, M.; Imamura, M.; Fruchter, J.S.; Finkel, R.C.; Kohl, C.P.; Arnold, J.R.; and Reedy, R.C., 1972, *Proceedings Third Lunar Science Conference*, Vol. 2 (Suppl. 3, *Geochim. Cosmochim. Acta*), ed. D. Criswell et al., M.I.T. Press, Cambridge, Mass., 1719.

Waldmeier, M., 1955, *Ergebnisse und Probleme der Sonnenforschung*, 2d ed., Geest und Portig, Leipsig, East Germany.

_____ , 1961, *The Sunspot Activity in the Years 1610-1960*, Schulthess, Zurich, Switzerland.

Walker, R.M., 1975, *Ann. Rev. Earth and Planet. Sci. 3*, 99.

Webber, W.R.; Benbrook, J.R.; Thomas, J.R.; Hunting, H.; and Duncan, R., 1963, *Boeing Report D2-90469*.

Yokoyama, Y.; Sato, J.; Reyss, J.L.; and Guichard, F., 1973, *Proceedings Fourth Lunar Science Conference*, Vol. 2 (Suppl. 4, *Geochim. Cosmochim. Acta*), ed. W.A. Gose, Pergamon Press, Elmsford, N.Y., 2209.

VI. THE THEORY OF THE SOLAR VARIATION

In our discussion thus far, we present a picture of the solar output as determined from measurement and analysis, but if we are to understand the sun itself and the solar/terrestrial interaction, we must also have a theoretical framework to describe how the various solar outputs are produced. In this chapter, Gough summarizes the time scales for global processes that would be most likely to affect the total radiant output of the sun and discusses the possibility of variation in the solar luminosity. The physical processes discussed here apply to the solar interior and to the solar photosphere, but a theoretical description of the outer chromospheric and coronal layers of the sun and the causes of their radiative and particle emissions is not presented.

The current theory of the evolution of the solar interior postulates an increase of some 30% in the solar luminosity over the last 5 billion years, and the confrontation between this hypothesis and the results from paleoclimatic studies have been discussed in Chapters I and II of this volume. Besides discussing this and other theoretical possibilities for solar variation, Gough presents a hypothesis for an irregular variation in the solar luminosity resulting from a departure from nuclear equilibrium in the solar core. He suggests that such a departure might occur infrequently, at intervals on the order of 10^8 y.

This summary of the current hydrodynamical treatment of the solar interior and its evolution makes it clear that a magnetohydrodynamic theory for stellar structure is required if the occurrence of solar activity is to be theoretically anticipated, much less understood. Without consideration of the influence and generation of magnetic fields in a star, we are unable to find a time scale of variation to match the observed 11-y periodicity in solar activity in the outer shell of the sun. In view of the difficulty of a full magnetohydrodynamical treatment of the stellar interior and its evolution, it is important to understand the extent to which the theoretical discussion of solar activity can be limited to the very outer layers of the star and not be forced to extend deep into the solar interior. If solar activity can be considered as caused principally by differential rotation and the generation of magnetic fields in an outer shell, current theory for the deep solar core can perhaps remain largely intact with only a modification in the outer boundary condition. There is, however, another fundamental question concerning the origin of solar activity and variability: are these processes the result of changes in the energy output of the solar core or can they arise from a variation of the energy transport by the solar convection zone over time scales of decades to centuries?

Given the types of variation in the solar output established empirically in earlier chapters, the theoretical framework for discussion of this variability obviously rests on our understanding of the solar dynamo and its relation to the observed differential rotation of the solar atmosphere. With a theoretical picture for the growth and decay of solar magnetic fields and their systematic redistribution over the solar surface by large-scale flows, we hope to understand both the variability of the radiation and high-energy particles from active regions and the variability of the solar wind that appears to originate from the solar poles and outside active regions in quiet zones. One useful result from such theory would be a means by which we might describe and infer the topological evolution of the global magnetic field structure in the solar photosphere and up through the corona.

Oran R. White

THEORETICAL PREDICTIONS OF VARIATIONS OF THE SOLAR OUTPUT

Douglas Gough
*Institute of Astronomy & Department of Applied Mathematics and Theoretical Physics,
University of Cambridge.*

The principal energy output of the sun is in the form of electromagnetic radiation, mainly in the visible range of the spectrum. Energy is lost also via the solar wind, cosmic rays, and neutrinos. Measurements of the present values of the fluxes of all but the neutrinos and evidence for past variations in those fluxes are discussed in detail in the preceding contributions to this volume. It is the purpose of this article to review the predictions of the flux variations from the theory of stellar evolution. Attention will be restricted to the "main sequence" lifetime of the sun, when almost the entire energy output is believed to derive from the conversion of H to ^{4}He in the solar core. This is the period that has the most obvious relevance to the problems discussed in this volume. Prior to the sun's arrival on the main sequence, the earth was either in the process of formation or did not yet exist as a single entity. Once hydrogen has been exhausted from its innermost regions, the sun will rapidly evolve from the main sequence, becoming extremely distended and much more luminous than it is at present; conditions on earth will change drastically.

I. TIMESCALES

The timescales on which one might expect variations in the solar output to take place depend on the timescales characteristic of the processes that determine the sun's structure. The most obvious of these are listed below. The list is not exhaustive, and others will be mentioned later.

A. DYNAMICAL TIMESCALE

The dynamical timescale is the time in which the sun would readjust its structure if it were imagined to be perturbed from its state of hydrostatic equilibrium. Hydrostatic equilibrium is a balance principally between gravitational attraction and pressure gradient. Magnetic fields and Reynolds stresses make only small contributions to the balance of forces throughout all but the outermost layers of the sun. Consequently the dynamical timescale is both the characteristic free-fall time τ_{ff} under gravity through the solar radius R:

$$\tau_{ff} \simeq \sqrt{R/(GM/R^2)} \qquad (1)$$

and the propagation time τ_a of a pressure (acoustic) wave:

$$\tau_a \simeq R/c_a , \qquad (2)$$

where M is the solar mass, G the gravitational constant, and c_a is a mean sound speed. For the sun, these timescales are of the order of an hour. The sun is observed to maintain its physical dimensions to within a few parts in 10^5 over such a period, which indicates that the assumption of an exact balance of forces used in constructing theoretical solar models is a good approximation. This implies a relation between the gravitational energy Ω of the sun and the total thermal energy U, which can be derived from the virial theorem and expressed approximately as (e.g., Schwarzschild, 1958)

$$U + \Omega = \frac{1}{2}\Omega . \qquad (3)$$

Low-amplitude oscillations of the solar surface have been observed, indicating that small deviations from hydrostatic equilibrium do occur. These are discussed in Section III.

B. THERMAL TIMESCALE

The thermal diffusion timescale is

$$\tau_{th} \simeq R^2/<\kappa> , \qquad (4)$$

where $\kappa = 4acT^3/3\chi\rho^2 c_p$ is the thermal diffusivity and the angular brackets denote taking an appropriate average (Henyey and l'Ecuyer, 1969). Here a is the radiation density constant, c the velocity of light, c_p the specific heat at constant pressure, χ the opacity, T the temperature, and ρ the density. Estimating $<\kappa>$ from theoretical solar models yields $\tau_{th} \simeq 10^7$ y.

It is true of nearly all stars that τ_{th} is much greater than the dynamical timescale. Because of this inequality it is believed that the sun contracted to the main sequence more or less in hydrostatic equilibrium, liberating gravitational energy preferentially in the central regions. About half of this was diffused to the surface and radiated into space, thus maintaining the balance of equation (3). This is sometimes called the Kelvin-Helmholtz contraction phase; observations of young stars provide supporting evidence for the existence of such a phase. The timescale of contraction is

$$\tau_{KH} \simeq U/L , \qquad (5)$$

the time to radiate the thermal energy of the star, where L is the stellar luminosity. Since the whole process is diffusively controlled, this time is numerically equal to τ_{th}. It follows that the timescale for photon diffusion from center to surface, which is simply the time to diffuse the total energy U_r of the radiation, is (U_r/U) τ_{th}. For the sun this is about 10^{-3} τ_{th}.

During Kelvin-Helmholtz contraction, the thermal energy not radiated into space remains to heat the star. When the central regions of the sun became hot enough and dense enough for thermonuclear reactions to provide the energy that was being lost from the surface, gravitational contraction halted. The sun had reached the main sequence, and it continued its evolution on the longer nuclear timescale τ_n. During the main sequence evolution, the gross structure of the sun has not changed appreciably so the values of τ_{th} and τ_{KH} are still approximately equal.

C. NUCLEAR TIMESCALES

The dominant timescale controlling the sun's main sequence evolution is that which characterizes the rate of conversion of hydrogen to helium. Taking the present luminosity of the sun $L_\odot$ to be typical, this is

$$\tau_n \simeq XME_n/L_\odot , \qquad (6)$$

where X is the mass fraction of hydrogen and $E_n \simeq 0.007c^2$ is the energy liberated per unit mass by the hydrogen to helium reaction. This time is about 10^{11} y. According to the generally accepted theory, the main sequence lifetime τ_{MS} is somewhat shorter than this: $\tau_{MS} \simeq 10^{10}$ y, because only about 10% of the sun's hydrogen content will have been consumed by the time hydrogen is exhausted from the central core. This result depends on the assumption of negligible material mixing between the core and the surrounding envelope during the main sequence phase. The age of the sun is about 4.8×10^9 y (Fowler, 1972), about half its expected main sequence lifetime.

The dominant nuclear reactions in the solar core are believed to be those of the proton-proton chain, which may be represented by

$$p(p,\beta^+\nu)D(p,\gamma)^3He(^3He, 2p)^4He$$
$$| \qquad |$$
$$p(pe^-,\nu)D \quad ^3He(^4He,\gamma)^7Be(e^-,\nu)^7Li(p,^4He)^4He \qquad (7)$$
$$|$$
$$^7Be(p,\gamma)^8B(\beta^+\nu)^8Be*(^4He)^4He .$$

The two reactions producing deuterium and the three chains destroying ^3He have been written in decreasing order of importance, so far as their contribution to the energy production in the sun is concerned. The deuterium-producing reactions are the slowest, and therefore control the rate of the entire chain, on the timescale τ_n. Within the central 20% by mass, where 75% of the energy is generated, the characteristic times of the other reactions are all very much shorter than τ_n, and so the abundances of the elements produced in the chain are usually assumed to be adjusted to their equilibrium values. It can be shown that this equilibrium is stable to perturbations in the abundances at constant ρ and T. Table 1 shows the characteristic times of the reactions of the chain in a model of the present sun.

The assumption that the chain is in equilibrium is valid, provided the sun is evolving on a timescale of τ_n. But if variations on a shorter time occur, it may be necessary to consider deviations from nuclear equilibrium.

TABLE 1

Nuclear reactions of the pp chain. Q and Q_ν are the mean thermal and neutrino energies liberated per reaction. The temperature exponents $\eta = \partial \ln \mathcal{R}/\partial \ln T$, where $\mathcal{R}$ is the reaction rate, are evaluated at the solar center and increase weakly outwards. The characteristic reaction times are evaluated for a model of the present sun at the mass ratios indicated, where m is the Lagrangian mass coordinate and M is the mass of the sun. About 75% of the total luminosity is generated in m/M<0.2, 95% in m/M<0.4.

Reaction	Q (Mev)	Q_ν (Mev)	η	Reaction timescale (yr)		
				m/M = 0	0.2	0.4
$p(p,\beta^+\nu)D$	1.19	0.42	4	7×10^9	4×10^{10}	2×10^{11}
$p(pe^-,\nu)D$	.001	1.44	4	2×10^{12}	3×10^{13}	2×10^{14}
$D(p,\gamma)^3He$	5.49		4	4×10^{-8}	3×10^{-7}	2×10^{-6}
$^3He(^3He,2p)^4He$	12.86		16	6×10^5	3×10^7	8×10^8
$^3He(^4He,\gamma)^7Be$	1.59		17	6×10^5	1×10^9	3×10^{11}
$^7Be(e^-,\nu)^7Li$	0.06	0.81	-1/2	2×10^{-1}	5×10^{-1}	7×10^{-1}
$^7Li(p,^4He)^4He$	17.35		11	3×10^{-5}	3×10^{-3}	8×10^{-2}
$^7Be(p,\gamma)^8B$	0.14		13	9×10^1	2×10^4	1×10^6
$^8B(\beta^+\nu)^8Be^*$	10.87	7.2	0	3×10^{-8}	3×10^{-8}	3×10^{-8}

D. ROTATION PERIOD

The solar surface rotates with a period that is about 25 d at the equator and more than 35 d at the poles. Doppler velocity measurements seem to yield an angular velocity lower than that inferred from sunspot motion, probably indicating that the angular velocity ω increases with depth in the region of the photosphere. It has been argued by Dicke (1964, 1970) that this increase continues to the center, where the rotation period is about 2 d. Recent measurements of the oblateness of the sun's figure (Hill and Stebbins, 1975) render this hypothesis unlikely and are consistent with an angular velocity of the same order of magnitude as the surface value throughout the sun. Such an angular velocity implies that at the photosphere the ratio of centrifugal to gravitational accelerations $\lambda \equiv R^3 \omega^2 / GM$ is about 2×10^{-5}.

E. CONVECTION TIMESCALES

Except near the top of the convection zone just beneath the photosphere, the superadiabatic temperature gradient is very small and acceleration due to buoyancy is much less than the gravitational acceleration. Typical convective velocities w are therefore best estimated from the convective heat flux, which can be shown to be of order ρw^3.

Since almost all the heat flux is carried by convection throughout most of the convective zone, $4\pi r^2 \rho w^3 \simeq L$, where r is a radial distance coordinate. If ℓ is the characteristic size of a convective eddy, usually assumed to be of the order of a pressure scale height, the characteristic turnover time is of order

$$\tau_c \simeq \ell/w \simeq \ell(4\pi r^2 \rho / L)^{1/3} \quad . \tag{8}$$

This time is a few minutes near the top of the convection zone and increases downwards to a few weeks near the bottom, where it is comparable with the rotation period.

F. OTHER TIMESCALES

The Prandtl number, the ratio of kinematic viscosity to thermal diffusivity, is about 10^{-6} at a median point in the sun; the characteristic viscous diffusion time over a distance comparable with the solar radius is therefore of order $10^6 \tau_{th} \simeq 10^{13}$ y, which is much greater than the age of the sun. For this reason viscosity is normally ignored, though it may be important above the photosphere where the density is very low or in thin internal boundary layers, if they exist.

The different degrees of rotational distortion of the isobaric and equipotential surfaces leads, in the absence of a magnetic field, to fluid motion in meridional planes: the Eddington-Sweet circulation. If rotation is almost uniform, the time τ_{ES} for such a motion to transport material from the center to the surface of the sun exceeds $\lambda^{-1} \tau_{th}$, which is of order 10^{12} y now and is much greater than the age of the sun. Material mixing by the Eddington-Sweet circulation is therefore probably negligible at present. Material diffusion is also small, since diffusion coefficients are of the same order as the kinematic viscosity. It has been argued (Roxburgh, 1964; Mestel, 1965; Mestel and Moss, 1976) that a small seed magnetic field with a poloidal component would be distorted by the combined influence of the rotation and meridional circulation in such a way as to almost quench the circulation everywhere except near the very surface of the star.

The magnetic diffusivity of the solar material is somewhat greater than the kinematic viscosity. If there were no differential rotation or turbulence, a large-scale primeval magnetic field would decay on a timescale comparable with the age of the sun and would not, therefore, have disappeared.

The loss of angular momentum from the surface layers by the solar wind torque via the magnetic field influences the angular velocity distribution and so modifies the meridional circulation. In analogous cases in the laboratory, the circulation is so great that angular momentum transport takes place in a time much shorter than the viscous diffusion time (Benton and Clark, 1974). This is likely to be the case for the sun, too (Howard, Moore, and Spiegel, 1967; Bretherton and Spiegel, 1968; Sakurai, 1972, 1975), but since laminar circulation is likely to be unstable, the situation is too complicated for a reliable estimate of the timescale for angular momentum transport to have been computed. It is likely to be much less than the age of the sun.

The characteristic timescale of mass loss due to the solar wind is presently about 10^{13} y.

II. THE MAIN SEQUENCE HISTORY OF THE SUN: EVOLUTION ON A NUCLEAR TIMESCALE

Solar models are usually evolved on the main sequence assuming hydrostatic support and thermal balance; the latter is justified because the thermal diffusion time τ_{th} is much shorter than the characteristic evolution time. Rotation is normally ignored, so the models are spherically symmetrical. It is usual to assume that initially the chemical composition is homogeneous, either because the sun condensed from a chemically homogeneous gas cloud or because prior to reaching the main sequence it went through a fully turbulent phase during which any inhomogeneities were destroyed. Both these assumptions have been questioned (Larson, 1969; Wheeler and Cameron, 1975; Hoyle, 1975). Furthermore, it is normal to assume that no significant mass loss or accretion takes place during the main-sequence evolution. Models evolved under these assumptions are found to be convectively stable in the inner 80% or so by radius, and since diffusion and Eddington-Sweet circulation appear to produce negligible mixing, the products of the nuclear reactions are assumed to remain in situ.

A. PHOTON LUMINOSITY AND EFFECTIVE TEMPERATURE

As hydrogen is burned to helium in the core, the number of particles per unit mass decreases. In order to maintain sufficient pressure to hold up the envelope, the central regions contract and heat up, thereby increasing the rate of thermonuclear energy generation. This energy is emitted at the surface almost entirely as electromagnetic radiation, and is thus set equal to the photon luminosity L. The photosphere radiates approximately as a black body, so it is meaningful to set

$$L = \pi a c R^2 T_e^4 \quad , \tag{9}$$

where T_e, the effective temperature of the star, represents the material temperature in the region where most of the radiation is emitted.

The variation of L with time t, taking t = 0 to be at the beginning of the main-sequence evolution, can be approximated by

$$L = L_\odot [1 + \alpha(1 - t/t_\odot)]^{-\beta} \tag{10}$$

with α = 0.4 and β = 1, and where $L_\odot$ and $t_\odot$ are the present luminosity and age of the sun. Thus the e-folding time for the luminosity, $\tau \simeq 2.5\,(L_\odot/L)t_\odot$, is comparable with, though somewhat greater than, the present age of the sun. The theory predicts that this trend will continue until hydrogen is more or less exhausted at the center. Note that at the beginning of the main sequence, which corresponds approximately to the time at which the earth was formed, the solar luminosity was about 70% of its value today.

Although the core contracts as the sun evolves, the radius of the entire sun does not necessarily decrease, because the increasing luminosity has a tendency to cause expansion of the envelope which transports it. The precise details depend on the theory adopted for convective transport in the outer layers, but increases since t = 0 of about 15% in R and 1 - 3% in T_e are typical.

It should be pointed out that the theory does not really predict the present luminosity $L_\odot$ and effective temperature T_e of the sun. Aside from uncertainties in the nuclear reaction rates and the equation of state and opacity of stellar material, we do not know within quite broad limits what the chemical composition of the sun is (Ross and Aller, 1976). Furthermore we cannot reliably describe the convection zone in the outer envelope or assume with certainty that no material mixing occurs. The composition at t = 0 is defined by X and Y, the relative abundances by mass of H and He, which are usually assumed to be uniform. The remaining elements have a total abundance Z = 1-X-Y which is probably only about 1 or 2%; their presence hardly affects the equation of state but it does provide the main source of opacity. Convective heat transport is usually described within the framework of mixing-length theory and depends on at least one undetermined parameter, say ε. The solar age $t_\odot$ is rarely questioned. The theory attempts to rationalize the data: for any plausible Z, values of X and ε can be chosen that reproduce $L_\odot$ and T_e at t = $t_\odot$. Thus one is left with at least a single infinity of solutions even if X and Y are assumed to be initially uniform.

B. NEUTRINO LUMINOSITY

The measure of the upper bound to the flux of neutrinos reaching the earth (Davis, Harmer, and Hoffman, 1968) has cast considerable doubt on the theory outlined above, because it is less than that predicted by the models (e.g., Bahcall and Sears,

1972). It has been proposed that neutrinos decay before reaching the earth, but most workers seem to regard this as unlikely. Thus, although the neutrino output of the sun is of little direct significance to the earth, it has become an important quantity for diagnosing solar structure.

Because of the discrepancy between most theoretical predictions of the neutrino luminosity L_ν and the observations, one can have little faith in any estimate of the variation of L_ν with time. Most stellar physicists believe that the theory is not seriously wrong, because it appears to explain adequately many of the gross features of the observations of other stars; the principal neutrino-producing reactions are extremely sensitive to temperature, and seemingly modest changes to conditions in the solar core might significantly change the predicted neutrino flux. Moreover, Davis' most recent measurements (Bahcall and Davis, 1976) are somewhat higher than the mean, and if it turns out that these are more reliable than the values quoted in the past, the problem may be removed. Thus, the increases in temperature and density on the main sequence that give rise to the variation in luminosity described by Equation (10) are generally accepted. This would imply that L_ν also increases, with an e-folding time probably somewhat shorter than τ.

C. SOLAR WIND AND COSMIC RAYS

Much progress has been made in understanding the dynamics of the solar wind far from the sun (Holzer and Axford, 1970), but there is no reliable theoretical estimate of the strength of the wind in terms of the interior structure of the sun. The energy balance of the corona (the existence of which is essential to the wind) is still not understood (Stein and Leibacher, 1974), though it is generally believed that the corona is maintained by dissipation of mechanical waves propagating outward from the convection zone. It is not unlikely that the propagation and dissipation rates depend on the strength of the magnetic field. The magnetic field is believed to be generated by dynamo action, which is controlled by solar rotation. Solar rotation is affected by the loss of angular momentum via the solar wind, and that in turn depends on the mass flux of the wind and the magnetic coupling between the wind and the sun itself. Thus solar rotation, magnetic field, and solar wind are intimately connected in a complicated way that has not been unraveled.

Since angular momentum is being lost via the wind, it seems plausible that the surface angular velocity of the sun is decreasing with time, at least on the evolution timescale τ. This conclusion is consistent with, for example, Kraft's (1967) observations of the surface rotation rates of young solar-type stars in the Pleiades and Hyades, which are some ten times greater than the rotation rate of the sun. It also seems plausible that the magnetic field, and hence the solar wind flux, should decrease as the rotation rate decreases; however, Mestel and Moss (1976) have pointed out that there may be a tendency for rapid rotation to submerge the field, possibly leading to a surface field that increases with decreasing angular velocity.

The changing vigor of the convection, which is presumed to generate the waves maintaining the corona, probably has a relatively minor influence. Spiegel (1968) has suggested simple functional forms for the time dependence of these quantities that rationalize the data.

Less can be said of the cosmic rays. High-energy particles are known to be associated with violent surface phenomena such as flares, which are no doubt intimately connected with hydromagnetic instabilities. If the sun's surface magnetic field has generally decreased during the main-sequence evolution, so presumably have the frequency and intensity of cosmic ray bursts.

III. VARIATIONS ON A DYNAMICAL TIMESCALE

It has been known for some time that the surface of the sun oscillates on a timescale of about 5 min (Leighton, Noyes, and Simon, 1962). More recently, longer period oscillations have been detected with periods up to about an hour (Deubner, 1972; Kaufman, 1972; Kobrin and Korshunov, 1972; Fossat and Ricort, 1973), but these are not as prominent as the 5-min oscillations. The oscillations have been studied mainly through doppler measurements using lines in the visible spectrum; variations in the visible intensity are not normally seen, though intensity variations in radio emission have been observed by Kaufman and by Kobrin and Korshunov. Where horizonal scales have been measured (Deubner, 1975; Fossat and Ricort, 1975) the oscillation periods are found to increase with wavelength. This suggests nonradial, standing acoustic waves, an interpretation suggested originally by Leighton, Noyes, and Simon, but discussed even earlier by Whitney (1958) in connection with the solar granulation. Deubner (1975) has measured the power distribution in wavenumber and frequency of the 5-min oscillation, and found that it agrees well with the theoretical diagnostic diagrams computed by Ulrich (1970) and Ando and Osaki (1975). The acoustic modes have significant amplitude only in the outer 10^4 km of the sun and appear to be driven principally by the κ mechanism, which is responsible for driving the radial pulsations of Cepheid variables (Wolff, 1972; Ando and Osaki, 1975).

A spectrum of oscillations in an apparent diameter of the sun has recently been reported by Hill, Stebbins, and Brown (1975), also with periods between about 5 min and 1 h. The spectrum appears to be discrete, with frequencies that agree with theoretically determined values for radial pulsations and nonradial pulsations corresponding to low values of the principal order l of the spherical harmonic that describes the angular dependence of the modes (Scuflaire et al., 1975; Christensen-Dalsgaard and Gough, 1976). Of course, the shorter period oscillations might just be the modes with high l observed by Deubner and by Fossat and Ricort; the longer-period modes could be gravity waves, but it seems unlikely that low-order modes would be excluded from such a rich spectrum. These modes penetrate much more deeply into the sun than those discussed by Ulrich, Wolff, and Ando and Osaki. No attempt has yet been made to measure the spatial dependence of the oscillations.

An even longer-period (160 min) global oscillation has been reported by Severny, Kotov, and Tsap (1976) and by Brookes, Isaak, and van der Raay (1976). Both sets of observations involved measuring components of the velocity field with large-scale spatial structure. Severny, Kotov, and Tsap observed the difference in doppler shift between light from the equatorial region of a vertical strip across the solar disk and that from the polar regions; Brookes, Isaak, and van der Raay measured the shift of a sodium line in the light from the entire solar disk relative to a laboratory source. As with the radius measurements, no spatial structure has yet been observed. The velocity amplitude of the oscillations is smaller by a factor of ten than that of the higher frequency oscillations observed by Deubner and by Fossat and Ricort. This may be because the measurements average over many horizontal wavelengths of a nonradial mode of oscillation with quite high l, or it may be simply that this oscillation is a mode with low l and really does have low amplitude. It is at least clear that the 160-min oscillation can hardly be an acoustic mode, because no acoustic mode of so low a frequency can be sustained by a stable, self-gravitating gaseous body of the sun's mass and size that is in hydrostatic equilibrium. The most likely candidate is a high-order gravity mode.

The interpretation of the recent oscillation data has been questioned by Grec and Fossat (1976). They claim that a combination of instrumental noise and fluctuations in the transparency and refractive index of the earth's atmosphere could be responsible for the results. Careful control experiments have yet to be performed.

Dissipation of coherent solar oscillations in the sun's atmosphere would distort the isophotes of the solar image. This may explain the excess equatorial brightness that Hill and Stebbins (1975) believe to be the cause of the high measure of the solar oblateness obtained by Dicke and Goldenberg (1974). For a non-rotating spherically symmetrical star there is a degeneracy in the frequencies of nonradial oscillations with the same value of l. The degeneracy is split by rotation, and modes with the same l oscillating together would then beat with a period that depends on an average of the angular velocity of the star. There is some indirect evidence that this is occurring in the sun, for Dicke (1976) has recently reported that his oblateness measures vary with a period of 12.2 d. From this, one could deduce a mean angular velocity of the sun, if only the modes that are presumed to be beating could be identified. The most likely candidates are either two dipole modes or two even-parity, nonaxially symmetric quadrupole modes, which would dissipate preferentially in the equatorial regions and therefore produce an apparent oblateness. These modes yield mean rotation periods for the sun between about 8 and 24 d, which suggests that the interior rotation is somewhat faster than the surface rotation.

The amplitudes of the oscillations that seem to have been observed are all too low to affect the structure of the sun appreciably. The energy they transport, and the stresses they produce are negligible. However, like the neutrino luminosity, the frequencies constitute a new class of data that is potentially useful for diagnosing the internal structure of the sun. Linear theory is no doubt adequate for calculating the frequencies, though nonlinear resonant interactions between modes may be an important factor in determining which of the modes are excited to observable amplitudes. In particular, it may explain why Severny and Brookes and their co-workers seem to have observed an isolated mode in the quite densely spaced g-mode spectrum. If this could be established, the resonance conditions would add further constraints to the structure of solar models.

No careful theoretical estimates of the amplitudes of the oscillations have yet been made. It is not even known what drives the modes, many of which are stable, according to linear theory. Direct coupling via the Reynolds stresses produced by the convection seems to be the most likely possibility, but it will probably be some time before this is confirmed or refuted. Goldreich and Keeley (1976) have made rough estimates of this coupling and have predicted oscillation amplitudes considerably lower than those reported by the observers.

IV. THE SOLAR CYCLE

The sun's magnetic field reverses every 11 y. Magnetic phenomena such as sunspots and flares vary in magnitude and frequency with the same periodicity. This period does not correspond to any of the natural timescales listed in Section II, and its origin seems to be a mystery.

Because the solar magnetic field varies on a timescale very much shorter than the diffusion time in the deep interior, it is often thought to be maintained by dynamo action in the convection zone. Many dynamos are oscillatory; their periods exceed the characteristic timescales of the motions producing the field because diffusion plays an important role in their generation. It is not implausible, therefore, that the motion in the convection zone, with turnover times of up to several weeks, might produce the desired effect. Moreover, the rotation of the sun gives helicity to the convective motion, and helicity is known to be important for the generation of magnetic fields. Both Babcock (1961) and Leighton (1969), have used simple models to describe how convection and rotation might interact with the magnetic field to produce the solar cycle. It cannot be said that the 22-y complete period is a firm prediction of the theories, but the value is made to seem plausible. More detailed models have been constructed since, but a clear picture of the most relevant dynamics has not yet emerged. The recent reviews by Gubbins (1974) and Stix (1975) discuss the matter in some detail.

The dynamo models do not have memory lasting over a large number of cycles.

Dicke (1970) has pointed to evidence that the phase of the solar cycle is maintained more accurately than one might expect from the observed fluctuations in period if the convective dynamo were the sole mechanism for reversing the large-scale field; a careful statistical analysis of this proposition has not been made. Dicke inferred that there must be an underlying clock controlling the cycle, such as torsional oscillations of the core (cf. Walén, 1946). This would require the poloidal component of the magnetic field inside the sun to be of the order of several thousand gauss, comparable with the fields in sunspots and too small to affect the hydrostatic structure of the sun significantly. There is some difficulty with this idea, however, because one would expect the restoring force arising from Coriolis effects to dominate the torsional dynamics, yielding inertial oscillations with periods comparable to the rotation period.

An alternative hypothesis, made recently by Wolff (1976), is that the solar cycle is driven by nonlinear interactions between the low-amplitude oscillations discussed in Section III and large-scale circulation currents. The theory makes no detailed analysis of the interactions, but does find interesting correspondences between estimated beat frequencies of high-order gravity modes and both the 22-y cycle and longer timescales apparent in the sunspot data. The detailed quantitative results should be treated with some caution because the asymptotic formula Wolff used for the periods is valid only for extremely high overtones. Nevertheless the idea deserves further pursuit.

V. VARIATIONS ON A THERMAL TIMESCALE

The possibility that low-order gravity modes (g modes) might sometimes be excited to quite high amplitude in the solar interior has received some attention in recent years, largely in connection with the solar neutrino problem discussed in Section II. But such modes perhaps have greater importance as a possible cause of terrestrial glaciation. Unlike the acoustic modes and the high-overtone g mode suggested as being responsible for the 160-min solar oscillation, low-overtone g modes with low l have appreciable amplitude in the central regions of the sun. There they induce fluctuations in the energy-generation rates, mainly in response to temperature fluctuations associated with the motion. These are in the correct phase to drive the oscillations and render them unstable. Crucial to this mechanism is the fact that the oscillation periods (of order 1 h) are very much less than the characteristic equilibration time for the ^{3}He -^{3}He reaction (see Table 1) and of course much less than the timescale for material diffusion. Advection of ^{3}He by the motion perturbs the equilibrium of the pp chain, and the H and ^{3}He burning reactions decouple. The sensitivity of the energy-generation rate to the temperature fluctuations associated with the motion is essentially an average of the temperature dependences of the H and ^{3}He reactions, weighted by the proportion of heat that each produces.

This means that the pp chain is much more temperature-sensitive on the timescale of the oscillations than it is on the timescale of the main-sequence evolution, provided ^{3}He has attained its equilibrium abundance. Computations by Christensen-Dalsgaard, Dilke, and Gough (1974), Boury *et al.* (1975), and Shibahasi, Osaki, and Unno (1975) suggest that once ^{3}He has reached its equilibrium abundance throughout most of the energy-generating core, the low-order g modes become unstable. The calculations predict that when the sun arrived on the main sequence it was not unstable to g modes, but that after about 2×10^8 y, ^{3}He achieved its equilibrium abundance throughout a sufficient region of the core to render the lowest order gravity mode g_1 (l = 1), with a period of about 1.7 h,) unstable.

Why is the g_1 (l = 1) mode not observed with large amplitude now? Boury *et al.* suggest that the sun has changed its structure so much on the main sequence that it is no longer self-excited. On the other hand, Dilke and Gough propose that the oscillations give way to direct convection, which mixes up the solar core and destroys nuclear equilibrium. The core then becomes quiescent again for another few hundred million years until the equilibrium abundance of ^{3}He is once more achieved over a sufficiently extensive proportion of the core for the whole process to start again. Unno (1975) has pointed out that the modification to the distribution of molecular weight in the solar interior produced by such a mixing tends to trap subsequent g modes of the type that trigger that mixing in the very region where they are driven and so enhances the instability in the later stages of the sun's main-sequence evolution. This phenomenon has been confirmed numerically by Gabriel *et al.* (1976).

A consequence of a comparatively sudden mixing is that the core is enriched with H and ^{3}He, and thermonuclear generation increases faster than the energy can escape. The core expands and the temperature decreases, in accord with the virial theorem [Equation (3)] and to the extent that nuclear reaction rates then drop below their original values. This occurs in about 10^6 y, which is less than the thermal diffusion time for the entire star; the envelope is therefore pushed out by the expanding core almost adiabatically; its cooling causes a temporary reduction in the luminosity of the sun. On a timescale of $\tau_{th} \simeq 10^7$ y, the thermal wave produced by the initial burst of thermonuclear energy reaches the surface, and the sun then settles back into thermal balance.

The computations predicting the g-mode instability are approximate and must be viewed with caution (Christensen-Dalsgaard and Gough, 1975). But even if such modes are excited, it has not been demonstrated that they initiate turbulent mixing in the solar core. Dilke and Gough (1972) have argued by analogy, following Veronis' (1965) explanation of finite amplitude thermohaline convection. Ulrich (1974) made a rough estimate of whether to expect mixing from a simplified laminar flow in the sun and concluded it was unlikely. The issue is unresolved. If the process does take place, one would expect the solar luminosity to suffer periods of variation on a timescale τ_{th}, separated by intervals of a few hundred million years during which there is a smooth slow rise, similar to that described by Equation (10). The amplitude of the luminosity variations would depend on how much of the core is mixed. Dilke and Gough presumed that the mixing would be controlled by the convection, which they expected to be confined to the inner 20 - 25% by mass of the core, where thermonuclear energy is generated. They found an initial decrease of about 5% lasting about 4×10^6 y, followed by a period in which the luminosity is about 5% higher than normal and lasting about twice as long. Gabriel et al. argued that the characteristics of the linear oscillatory g mode should determine the extent of the zone that is subsequently mixed. They assumed mixing out to the node in the g_1 ($l = 1$) mode (about 83% by mass) and found that the luminosity then varied with an amplitude of about 20%. (cf. also Rood, 1972; Ezer and Cameron, 1972.)

Another consequence of the transient mixing phases is a temporary reduction in the neutrino luminosity. This occurs partly because the central temperature is decreased, which reduces the rate of the temperature-sensitive ^{7}Be (p, γ) ^{8}B reaction, and partly because ^{7}Be is mixed away from the center where the ^{8}B-producing reaction competes most favorably with the electron capture by ^{7}Be. According to Dilke and Gough, mixing 25% by mass of the core reduces L_ν almost to within the limit set by observation; mixing more of the mass reduces L_ν still further (Rood, 1972; Ezer and Cameron, 1972).

It has been suggested that variations in the photon luminosity resulting from the proposed core mixing induce terrestrial glaciation. The interval between the events, which is set by the ^{3}He equilibration time throughout the bulk of the energy-generating core, is in good agreement with the period between the major glacial epochs. Furthermore, the duration of the luminosity fluctuation is of the same order as the duration of the major ice ages, and the amplitude corresponds to what climatologists and theoretical meteorologists believe is required to induce significant changes in the climate. Nevertheless, such an interpretation leaves several phenomena unexplained; among these is the steady decrease in mean sea temperature during the 10^7 y preceding the onset of the Pleistocene epoch. On the other hand, Sagan and Young (1973) have pointed out that glaciation on Mars and the earth may have taken place simultaneously, which suggests extraterrestrial control. If one accepts a correlation between climate and photon luminosity, then the earth's climatic history can be used to date the solar luminosity transients. The beginning of the Pleistocene period only two or three million years ago implies that the sun is not presently in thermal balance; it could explain why the neutrino luminosity is not observed to be as high as the values predicted theoretically from equilibrium solar models. It also implies that the solar photon luminosity will temporarily rise above its present value in the next few million years.

Material mixing in the solar interior can influence the long-term evolution of the sun. The descent of unburned hydrogen into the central regions from above prolongs the main-sequence phase. There is comparatively little effect if the region mixed is no larger than the region where most of the nuclear reactions are taking place, but once the mixing extends into the envelope, the time variation in luminosity L is reduced. The most extreme case would be if the entire sun were mixed: the luminosity would then still increase approximately according to Equation (10) until hydrogen is almost depleted in the center, but with $\alpha = 0.5$ and $\beta = 0.9$. The zero-age main-sequence luminosity would then have been 75% of the luminosity today. If such mixing has taken place, it has presumably occurred in other solar-type stars, too, and would require a revision in the estimated ages of globular clusters. Shaviv and Salpeter (1971) have shown that if the inner 40% by mass were mixed, age estimates would be increased by nearly a factor of two, but that 20% mixing would make very little difference.

VI. VARIATIONS ON THE ROTATION AND CONVECTION TIMESCALES

Variations in convective heat flux on the eddy turnover timescale, or the scale of the rise time of buoyant thermals, are observed in the laboratory (Thomas and Townsend, 1957; Deardorff and Willis, 1967; Krishnamurti, 1973; Busse and Whitehead, 1974) and in the earth's atmosphere (Priestley, 1959). Though conditions are very different in the solar convection zone, it seems likely that similar variations should occur there too. Of course, the solar granulation is observed to vary on a timescale of about 5 min, but since there are $10^5 - 10^6$ granules on the solar surface, the statistical fluctuations they produce in the total energy output of the sun are very small. Deeper in the convection zone, the length and time scales of the energy-transporting convective motions increase. The so-called giant cells, if they exist, are presumably vestiges of such motions. As Gilman (1976) has pointed out, since there are only a few such cells on the solar disk, variations in efficacy and number might produce measurable changes in the luminosity of the sun on a timescale of weeks or months. Moreover, any inhomogeneities in surface heat flux produced by the giant cells would rotate with the sun and cause fluctuations in the solar constant.

It is difficult to estimate the magnitude of the luminosity variations that might be produced in this way. Mixing-length models of the solar envelope predict temperature fluctuations of only a few parts per million in the bottom few pressure scale heights of the convection zone, so the lower part of the convection zone does not provide a particularly nonuniform base for the smaller scale eddies above. Large-scale surface inhomogeneities are more likely to be produced by fluid moving from the bottom of the convection zone to the photosphere, but to estimate their amplitude requires knowledge of the flow field. We must await convection theories more sophisticated than those presently available for solar modeling, such as that being developed by Latour et al. (1976) to model A stars, before a reliable theoretical description on this phenomenon can be produced.

Inertial oscillations of the sun have not been studied. Their periods are of the order of the rotation period. Their existence could be detected through measurements of variations in the angular velocity of the solar surface. Since the sun is compressible, these torsional oscillations induce differential rarefaction via the centrifugal forces, but since their amplitude must be considerably less than the solar angular velocity, the relative changes in structure must be less than $\lambda = R^3 \omega^2 / GM$. Consequently any luminosity changes would be below the present observational limits, though one might expect them to modulate the solar activity.

VII. EXTERNAL INFLUENCES

It was suggested by Hoyle and Lyttleton (1939) that the gravitational energy released by infalling matter as the sun passes through interstellar gas clouds could temporarily increase the solar luminosity. Gas is accreted by being focused into a wake behind the sun. Collisions in the wake dissipate the kinetic energy of motion perpendicular to the direction of motion of the sun. Thus the angular momentum of the gas relative to the sun is destroyed, and the gas then falls into the sun.

The existence of the solar wind renders this process unlikely. Interstellar gas of typical density is deflected by the outflow and prevented from falling in by the pressure of the wind. However, dust grains can fall through the wind, though their angular momentum relative to the sun is less easily dissipated. Lyttleton (1970) has argued that such a process accounts for the comets. More recently, McCrea (1975 a,b) has suggested that the accretion of dust might even be efficient enough to cause substantial variations in solar luminosity L, reviving Simpson's (1937) idea that the temporary increase in L would induce terrestrial glaciation. Accretion would take place as the sun passed through the dust lanes in the spiral arms of the galaxy, which occurs at intervals of about 10^8 y and corresponds to the intervals between the major glacial epochs. The duration of each passage through a dust lane is of order 10^7 y. The sun's present position near the inner edge of the Orion arm suggests that accretion has taken place during the last 10^7 years and caused the Pleistocene glaciation.

Most climatologists and meteorologists agree that variations of a few percent in the solar constant would have a drastic effect on the earth's climate, though current opinion seems to be that a decrease in L is required for glaciation. A change of that magnitude caused by accretion would require gas clouds with densities equivalent to at least 10^5 hydrogen atoms per cubic centimeter. Although this is a factor 10^4 greater than normal gas clouds, McCrea points out that such dense clouds have been observed (Lequeux, 1972). If such clouds occupy about 4% as much space as ordinary clouds, which themselves occupy about 4% of the space near the galactic plane, then enhancements of the solar luminosity might occur at intervals of somewhat less than a million years during our passage through a spiral arm. Lequeux's estimates of the sizes of the dense gas clouds yield an estimate of 50,000 y for the duration of the luminosity enhancements; during this time the sun would accrete matter at the rate of one medium-sized comet every 10 s. However, there appears to be no gas cloud large enough and close enough to the solar system to have been responsible for the most recent advance of the ice sheets (Dennison and Mansfield, 1976).

Another suggested implication of McCrea's hypothesis arises from the fact that the chemical composition of the dust is very different from the mean composition of the sun. Had the sun started life with a very low heavy-element abundance Z, the convection zone would be so thin that its mass might be hardly greater than the total mass accreted. The accreted mass might not then have been greatly diluted, and the heavy-element abundance at the surface would now be quite different from that in the interior. As Joss (1974) has pointed out, the central temperature of the sun would then be lower than the value predicted by models with homogeneous Z, and could account for the neutrino discrepancy. However, this result seems unlikely, for one would expect the surface material with high molecular weight to mix into the interior by the fingering process (cf. Turner, 1973). Furthermore, the low observed abundances of lithium and boron suggest that mixing of surface material well into the interior has taken place. The accretion does not perceptibly affect the long-term variation of L given by Equation (10).

The influence of ordinary gas clouds on the cosmic ray flux at the earth has been discussed recently by Begelman and Rees (1976). They point out that when the earth passes through a gas cloud with density of order 100 particles per cubic centimeter the upstream interface between the deflected solar wind and the interstellar medium is within 1 AU of the sun. Consequently the earth's orbit is partially in the gas cloud. Since the sun's velocity through the gas cloud is only about 20 km/s, a factor 20 less than the present solar wind velocity at the earth, the particles incident on the earth while it is immersed in the cloud are less numerous and less energetic than usual. The net solar output is probably hardly affected. Begelman and Rees point out that correlations between low sea temperatures and high magnetic field strengths, and by inference low incident cosmic ray flux, suggest that the decreased flux associated with the passage of the sun through a cloud induces glaciation. Though they consider the details of McCrea's accretion hypothesis unlikely, they accept that the ice ages may be a result of the passage of the sun through the spiral arms of the galaxy.

VIII. DISCUSSION

The only firm theoretical prediction of the variation of the solar output is the gradual rise in the photon luminosity described by Equation (10). Such a rise is an unavoidable consequence of the increase in mean atomic weight of stellar material as nuclear reactions convert H to He, whether or not internal mixing has taken place, provided textbook physics is correct. Dynamical oscillations of the entire sun have been predicted to have occurred in the past, but the only modes with a clear excitation mechanism in the present sun are the high-order acoustic modes, whose amplitude is confined to the outer regions. Observations suggest that low-order modes are excited, too; their frequencies have been reproduced theoretically, but their source of energy has yet to be explained. Their existence was not confidently predicted before they were observed. The solar cycle has been inferred from documentation of sunspots and aurorae, but theoretical discussions of the periodicity are little more than rationalizations of the data. On the other hand, there are theoretical predictions of luminosity transients on a longer timescale for which there is almost no observational evidence. Other variations, on a thermal timescale, resulting from a conjectured secular instability, have been investigated theoretically (Rosenbluth and Bahcall, 1973; Schwarzschild and Harm, 1973) with negative results. Variations resulting from oscillations in the nuclear rates of the kind found in certain classes of chemical reactions (Faraday Society, 1974) also seem not to occur (Aizenman and Perdang, 1973), because there are no feedback loops in the pp chain that can modify chemical abundances on a timescale shorter than τ_n.

The photon flux is the only solar output that astrophysicists have much confidence in predicting. Theoretical solar models predict the neutrino flux, but these seem to conflict with observation. There are no predictions of solar wind and cosmic ray fluxes that yield reliable estimates of how these quantities would vary as the structure of the sun changes.

Potentially the most powerful way of finding observational evidence for past variations in the solar output is to consider their influence on the planets, and on the earth in particular. Climatologists have discussed the possibility that variations in the solar constant of order 5% would cause terrestrial glaciation (Budyko, 1969; Sellers, 1969; Parkin, 1976) but these are all in the context of equilibrium climate states. It is not entirely certain whether insolation should increase or decrease to induce an ice age, though modern theories favor the latter. The climate models do not take into account the general rise in the solar luminosity over the past 4.8×10^9 y from about 70% of its present day value. Indeed this point has received very little attention. The earth's climate appears to have been relatively stable to substantial changes in the solar luminosity on the nuclear timescale (Sagan and Mullen, 1972; Margulis and Lovelock, 1974). If variations in L on a timescale of millions of years increase glaciation, the response must be transient (Dilke and Gough, 1972; Margulis and Lovelock, 1974), which implies the existence of natural terrestrial timescales longer than 10^6 y. It is presumed that a change in the solar constant triggers a response in the climate that does not necessarily follow the detailed time variations of L.

If climatic variations are an indicator of solar change, then one is tempted to look for variations in the sun's outputs on other timescales. Short-term variations in the climatic record seem to be well correlated with insolation variations arising from the relative motion of the sun and earth (Milankovich, 1920; Emiliani, 1955; van den Heuvel, 1966) but there seem to be difficulties in explaining the variations of Quaternary glaciation on timescales of 10^5 y in this way. Ulrich (1975) has pointed out that the thermal cooling time of the sun's convective envelope is of the right order of magnitude, but there is no obvious reason why the mean thermal balance of just the convection zone should be disturbed.

Finally, it should be pointed out that our ideas about the main sequence evolution of the sun would be changed if it were found that the physics that has been used were wrong. For example several theories of gravitation (Brans and Dicke, 1961; Hoyle and Narlikar, 1973; Dirac, 1974) can be regarded to imply that the constant of gravitation G is decreasing with time. In most cases the time dependence is of the form

$$G/G_0 = (\Lambda H t')^{-n} \equiv (t'/T)^{-n} , \qquad (11)$$

where G_0 is the present value of G, H is the Hubble constant, $t' = t + T - t_\odot$ is time measured from an origin at a time T before present, and Λ and n are constants of order unity. In the flat space cosmology of Brans and Dicke Λ and n are related to a coupling constant ω by

$$\Lambda = (4 + 3\omega)/(2 + 2\omega) , \quad n = 2/(4 + 3\omega) . \qquad (12)$$

Dicke (1968) has favored $\omega \simeq 5$, which implies $n \simeq 0.1$; in the Hoyle-Narlikar theory n is typically ½, and Dirac predicts n = 1. The luminosity of a stellar model increases steeply with G, as was pointed out by Teller (1948), in a manner given approximately by

$$L \propto G^\gamma , \qquad (13)$$

with $\gamma \simeq 7$. Unless the decrease in G with time is very weak, this dominates the effect of the changing composition and leads to a decrease of L with t (Pochoda and Schwarzschild, 1964; Ezer and Cameron, 1965; Roeder and Demarque, 1966; Shaviv and Bahcall, 1969). The luminosity variation is approximated by the formula

$$L/L_\odot = \left\{1 + 2T[1-(t'/T)^{1-n\gamma}]/[5(1-n\gamma)t_\odot]\right\}^{-1} (t'/T)^{-n\gamma} , \qquad (14)$$

whose form can be derived by homology arguments. Typically the decrease in L implied by this formula is greater than the variation when G is constant (n = 0); the decrease in the solar constant is even greater, since the radius of the earth's orbit varies in inverse proportion to G. Moreover, higher L in the past would mean that more hydrogen has been consumed, and the structure of the sun now would resemble that of a star that had evolved with constant G but for a longer time. The central temperature would therefore be greater than that of the usual models and consequently the predicted neutrino flux would be higher. But whether G actually is varying is a matter that is not settled. Recent measurements of the gravitational deflection of radio waves by the sun (Fomalont & Sramek, 1976) are in very good agreement with general relativity, and in the context of the Brans-Dicke theory give a lower bound to ω of about 23. This yields so weak a variation in G that the solar constant increases with time, from a value at t = 0 of about 85% of its current value if H is taken to be 70 KM s^{-1} M parsec $^{-1}$. Van Flandern (1975) has concluded from an analysis of lunar occultations that G probably decreases with a timescale of about 10^{10} y, though Lyttleton (1976) has shown that the observations might be explained by a decreasing moment of inertia of the earth. Planetary ranging will no doubt settle the matter in due course.

It is evident, therefore, that at present there is no unquestioned theoretical prediction of the variation of any of the solar outputs.

ACKNOWLEDGMENTS

I thank J. Christensen-Dalsgaard, P.A. Gilman, and R.K. Ulrich for commenting on a draft typescript.

REFERENCES

Aizenman, M.L., and Perdang, J., 1973, *Astr. and Ap. 28*, 327.
Ando, H., and Osaki, Y., 1975, *Publ. Astr. Soc. Japan 27*, 581.
Babcock, H.W., 1961, *Ap. J. 133*, 572.
Bahcall, J.N., and Davis, R., 1976, *Science 191*, 264.
_____, and Sears, R.L., 1972, *Annu. Rev. Astr. and Ap. 10*, 25.
_____; Huebner, W.F.; Magee, N.H.; Merts, A.L.; and Ulrich, R.K., 1973, *Ap. J. 184*, 1.
Begelman, M., and Rees, M.J., 1976, *Nature 261*, 298.
Benton, E.R., and Clark, A., 1974, *Annu. Rev. Fluid Mech. 6*, 257.
Boury, A.; Gabriel, M.; Noels, A.; Scuflaire, R.; and Ledoux, P.; 1975, *Astr. and Ap. 41*, 279.
Brans, C., and Dicke, R.H., 1961, *Phys. Rev. 124*, 925.
Bretherton, F.P., and Spiegel, E.A., 1968, *Ap. J. (Letters) 153*, L77.
Brookes, J.R., Isaak, G.R., and van der Raay, H.B., 1976, *Nature 259*, 92.
Budyko, M.I., 1969, *Tellus 21*, 611.
Busse, F.H., and Whitehead, J.A., 1974, *J. Fluid Mech. 66*, 67.
Christensen-Dalsgaard, J., and Gough, D.O., 1975, *Mém. Soc. Roy. Sci. Liège, Sér. 6, 8*, 309.
_____, and _____, 1976, *Nature 259*, 89.
_____, Dilke, F.W.W., and Gough, D.O., 1974, *M.N.R.A.S. 169*, 429.
Davis, R., Harmer, D.S., and Hoffman, K.C., 1968, *Phys. Rev. Letters 20*, 1205.
Deardorff, J.W., and Willis, G.E., 1967, *J. Fluid Mech. 28*, 675.
Dennison, B., and Mansfield, V.N., 1976, *Nature 261*, 32.

Deubner, F.-L., 1972, *Solar Phys. 22*, 263.
——————, 1975, *Astr. and Ap. 44*, 371.
Dicke, R.H., 1964, *Nature 202*, 432.
——————, 1968, *Ap. J. 152*, 1.
——————, 1970, *IAU Colloquium on Stellar Rotation*, ed. A. Slettebak, D. Reidel, Dordrecht, 289.
——————, 1976, *Solar Phys.*, in press.
——————, and Goldenberg, H.M., 1974, *Ap. J. Suppl. 27*, 131.
Dilke, F.W.W., and Gough, D.O., 1972, *Nature 240*, 262.
Dirac, P.A.M., 1974, *Proc. Roy. Soc. A 338*, 439.
Emiliani, C., 1955, *J. Geol. 63*, 538.
Ezer, D., and Cameron, A.G.W., 1965, *Canad. J. Phys. 44*, 593.
—————— and ——————, 1972, *Nature Phys. Sci. 240*, 18.
Faraday Society, 1974, *Symposium No. 9*, Chemical Society, London.
Fomalont, E.B., and Sramek, R.A., 1976, *Phys. Rev. Letters 36*, 1475.
Fossat, E., and Ricort, G., 1973, *Solar Phys. 28*, 331.
—————— and ——————, 1975, *Astr. and Ap. 43*, 243.
Fowler, W.A., 1972, *Cosmology, Fusion and Other Matters*, ed. F. Reines, Colorado Associated University Press, Boulder, 67.
Gabriel, M.; Noels, A.; Scuflaire, R.; and Boury, A., 1976, *Astr. and Ap. 47*, 137.
Gilman, P.A., 1975, *Proceedings IAU Symposium No. 71, Basic Mechanisms of Solar Activity*, ed. V. T. Bumba and J. Kleczek, D. Reidel, Dordrecht, 207.
Goldreich, P., and Keeley, D., 1976, *Ap. J.*, in press.
Grec, G., and Fossat, E., 1976, *Astr. and Ap.*, in press.
Gubbins, D., 1974, *Rev. Geophys. Space Phys. 12*, 137.
Henyey, L., and L'Ecuyer, J., 1969, *Ap. J. 156*, 549.
Hill, H.A., and Stebbins, R.T., 1975, *ibid. 200*, 471.
——————, ——————, and Brown T.M., 1975, *Proceedings Fifth International Conference Atomic Masses and Fundamental Constants*, Gordon & Breach, New York.
Holzer, T.E., and Axford, W.I., 1970, *Annu. Rev. Astr. and Ap. 8*, 31.
Howard, L.N., Moore, D.W., and Spiegel, E.A., 1967, *Nature 214*, 1297.
Hoyle, F., 1975, *Ap. J. (Letters) 197*, L127.
——————, and Lyttleton, R.A., 1939, *Proc. Cambridge Phil. Soc. 35*, 405.
——————, and Narlikar, J.V., 1973, *Action at a Distance in Physics and Cosmology*, Freeman, San Francisco.
Joss, P., 1974, *Ap. J. 191*, 771.
Kaufman, P., 1972, *Solar Phys. 23*, 178.
Kobrin, M.M., and Korshunov, A.I., 1972, *ibid. 25*, 339.
Kraft, R.P., 1967, *Ap. J. 150*, 551.
Krishnamurti, R., 1973, *J. Fluid Mech. 60*, 285.
Larson, R.B., 1969, *M.N.R.A.S. 145*, 271.
Latour, J.; Spiegel, E.A.; Toomre, J.; and Zahn, J.-P., 1976, *Ap. J., 207*, 233.

Ledoux, P., 1951, *ibid. 114*, 373.
Leighton, R.B., 1969, *ibid. 156*, 1.
──────────, Noyes, R.W., and Simon, G.W., 1962, *ibid. 135*, 474.
Lequeux, J., 1972, *On the Origin of the Solar System*, ed. H. Reeves, Gordon & Breach, London, 118.
Lyttleton, R.A., 1970, *Observatory 90*, 178.
──────────, 1976, *Moon*, in press.
McCrea, W.H., 1975a, *Nature 225*, 607.
──────────, 1975b, *Observatory 95*, 239.
Margulis, L., and Lovelock, J.E., 1974, *Icarus 21*, 471.
Mestel, L., 1965, *Stars and Stellar Systems*, Vol. 8, ed. L.H. Aller and D.B. McLaughlin, University of Chicago Press, 465.
────────── and Moss, D.M., 1976, *M.N.R.A.S.*, in press.
Milankovich, M., 1920, *Theorie Mathematique des Phénomènes Thermiques Produits par la Radiation Solaire*, Gauthier-Villars, Paris.
Noels, A.; Gabriel, M.; Boury, A.; Scuflaire, R.; and Ledoux, P., 1975, *Mém. Soc. Roy. Sci. Liège, Sér. 6, 8*, 317.
Parkin, D.W., 1976, *Nature 260*, 28.
Pochoda, P., and Schwarzschild, M., 1964, *Ap. J. 139*, 587.
Priestley, C.H.B., 1959, *Turbulent Transfer in the Lower Atmosphere*, University of Chicago Press, Chicago.
Roeder, R.C., and Demarque, P.R., 1966, *Ap. J. 144*, 1016.
Rood, R., 1972, *Nature Phys. Sci. 240*, 178.
Rosenbluth, M.N., and Bahcall, J.N., 1973, *Ap. J. 184*, 9.
Ross, J.E., and Aller, L.H., 1976, *Science 191*, 1223.
Roxburgh, I.W., 1964, *M.N.R.A.S. 128*, 157.
Sagan, C., and Mullen, G., 1972, *Science 177*, 52.

────────── and Young, A.T., 1973, *Nature 243*, 459.
Sakurai, T., 1972, *Publ. Astr. Soc. Japan 24*, 153.
──────────, 1975, *M.N.R.A.S. 171*, 35.

Schwarzschild, M., 1958, *The Structure and Evolution of the Stars*, Princeton University Press, Princeton, N.J.
────────── and Härm, R., 1973, *Ap. J. 184*, 5.
Scuflaire, R.; Gabriel, M.; Noels, A.; and Boury, A., 1975, *Astr. and Ap. 45*, 15.
Sellers, W.D., 1969, *J. Appl. Meteorol. 8*, 392.
Severny, A.B., Kotov, V.A., and Tsap, T.T., 1976, *Nature 259*, 87.
Shaviv, G., and Bahcall, J.N., 1969, *Ap. J. 155*, 135.
────────── and Salpeter, E.E., 1971, *ibid. 165*, 171.
Shibahashi, H., Osaki, Y., and Unno, W., 1975, *Publ. Astron. Soc. Japan 27*, 401.
Simpson, G.C., 1937, *Proc. Roy. Inst. Great Britain 30*, 125.
Spiegel, E.A., 1968, *Highlights of Astronomy*, ed. L. Perek, D. Reidel, Dordrecht, 261.
Stein, R.F., and Leibacher, J., 1974, *Annu. Rev. Astr. and Ap. 12*, 407.

Stix, M., 1975, *Proceedings IAU Symposium No. 71*, D. Reidel, Dordrecht.
Teller, E., 1948, *Phys. Rev. 73*, 801.
Thomas, D.B., and Townsend, A.A., 1957, *J. Fluid Mech. 2*, 473.
Turner, J.S., 1973, *Buoyancy Effects in Fluids*, Cambridge University Press, Cambridge, England.
Ulrich, R.K., 1970, *Ap. J. 162*, 993.
────────, 1974, *ibid. 188*, 369.
────────, 1975, *Science 190*, 619.
Unno, W., 1975, *Publ. Astron. Soc. Japan 27*, 81.
van den Heuvel, E.P.J., 1966, *Geophys. J. 11*, 323.
Van Flandern, T.C., 1975, *M.N.R.A.S. 170*, 333.
Veronis, G., 1965, *J. Marine Res. 23*, 1.
Walén, C., 1946, *Askiv. Mat. Astr. Fys. 33A*, No. 18.
Wheeler, J.C., and Cameron, A.G.W., 1975, *Ap. J. 196*, 601.
Whitney, C., 1958, *Smithsonian Contrib. Ap. 2*, 365.
Wolff, C.L., 1972, *Ap. J. (Letters) 177*, L87.
────────, 1976, *Ap. J. 205*, 612.

Comment

In connection with the possibility of the solar system colliding with an interstellar dust cloud, an event which might produce changes in the solar wind or interplanetary dust at 1 AU, I want to comment on the evidence apparently and potentially available in studies of lunar samples and meteorites. More extensive discussion is given in my review paper on the subject (Walker, 1975). A typical scoop of lunar dust in a "mature" soil sample contains many crystals that have been at or very near the lunar surface. This is true even for crystals from the deepest portions of the core samples that extend down to 3 m in the lunar crust. From the isotopic measurements along one such core, it is known that the crystals were exposed to the sun at least a billion years ago. Typically, the lunar soil was built up from relatively thin layers of fresh material that were then reworked by micrometeoroid bombardment before being covered over by another layer (or being excavated completely and thrown out somewhere else). The exposure of any given grain at the surface is thus a statistical process. Estimates for typical "true" surface exposure ages for these grains range from 10^3 to 10^5 y. In the past, most studies of lunar soils have been of a statistical nature, and measurements have typically been made of the total solar wind contribution, the average solar flare track densities, the percentage of grains with impact pits, etc. For a number of soil samples, including one that was known to have been deposited as a fresh layer 50 million years ago, our group has found no difference in the percentage of grains with microcrater densities greater than 10^6 cm^{-2} or in the statistical ratio of the number of microcraters to the expected solar flare track densities. However, the crater/track ratios vary considerably from grain to grain, presumably because of the larger penetration depths of solar flare particles as opposed to micrometeoroids. There is thus no evidence that any very large dust fluxes have been encountered but they cannot be totally excluded.

For some time we have also been working to develop techniques that permit simultaneous measurement of the solar wind contribution and the microcraters on individual crystals. Detailed simulation studies on implanted ions in various crystals have convinced us of the basic validity of using an ion microprobe, coupled with laboratory-implanted "marker" isotopes, to study the solar wind effects in single grains. We recently reported surface enhancements in a lunar crystal of the elements Fe, Mg, Cr, Ti, P, and C. The beauty of this technique is that the penetration distance of solar wind ions (about 300 Å) is close to the dimensions of the craters that we measure (greater than or equal to 100 Å). Thus, any passage through an interstellar dust cloud that results in an increased flux of crater-producing particles should manifest itself in a greatly altered crater/solar wind ratio. The techniques are far from simple, and much more work remains to be done to extract the full information carried by the lunar samples.

Gas-rich meteorites also contain individual crystals that exhibit microcraters and solar flare effects. It is the prevailing prejudice that these objects were assembled in the time interval from 4 to 4.6 $\times$ 10^9 years ago, although this remains to be proved. The time of compaction is a problem that can be solved, albeit with great difficulty. It thus appears to us that with refined and careful experimental techniques just now being developed, it will be possible for us to set stringent limits on the size and density of interstellar dust clouds encountered by the solar system in its motion through the galaxy.

R.M. Walker
McDonnell Center for Space Sciences

References

Walker, R.M., 1975, *Ann. Rev. Earth and Planet. Sci. 3*, 99.

VII. STELLAR VARIABILITY

Given the solar mass, the measured solar luminosity, the chemical composition, and a measure of the shape of the solar spectrum, the theory of stellar interiors and their evolution allows us to place the sun on the evolutionary track described by similar stars. Consequently, a combination of empirical boundary conditions and a fundamental astrophysical theory sets the sun in the hierarchy of all stellar objects that we observe. In order to understand both the past and future of the sun, we can use other stars and the theory as guides. Here Linsky critically examines the empirical and analytical framework that connects our picture of the sun with that of other stars, and gives his view of the state of knowledge of the empirical determination of the output of different types of stars.

With the discovery that other stars have chromospheres, coronae, winds, and chromospheric activity, the variation of these well-observed solar phenomena with stellar type can, perhaps, give us clues to the long-term evolution of both solar activity and the outer layers of the sun. The established variation of stellar activity with stellar rotation rate has already emphasized the importance of rotation in the general problem of the production of stellar activity and magnetic fields. The opportunity to strengthen these connections between the sun and other stars rests squarely on our ability to analyze stellar data through a physically realistic theory of stellar spectra. Linsky describes the application of spectral diagnostics developed in solar work to the analysis of stellar spectra. The future yield of this analytic approach depends on the availability of high-resolution stellar spectra obtained through the development and application of more sensitive spectrophotometric techniques to stellar observation.

In the development of the theory of stellar evolution, the placement of the sun in the set of evolutionary tracks for the different types of stars is a basic step in understanding the global properties of the sun. But now, when we can begin to see real similarities in smaller-scale properties of the sun and other stars, we can extend our knowledge of the physics of detailed solar processes to these other objects, where the direct observation of surface detail is not possible. Stellar and solar astrophysics are becoming more strongly coupled through such current research, and information will flow in both directions between these two fields. An accurate description of the solar output gives an essential empirical boundary condition for comparison with the output of other stellar objects.

Oran R. White

THE SOLAR OUTPUT AND VARIABILITY VIEWED IN THE BROADER CONTEXT OF STELLAR ACTIVITY

Jeffrey L. Linsky
Joint Institute for Laboratory Astrophysics
National Bureau of Standards and University of Colorado

The sun is most likely a unique star only by virtue of its proximity, but the consequences of its closeness are far-reaching indeed. Because of its proximity the sun appears to be 10^{10} times as bright visually as the brightest stars and subtends a solid angle 10^9 that of the largest stars. Also the solar radiative flux is essentially unattenuated by intervening matter (except for the terrestrial atmosphere), whereas interesting stellar emissions shortward of 912 Å and in the cores of ultraviolet resonance lines are heavily attenuated. As a consequence, many phenomena routinely studied on the sun have not been observed in stars where they are presumably present, and the opportunity for cross-fertilization between solar and stellar research has heretofore not presented itself despite the similarity of physical processes and diagnostic techniques available to understand them. The phenomena observed on the sun thus tend to be considered solar phenomena, rather than examples of stellar phenomena seen on the sun.

I. PERSPECTIVE

In classical terms, stellar atmospheres are usually assumed to be homogeneous, plane-parallel, and static envelopes which are in local thermodynamic equilibrium (LTE), radiative and/or convective equilibrium, and hydrostatic equilibrium, and which have no magnetic fields. In recent years, stellar models have been constructed that incorporate radiatively driven winds, extended geometries, and departures from LTE. What I refer to as solar-type phenomena are those that depart further from these assumptions through considerations of nonradiative heating, magnetic fields, and small-scale or dynamic structures governed by nonradiative heating and magnetic effects. Praderie (1973) and Doherty (1973) have reviewed the observational basis for stellar chromospheres and for nonradiative heating therein, as it was known in 1972. At that time the only evidence for solar-type phenomena in stars was evidence of chromospheres, of flares in dMe stars, and of stellar winds in supergiants and M-type stars.

More recently, evidence has been presented for other solar-type phenomena in stars, as will be described below; and theoretical and semiempirical models for stellar chromospheres, coronae, and flares have been computed. This recent activity has resulted from new satellite and rocket instruments, able to observe cool stars (spectral classes F-M) in the ultraviolet and X-ray portions of the spectrum, and also from the development of spectroscopic diagnostic techniques to interpret spectral features, such as the Ca II H and K lines, that are observed frequently but have not yet been properly analyzed. This trend should accelerate in the next few years as more powerful space experiments become available (International Ultraviolet Explorer, HEAO-B, Large Space Telescope), as ground-based spectra of fainter stars are obtained, and as the diagnostic techniques are further developed and applied systematically to a wide range of stars. For this reason it is useful to assess where we stand in this field, to delineate some important problems to be solved, and to suggest how they might be solved.

It is important to emphasize that the study of solar-type phenomena in the sun and stars is a two-way street. Perhaps it is more obvious that our understanding of solar-type phenomena on the sun is important to the study of similar stars. Our experience in studying the sun naturally suggests what phenomena to search for in stars and how best to search for them. For example, the λ 1175-1600 Å emission line spectrum of Capella (G8III+F?) is similar to that of a solar plage (Vitz et al., 1976; Haisch and Linsky, 1976); thus the stellar spectrum may reasonably be interpreted in terms of a transition region similar to that in a solar plage. Conversely, when a stellar ultraviolet emission line spectrum differs qualitatively from that seen in the sun, as in the case of Arcturus (K2 III) (Weinstein, Moos, and Linsky, 1976), then the atmospheric structure of the star must differ considerably from that of the sun.

The opposite case—what we can learn about the sun by studying stars—may not be as obvious, but it is important and has several aspects.

First, we have only a tiny temporal baseline (compared to the age of the sun) from which to study the variability of solar radiative output and phenomena that modulate this output. One way to study long-term variability is to study stars similar to the sun but of different ages. In particular, it is important to ascertain whether the statistical trend of decreasing chromospheric activity with age seen by Wilson (1963), Wilson and Skumanich (1964), and Skumanich (1972) is applicable to the solar chromosphere and corona.

Second, we will probably not understand many solar phenomena in detail unless we can investigate the effects of changing basic parameters such as gravity, convective generation of wave modes, and background radiation fields. This may only be possible by studying stars.

Third, we are still quite ignorant about the underlying causes of solar variability, presumably because these phenomena are not readily apparent owing to their long time scale or small amplitude near the surface of the sun. I have in mind here large-scale circulation patterns, causes of sunspots and the solar magnetic cycle, and the global properties of the solar wind. In stars with more active outer atmospheres, these phenomena may be more apparent and easier to study.

Until now I have spoken of the sun as one star with a wide range of phenomena and structures. For our purposes it might be more instructive to speak of the sun as many stars coexisting in the same gravitational field with the same effective temperature and chemical composition. Below we will identify a number of different solar chromospheres, transition regions, and coronae identifiable by their emission spectra produced by differing pressure and temperature structures. These differences in physical properties, in turn, can probably be traced back to characteristic magnetic field structures. It is likely that these differences in structures on the sun are prototypes for the range of phenomena in cool stars, and the best way to understand the range of stellar "activity" is first to understand the range of solar "activity." We will endeavor below to quantify the word "activity."

Finally, I mention that, most conveniently, the sun has a near twin close by. This twin is α CEN A at a distance of only 1.33 parsec and an apparent visual magnitude of m_v = -0.01. Table 1 summarizes the parameters for the two stars taken from Ayres et al. (1976). It is important to note that αCEN A appears to be twice as old as the sun, on the basis of its location in the Hertzsprung-Russell diagram, and thus may be indicative of the direction in which the sun is evolving. Consequently, this star should be studied in detail for evidence of solar-type phenomena, as described in the next section, and for variability in its radiative output.

Table 1
STELLAR PARAMETERS

	α Cen A		Sun
Spectral Type	G2 V		G2 V
Mass/Mass (θ)	1.1		1.0
L/L (θ)	1.6		1.0
Age (yr)	10×10^9		5×10^9
Assumed Metal abundance	1.0	2.0	1.0
T_{eff} (K)	5700	5770	5770
log g (cm s^{-1})	4.25	4.27	4.44

II. A SURVEY OF SOLAR-TYPE PHENOMENA IN THE SUN AND EVIDENCE FOR SUCH PHENOMENA IN COOL STARS

We now consider in turn a number of solar-type phenomena, giving a description of the phenomena on the sun, some of the useful diagnostics for identifying and characterizing these phenomena, and evidence (if any) for these phenomena in cool stars.

A. NONRADIATIVE HEATING IN THE UPPER PHOTOSPHERE

Except in unusual circumstances, radiative equilibrium model atmospheres are characterized by a decrease in temperature with height. The addition of nonradiative heating of any sort to such an atmosphere will tend to make the decrease less steep. Then, at some height where the densities and radiative cooling rates are sufficiently small, the nonradiative heating will force the temperature to increase with height. For clarity we refer to the region where $dT/dh < 0$ as a nonradiatively heated "photosphere" and the region where $dT/dh > 0$ as a "chromosphere," but the question is more complex and interesting than stated here (Praderie, 1973; Underhill, 1973).

The identification of nonradiative heating in the upper portion of a photosphere is important because it is evidence for the violation of a classical assumption and because it can affect the emergent spectrum. In stars like the sun an increase in the upper photosphere temperature distribution at $10^{-2} < \tau_{5000} < 10^{-4}$ would manifest itself in significantly increased flux in the far infrared continuum (10-200 μm), ultraviolet continuum (1200-2000 Å; see Vernazza, Avrett, and Loeser, 1976), the inner wings of the strong resonance lines, and in the cores of some strong absorption lines.

The evidence for nonradiative heating in the solar upper photosphere is presently in dispute. Vernazza, Avrett, and Loeser (1976) have computed a photosphere-chromosphere model to best fit the existing infrared and ultraviolet continuum data, explicitly treating non-LTE effects in H I, Si I, C I, Mg I, and Al I. Their model has a temperature minimum of 4150 K located at $\tau_{5000} = 3 \times 10^{-4}$, and their data, they claim, are consistent with a minimum temperature in the range 4050-4250 K. This result is somewhat sensitive to the line-blanketing assumptions and absolute calibration errors.

The inner wings of the Ca II and Mg II resonance lines are also formed in the upper photosphere. Using the partial redistribution (PRD) technique (see Appendix A) for solving the transfer equation in the wings of these lines, Milkey and Mihalas (1974) and Shine, Milkey, and Mihalas (1975a) estimate a minimum temperature of 4400-4450 K; Ayres and Linsky (1976) derive detailed models with a temperature minimum of 4450 K located at a mass column density (measured inward) of 0.05 g cm^{-2}, corresponding to $\tau_{5000} = 2 \times 10^{-4}$. The latter result is sensitive somewhat to errors in the atomic abundances, line broadening, and absolute intensity calibration. The Ayres-Linsky models are hotter than the Vernazza-Avrett-Loeser model over the range $2 \times 10^{-4} < \tau_{5000} < 5 \times 10^{-2}$.

The important question is whether the semiempirical models are consistent with, or significantly hotter than, radiative equilibrium models. On the basis of a radiative equilibrium, line-blanketed model including a representative sample of non-LTE lines, Athay (1970) derived an upper photosphere boundary temperature of 4330 ± 150 K. Kurucz (1974) has computed a radiative equilibrium model blanketed with 1.7×10^6 lines with a boundary temperature of 4300 K. Compared to these radiative equilibrium models, the semiempirical model by Vernazza, Avrett, and Loeser appears cooler in the upper photosphere, possibly indicative of an unknown cooling process; and the Ayres-Linsky models appear hotter, indicative of nonradiative heating. The question is thus not yet resolved. An additional consideration is that in plages the inner wings of the Ca II lines are clearly brighter than for the quiet sun (Shine and Linsky, 1973), indicative of nonradiative heating or backwarming by chromospheric radiation (Underhill, 1973).

An important point is that in the upper photosphere the densities are sufficiently high that a temperature excess of 100-200 K over radiative equilibrium corresponds to a radiative loss (due mainly to H^-) that could be as large as 1×10^9 ergs cm^{-2} s^{-1} (Ayres, 1975). This corresponds to 2 percent of the solar luminosity and is far in excess of the radiative losses of 4.6×10^6 ergs cm^{-2} s^{-1} estimated by Athay (1976) for the solar chromosphere and corona.

Shine, Milkey, and Mihalas (1975b) have computed theoretical PRD models for solar-type stars with log g = 4.44 and 2.0. They find that for the lower gravity star the inner wings of strong chromospheric resonance lines are much darker despite the same (scaled) temperature distribution, demonstrating that PRD diagnostics must be employed to empirically derive an accurate temperature distribution for the upper photosphere.

So far only three stars have been studied for evidence of photospheric non-radiative heating by comparisons among the semiempirical temperature distributions obtained from PRD analyses of the Ca II line wings and line-blanketed radiative equilibrium model atmospheres. In two cases, Procyon (F5 IV-V) studied by Ayres (1975), and α CEN A studied by Ayres et al. (1976), there seems to be evidence for nonradiative heating. In the third case, α CEN B (KI V) studied by Ayres et al. (1976), the lack of evidence may be due to omission in the radiative equilibrium model of CO line-blanketing, which should be important for stars having an effective temperature near 5000 K (Johnson, 1973).

B. CHROMOSPHERES

As noted above, the question of how to define a chromosphere is an interesting one. I feel, however, that given the rudimentary state of our understanding of stellar chromospheres, it is premature to define a chromosphere very rigorously. Instead, I will adopt as a working definition for a chromosphere that region of a stellar atmosphere where $dT/dh > 0$ ($dT/dm < 0$) and where the energy balance is dominated by radiative and nonradiative (wave-dissipation) heating terms and radiative losses. Specifically excluded are conductive and stellar wind terms characteristic of transition regions and coronae. Praderie (1973) has argued that a net mass flux is necessary for nonradiative heating and is thus required for a chromosphere. In the sun the chromosphere covers the approximate temperature range from 4500 to 30,000 K, but it is incorrect to identify any stellar atmospheric region having this range of temperatures as a "chromosphere" without first considering the energy balance equation.

At the present time no detailed models of solar or stellar chromospheric regions have been constructed on a completely theoretical basis, that is, purely on the basis of computed wave dissipation and energy-balance considerations. Instead, chromospheric models have been constructed semiempirically to match one or more observed spectral lines or continua. Such spectral features are referred to as diagnostics. Diagnostics are useful to the extent that they uniquely define some physical parameters of the atmosphere such as the distribution of temperature, pressure, and velocities, or a background radiation field. Clearly, the accuracy of the semiempirical models is intimately tied to the usefulness of the diagnostics and the accuracy of the physical basis for these diagnostics.

In Appendix A we consider in detail the various chromospheric diagnostics in use, their physical basis, the data available, and models of solar and stellar chromospheres computed by means of these various diagnostics. The reader is also referred to the earlier review by Praderie (1973). Here we give a broad outline of the various diagnostics, their usefulness, and what is being learned about solar and stellar chromospheres with them. The various available diagnostics are readily divided into a number of categories:

(1) Collisionally Dominated Resonance Lines. In this category are such lines as the H (λ 3968) and K (λ 3933) lines of Ca II, the h (λ 2803) and k (λ 2796) lines of Mg II, and the Lyman-alpha (λ 1216) line of H I. Under conditions that exist in solar chromospheres and probably also in a wide range of stellar chromospheres, these lines are collisionally dominated (Thomas and Athay, 1961) and of sufficient optical thickness that their cores are thermalized above the temperature minimum (Avrett and Hummer, 1965). Thus the emission cores of these lines are useful diagnostics of the lower and middle chromosphere temperature and density structure. Milkey and Mihalas (1973) showed the importance of incorporating coherency effects into the transfer equation for these lines; and Shine, Milkey, and Mihalas (1975a) were able to resolve the long-standing question of limb darkening in the Ca II lines by using this new PRD formulation. Essentially all of the recent work in constructing semiempirical models of stellar chromospheres as described below is based on these lines by using the PRD formulation. Other potentially useful lines in this category are Ca I (λ 4226), Na I (λ 5890,

λ 5896), Mg I (λ 2852), C II (λ 1334, λ 1335), and Si II (λ 1260, λ 1264).

(2) Resonance Lines of Uncertain Formation. The resonance lines of He I (λ 584) and He II (λ 304) are potentially useful diagnostics of chromospheric structure, but have not been utilized since the process of formation, even in the sun, is in dispute. Various authors have proposed excitation by electron collisions, by recombination following photoionization by the coronal XUV radiation field, and by mixing- and diffusion-type processes. Depending on the method of formation, the strengths of these emission lines could be diagnostics of the coronal radiation field or of the upper chromosphere and transition-region temperature structure.

(3) Collisionally Dominated Subordinate Lines. To fall in this category, a line must respond to a change in chromospheric temperature structure or density by changing in core intensity. The infrared triplet lines of Ca II (λ 8542, λ 8498, λ 8662) fall in this category for the solar chromosphere and many stellar chromospheres, while the Balmer-alpha line in dMe stars and subordinate lines of He I (λ 10830, λ 5876) and He II (λ 1640) may be useful under certain conditions (see Appendix A). These lines have not been utilized (except λ 8542) in conjunction with the Ca II resonance lines.

(4) Spectral Lines Responsive to Radiation Fields. The central intensities of a number of lines are sensitive to continuum or line radiation fields and thus can be used as diagnostics for these radiations, but not directly for other chromospheric properties. Examples include the Balmer-alpha and epsilon lines of H I which are sensitive to the Balmer and Paschen radiation fields, the O I resonance lines (λ 1302, λ 1304, λ 1306), which are pumped by the H I Lyman-beta line, and subordinate lines of He I (λ 10830, λ 5876) and He II (λ 1640), which may be excited by the coronal XUV radiation field.

(5) Other Diagnostics. In this category are the Balmer continuum (which is in emission in T Tauri stars and occasionally in solar flares), the hydrogen free-free continuum in the far infrared, and the Si bound-free continuum (λ < 1525 Å) and other continua in the ultraviolet. These features are sensitive to the temperature distribution at the base of the solar chromosphere (Vernazza, Avrett, and Loeser, 1973;1976). Also the CO fundamental vibration-rotation band in K giants may show emission owing to a chromosphere (Heasley and Milkey, 1976).

Table 2 summarizes the solar and stellar chromospheric models that have been constructed using the above diagnostics, mainly category (1). The table includes the minimum temperature (T_{min}) and the pressure and mass column density at the location of the temperature minimum. As originally suggested by Thomas and Athay (1961), the temperature rises very steeply when the H I Lyman continuum

Table 2
CHROMOSPHERIC MODELS

Structure	Paper	Diagnostic	T_{min} (°K)	$P(T_{min})$ (dynes/cm^2)	$m(T_{min})$ (g/cm^2)	$m(\tau_{LyC}=1)$ (g/cm^2)
Quiet Sun	Vernazza et al. (1973)	Continua	4100	5 x10^2	0.02	4 x10^{-6}
	Vernazza et al. (1976)	Continua	4150	1.4x10^3	0.05	
	Shine et al. (1975a)	Ca II PRD	~4450			
	Ayres and Linsky (1976)	Ca II, Mg II PRD	4450	1.6x10^3	0.06	5.6x10^{-6}
α CEN A (G 2V)	Ayres et al. (1976)	Ca II PRD	4490	3.0x10^2	0.016	5 x10^{-6}
α CEN B (K1 V)	Ayres et al. (1976)	Ca II PRD	3730	1.9x10^3	0.063	3 x10^{-6}
Procyon (F5 IV-V)	Ayres et al. (1974) Ayres et al. (1976)	Ca II Mg II PRD	5200	1.1x10^3	0.11	1 x10^{-5}
Arcturus (K2 III)	Ayres and Linsky (1975a)	Ca II, Mg II PRD	3150	9 x10^2	1.8	3 x10^{-5}
Solar Plage	Shine and Linsky (1974)	Ca II CRD	4400	8 x10^2	0.03	1 x10^{-4}
Solar Flare (3)	Machado and Linsky (1975)	Ca II PRD	5030	1.1x10^4	0.4	3 x10^{-3}
T Tau (dG5e)	Dumont et al. (1973)	Hα		4 x10^4	40	

becomes optically thin owing to the loss of a major source of cooling. As a result, the mass column density at $\tau(\lambda = 911 \text{ Å}) = 1$, $m_o = m(\tau_{Lyc} = 1)$, is a convenient benchmark for the top of the lower chromosphere.

Table 3 summarizes parameters for various structures in the solar chromosphere. These parameters are based on the chromospheric models described in Table 2 and some guesswork where models are not yet available. The various structures are arranged in order of increasing chromospheric "activity" as measured by the emission strength of the K line. Note that this sequence is also a sequence of increasing P_o ($\tau_{Lyc} = 1$) and dT/dh between the temperature minimum and the location where $\tau_{Lyc} = 1$. The reason for the increase in emission with increasing P_o and dT/dh is that at each K-line optical depth the temperature T_e (τ_K) and Planck function B_ν (τ_K) are larger, which increases the ionization of hydrogen and the metals so that the Ca II line source functions are more thermalized by collisions.

Table 4 summarizes the evidence for stellar chromospheres in different classes of stars and according to different diagnostics. The symbols are N (no emission), W (weak emission), S (strong emission), and VS (very strong emission). The stars are also listed approximately in order of increasing visibility of the chromospheric diagnostics, but visibility is not synonymous with absolute flux as described below because the background flux is a strong function of stellar effective temperature.

We consider now what can be said concerning the energy balance in solar and stellar chromospheres. Athay (1976) has recently summarized the various radiative losses from the spatially averaged solar chromosphere. He estimates a total loss of about 4×10^6 ergs cm^{-2} s^{-1} in which the dominant contributors are Lyman alpha in the upper chromosphere, the Balmer continuum, Balmer-alpha, Mg II, Ca II,

Table 3
SOLAR CHROMOSPHERES

STRUCTURE	SCALE (Mm)	$P_o(\tau_{Lyc}=1)$ (dynes/cm^2)	dT/dh ($^\circ$K/km)	B (gauss)	LIFETIME (s)
UMBRA	15	?	?	3300	10^6-10^7
PENUMBRA	10	?	?	1900	10^6-10^7
PROMINENCE	100	0.02	small	5 -50	10^6-10^7
SPICULE	1x10	0.15	small	25 -50	6 x10^2
QUIET REGION	300x2	0.15	2		
NETWORK	1	?	?	10^2-10^3	10^5
CELL	30	?	?	~1	10^5
PLAGE	60x1	2	3.0	10^2-10^3	10^6-10^7
FLARE (SF-3)	10x0.6	8-80	3.5-4.3	10^3	10^2

Table 4
EVIDENCE FOR STELLAR CHROMOSPHERES

CLASS	EXAMPLES	CaII	MgII	Lα	Hα	Hϵ	Bal. Cont.	EUV Lines	HeI	HeII
A STARS	γBOO	W			N	N	N		N	N
EVOLVED F-M IV-V	αCMI	W	S	S	N	N	N	?	W	N
	QUIET SUN	W	S	S	N	N	N	S	?	N
YOUNG F-M V	ϵERI	VS	VS	VS					S	
	ACTIVE SUN	VS	VS	VS	N	N	N	VS	S	N
G-M GIANTS	αBOO, βGEM	S	VS	VS	N	W	N	W	W	N
G-M SUPERGIANTS	αORI	S	VS	VS	N		N		W	N
LONG PERIOD VAR.	oCET	VS			S					
dG$_e$-dM$_e$	GM AUR	VS			S	S			S	
SPEC. BINARIES	αAUR, λAND	VS	VS	VS				VS	VS	
T TAU	T TAU	VS			S	W	S		S	
TURNED ON	FU ORI,V1057 CYG	VS			S	S			S	
NOVAE (Principal)	T CRB,T PYX	VS			S	S			W	
(Nebular)					S	S			S	S
FLARE STARS	SUN	VS	VS	VS	VS	S	W	VS	S	S
	UV CYG,AD LEO	VS			VS	S	S		W	N
	SS CYG,U GEM	VS			VS	S	S		VS	S

and Fe I in the middle chromosphere, and H⁻ in the low chromosphere.

At present we cannot completely estimate the radiative losses from the solar chromosphere, much less from stars, but we can compare the losses in a few important lines formed at different chromospheric layers. Table 5 summarizes the data now available for the Lyman alpha, Mg II, and Ca II lines. These data are of necessity very heterogeneous. One unresolved question is whether the chromospheric losses in the Ca II and Mg II lines are measured by the emergent flux between, say, k_{1R} and k_{1V} (e.g., Ayres, Linsky, and Shine, 1974; Ayres and Linsky, 1975a) or by the flux above an estimated "photospheric" absorption line profile (e.g., Dravins, 1976; Blanco et al., 1974; 1976; Kondo et al., 1976b; Kondo, Morgan, and Modisette, 1976b; Kondo et al., 1976a). Despite the heterogeneity of the data, several trends clearly appear:

(1) There is a strong trend in all the lines of decreasing emission with decreasing stellar effective temperature (later spectral type).

(2) The emission does not appear to depend on stellar luminosity.

(3) The sun does not appear to be anomalous to this sample of 17 stars.

(4) The Mg II lines are the strongest chromospheric emitters, as originally suggested by Kandel (1967). The star α CMI may not be an exception to this rule since the Ca II and Mg II line fluxes were estimated differently as described above. Thus it may be possible to roughly estimate the total chromospheric radiative loss as five times the Mg II flux (the solar ratio).

Dravins (1976) and Blanco et al. (1974; 1976) have measured Ca II K line fluxes (above the estimated "photospheric" absorption line profile) for a wide range of late-type stars. Dravins finds that the K line absolute fluxes measured this way are approximately constant between spectral classes F0 and K0. An important result obtained by Blanco et al. (1974; 1976) is that the ratio of the K line flux to the integrated Planck function goes through a maximum at T_{eff} = 4500 K for giants and at 5000 K for main-sequence stars. This trend is not found in the theoretical mechanical flux calculations of de Loore (1970) based on the mixing-length theory of convection. As a consequence, either the calculations are in error or the fraction of the total mechanical flux available that is eventually emitted in the Ca II lines depends on spectral type. There is no evidence in Table 5 that the latter is true.

In addition to de Loore (1970), Kuperus (1965), Ulmschneider (1967), and Nariai (1969), among others, have computed the mechanical energy flux generated by convective zones and available to heat the outer atmospheres of stars. Athay (1976) and de Jager (1976) have recently discussed this work and have pointed out the large uncertainties in the mechanical flux generation that results from uncertainties in the mixing-length theory. There is the further point that by far the largest amount of this energy may be dumped in the photosphere (cf. § IIA) and not in the chromosphere or corona, so that it is premature to compare these calculations with the estimated chromospheric radiative losses.

Table 5
STELLAR CHROMOSPHERE RADIATIVE LOSSES

Star	Spectral Type	Assumed Angular Diameter	Assumed Interstellar n_N (cm^{-3})	Lα	Mg II h+k	Ca II H+K	Other
β CAS	F2 IV	0.0019			4.2×10^6(1)		
α CMI	F5 IV-V	0.0057	0.03	1.7×10^5(2)	1.5×10^6(1,3)	3.8×10^6(3)	
α PER	F5 Ib	0.0030			2.6×10^6(4)		
Average Sun	G2 V			2.5×10^5(5)	1.2×10^6(5)	8×10^6(6)	4×10^6(7)
Solar Plage	G2 V			3.1×10^6(5,8)	3.6×10^6(5,9)	4×10^6(5,10)	
α AUR	G8 III+F?	0.0094	0.01	4.2×10^5(11)	1.2×10^6(12)		
ε GEM	G8 Ib	0.0085			2.1×10^5(13)		
α UMA	K0 III	0.0081			1.5×10^5(12)		
β GEM	K0 III	0.0105	0.10	4.1×10^4(14)	1.2×10^5(5,12)		
ε ERI	K2 III	0.0026	0.10	6.0×10^5(14)	1.1×10^6(14)	4.3×10^5(15)	
α BOO	K2 III	0.022	0.10	5.2×10^4(14)	1.7×10^5(13,14)	1.3×10^5(16)	
ε PEG	K2 Ib	0.014			1.0×10^5(13,14)		
α HYA	K3 II-III	0.014			4.8×10^4(12)		
β UMI	K4 III	0.014			3.6×10^4(12)		
α TAU	K5 III	0.022	0.10	2.6×10^4(14)	6.9×10^4(1,5,13)		
β AND	M0 III	0.019			1.1×10^5(13)		
α SCO	M1 Ib	0.042			4.8×10^4(17)		
α ORI	M2 Iob	0.060			1.7×10^4(1,13,17)		

(1) Kondo et al. (1976a); (2) Evans et al. (1975); (3) Ayres et al. (1974); (4) Kondo et al. (1976b); (5) McClintock et al. (1975b); (6) Ayres and Linsky (1976); (7) Athay (1976); (8) Dupree et al. (1973); (9) Lemaire and Skumanich (1973); (10) Shine and Linsky (1974); (11) Dupree (1975); (12) Kondo et al. (1976a); (13) Doherty (1972a); (14) McClintock et al. (1975a); (15) Blanco et al. (1974); (16) Ayres and Linsky (1975a); (17) Bernat and Lambert (1976).

Given the abundance of diagnostics available, the large and growing set of observations, and the few pioneering model chromospheres computed; it is important to state some realistic goals for future studies of chromospheres.

Specific lines such as the Balmer-alpha line of H I, the λ 5876 and λ 10830 lines of He I, and the λ 1640 and λ 4686 lines of He II will be of great use, when these diagnostics are better understood, in probing regions of the chromosphere above where the Ca II and Mg II lines are formed.

It is important to determine the region of the Hertzsprung-Russell diagram where chromospheres exist, including both the high and low temperature limits, if any.

The difference between active and quiet chromospheres in the sun and stars in terms of temperature and density structures as well as energy balance remains to be studied. One question to be resolved is whether atmospheric extension produces the bright emission spectra in spectroscopic binaries and in T Tauri stars—and if so, why.

It is important to know the extent to which the solar chromosphere models are reliable prototypes, in terms of the general temperature structures and energy balance, for the range of stellar chromospheres.

C. TRANSITION REGIONS

The term "chromosphere-corona transition region" or "transition region," TR for short, applies to the region in the sun where the energy balance is determined by conductive heating from the corona, radiative losses, and possibly also dynamical terms. Typically the temperature range is from 3×10^4 to 1×10^6 K. For the present we apply this term also to stellar structures where the energy balance equation appears roughly to apply, but the energy balance may be more complex than usually assumed, even for the solar TR.

Pottasch (1964) and Athay (1966) originally showed that the solar TR temperature gradient is consistent with thermal conduction heating from the corona. Subsequent semiempirical (e.g., Dupree, 1972) and theoretical models (Moore and Fung, 1972; Shmeleva and Syrovatskii, 1973) have either derived or assumed a thermal conduction temperature distribution. Withbroe and Gurman (1973) then showed that a sequence of TR models for coronal holes, quiet sun, and active regions is a sequence of increasing temperature gradient and downward conductive flux. Since the solar TR appears typically to be thin compared to a pressure scale height, it appears reasonable to set $P_{TR} = P_o$, the pressure at the top of the chromosphere. Table 6 summarizes a number of solar TRs and their properties.

Table 6
SOLAR AND STELLAR TRANSITION REGIONS

	Structure	P_o (dynes/cm^2)	$n_e T_e$ (K cm^{-3})	F_c (ergs/cm^2/s)	Thickness (km) (40,000–400,000°K)
	Sunspot	?		?	~30,000
	Prominence	0.02	8×10^{13}	3×10^4	5,400
	Hole	0.09	3.6×10^{14}	1.2×10^6	1,400
	Spicule	0.15	6.0×10^{14}	1.2×10^6	135
SUN	Quiet Region	0.15	6.0×10^{14}	1.2×10^6	135
	Cell	?		?	?
	Network	?		?	?
	Active Region	1.4	5.4×10^{15}	6×10^6	30
	Flare (SF-3)	8–80	$3 \times 10^{16} - 3 \times 10^{17}$	large	small
αCMI (Evans et al. 1975)		0.006–0.05	$2.0 \times 10^{13} - 1.7 \times 10^{14}$		
αAUR (Haisch and Linsky 1976)		1.5	5.6×10^{15}	3.3×10^6	

Before proceeding we summarize the assumptions that are included in the models just described: (1) plane-parallel geometry, (2) constant pressure, (3) no magnetic fields, and (4) an energy balance consisting solely of conductive heating and radiative losses. These assumptions are now being questioned from several directions. One deduction from (4) is that $dT/dz \sim P$ for any assumed temperature dependence of the radiative cooling function (e.g., Haisch and Linsky, 1976). Withbroe and Gurman (1973) find empirically, however, that in the sun $dT/dz \sim P^{1.8}$. This discrepancy could be resolved by an additional local heating term such as wave dissipation (Munro and Withbroe, 1972; Boland et al., 1973; Kopp, 1972), an additional loss term due to the solar wind (Noci, 1973), or a non-plane-parallel geometry defined by magnetic flux tubes (Kopp and Kuperus, 1968). An additional consideration suggested by McWhirter, Thonemann, and Wilson (1975) is that the pressure exerted by sound waves may be comparable in magnitude to the hydrostatic pressure (cf. Flower and Pineau des Forets, 1974).

One typically determines TR models by inferring the run of emission measure ($\int n_e^2 \, dh$) with temperature from the intensities of collisionally excited resonance lines in the ultraviolet. Such lines as Si III (λ 1206), Si IV (λ 1394 and λ 1403), C III (λ 977), N V (λ 1238 and λ 1242), Ne VII (λ 465), and Mg X (λ 610 and λ 625) have been very useful. Also ratios of lines in the beryllium isoelectronic sequence have been proposed as electron-density diagnostics (Munro, Dupree, and Withbroe, 1971). The use of such diagnostics may lead to considerable error, however, when the atomic rates are poorly determined, where photoionization is ignored (Nussbaumer and Storey, 1976), when the geometry is unknown, or when local time-independent ionization balance is invalid owing to a wind or to diffusion (Shine, Gerola, and Linsky, 1975; Kjeldseth Moe, 1976).

The identification of transition regions in stars requires the measurement of line intensities from several ions so as to determine whether thermal conduction from a corona plays a role in the energy balance. Two stars have been analyzed to date. From measurements of the Lyman-alpha, Si III (λ 1206), and O VI (λ 1032) lines in α CMi; Evans, Jordan, and Wilson (1975) showed that a solar-like TR may set in at a temperature of 100,000 K, and that this TR is thicker and less dense ($1.7 \times 10^{14} \geq n_e T_e \geq 2.0 \times 10^{13}$) than the solar TR ($n_e T_e \approx 6 \times 10^{14}$). Observations of lines of Si II-IV, C II-III, N V, and O VI in α AUR by Vitz et al. (1976) and Dupree (1975) have led to a TR model in the α AUR primary star with $n_e T_e = 5.6 \times 10^{15}$ (Haisch and Linsky, 1976; Dupree and Baliunas, 1976), similar to pressures in solar active regions (Dupree et al., 1973). Haisch and Linsky (1976) state that these lines are formed in a TR rather than a corona because the line flux ratios are similar to those in the sun. Also the wind speed is about 0.2 times the sound speed at the O VI (T $\approx$ 300,000 K) level, and this flow should affect the energy balance.

D. CORONAE AND WINDS

The term "corona" has come to imply that region of the solar outer atmosphere where $T > 1 \times 10^6$ K. A far better definition might be given in terms of an energy balance between nonradiative heating due to wave dissipation or magnetic annihilation (Tucker, 1973; Sheeley et al., 1975) and losses due to radiation, wind, and conduction into the TR and out to space. The sun has at least four coronae, and there is a close correlation of coronal base pressure and temperature (see Table 7). As pointed out by Adams and Sturrock (1975), among others, the divergence of the magnetic field lines in coronal holes increases the solar wind energy loss, resulting in lower coronal temperatures and densities. Conversely, closed magnetic field structures, which characterize active regions, result in decreased energy loss by the wind and higher densities and temperatures. Thus the magnetic field geometry plays a crucial role in the solar corona and presumably in stellar coronae as well. As mentioned above, purely theoretical models of stellar coronae are highly uncertain owing to uncertainties in the mechanical flux generation rates. As a result we concentrate on observations and semiempirical models.

One can argue that if a transition region has been identified in a star, then a corona must exist surrounding the TR; accordingly, coronae must exist in α CMI and α AUR. But only the coronal base pressure and a minimum value of the coronal temperature can be determined in this way. For more information the detection of a coronal line is necessary.

The coronal forbidden red and green lines of [Fe X] and [Fe XIV] have not been detected in any nonexploding star, presumably because of their relative faintness when seen against the stellar disk. An exception is the star AS 295 B (Herbig and Hoffleit, 1975) which shows these lines together with [Fe XI] λ 7891 and [Ar X] λ 5533, suggesting a corona of at least 1.25×10^6 K. This star may have been a recent nova, however, and novae often show a coronal spectrum (McLaughlin, 1960). Another diagnostic of coronae in the visible may be continuum polarization, which in white dwarfs may be due to cyclotron emission (Ingham, Brecker, and Wasserman, 1976).

In the ultraviolet, Gerola et al. (1974) have observed the O V (λ 1218) intercombination line in β GEM, which they state is far too strong to be formed in a

Table 7
SOLAR CORONAE

STRUCTURE	P_o (dynes/cm^2)	T_{COR} (°K)
HOLE	0.09	1.0×10^6
QUIET REGION	0.15	1.6×10^6
ACTIVE REGION	1.4	$2-8 \times 10^6$
FLARE (SF-3)	8-80	$> 8 \times 10^6$

TR. Instead, they suggest that the line is formed in a cool corona at $T \approx 260{,}000$ K. This observation, however, has not been confirmed and could be spurious. In α CMI (Evans, Jordan, and Wilson, 1975) and α AUR (Dupree, 1975; Vitz et al., 1976), and O VI (λ 1032) line has been observed; but in both cases the line may be formed in a TR rather than a corona. From the upper limits of the ultraviolet emission line, assuming that the coronal base pressure equals the pressure at the top of the chromosphere, Weinstein, Moos, and Linsky (1976) conclude that the coronal temperature of α BOO is unlikely to be in the range 20,000-350,000 K. They suggest however that the coronal pressure may be overestimated owing to a thick transition region, in which case a wide range of coronal temperatures is possible. Riegler and Garmire (1975) have searched for 140-430 Å emission from four stars including o^1 CMA (K3 Iab) without success.

Recent X-ray observations of several stars have revealed positive evidence for coronae. The star α AUR has been observed at 0.25 keV (Mewe et al., 1975; 1976) and with a 0.2-1.6 keV broadband channel (Catura, Acton, and Johnson, 1975). These observations do not imply a unique coronal temperature, but rather a range of possible temperatures and pressures. In particular, if we assume a pressure of 1.5 dynes cm^{-2} from the Haisch and Linsky (1976) TR model, then the Mewe et al. (1975) data imply a coronal temperature of 2×10^5 or 10^8 K for a homogeneous isothermal corona. Upper limits on the X-ray flux from a number of late-type stars have been given by Mewe et al. (1975;1976), Margon, Mason, and Sanford (1974), Cruddace et al. (1975), and Vanderhill et al. (1975). The white dwarfs α CMA B (Mewe et al., 1975; 1976) and possibly HZ 43 (Hearn et al., 1976; Lampton et al., 1976) have been detected at soft X-ray wavelengths and in the extreme ultraviolet, but the emission may be photospheric rather than coronal (Shipman, 1976; Durisen, Savedoff, and Van Horn, 1976).

Oster (1975) has reviewed radio observations relevant to the question of stellar coronae and has pointed out that no coronae similar to the quiet solar coronae have yet been discovered. On the other hand, nonthermal and highly variable emission has been detected from α ORI and α SCO (cf. Oster, 1971), and UV Ceti-type flare stars have been detected during flares (see below). Altenhoff et al. (1976) have recently published observations and upper limits on a number of stars.

The detection of winds would provide indirect evidence for stellar coronae in late-type stars even if there is no direct spectroscopic evidence for ions at coronal temperatures. A severe problem in using this method, however, is that if downward motions of chromospheric or TR material are correlated with brighter emission, as is apparently true for the sun, then a net red shift in a stellar emission line will give a false mass loss signal (Doschek, Feldman, and Bohlin, 1976). Dupree (1975) measured a blue shift of 20 ± 7 km s^{-1} in the O VI (λ 1032) line of α AUR which corresponds to a mass loss of $(1.2 \pm 0.4) \times 10^{-8}$ solar masses y^{-1} (Haisch and Linsky, 1976) if the blue shift accurately measures the net doppler motion. With the same caveat, the asymmetric Mg II and Ca II lines of α BOO (Chiu et al., 1976) imply a mass loss of 8×10^{-9} solar masses y^{-1}, but this loss may be only occasional.

Hills (1973) has proposed that the integrated soft X-ray emission from a corona is related to the stellar mass loss. On the basis of this theory Cruddace et al. (1975) and Margon, Mason, and Sanford (1974) have estimated upper limits to the mass loss for several bright late-type stars.

Hearn (1975) has developed a theory to predict coronal temperature and mass loss given the stellar mass, radius, and coronal base pressure. The theory is based on the assumption that the most likely coronal configuration is one in which the total loss for a given base pressure is a minimum. The physical basis underlying the minimum flux assumption is unclear because it is unstable. A slight decrease in temperature increases the total energy loss from the corona, leading to a further decrease in temperature and thermal runaway. This point needs to be studied, but in the meantime the minimum flux assumption leads to a straightforward method of computing coronal properties which yields realistic values for the sun. Using this theory with some modification, Haisch and Linsky (1976) estimate for α AUR a coronal temperature of 1.2×10^6 K and mass loss of 2×10^{-8} solar masses y^{-1}, close to the value derived using the wavelength shift of the O VI (λ 1032) line. Mullan (1976) has computed coronal models for dwarfs and giants using the minimum flux theory and the assumption that the fraction of the total stellar luminosity used to heat the corona is the same as for the sun. He finds for main-sequence stars that the coronal temperature should increase with effective temperature. In G and K giants (like α AUR and α BOO) he estimates coronal temperatures well under 10^6 K and detectable winds. Also he suggests that the transition region in α BOO may be thick, as independently suggested by Weinstein, Moos, and Linsky (1976).

E. MAGNETIC FIELDS AND STELLAR CYCLES

Since solar chromospheric activity is related to the solar magnetic cycle, several observers have searched for stellar cycles by attempting to measure the longitudinal magnetic fields of late-type stars. In their search Severny (1970) and Borra and Landstreet (1973) report a measurable magnetic field only in γ CYG (F8 Ib), and this field appears variable. Boesgaard (1974) has observed 8 F0-K0 dwarfs thought to be young on the basis of strong Ca II emission, high lithium content, or high rotation velocities. In this sample, γ VIR N (F0 V) appears to have a small variable magnetic field, and ξ BOO A (G8 V) and 70 OPH (K0 V) show marginal evidence for fields. Further work to confirm these not wholly convincing observations and expand our information on magnetic fields in late-type stars would be very helpful.

Since 1967, Olin Wilson has been systematically observing main-sequence stars to search for variability in the Ca II H and K line emission as evidence of stellar cycles. Stellar rotation would also be evidenced by the appearance of plages on the disk. His approach (described in Wilson, 1968) is to measure the flux in 1 Å bands centered on H and K and divide the sum by the flux in two continuum bands separated by about 250 Å on either side of the H and K region. The results of this work are to be published in 1977. Wilson has now found several stars of a spectral type later than G0 V with completed cycles. Figure 1 is one such example (shown with his permission), of the star HD 32147, a K5 dwarf. Wilson finds no evidence for periodicity in stars hotter than spectral type G0 V or in cooler stars with very strong H and K emissions. These latter, presumably younger, stars are very "noisy" in Ca II emission without much evidence for periodicity.

F. FLARES, STARSPOTS, AND PLAGES

A complete description of stellar flares is beyond the scope of the present paper and the reader is referred to the excellent reviews of Ambartsumian and Mirzoyan (1975), Kunkel (1975), and Gershberg (1975). We will dwell here briefly on only a few possible analogues of solar phenomena in flare stars.

There is considerable evidence that flares in UV Ceti-type stars are similar to the chromospheric aspects of solar flares. As summarized by Gershberg (1975), the similarities include light curves in emission lines, confinement to small areas in

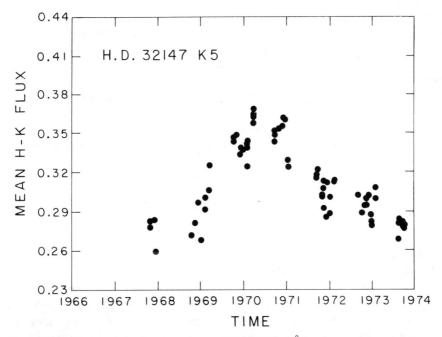

Fig. 1 *Time variation of the flux from the star HD 32147 in 1-Å bands centered on the H and K lines, divided by the flux in two continuum bands separated by about 250 Å on either side of the H and K spectral region (cf. Wilson, 1968). These are unpublished observations kindly supplied by Olin Wilson.*

the disk, sympathetic flares, the emission line spectra, plasma temperatures and densities, and motions in excess of 100 km s^{-1}. However, Kunkel (1970) finds that the energies in these flares can exceed by at least a factor of 100 the total energy of solar class 3$^+$ flares. Hard X-ray emission has been predicted (Grindlay, 1970) and 1-7 keV X-rays have been detected (Heise et al., 1975) from YZ CMI.

Many flare stars (BY DRA is the prototype here) show quasiperiodic brightness variations suggestive of cool starspots rotating on and off the disk. These spots may be 500-1500 K cooler than the stellar photosphere (Torres and Ferroz-Mello, 1973; Bopp and Evans, 1973), but Vogt (1975) has suggested bright plage regions as an alternative (cf. Fix and Spangler, 1976). An interesting point made by Gershberg (1975) is that stellar flares, like solar flares, tend to occur when the continuum flux is low (i.e., when the starspots are on the disk); thus the starspots may be the seat of flare activity (cf. Bopp, 1974). On theoretical grounds, Mullan (1974) has estimated magnetic fields of $\sim 2 \times 10^4$ gauss for these spots, and he suggests that the energy for stellar flares is the missing energy in the starspots (Mullan, 1975a). An unequivocal measurement of the magnetic fields in these spots would confirm or disprove the theory (cf. Anderson, Hartmann, and Bopp, 1976). There is also evidence (Kunkel, 1975) that, as in the sun, the lifetime for individual active regions is months and that long-term cycles on the order of one or several decades (like the solar cycle) may exist. Mullan (1975b) has proposed that W UMA-type stars also have magnetic starspots and cycles.

There is evidence for nonthermal processes during stellar flares. Flares may be heated by fast electrons (Bopp and Moffett, 1973). Spangler and Moffett (1976) and Spangler, Rankin, and Shawhan (1974) conclude on the basis of polarization and intensity measurements that the radio emission is nonthermal, possibly as a result of a coherent process (Robinson, Slee, and Little, 1976). Also the radio emission lags the optical emission, suggestive of a disturbance propagating outward in the stellar corona.

III. BASIC QUESTIONS AND POTENTIAL EXPERIMENTS

We conclude this survey of solar-type phenomena in the sun and late-type stars by posing a number of fundamental questions and suggesting experiments, both observational and theoretical, that may help in answering these questions.

What is stellar activity? I think that we have skirted this fundamental question long enough. It appears to me that the most productive approach is to ask why the sun has so many chromospheres, transition regions, and coronae, loosely described as varying "activity," coexisting in the same gravitational field. To answer this we should first assess the energy balance in these various structures semiempirically and then question how it is that the various magnetic field structures lead to the various energy balance regimes. With this insight, we can productively begin to inquire into stellar activity.

What are the various energy balance options that occur in the outer atmospheres of late-type stars? This question may be resolved into questions concerning nonradiative heating, radiative losses, winds, and conduction. I suggest that purely theoretical approaches, computing the generation and dissipation of mechanical energy, are not sufficiently accurate to yield very much insight into this question. Instead, it may be more productive to derive semiempirically the radiative losses and temperature structures in the various outer layers (including photospheres) of different kinds of stars and to use this information as a guide for future theoretical calculations. This suggests a major effort to obtain high-resolution spectra in the ultraviolet, visible, and X-ray portions of the spectrum. Even more important is the development and application of diagnostics to obtain reliable information and models from these data (Linsky, 1976).

What might be the long-term variation in solar activity and the solar radiative output? One approach is to seek out statistical effects, such as the decline of chromospheric emission with age (Skumanich, 1972; Blanco et al., 1974; Blanco, Catalano, and Marilli, 1976) and the properties of stellar cycles. Another approach is to study closely pairs of stars which are identical except for their outer atmospheres and estimated ages. From the differences in the temperature structure and energy balance of such star pairs, one may obtain insight into evolutionary effects. In particular, the α CEN A - sun pair should be closely studied.

Finally, what is the significance for its outer atmosphere of whether a star is a part of a binary or multiple system? Observationally we know that close spectroscopic binaries tend to have bright chromospheric and probably also coronal emission spectra (α AUR may be the prototype). It is commonly thought that when stars form they lose excess angular momentum by forming either a planetary system or a multiple star system. If so, the internal angular momentum distribution and convective envelopes of stars may depend on their past history in complex ways. Also, the existence of nearby stars will produce Roche lobes about stars which could be significantly filled, and tidal coupling may alter the generation and transport of mechanical energy in stars (Young and Koniges, 1976). This is a particularly intriguing question.

ACKNOWLEDGMENTS

This work was supported by grants NGR-06-003-057 and NAS5-23274 from the National Aeronautics and Space Administration to the University of Colorado. I am extremely grateful to Olin Wilson for permission to reproduce Figure 1 and to report on his work prior to publication. I also wish to thank Walter Kelch for a careful reading of the manuscript.

APPENDIX A: CHROMOSPHERIC DIAGNOSTICS

In this appendix we consider in detail the various chromospheric diagnostics that are available—their physical bases and the chromospheric models that have been constructed using them.

Ca II

The Ca II resonance lines H (λ 3968) and K (λ 3933) have proved to be the most useful chromospheric diagnostics because Ca II is the most abundant ion in the chromosphere with resonance lines in the visible spectrum and because these lines are collisionally excited and thermalized in the low chromosphere (Linsky and Avrett, 1970). As a result the positive chromospheric temperature gradient (dT/dh > 0) produces emission cores in these lines.

Until recently, chromospheric models were derived by matching the observed line profiles with theoretical profiles computed in the complete redistribution (CRD) approximation (e.g., Linsky and Avrett, 1970). This approach was able to match line profiles at the center of the solar disk, but was unable to account for limb darkening of the whole line profile (Zirker, 1968; Athay and Skumanich, 1968). Milkey and Mihalas (1973) and Shine, Milkey, and Mihalas (1975a) subsequently showed that taking account of coherency effects in the inner line wings (via the partial redistribution or PRD formulation) is more realistic on physical grounds and can account for limb darkening as well. This approach is now in general use, although CRD is entirely adequate to explain the H and K line cores (but not the H_1 and K_1 minima features) and the entire profiles of the subordinate triplet lines at λ 8498, 8542, and 8662 (Shine, Milkey, and Mihalas, 1975b).

A number of solar chromospheric models have now been constructed that are based on the Ca II lines (see Table 2) to describe the quiet sun (Ayres and Linsky, 1976), plage regions (Shine and Linsky, 1974), and the chromospheric portions of flares (Machado and Linsky, 1975). These models are parameterized in terms of the temperature minimum (T_{min}) and pressure [$P(T_{min})$], or the mass column density [$m(T_{min}) = P(T_{min})/g$] at the base of the chromosphere, and the mass column density at the layer where the hydrogen Lyman continuum (λ 911) has unit optical depth, $m_0 = m(\tau_{LyC} = 1)$. Since the temperature corresponding to m_0 is about 8300 K for quiet and active regions on the sun (Noyes and Kalkofen, 1970), and since T(m) rises very steeply above this layer (Thomas and Athay, 1961; Vernazza, Avrett, and Loeser, 1976), these solar models and the stellar models described below assume this upper chromosphere temperature distribution. Between $m(T_{min})$ and m_0 these models generally assume for convenience that dT/d log m is a constant, but there is no independent evidence for this and Vernazza, Avrett, and Loeser (1976) suggest a nonlinear structure for the quiet sun. It is possible that the assumption that dT/d log m is constant leads to errors in the derived chromospheric microturbulent velocities.

As noted above, the sun has a number of chromosphere structures. We list these in Table 3, which gives the horizontal (or both horizontal and vertical) scales in megameters, typical magnetic fields, and lifetimes. Shine and Linsky (1974) are able to explain enhanced H and K emission in plages simply by increasing m_0 and thus P_0 about a factor of 10, which steepens the average value of dT/dh in the chromosphere about 50%. This change in the model parameters results in enhanced emission because at each K line optical depth $T_e(\tau_K)$ and $B_\nu(\tau_K)$ are larger, which increases the ionization of hydrogen and the metals so that the Ca II line source functions are more thermalized. Machado and Linsky (1975) find that H and K emission in solar flares can similarly be explained by further increasing P_0 to 8 dynes cm^{-2} for subflares and on up to 80 dynes cm^{-2} for Class 3 flares. This increase in P_0 and K line emission is presumably due to enhanced heating. The different solar chromospheres are listed in Table 3 in order of increasing P_0 according to computed models; where models have not yet been computed, the values are estimated.

A considerable literature now exists concerning stellar observations in the H and K lines, and Bidelman (1954) has compiled a very useful bibliography of the various stellar types that exhibit emission in these lines. H and K emission is occasionally seen in F stars, is usually seen in G stars, and is essentially ubiquitous in K and M stars. The earliest stars seen with Ca II emission include the F0 V star γ Vir N (Warner, 1968), and the F0 Ib star α CAR (Warner, 1966); occasionally Ca II is seen in the A7 III star γ BOO (Le Contel et al., 1970). One important unanswered question is whether chromospheres exist in A-type stars, where it is commonly assumed that they should not occur because of the thinness of the convective zones in these stars. A search is presently under way by Chiu, Linsky, and Maran for chromospheric filling in of the K-line core in A-type stars.

A second question is whether chromospheres extend to the coolest low-gravity stars, the M giants and supergiants. Dyke and Johnson (1969), Jennings and Dyke (1972), and Jennings (1973) find an inverse correlation in these stars between K-line emission and polarization and infrared excess, both indicative of grains in a cool outer atmosphere. They interpret this empirical result as suggesting that when the grain density is high, the grains cool the gas and dissipate the nonradiative energy input to the outer atmosphere, which would otherwise be cooled by emission lines like H and K.

Table 4 lists various groups of stars or structures on the sun in rough order of increasing chromospheric emission. Listed under each diagnostic is an indication of whether the diagnostic appears weak or strong as measured against the background continuum, with the symbols N (not present), W (weak), S (strong), VS (very strong), and blank (unknown). H and K emission is seen in a number of interesting star types including T Tauris (e.g., Kuhi, 1965), classical cepheids on the rising branch (e.g., Kraft, 1960; Hollars, 1974), carbon stars (Richer, 1975), eclipsing variables (Odgers and Wright, 1965), and is especially strong in close spectroscopic binaries (Eilek and Walker, 1976; Young, 1976). The emission is often variable even in single stars that do not show continuum variations (Griffin, 1964; Deutsch, 1967; Liller, 1968; Chiu et al., 1976), indicating that, like the sun, stellar chromospheric structures change with time and rotate on and off the disk. But in stars that are cooler and have lower gravity than the sun, the changes are of greater magnitude and the structures of larger size (Schwarzschild, 1975; Chiu et al., 1976).

Wilson and Bappu (1957) began a systematic study of H and K emission in cool stars at a dispersion of 10 Å mm^{-1}. They characterized their K-line intensities on a scale, the so-called Wilson-Bappu intensities (WBI), from 0 (no detectable emission) to 5 (emission stronger than the continuum). On this scale, the quiet sun would be 0, plages 1-2, and solar flares 4-5. They also characterized the line widths from direct measurements of the plates at this dispersion by means of a width index W (km s^{-1}), indicative of the width somewhere between the K_2 emission peak and the K_1 minimum beyond the emission peaks (cf. Wilson, 1976). Wilson (1968) has extended this observing program and now obtains photoelectric observations of the H- and K-line cores. Warner (1969) has also applied the program to southern-hemisphere stars.

Several interesting empirical correlations have come from this program. Wilson and Bappu (1957) found that $W \sim L^{1/6}$ (cf. Wilson, 1966a). Unfortunately the great interest in interpreting and extending this width-luminosity relationship (the so-called Wilson-Bappu effect) has so dominated the study of stellar chromospheres that other, possibly more interesting, empirical results have not been studied to the extent they merit. The fundamental question in interpreting the Wilson-Bappu effect is whether the width W refers to the doppler core or the damping wing. In the former case one is led to the conclusion that chromospheric turbulent velocities increase with stellar luminosity (Goldberg, 1957; Fosbury, 1973; Reimers, 1973; Wilson, 1957b). An alternative approach is that W refers to a feature in the damping part of the line profile, in which case Ayres, Linsky, and Shine (1975) showed that $W \sim L^{1/6}$ naturally follows from the increase in mass column density above the temperature minimum needed to retain the same H^- optical depth as the gravity of a star decreases (cf. Lutz, Furenlid, and Lutz, 1973). This question remains unresolved. In the meantime other width-luminosity relations have been found for Balmer-alpha (Kraft, Preston, and Wolfe, 1964; Fosbury, 1973; Reimers, 1973), the Mg II resonance lines (Kondo, Morgan, and Modisette, 1976a; McClintock et al., 1975a), and Lyman-alpha (McClintock et al., 1975a).

Wilson (1963; 1966a), Wilson and Skumanich (1964), and Wilson and Woolley (1970) have found statistically that the relative strength of the Ca II K_2 emission decreases with age as stars evolve off the main sequence, and Skumanich (1972) has derived an evolutionary time scale for this decay. Similar decreases with age in the rotational velocity of main-sequence stars (Wilson, 1966b; Kraft, 1967) and lithium depletion with age (e.g., Herbig, 1965) have suggested the picture (Wilson,

1966b; Kraft, 1967; Demarque and Roeder, 1967) that stars with well-developed convective envelopes (spectral types F and later) have chromospheres, winds, and magnetic fields as a result of dynamo processes. The loss of angular momentum by the wind in time (Brandt, 1966) then brakes the star, decreasing the dynamo and resultant magnetic fields, the dissipation of acoustic energy (somehow related to the magnetic field), and ultimately the Ca II emission. This picture has considerable plausibility, especially since magnetic field strength and Ca II emission are well correlated in the sun (e.g., Frazier, 1971), but the picture has never really been tested. In particular, the correlation of decreasing Ca II emission with age may only be true statistically as α CEN A and the sun have similar Ca II emission but α CEN A is probably twice as old (Ayres et al., 1976). Also, high-luminosity evolved stars often have strong emission, and the strength of emission in spectroscopic binaries appears to be a function of the filling of one star's Roche lobe rather than of age (Young, 1976).

Stellar chromospheric models have now been constructed for α CMI (F5 IV; Ayres et al., 1974), α BOO (K2 IIIp; Ayres and Linsky, 1975a), α CEN A (G2 V; Ayres et al., 1976), and α CEN B (K1 V; Ayres et al., 1976). One important trend seen in these models is that despite the 2250 K range in effective temperature and the factor-of-600 range in gravity among these quiet chromosphere stars, the values of m_0 are all within a factor of 3 and 10^{-5} g cm^{-2}. It is important to determine what causes this small range in m_0 and whether m_0 is much larger in active chromosphere stars than it is in solar plages and flares.

In addition to the H and K lines, five subordinate lines of Ca II are potentially useful chromospheric diagnostics. The so-called infrared triplet lines λ 8498, 8542, and 8662 couple the 4 ^{2}P states (upper states of H and K) with the 3 ^{2}D metastable states. These lines are less opaque than H and K, but they are also collisionally dominated and are formed in the solar chromosphere (Vernazza, Avrett, and Loeser, 1976). Spectroheliograms in these lines (Title, 1966) show the chromospheric network and plages as bright regions, and self-reversed emission features appear in the line cores in strong plages (Shine and Linsky, 1973) and flares (Machado and Linsky, 1975). Shine and Linsky (1974) have shown that plage models that match the H- and K-line profiles also match the infrared triplet line profiles, including the observation that the weakest of the triplet lines (λ 8498) is the first to exhibit a clear emission feature as one goes from weak to strong plages. Thus these lines should be good chromospheric diagnostics. Also Shine, Milkey, and Mihalas (1975a, b) have shown that CRD diagnostics are entirely adequate in analyzing these lines, at least in solar-type stars.

The infrared triplet lines are typically in absorption in late-type stars, the exceptions being long-period variables near maximum light (Kraft, 1957; Merrill, 1934; 1961), T Tauri stars, and the peculiar binary AX MON (Wallerstein, 1971; Herbig and Rao, 1972; cf. Andrillat and Swings, 1976). Anderson (1974) has searched for emissions in the core of λ 8498 without success, but he observed only one active chromosphere star, βDRA (G2 II), and the λ 8498 profile of this star is shallow, presumably filled in by weak chromospheric emission. Linsky, Hunten, and Sowell (1976) are searching for core emission or filled-in profiles of the λ 8542 line, with emphasis on active chromosphere stars including spectroscopic binaries, to establish the usefulness of this line as an empirical test for active stellar chromospheres.

The $4\,^2S$ ground state of Ca II is connected to the metastable $3\,^2D$ states by quadrupole transitions at λ 7291 and λ 7324. These very weak lines, which Lambert, Mallia, and Warner (1969), and Schorn, Young, and Barker (1975) have observed in the sun, will be useful diagnostics of chromospheres in low-gravity, late-type stars.

Mg II

The Mg II resonance lines h (λ 2803) and k (λ 2796) are similar to H and K in that they are collisionally excited emission lines that are thermalized in the solar chromosphere. In the quiet sun they appear as stronger emission lines with peak residual intensities of, typically, 0.3, compared to 0.07 for H and K, in part because they are measured against a darker background and in part because they are thermalized higher in the chromosphere at slightly hotter temperatures. This is because magnesium is about 14 times more abundant than calcium; its ionization potential is larger (15.03 eV for Mg II compared to 11.87 eV for Ca II); Lyman-alpha photoionizes Ca II but not Mg II from the metastable $3\,^2D$ states. In absolute flux units, however, the k-line emission core is only slightly stronger than K (2.24×10^5 ergs cm^{-2} s^{-1} sr^{-1} for k in a ±0.60-Å bandpass compared to 1.56×10^5 ergs cm^{-2} s^{-1} sr^{-1} for K in a ±0.275-Å bandpass; Ayres and Linsky, 1976). Also the Mg II lines appear bright in plage regions and the chromospheric network (Fredga, 1969; Lemaire and Skumanich, 1973).

As for the Ca II resonance lines, h and k should be synthesized using PRD codes rather than CRD codes (Milkey and Mihalas, 1974; Milkey, Ayres, and Shine, 1975). An important difference in synthesizing the Mg II lines is that, unlike Ca II, there are no intermediate metastable states between the resonance line upper and lower states and thus no finite minimum value of the incoherence fraction. As a result the inner wings of the Mg II lines are more nearly pure coherent scattering and thus can be darker. Milkey, Ayres, and Shine (1975) show that the inner wings of the Mg II lines in solar-type stars should darken considerably as the stellar gravity decreases.

Milkey and Mihalas (1974) have computed PRD Mg II line profiles using the HSRA model atmosphere (Gingerich et al., 1971). They were able to closely match the quiet sun line cores observed by Lemaire and Skumanich (1973), but with the HSRA model (T_{min} = 4170 K) they naturally compute k-line inner wings which are lower than observed. As a result they suggest that T_{min} must be raised considerably. Previously Lemaire and Skumanich (1973) have computed Mg II for several chromosphere models using CRD diagnostics and have compared these theoretical profiles with their observations of quiet and plage regions.

Using PRD diagnostics, Ayres and Linsky (1976) have derived a quiet chromosphere model by matching computed profiles with the observed profiles obtained by Kohl and Parkinson (1976) at two positions on the disk. The Ayres-Linsky Mg II model is slightly hotter in the upper photosphere than their Ca II model, and both models are characterized by T_{min} = 4450 K. The Mg II lines thus appear to be useful chromospheric diagnostics for deriving models for different chromospheric structures; this can be done with the aid of OSO-8 observations, for example.

Many stellar observations in the Mg II lines are now being acquired by space experiments. Doherty (1972a) has given integrated Mg II line fluxes for eight late-type stars obtained from the 25-Å photometry of OAO-2. Gurzadyan (1975) has also observed several late-type stars at 25-Å resolution. Kondo et al. (1972) and Kondo, Morgan, and Modisette (1975) have observed a number of late-type stars at 0.25-Å resolution with their balloon-borne stellar spectrometer. They found Mg II emission in stars as early as βCAS (F2 IV), and in α ORI (M2 Iab) they noted asymmetrical emission in the k line which Modisette, Nicholas, and Kondo (1973) have interpreted as due to Fe I (λ 2795) absorption, possibly circumstellar. Observations with a new, higher resolution balloon-borne spectrometer are now underway.

Several groups have been using the Princeton spectrometer on *Copernicus* for Mg II chromospheric observations. In a study of O- and B-type stars, Kondo, Modisette, and Wolf (1975) and Kondo, Morgan, and Modisette (1976b) have observed emission in γARA (B1 Vek) and βCEN (B1 II) and possible emission in λ ORI (O8 III), α VIR (B1 IV), α GRU (B7 IV), and βLIB (B8 V). If confirmed, these observations may be evidence for chromospheres in stars hotter than heretofore thought, although the heating mechanism(s) for these chromospheres could be different from that of the sun. Evans, Jordan, and Wilson (1975) have obtained Mg II profiles for α CAR (F0 Ib) and α CMI (F5 IV-V). Dupree and Baliunas (1976) have reported on strong Mg II emission from the α AUR primary (G8 III) and Baliunas, Dupree, and Lester (1976) find asymmetrical Mg II emission lines with no self-reversal and four times the solar surface flux from the spectroscopic binary λ AND (G8 III-IV). Moos et al. (1974) and McClintock et al. (1975a, b) have been studying K-type stars of all luminosity classes. They have found Mg II stellar surface fluxes that range from 0.05 to 0.16 that of the sun for giants and supergiants to the solar value for ε ERI (K2 V). They also find asymmetric emission in α BOO (K2 IIIp) and possibly also in α TAU (K5 III), indicative of significant mass loss. Finally, Bernat and Lambert (1976), in their study of the M stars α ORI and α SCO, find the k line asymmetric and h line symmetric (cf. Kondo et al., 1972), suggestive of circumstellar absorption by Fe I and Mn I as proposed by Modisette, Nicholas, and Kondo (1973).

Mg II observations have now been utilized in constructing stellar chromospheric models for α CMI (Ayres, Linsky, and Shine, 1974) using the Kondo *et al.* (1972) data and for α BOO (Ayres and Linsky, 1975a) using the Moos *et al.* (1974) data.

H I

A number of spectral features of hydrogen are potentially useful chromosphere diagnostics including Lyman-alpha, Balmer-alpha, Balmer-epsilon, and the Lyman and Balmer continua. The Lyman-alpha (Lα) line, like the Ca II and Mg II resonance lines, is collisionally dominated and thermalized in the solar chromosphere. Due to the larger abundance of hydrogen, however, Lα is formed higher in the chromosphere and, as suggested by Vernazza, Avrett, and Loeser (1973;1976), the core of Lα is formed in a plateau of about 160 km width and a temperature near 20,000 K. The existence of this plateau remains to be confirmed by analyses of other lines of other species. Milkey and Mihalas (1973), following an earlier suggestion of Vernazza, have shown that PRD diagnostics are needed to analyze the inner wings of Lα.

In the sun, spectroheliograms of Lα clearly show the bright plage and chromospheric network regions (Prinz, 1973) and in the preliminary OSO-8 data they also weakly show polar coronal holes. Basri *et al.* (1976) point out, however, that in images with high spatial resolution Lα does not show precisely the same chromospheric structure as other lines, such as Balmer-alpha and Ca II, presumably owing to its higher level of formation.

Despite its clear usefulness as a chromospheric diagnostic, Lα has not been effectively employed yet in model building. The exceptions have been the early analysis of Morton and Widing (1961) and the CRD analysis of OSO-4 and OSO-6 quiet chromosphere profiles by Vernazza, Avrett, and Loeser (1973;1976) that led to the suggestion of a 20,000 K plateau noted above. Basri *et al.* (1976) have recently begun a PRD analysis of the Lα wings and core using the one-arcsec-resolution Brueckner spectra. They are interested in deriving temperature structures over the range 6,000-25,000 K in active and quiet chromosphere structures.

Stellar observations of Lα are feasible only for stars nearer than about 30 parsec due to strong interstellar absorption. Even for the nearby stars it is difficult to disentangle interstellar absorption from stellar or circumstellar self-reversal at line center. Evans, Jordan, and Wilson (1975) have obtained *Copernicus* spectra of Lα in α CMI with a line surface flux comparable to the sun. Dupree (1975), Dupree and Baliunas (1976), and Vitz *et al.* (1976) find the line surface flux in α AUR to be about 1.5 times stronger than the sun. Baliunas, Dupree, and Lester (1976) report that in the spectroscopic binary λ AND the line surface flux is three times that of the sun and the red emission peak is stronger than the blue, suggestive of strong mass loss. The α BOO line flux has now been observed by several rockets and *Copernicus* (Rottman *et al.*, 1971; Moos and Rottman, 1972; Moos *et al.*, 1974; McKinney, Moos, and Giles, 1976; Weinstein, Moos, and Linsky, 1976) with clear evidence for variability. The α BOO Lα profile, like that of λ AND, exhibits a stronger red than blue emission peak, suggestive of mass loss. McClintock et al. (1975a, b) have measured Lα line fluxes in several other K-type stars. No Lα measurements of M-type stars have yet been reported, and none of the data just described have yet been analyzed.

The Balmer-alpha (Hα) line is also chromospheric and accessible to ground-based observing but is more difficult to use as a diagnostic of chromospheric densities and temperatures as it is typically photoionization- rather than collision-dominated in the sun (Gebbie and Steinitz, 1974). Hα spectroheliograms do show significant intensity variations, however, with the chromospheric network, plage regions, and especially flares appearing bright. Gebbie and Steinitz (1973; 1974) proposed that intensity contrasts in the quiet sun are due to lateral changes in the shape of the absorption profile, while in plages and flares the electron densities may be sufficiently large that the line is collision-dominated. The use of Hα as a spectroscopic diagnostic has not been vigorously pursued because of the typical photoionization-dominated nature of the line, except for M dwarfs (Fosbury, 1974), and because solutions of the transfer equation with the R_{IV} redistribution function (Mihalas, 1970) have not yet been attempted. Computations of model atmospheres on the basis of eclipse line intensities have been done (e.g., Thomas and Athay, 1961). Also, Schoolman (1972) has computed Hα CRD line profiles for various atmospheric models and Canfield (1974) has done so for models of flares.

Hα profiles have been obtained for 33 late-type stars by Fosbury (1973), who extended the work of Kraft, Preston, and Wolfe (1964) and Lo Presto (1971), and have related line widths to stellar luminosity, chromospheric velocity fields, and acoustical energy flux. Dumont et al. (1973) have shown that for 20 T Tauri stars the Hα emission flux is consistent with a chromospheric origin, the observed Paschen and Balmer continua, and a photoionization-dominated source function. Herbig and Rao (1972) have compiled a list of Hα emission-line stars. Further studies of this line as a diagnostic in active chromosphere stars might be quite productive.

The Balmer-epsilon (H ε) line is located +1.6 Å from the core of the Ca II H line and thus is measured against a very dark background in cool stars. In the sun, Hε is in absorption on the disk but in emission in flares. In late-type stars and especially in K and M giants (Wilson, 1957a) Hε is often in emission while other members of the Balmer series are in absorption, suggesting that the dark background against which it is measured is an important aspect of Hε appearing in emission. Hε emission is also seen in active chromosphere stars (spectroscopic binaries, dMe, T Tauris, flare stars), long-period variables, and novae.

Fosbury (1974) argues that Hε should be photoionization-dominated in late-type stars such as the sun and K giants, but in M dwarfs chromospheric electron densities may be sufficiently high to produce a collision-dominated line whose emission is indicative of a chromosphere temperature rise. He concludes that the

Hε emission clearly seen in α BOO by Griffin (1968) results from the line excitation temperature (due to the hydrogen-continuum radiation fields) exceeding the local radiation temperature (dark H line wing) and is thus not indicative of a chromosphere. Ayres and Linsky (1975b) have made detailed calculations of the Hε line profile for the sun and α BOO. They find that the line, as expected, is insensitive to the chromospheric model for the sun; but the contrary is true for α BOO because the relative contribution to $\tau(H\varepsilon)$ by the chromosphere is large in stars with cool photospheres. Thus, although Hε remains photoionization-dominated (controlled by the Balmer-continuum radiation field) in K giants, it is largely formed in the chromosphere and is indirectly sensitive to chromospheric properties. It is therefore a good candidate for a chromospheric diagnostic in these stars and M dwarfs.

Since the Lyman continuum is probably not readily observable in any star other than the sun owing to interstellar absorption, we conclude this discussion of hydrogen chromospheric diagnostics by considering emission in the Balmer and Paschen continua. In the sun the only features that show this emission are white-light flares. A number of suggestions have been made that this emission is photospheric because of heating by flare X-rays (Rust and Hegwer, 1975; Somov, 1975) or by nonthermal ions and electrons (Svestka, 1973; Najita and Orrall, 1970). Alternative suggestions are that the emission is indeed chromospheric, originating either in a layer where T = 8500 K, which has been pushed down to a region of high density by the flare (Machado and Rust, 1974), or in a layer where $T \approx 20{,}000$ K and the Lyman lines are formed (Machado, 1976).

Evidence for chromospheric hydrogen bound-free emission has not yet been presented for most types of late-type stars. Possible exceptions include the T Tauri stars (Kuhi, 1974; Dumont et al., 1973; but see Kuan, 1975), Herbig-Haro objects (Bohm, Schwartz, and Siegmund, 1974), and flare stars (e.g., Kunkel, 1970; Moffett, 1975).

He I and He II

We treat these two ions together because their similar energy level diagrams and spectra are likely produced by similar mechanisms and should be amenable to similar diagnostics. The bound-free continuum ($\lambda < 504$ Å for He I and $\lambda < 228$ Å for He II) and resonance lines ($\lambda\,584$ for He I and $\lambda\,304$ for He II) have been studied in the sun, but they may not be observable except for the very closest stars. The visible and infrared lines of He I ($\lambda\,5876$, $\lambda\,10830$) and He II ($\lambda\,4686$) have been observed, however, in stars and should prove to be very interesting diagnostics because of the large excitation energies of these levels. Also the $\lambda\,1640$ (Hα) line of He II may be a useful diagnostic.

Three mechanisms have been proposed for the formation of the ultraviolet spectra of these ions: (a) excitation and ionization by electron collisions (Athay, 1966); (b) recombination following photoionization by coronal XUV emission and the λ 304 line in the case of He I (Hirayama, 1972; Zirin, 1975); (c) enhancement of collisional excitation by mixing of hot and cold plasma (Jordan, 1975), for example, by diffusion (Shine, Gerola, and Linsky, 1975).

In quiet-sun regions the color temperature of He II Lyman continuum is about 13,000 K (Linsky et al., 1976) and the He I continuum 11,500 K (Vernazza and Reeves, 1976). These cool temperatures are indicative of formation in the chromosphere by the second mechanism, and detailed computations by Linsky et al. (1976) and Avrett, Vernazza, and Linsky (1976) confirm this. The resonance lines, however, are observed to be much stronger than computed using mechanisms (a) and (b) (Jordan, 1975; Avrett, Vernazza, and Linsky, 1976), unless considerably more material exists in the solar transition region than presently accepted models allow (Milkey, Heasley, and Beebe, 1973; Dupree, 1972). This discrepancy has led to suggestions of mixing or diffusion, but it could also imply errors in our transition region models. In any case, the λ 304 and λ 584 lines are collisionally dominated in the quiet sun, but they may be formed over a wide temperature range in the chromosphere and transition region (Avrett, Vernazza, and Linsky, 1976). In late-type giants, however, the recombination mechanism may dominate because of low chromospheric densities, in which case the He I and He II spectra may be diagnostics of coronae rather than of chromospheres.

The formation of the subordinate lines λ 5876 (D_3) and λ 10830 of He I and λ 4686 of He II has not yet been clarified because the necessary calculations including all the above mechanisms have yet to be performed. Milkey, Heasley, and Beebe (1973) state that λ 10830 should be a pure scattering line in the quiet sun and that the He I triplet states should be populated by both mechanisms (a) and (b). In Zirin's (1975) analysis there is sufficient coronal radiation at $\lambda < 504$Å to account for the observed quiet-sun λ 10830 absorption, but Linsky et al. (1976) point out that energetically this mechanism cannot work because most of this coronal radiation will be absorbed by H I rather than He I.

Spectroheliograms show λ 10830 and λ 5876 absorption in plages and the chromospheric network and decreased absorption in supergranule cells and coronal holes (e.g., Harvey et al., 1975). The He II λ 4686 line may also appear as weak absorption above plages. Zirin (1975) interprets the appearance of dark bands in λ 5876 and λ 4686 at 1150 km and 1500 km above the disk, respectively, as evidence for excitation of He I and He II low in the chromosphere and thus for mechanism (b). The shape of the He II Hα line profile λ 1640 in the quiet sun, however, is very suggestive of formation by collisional rather than recombination processes (Feldman et al., 1975; Kohl, 1976). It should therefore be clear that until the formation mechanisms for the He I and He II subordinate lines are sorted out, these lines will not be useful diagnostics.

A considerable number of stellar observations in the λ 10830 line have been obtained by Vaughan and Zirin (1968), Zirin (1976), and Andrillat and Swings (1976). They find variability in many stars, and often the line is seen in emission or with emission components. One interesting result is that λ 10830 absorption or emission is not tightly correlated with K-line emission as would be expected if λ 10830 were a coronal diagnostic [mechanism (b)]. Spectroscopic binaries typically exhibit very strong λ 10830 absorption, as they do K-line emission. M giants and supergiants usually do not show λ 10830 absorption or emission, possibly indicating weak or nonexistent coronae in these stars.

The λ 5876 line is typically weaker in stars than in the sun but a few observations have been presented (Pasachoff and Lepler, 1972). This line is seen in spectroscopic binaries like λ AND (Pasachoff and Lepler, 1972), dMe stars (Worden and Peterson, 1976), novae (McLaughlin, 1960), flare stars (Joy, 1960), and presumably T Tauri stars. Observations of the λ 4686 line have been reported for novae and flare stars. No chromospheric models have yet been attempted on the basis of stellar He I or He II data.

O I

Neutral oxygen is interesting because it is an example of a fluorescence spectrum; that is, a spectrum largely excited by a strong emission line in another species. Bowen (1947) pointed out that the H I Lyman-beta (Lβ) line is coincident with an O I transition from the ground state to the 3d $^3D^o$ level, which then cascades to give lines at λ 11287, λ 8446, and the resonance lines λ 1302, λ 1305, and λ 1306. Strong λ 11287 and λ 8446 emission is indeed often seen in H II regions, Be stars, and novae as well as various peculiar stars (Andrillat and Swings, 1976).

In the sun, λ 11287 and λ 8446 appear in absorption and the λ 1302 triplet is in emission, as are the λ 1355 and λ 1358 intercombination lines (Dupree *et al.*, 1973). In α AUR the resonance triplet lines have about eight times the solar surface brightness (Vitz *et al.*, 1976) and in α BOO (Weinstein, Moos, and Linsky, 1976) the resonance triplet and the λ 1355 and λ 1358 lines are about equal to the solar surface brightness and are anomalously bright compared to other lines formed in the lower chromosphere.

The anomalous character of the O I lines in α BOO results from the Bowen fluorescence mechanism. Haisch *et al.* (1976) calculated the O I spectrum in α BOO using the Ayres and Linsky (1975a) chromospheric model. They find that Lyman-beta pumping enhances the resonant triplet flux a factor of 25 above the collisional excitation value, can explain the ultraviolet line fluxes, and can account for measured equivalent widths in λ 8446 and other near infrared lines in the Griffin (1968) atlas. Shine *et al.* (1976) find similar factor-of-30 enhancements in the O I resonant triplet in the quiet sun. Clearly the O I ultraviolet and near-infrared lines are diagnostics for the Lyman-beta flux and thus indirectly for chromospheric properties near $T = 20,000$ K.

OTHER SPECTRAL LINES

Other potentially useful chromospheric diagnostics include the resonance lines of

C II at 1334 and 1335 Å and Si II at 1260 and 1264 Å. These lines have been measured in α AUR (Vitz et al., 1976) and are being studied by Shine et al. (1976) in the sun. Heasley and Milkey (1976) have suggested that emission in the cores of the CO fundamental vibration-rotation bands should be a diagnostic of chromopheres in late-type giants.

ULTRAVIOLET AND INFRARED CONTINUA

In the sun the continuum originates in the chromosphere for $\lambda < 1525$Å (the Si I edge) and $\lambda > 150\mu m$ (hydrogen free-free opacity) (Vernazza, Avrett, and Loeser, 1976), and in other stars we might expect to see chromospheric emission at somewhat different wavelengths, depending on the atmospheric models. The Wisconsin experiment on OAO-2 observed a number of late-type stars in broad-band channels down to 1900 Å with no clear evidence for chromospheric emission (Doherty, 1972b). Parsons and Peytremann (1973), however, show evidence for chromospheric emission in HR 2786, a G0 II star, in the U_3 (1620 Å) channel of the Smithsonian experiment on OAO-2. Evidence of chromospheric emission in other cool stars may be available in the OAO-2 Celescope data. At 20 μm no excess emission clearly ascribable to a chromosphere has been detected (Morrison and Simon, 1973) other than a 0.3 magnitude excess for α LYR (A0 V), which the authors tentatively call chromospheric.

REFERENCES

Adams, W.M., and Sturrock, P.A., 1975, Ap. J. 202, 259.

Altenhoff, W.J.; Braes, L.L.E.; Olnon, F.M.; and Wendker, H.J., 1976, Astr. and Ap. 46, 11.

Ambartsumian, V.A., and Mirzoyan, L.V., 1975, Variable Stars and Stellar Evolution, ed. V.E. Sherwood and L. Plaut, D. Reidel, Dordrecht, 1.

Anderson, C.M., 1974, Ap. J. 190, 585.

──────────, Hartmann, L.W., and Bopp, B.W., 1976, Ap. J. (Letters) 204, L51.

Andrillat, Y., and Swings, J.P., 1976, ibid. 204, L123.

Athay, R.G., 1966, Ap. J. 145, 784.

──────────, 1970, ibid. 161, 713.

──────────, 1976, The Solar Chromosphere and Corona: Quiet Sun, D. Reidel, Dordrecht.

────────── and Skumanich, A., 1968, Solar Phys. 4, 176.

Avrett, E.H., and Hummer, D.G., 1965, M.N.R.A.S. 130, 295.

──────────, Vernazza, J.E., and Linsky, J.L., 1976, Ap. J. (Letters) 207, L19.

Ayres, T.R., 1975, Ap. J. 201, 799.

———— and Linsky, J.L., 1975a, ibid. 200, 660.

———— and ————, 1975b, ibid. 201, 212.

———— and ————, 1976, ibid. 205, 874.

————, ————, and Shine, R.A., 1974, ibid. 192, 93.

————, ————, and ————, 1975, Ap. J. (Letters) 195, L121.

————; ————; Rodgers, A.W.; and Kurucz, R.L., 1976, Ap. J., in press.

Baliunas, S., Dupree, A.K., and Lester, J.B., 1976, Bull. AAS 8, 353.

Basri, G.; Bartoe, J.-D. F.; Brueckner, G.; Linsky, J.L.; and Van Hoosier, M.E., 1976, Bull. AAS 8, 331.

Bernat, A.P., and Lambert, D.L., 1976, Ap. J. , in press.

Bidelman, W.P., 1954, Ap. J. Suppl. 1, 214.

Blanco, C., Catalano, S., and Marilli, W., 1976, Astr. and Ap. 48, 19.

————; ————; ————; and Rodono, M., 1974, ibid. 33, 257.

Boesgaard, A.M., 1974, Ap. J. 188, 567.

Böhm, K.-H., Schwartz, R.D., and Siegmund, W.A., 1974, ibid. 193, 353.

Boland, B.C.; Engstrom, S.F.T.; Jones, B.B.; and Wilson, R., 1973, Astr. and Ap. 22, 161.

Bopp, B.W., 1974, Ap. J. 193, 389.

———— and Evans, D.S., 1973, M.N.R.A.S. 164, 343.

———— and Moffett, T.J., 1973, Ap. J. 185, 239.

Borra, E.F., and Landstreet, J.L., 1973, Ap. J. (Letters) 185, L139.

Bowen, I.S., 1947, Pub. Astr. Soc. Pacific 59, 196.

Brandt, J.C., 1966, Ap. J. 144, 1221.

Canfield, R.C., 1974, Solar Phys. 34, 339.

Catura, R.C., Acton, L.W., and Johnson, H.M., 1975, Ap. J. (Letters) 196, L47.

Chiu, H.Y.; Adams, P.J.; Linsky, J.L.; Basri, G.S.; Maran, S.P.; and Hobbs, R.W., 1976, Ap. J., in press.

Cruddace, R.; Bowyer, S.; Malina, R.; Margon, B.; and Lampton, M., 1975, Ap. J. (Letters) 202, L9.

de Jager, C., 1976, Mém. Soc. Roy. Sci. Liège 9, 369.

de Loore, C., 1970, Astr. Space Sci. 6, 60.

Demarque, P., and Roeder, R.C., 1967, Ap. J. 147, 1188.

Deutsch, A.J., 1967, Pub. Astr. Soc. Pacific 79, 431.

Doherty, L.R., 1972a, Ap. J. 178, 495.

————, 1972b, ibid. 178, 727.

—————————, 1973, *Stellar Chromospheres*, ed. S.D. Jordan and E.H. Avrett, NASA SP-317, 99.

Doschek, G.A., Feldman, U., and Bohlin, J.D., 1976, *Ap. J. (Letters) 205*, L177.

Dravins, D., 1976, *Basic Mechanics of Solar Activity*, ed. J. Kleczek, in press.

Dumont, S.; Heidmann, N.; Kuhi, L.V.; and Thomas, R.N., 1973, *Astr. and Ap. 29*, 199.

Dupree, A.K., 1972, *Ap. J. 178*, 527.

—————————, 1975, *Ap. J. (Letters) 200*, L27.

————————— and Baliunas, S., 1976, *Bull. AAS 8*, 397.

—————————; Huber, M.C.E.; Noyes, R.W.; Parkinson, W.H.; Reeves, E.M.; and Withbroe, G.L., 1973, *Ap. J. 182*, 321.

Durisen, R.H., Savedoff, M.P., and Van Horn, H.M., 1976, *Ap. J. (Letters) 206*, L149.

Dyke, H.M., and Johnson, H.R., 1969, *Ap. J. 156*, 389.

Eilek, J.A., and Walker, G.A.H., 1976, *Pub. Astr. Soc. Pacific 88*, 137.

Evans, R.G., Jordan, C., and Wilson, R., 1975, *M.N.R.A.S. 172*, 585.

Feldman, U.; Doschek, G.A.; Van Hoosier, M.E.; and Tousey, R., 1975, *Ap. J. (Letters) 199*, L67.

Fix, J.D., and Spangler, S.R., 1976, *Ap. J. (Letters) 205*, L163.

Flower, D.R., and Pineau des Forets, G., 1974, *Astr. and Ap. 37*, 297.

Fosbury, R.A.E., 1973, *ibid. 27*, 141.

—————————, 1974, *M.N.R.A.S. 169*, 147.

Frazier, E.N., 1971, *Solar Phys. 21*, 42.

Fredga, K., 1969, *ibid. 9*, 358.

Gebbie, K.B., and Steinitz, R., 1973, *ibid. 29*, 3.

————————— and —————————, 1974, *Ap. J. 188*, 399.

Gerola, H.; Linsky, J.L.; Shine, R.A.; McClintock, W.; Henry, R.C.; and Moos, H.W., 1974, *Ap. J. (Letters) 193*, L107.

Gershberg, R.E., 1975, *Variable Stars and Stellar Evolution*, ed. V.E. Sherwood and L. Plaut, D. Reidel, Dordrecht, 47.

Gingerich, O.; Noyes, R.W.; Kalkofen, W.; and Cuny, Y., 1971, *Solar Phys. 18*, 347.

Goldberg, L., 1957, *Ap. J. 126*, 318.

Griffin, R.F., 1964, *Observatory 83*, 255.

—————————, 1968, *A Photometric Atlas of the Spectrum of Arcturus*, Cambridge Philosophical Society, London.

Grindlay, J., 1970, *Ap. J. 162*, 187.

Gurzadyan, G.A., 1975, *Pub. Astr. Soc. Pacific 87*, 289.

Haisch, B.M., and Linsky, J.L., 1976, *Ap. J. (Letters) 205*, L39.

_____; _____; Weinstein, A.; and Shine, R., 1976, *Bull. AAS 8*, 331.

Harvey, J.W.; Krieger, A.S.; Timothy, A.F.; and Vaiana, G.S., 1975, *ibid. 7*, 358.

Hearn, A.G., 1975, *Astr. and Ap. 40*, 355.

Hearn, D.R.; Richardson, J.A.; Bradt, H.V.D.; Clark, G.W.; Lewin, W.H.G.; Mayer, W.F.; McClintock, J.E.; Primini, F.A.; and Rappaport, S.A., 1976, *Ap. J. (Letters) 203*, L21.

Heasley, J.N., and Milkey, R.W., 1976, *ibid. 205*, L43.

Heise, J.; Brinkman, A.C.; Schrijver, J.; Mewe, R.; Gronenschild, E.; den Boggende, A.; and Grindlay, J., 1975, *ibid. 202*, L73.

Herbig, G.H., 1965, *Ap. J. 141*, 588.

_____ and Hoffleit, D., 1975, *Ap. J. (Letters) 202*, L41.

_____ and Rao, N.K., 1972, *Ap. J. 174*, 401.

Hills, J.G., 1973, *Ap. Letters 14*, 69.

Hirayama, T., 1972, *Solar Phys. 24*, 310.

Hollars, D.R., 1974, *Ap. J. 194*, 137.

Ingham, W.H., Brecker, K., and Wasserman, I., 1976, *ibid. 207*, 518.

Jennings, M.C., 1973, *ibid. 185*, 197.

_____ and Dyke, H.M., 1972, *ibid. 177*, 427.

Johnson, H.R., 1973, *ibid. 180*, 81.

Jordan, C., 1975, *M.N.R.A.S. 170*, 429.

Joy, A.H., 1960, *Stellar Atmospheres*, ed. J.L. Greenstein, University of Chicago Press, Chicago, 653.

Kandel, R., 1967, *Ann. Ap. 30*, 999.

Kelch, W.L., and Milkey, R.W., 1976, *Ap. J.*, in press.

Kjeldseth Moe, O., 1976, *Bull. AAS 8*, 331.

Kohl, J.L., 1976, *Ap. J.*, in press.

_____ and Parkinson, W.H., 1976, *ibid. 205*, 599.

Kondo, Y., Modisette, J.L., and Wolf, G.W., 1975, *ibid. 199*, 110.

_____, Morgan, T.H., and Modisette, J.L., 1975, *Ap. J. (Letters) 196*, L125.

_____, _____, and _____, 1976a, *Ap. J. 207*, 167.

_____, _____, and _____, 1976b, *ibid.*, in press.

_____; Duval, J.E.; Modisette, J.L.; and Morgan, T.H., 1976a, *ibid.*, in press.

_____; Guili, R.T.; Modisette, J.L.; and Rydgren, A.E., 1972, *ibid. 176*, 153.

_____; Modisette, J.L.; Dufour, R.J.; and Whaley, R.S., 1976b, *ibid. 206*, 163.

Kopp, R.A., 1972, *Solar Phys. 27*, 373.

———— and Kuperus, M., 1968, *ibid. 4*, 212.

Kraft, R.P., 1957, *Ap. J. 125*, 336.

————, 1960, *Stellar Atmospheres*, ed. J.L. Greenstein, University of Chicago Press, Chicago, 401.

————, 1967, *Ap. J. 150*, 551.

————, Preston, G.W., and Wolfe, S.C., 1964, *ibid. 140*, 237.

Kuan, P., 1975, *ibid. 202*, 425.

Kuhi, L.V., 1965, *Pub. Astr. Soc. Pacific 77*, 253.

————, 1974, *Astr. and Ap. Suppl. 15*, 47.

Kunkel, W.E., 1970, *Ap. J. 161*, 503.

————, 1975, *Variable Stars and Stellar Evolution*, ed. V.E. Sherwood and L. Plaut, D. Reidel, Dordrecht, 15.

Kuperus, M., 1965, *Recherches Astr. Obs. Utrecht 17*, 1.

Kurucz, R.L., 1974, *Solar Phys. 34*, 17.

Lambert, D.L., Mallia, E.A., and Warner, B., 1969, *ibid. 7*, 11.

Lampton, M.; Margon, B.; Paresce, F.; Stern, R.; and Bowyer, S., 1976, *Ap. J. (Letters) 203*, L71.

Le Contel, J.M.; Praderie, F.; Bijaoui, A.; Dantel, M.; and Sareyan, J.P., 1970, *Astr. and Ap. 8*, 159.

Lemaire, P., and Skumanich, A., 1973, *ibid. 22*, 61.

Liller, W., 1968, *Ap. J. 151*, 589.

Linsky, J.L., 1976, *Ap. Letters 17*, 1.

———— and Avrett, E.H., 1970, *Pub. Astr. Soc. Pacific 82*, 169.

————, Hunter, D.M., and Sowell, R., 1976, in preparation.

————; Glackin, D.L.; Chapman, R.D.; Neupert, W.M.; and Thomas, R.J., 1976, *Ap. J. 203*, 509.

Lo Presto, J.C., 1971, *Pub. Astr. Soc. Pacific 83*, 674.

Lutz, T.E., Furenlid, I., and Lutz, J.H., 1973, *Ap. J. 184*, 787.

Machado, M.E., 1976, *Solar Phys.*, in press.

———— and Linsky, J.L., 1975, *ibid. 42*, 395.

———— and Rust, D.M., 1974, *ibid. 38*, 499.

Margon, B., Mason, K.O., and Sanford, P.W., 1974, *Ap. J. (Letters) 194*, L75.

McClintock, W.; Henry, R.C.; Moos, H.W.; and Linsky, J.L., 1975a, *Ap. J. 202*, 733.

————; Linsky, J.L.; Henry, R.C.; Moos, H.W.; and Gerola, H., 1975b, *ibid. 202*, 165.

McKinney, W.R., Moos, H.W., and Giles, J.W., 1976, *ibid. 205*, 848.

McLaughlin, D.B., 1960, *Stellar Atmospheres*, ed. J.L. Greenstein, University of Chicago Press, Chicago, 585.

McWhirter, R.W.P., Thonemann, P.C., and Wilson, R., 1975, *Astr. and Ap. 40*, 63.

Merrill, P.W., 1934, *Ap. J. 79*, 183.

—————————, 1961, *Stellar Atmospheres*, ed. J.L. Greenstein, University of Chicago Press, Chicago, 521.

Mewe, R.; Heise, J.; Gronenschild, E.; Brinkman, A.C.; Schrijver, J.; and den Boggende, A., 1975, *Ap. J. (Letters) 202*, L67.

————————— ; ————————— ; ————————— ; ————————— ; ————————— ; —————————, 1976, *Astr. Space Sci. 42*, 217.

Mihalas, D., 1970, *Stellar Atmospheres*, Freeman, San Francisco.

Milkey, R.W., and Mihalas, D., 1973, *Ap. J. 185*, 709.

————————— and —————————, 1974, *ibid. 192*, 769.

—————————, Ayres, T.R., and Shine, R.A., 1975, *ibid. 197*, 143.

—————————, Heasley, J.N., and Beebe, H.A., 1973, *ibid. 186*, 1043.

Modisette, J.L., Nicholas, R.E., and Kondo, Y., 1973, *ibid. 186*, 219.

Moffett, T.J., 1975, *Ap. J. Suppl. 29*, 1.

Moore, R.L., and Fung, P.C.W., 1972, *Solar Phys. 23*, 78.

Moos, H.W., and Rottman, G.J., 1972, *Ap. J. (Letters) 174*, L73.

—————————; Linsky, J.L.; Henry, R.C.; and McClintock, W., 1974, *ibid. 188*, L93.

Morrison, D., and Simon, T., 1973, *Ap. J. 186*, 193.

Morton, D.C., and Widing, K.G., 1961, *ibid. 133*, 596.

Mullan, D.J., 1974, *ibid. 192*, 149.

—————————, 1975a, *ibid. 200*, 641.

—————————, 1975b, *ibid. 198*, 563.

—————————, 1976, *ibid.*, in press.

Munro, R.H., and Withbroe, G.L., 1972, *ibid. 176*, 511.

—————————, Dupree, A.K., and Withbroe, G.L., 1971, *Solar Phys. 19*, 347.

Najita, K., and Orrall, F.Q., 1970, *ibid. 15*, 176.

Nariai, H., 1969, *Astr. Space Sci. 3*, 150.

Noci, G., 1973, *Solar Phys. 28*, 403.

Noyes, R.W., and Kalkofen, W., 1970, *ibid. 15*, 120.

Nussbaumer, H., and Storey, P.J., 1976, *Astr. and Ap.*, in press.

Odgers, G.J., and Wright, K.O., 1965, *J.R.A.S. Canada 59*, 115.

Oster, L., 1971, *Ap. J. 169*, 57.

—————, 1975, *Problems in Stellar Atmospheres and Envelopes*, ed. B. Baschek, W.H. Kegel, and G. Traving, Springer-Verlag, New York, 301.

Parsons, S.B., and Peytremann, E., 1973, *Ap. J. 180*, 71.

Pasachoff, J.M., and Lepler, E.C., 1972, *Bull. AAS 4*, 235.

Pottasch, S.R., 1964, *Space Sci. Rev. 3*, 816.

Praderie, F., 1973, *Stellar Chromospheres*, ed. S.D. Jordan and E.H. Avrett, NASA SP-317, 79.

Prinz, D.K., 1973, *Solar Phys. 28*, 35.

Reimers, D., 1973, *Astr. and Ap. 24*, 79.

Richer, H.B., 1975, *Ap. J. 197*, 611.

Reigler, G.R., and Garmire, G.P., 1975, *Astr. and Ap. 45*, 213.

Robinson, R.D., Slee, O.B., and Little, A.G., 1976, *Ap. J. (Letters) 203*, L91.

Rottman, G.J.; Moos, H.W.; Barry, J.R.; and Henry, R.C., 1971, *Ap. J. 165*, 661.

Rust, D.M., and Hegwer, F., 1975, *Solar Phys. 40*, 141.

Schoolman, S.A., 1972, *ibid. 22*, 344.

Schorn, R.A., Young, A.T., and Barker, E.S., 1975, *ibid. 43*, 9.

Schwarzschild, M., 1975, *Ap. J. 195*, 137.

Severny, A., 1970, *Ap. J. (Letters) 159*, L73.

Sheeley, N.R.; Bohlin, J.D.; Brueckner, G.E.; Purcell, J.D.; Scherrer, V.E.; and Tousey, R., 1975, *ibid. 196*, L129.

Shine, R.A., and Linsky, J.L., 1973, *Solar Phys. 25*, 357.

————— and —————, 1974, *ibid. 39*, 49.

—————, Gerola, H., and Linsky, J.L., 1975, *Ap. J. (Letters) 202*, L101.

—————, Milkey, R.W., and Mihalas, D., 1975a, *Ap. J. 199*, 724.

—————, —————, and —————, 1975b, *ibid. 201*, 222.

—————; Lites, B.W.; Chipman, E.G.; Roussel-Dupre, D.; Bruner, E.C., Jr.; Rottman, G.J.; Orrall, F.Q.; Athay, R.G.; and White, O.R., 1976, *Bull. AAS 8*, 331.

Shmeleva, O.P., and Syrovatskii, S.I., 1973, *Solar Phys. 33*, 341.

Shipman, H.L., 1976, *Ap. J. (Letters) 206*, L67.

Skumanich, A., 1972, *Ap. J. 171*, 565.

Somov, B.V., 1975, *Solar Phys. 42*, 235.

Spangler, S.R., and Moffett, T.J., 1976, *Ap. J. 203*, 497.

—————, Rankin, J.M., and Shawhan, S.D., 1974, *Ap. J. (Letters) 194*, L43.

Svestka, Z., 1973, *Solar Phys. 31*, 389.

Thomas, R.N., and Athay, R.G., 1961, *Physics of the Solar Chromosphere*, Interscience, New York.

Title, A., 1966, *Selected Spectroheliograms*, Mount Wilson and Palomar Observatories, Pasadena.

Torres, C.A.O., and Ferroz-Mello, S., 1973, *Astr. and Ap. 27*, 231.

Tucker, W.H., 1973, *Ap. J. 186*, 285.

Ulmschneider, P., 1967, *Z. Ap. 67*, 193.

Underhill, A.B., 1973, *Stellar Chromospheres*, ed. S.D. Jordan and E.H. Avrett, NASA SP-317, 53.

Vanderhill, M.J.; Borken, R.J.; Bunner, A.N.; Burstein, P.H.; and Kraushaar, W.L., 1975, *Ap. J. (Letters) 197*, L19.

Vaughan, A.N., and Zirin, H., 1968, *Ap. J. 152*, 123.

Vernazza, J.E., and Reeves, E.M., 1976, in preparation.

──────────, Avrett, E.H., and Loeser, R., 1973, *Ap. J. 184*, 605.

──────────, ──────────, and ──────────, 1976, *Ap. J. Suppl. 30*, 1.

Vitz, R.C.; Weiser, H.; Moos, H.W.; Weinstein, A.; and Warden, E.S., 1976, *Ap. J. (Letters) 205*, L35.

Vogt, S.S., 1975, *Ap. J. 199*, 418.

Wallerstein, G., 1971, *Pub. Astr. Soc. Pacific 83*, 77.

Warner, B., 1966, *Observatory 86*, 82.

──────────, 1968, *ibid. 88*, 217.

──────────, 1969, *M.N.R.A.S. 144*, 333.

Weinstein, A., Moos, H.W., and Linsky, J.L., 1976, *Ap. J.*, submitted.

Wilson, O.C., 1957a, *Ap. J. 126*, 46.

──────────, 1957b, *ibid. 126*, 525.

──────────, 1963, *ibid. 138*, 832.

──────────, 1966a, *Science 151*, 1487.

──────────, 1966b, *Ap. J. 144*, 695.

──────────, 1968, *ibid. 153*, 221.

──────────, 1976, *ibid. 205*, 823.

────────── and Bappu, M.K.V., 1957, *ibid. 125*, 661.

────────── and Skumanich, A., 1964, *ibid. 140*, 1401.

────────── and Woolley, R., 1970, *M.N.R.A.S. 148*, 463.

Withbroe, G.L., and Gurman, J.B., 1973, *Ap. J. 183*, 279.

Worden, S.P., and Peterson, B.M., 1976, *Ap. J. (Letters) 206*, L145.

Young, A., 1976, *Ap. J.*, in press.

——————— and Koniges, A., 1976, *ibid.*, in press.

Zirin, H., 1975, *Ap. J. (Letters) 199*, L63.

———————, 1976, *Ap. J.*, in press.

Zirker, J.B., 1968, *Solar Phys. 3*, 164.

APPENDIX I.
LIST OF PARTICIPANTS AND CONTRIBUTORS

PARTICIPANTS IN THE SOLAR OUTPUT WORKSHOP
AND CONTRIBUTORS TO THIS VOLUME

Richard G. Allen
Kitt Peak National Observatory
P.O. Box 26732
Tucson, Arizona 85726

James R. Asbridge
Group P4
Los Alamos Scientific Laboratory
Los Alamos, New Mexico 87544

Eugene H. Avrett
Center for Astrophysics
60 Garden Street
Cambridge, Massachusetts 02138

Samuel J. Bame
Group P4
Los Alamos Scientific Laboratory
Los Alamos, New Mexico 87544

Jean-David F. Bartoe
Naval Research Laboratory
Washington, D.C. 20375

Elmo C. Bruner
Lockheed Missiles and Space
  Company, Inc.
3251 Hanover Street
Palo Alto, California 94304

Paul J. Crutzen
National Center for
  Atmospheric Research
P.O. Box 3000
Boulder, Colorado 80307

Paul E. Damon
Department of Geosciences
University of Arizona
Tucson, Arizona 85721

Kenneth P. Dere
Naval Research Laboratory
Washington, D.C. 20375

Robert Dickinson
National Center for
  Atmospheric Research
P.O. Box 3000
Boulder, Colorado 80307

Richard F. Donnelly
NOAA/SEL Code R43
U.S. Department of Commerce
Boulder, Colorado 80302

APPENDIX

John A. Eddy
High Altitude Observatory
National Center for
  Atmospheric Research
P.O. Box 3000
Boulder, Colorado 80307

William C. Feldman
Group P4
Los Alamos Scientific Laboratory
Los Alamos, New Mexico 87544

Claus Fröhlich
Physikalisch-Meteorologisches
  Observatorium Davos
Weltstrahlungszentrum
CH-7370 Davos Platz
Oberwiesstrasse 4
Switzerland

Peter A. Gilman
High Altitude Observatory
National Center for
  Atmospheric Research
P.O. Box 3000
Boulder, Colorado 80307

Owen J. Gingerich
Center for Astrophysics
60 Garden Street
Cambridge, Massachusetts 02138

John T. Gosling
Group P4
Los Alamos Scientific Laboratory
Los Alamos, New Mexico 87544

Douglas O. Gough
Institute of Theoretical
  Astronomy
Madingley Road
Cambridge, United Kingdom

James D. Hays
Lamont-Doherty Geological
  Observatory
Palisades, New York 01964

Donald F. Heath
NASA/Goddard Space Flight
  Center
Code 912
Greenbelt, Maryland 20771

Leon J. Heroux
Ultraviolet Radiation Branch
Air Force Geophysics Laboratory
Hanscom AFB, Massachusetts 01731

Dieter Heymann
Department of Geology
Rice University
Houston, Texas 77001

John R. Hickey
The Eppley Laboratory, Inc.
12 Sheffield Avenue
Newport, Rhode Island 02840

Robert W. Hobbs
NASA/Goddard Space Flight
  Center
Greenbelt, Maryland 20771

Thomas E. Holzer
High Altitude Observatory
National Center for
  Atmospheric Research
P.O. Box 3000
Boulder, Colorado 80307

Donald M. Horan
Code 7125
Naval Research Laboratory
Washington, D.C. 20375

Appendix: PARTICIPANTS AND CONTRIBUTORS

Hugh H. Hudson
University of California
  at San Diego
Physics Department
P.O. Box 109
La Jolla, California 92037

Arthur J. Hundhausen
High Altitude Observatory
National Center for
  Atmospheric Research
P.O. Box 3000
Boulder, Colorado 80307

William L. Hunter
Code 7140
Naval Research Laboratory
Washington, D.C. 20375

Icko Iben, Jr.
University of Illinois
  at Urbana-Champaign
Department of Astronomy
  Observatory
Urbana, Illinois 61801

William W. Kellogg
National Center for
  Atmospheric Research
P.O. Box 3000
Boulder, Colorado 80307

Robert W. Kreplin
Naval Research Laboratory
Washington, D.C. 20375

Mukul R. Kundu
University of Maryland
Astronomy Program
Space Science Building
College Park, Maryland 20742

Dietrich Labs
Landessternwarte
69 Heidelberg 1
Köenigstuhl
Federal Republic of Germany

Louis J. Lanzerotti
Bell Laboratories
600 Mountain Avenue
Murray Hill, New Jersey 07974

Robert P. Lin
University of California
Space Sciences Laboratory
Berkeley, California 94720

Jeffrey L. Linsky
JILA
University of Colorado
Boulder, Colorado 80309

G. Wesley Lockwood
Lowell Observatory
Flagstaff, Arizona 86001

Julius London
Department of Astro-Geophysics
University of Colorado
Boulder, Colorado 80309

Robert P. Madden
National Bureau of Standards
U.S. Department of Commerce
Washington, D.C. 20234

William G. Mankin
National Center for
  Atmospheric Research
P.O. Box 3000
Boulder, Colorado 80307

James E. Manson
Ultraviolet Radiation Branch
Aeronomy Laboratory
Air Force Geophysics Laboratory
Hanscom AFB, Massachusetts 01731

John F. Meekins
Naval Research Laboratory
Washington, D.C. 20375

J. Murray Mitchell
NOAA
Rm. 625 GRMAX Bldg.
8060 13th Street
Silver Spring, Maryland 20910

O. Kjeldseth Moe
Naval Research Laboratory
Washington, D.C. 20375

Heinz Neckel
Hamburger Sternwarte
Hamburg-Bergedorf
Gojenbergsweg 112
Federal Republic of Germany

Gordon A. Newkirk Jr.
High Altitude Observatory
National Center for
  Atmospheric Research
P.O. Box 3000
Boulder, Colorado 80307

A. Keith Pierce
Kitt Peak National Observatory
P.O. Box 26732
Tucson, Arizona 85726

Raymond G. Roble
National Center for
  Atmospheric Research
P.O. Box 3000
Boulder, Colorado 80307

Gary J. Rottman
LASP
University of Colorado
Boulder, Colorado 80309

Gerhard Schmidtke
Institut für Physikalische
  Weltsramforschung
D79 Freiburg I. BR
Heidenhofstrasse 8
Federal Republic of Germany

Stephen H. Schneider
AAP/National Center for
  Atmospheric Research
P.O. Box 3000
Boulder, Colorado 80307

Fred I. Shimabukuro
Aerospace Corporation
P.O. Box 92957
Los Angeles, California 90009

Paul Simon
Institut d'Aéronomie Spatiale
  de Belgique
3, avenue Circulaire
1180 - Bruxelles
Belgium

Andre Skumanich
High Altitude Observatory
National Center for
  Atmospheric Research
P.O. Box 3000
Boulder, Colorado 80307

Appendix: PARTICIPANTS AND CONTRIBUTORS

Stephen T. Suess
NOAA/ERL
U.S. Department of Commerce
Boulder, Colorado 80302

Mathew P. Thekaekara*
Code 912
NASA/Goddard Space Flight Center
Greenbelt, Maryland 20771

Shelby G. Tilford
NASA Headquarters
Code ST
Washington, D.C. 20546

Adrienne F. Timothy
NASA Headquarters
Code ST
Washington, D.C. 20546

J. Gethyn Timothy
Center for Astrophysics
60 Garden Street
Cambridge, Massachusetts 02138

Roger K. Ulrich
University of California
Astronomy Department
Math-Science Building
Los Angeles, California 90024

Thomas C. Van Flandern
U.S. Naval Observatory
34th and Massachusetts Ave. N.W.
Washington, D.C. 20390

Anandu D. Vernekar
Institute of Fluid Dynamics
   and Applied Mathematics
University of Maryland
College Park, Maryland 20742

Alfred Vidal-Madjar
Centre National de la
   Recherche Scientifique
Laboratoire de Physique
   Stellaire et Planetaire
Boite Postale 10
91 Verrieres-le-Buisson
France

William J. Wagner
High Altitude Observatory
National Center for
   Atmospheric Research
P.O. Box 3000
Boulder, Colorado 80307

Arthur B. C. Walker
Stanford University
Institute for Plasma Research
   and Department of Applied Physics
Via Crespi
Stanford, California 94305

Robert M. Walker
Washington University
McDonnell Center for the
   Space Sciences
Box 1105
St. Louis, Missouri 63130

---

*We regret that our colleague Mathew Thekaekara died in November 1976.

Oran R. White
High Altitude Observatory
National Center for
 Atmospheric Research
P.O. Box 3000
Boulder, Colorado 80307

Richard C. Willson
Jet Propulsion Laboratory (183B-365)
4800 Oak Grove Drive
Pasadena, California 91103

Harold Zirin
California Institute of Technology
Building 354-33
Hale Observatory
Pasadena, California 91125

INDEX

A
Accretion effects, 466
Angular momentum loss, 456, 458
Atmospheric chemistry, 4, 13
Aurorae, 57

B
Brightness temperature, 166, 331, 334

C
Calcium spectra, 480, 482, 496
Calibration of spectra
  EUV, 239, 319
  infrared, 156-64
  Lyman-α, 221, 319
  optical, 171
  radio, 135
  standard s (NBS), 313
  UV, 196, 319
  X-ray (hard), 287-311, 319
  X-ray (soft), 261, 282, 319
Carbon^{14}, 21, 40, 65, 87, 415, 428-47
Carbon III spectrum (977Å), 234
Center-to-limb variation, 160, 176, 184, 342
Chromospheric-coronal transition region, 279, 341
Chromospheric models, 327
Chronology, radionuclides, 407
Circulation, atmospheric, 9
Climate
  oscillations, 20, 85-89
  sensitivity, 16
  trends, 86-89
  variability, 20, 445
Computed spectra, 339
Continental drift, climatic effects, 75-81
Convection, 455, 465
Corona
  at eclipse, 61, 63
  models, 279
Coronal holes, 38, 371

Cosmic rays, 39, 89, 383-447, 458

D
Deuterium, 230
deVries effect, 434
Diagnostics
  chromospheric, 496
  spectral, 487

E
Eclipses, solar, 61
Effective temperature, 457
Energetic particles, 39, 383-427
  composition, 394
  flux, 395
  galactic, 398, 408, 430
  modulation, 37, 398, 430
  propagation, 397
  spectra, 392
EUV
  spectroheliograms, 239
  spectrum, 33, 237
  variability, 246-58
Evolution, solar, 451, 456

F
Flares, 144, 221, 268, 301, 393, 408, 414, 416, 417, 431
  effects of, 15, 36, 37, 39, 387, 396
Flare stars, 493

G
Galactic cosmic rays, 430
Galactic rotation, 84, 466
Geomagnetic activity and variation, 6, 36, 66, 87, 397, 436, 441
Glaciation epochs, 75, 464, 466
Gleissberg cycle, 54, 70, 440
Gravitational constant, variability, 469

INDEX

H

Helium
  abundance relative to hydrogen, 359, 362, 372, 392, 411, 433
  isotopic abundance, 36
  spectra, 341, 504
Historical data, 52, 73-91, 440
Hydrogen spectra, 213-34, 482, 502

I

Infrared
  spectrum, 29, 151-68
  variability, 166
Inhomogeneities, solar, 160, 342
Insolation
  latitudinal distribution, 122
  variation, 122, 126
Instability, nuclear, 43, 462
Interplanetary magnetic field, 351-80
Interplanetary plasma, 351-80
Interplanetary shocks, 373
Interstellar dust clouds, 43, 466, 473
Ionosphere, 8
Isotopic ratios, 413

J

Joule heating, ionospheric, 7

L

Limb darkening, 160, 176, 184
Line blanketing, 176, 328, 481, 482
Line formation theory, 327-48, 480
Little Ice Age, 21, 65, 69, 107, 438, 443
Local thermodynamic equilibrium, 327-48
Luminosity, 457, 463, 467
Lunar samples, 405, 432, 473
Lyman alpha
  profile, 224-25
  spectroheliograms, 226, 227
  spectrum, 32, 213-34
  variability, 215-34

M

Magnesium spectra, 482, 500
Magnetic fields
  interplanetary, 351
  North-South component, 371
  radial component, 371
  solar, 36
  stellar, 492
Magnetosphere, 5
Maunder minimum, 56, 57, 440, 443
Mechanical heating, 48
Medieval climatic optimum, 69
Meteoritic samples, 405
Models
  chromosphere, 194, 337, 481
  corona, 134, 279
  photosphere, 329, 332, 480
  solar atmosphere, 327-48
  transition region, 341

N

Naked-eye sunspot observations, 60
NBS standards, 313
Network, chromospheric, 239, 342
Neutrino flux, 44, 457
Non-radiative heating, 48
Nuclear instability, 43, 463
Nuclear processes, 453, 462
Nucleosynthesis, early solar system, 419

O

Oblateness, solar, 455, 460
Optical
  spectrum, 29, 169-92
  variability, 190
Orbital variations, terrestrial, 49, 85, 117
Oscillations, 459

P

Paleoclimate studies, 21, 68, 73-90, 429-48
Participants, Solar Output Workshop, 517
Polar cap absorption, 389
Protons, 386, 408, 413, 431

R

Radiative losses, chromospheric, 487
Radio
  burst, 142
  quiet sun flux, 134
  s-component, 139
  spectrum, 28, 133-50
  10.7 cm flux correlations, 150, 218, 220, 245-54, 262, 266, 270, 285, 320
  variability, 147
Radiocarbon dating, 434
Radiometric scales, 94

Radiometric standards, 313
Redistribution, photon, 336, 480
Rotation, stellar, 458, 460

S

Samples
  dendrochronological, 428-47
  lunar, 405-27
  meteoritic, 405, 427
Sea core studies, 80
Solar activity, 36, 37, 51, 430, 434
Solar constant
  defined, 91
  measurements, 22, 91-109, 111-16
  value, 22, 104
  variation, 104-7, 114-16
Solar cycle, 11 year, 36, 51, 53, 461
Solar evolution, 451, 456
Solar interior, 41, 448-71
Solar magnetic cycle, 22-year, 38, 350, 461
Solar mass loss, 456
Solar particle events, 384
Solar spectrum
  EUV, 33
  general, 25, 129
  IR, 29
  radio, 28
  UV, 31
  visible, 29
  X-ray, 33, 34
Solar-type stars, 479
Solar wind, 351-82, 458
  abundances, 368
  ancient, 409
  bulk flow, 354, 357
  energy, 355
  fluid state, 355, 362, 364
  ions, 366, 368
  latitude variations, 370
  length scales, 361
  microstructure, 369
  sectors, 38, 370, 372
  streams, 37, 371
  time scales, 361
  variability, 374, 376
Space degradation, 319, 320
Spectral diagnostics, 487, 496-507
Spectral irradiance, 25, 131
  EUV, 237-58, 318
  infrared, 151-70

Lyman-α, 213-34, 318
  optical, 171-92
  radio, 133-50
  UV, 31, 193-211
  X-ray (hard), 287-311, 318, 321
  X-ray (soft), 261-78, 318
Spectrum of time project, 67
Spectrum synthesis, 339, 480
Spörer minimum, 62, 66, 443, 445
Starspots, 492
Stellar activity, 44, 477, 494
Stellar chromospheres, 45, 481, 484, 485, 487
Stellar coronae, 46, 490
Stellar magnetic fields, 492
  cycles, 492
  flares, 492
  plages, 492
Stellar photospheres, 480
Stellar transition regions, 488
Stellar variability, 44, 477, 492
Stellar winds and mass loss, 46, 490
Stratosphere, 13
Suess effect, 66, 434
Sunspot number, 16, 51, 430
  correlations, 140, 210, 220, 230, 245, 388, 389, 395, 398, 430, 437

T

Temperature, mean global, 17
Temperature minimum, solar, 166, 190, 332
Terrestrial absorption corrections, 24, 89, 91, 114, 154, 156, 173
Terrestrial response
  ionosphere, 8
  magnetosphere, 5
  stratosphere, 13
  troposphere, 16
Timescales
  convection, 455, 465
  dynamical, 451
  nuclear, 453
  rotation, 454
  thermal, 452, 462
Theoretical spectra, 339
Theory
  solar interior, 449-71
  solar variability, 449-71, 458
Troposphere, 16

U

UV absorption, 8, 13, 34, 193, 195
UV spectrum, 193
UV variability, 208, 323

V

Velocity fields, solar, 343
Visible spectrum; *see* Optical

W

Waves, 458, 459

X

X-ray
　bright points, 238, 319
　spectrum, 33, 34, 261, 287, 321
　variability, 268, 299, 309
XUV spectrum; *see* EUV and X-ray